Differential Geometrical Methods in Theoretical Physics

NATO ASI Series

Advanced Science Institutes Series

A Series presenting the results of activities sponsored by the NATO Science Committee, which aims at the dissemination of advanced scientific and technological knowledge, with a view to strengthening links between scientific communities.

The Series is published by an international board of publishers in conjunction with the NATO Scientific Affairs Division

A Life Sciences B Physics	Plenum Publishing Corporation London and New York
C Mathematical and Physical Sciences D Behavioural and Social Sciences E Applied Sciences	Kluwer Academic Publishers Dordrecht, Boston and London
F Computer and Systems Sciences G Ecological Sciences H Cell Biology	Springer-Verlag Berlin, Heidelberg, New York, London, Paris and Tokyo

Series C: Mathematical and Physical Sciences - Vol. 250

Differential Geometrical Methods in Theoretical Physics

edited by

K. Bleuler

and

M. Werner

Institute for Theoretical Nuclear Physics,
University of Bonn, Bonn, F.R.G.

Kluwer Academic Publishers

Dordrecht / Boston / London

Published in cooperation with NATO Scientific Affairs Division

Proceedings of the NATO Advanced Research Workshop
and the 16th International Conference on
Differential Geometrical Methods in Theoretical Physics
Como, Italy
24–29 August 1987

Library of Congress Cataloging in Publication Data

International Conference on Differential Geometric Methods in Theoretical Physics (16th : 1987 : Como, Italy)
Proceedings of the 16th International Conference on Differential Geometrical Methods in Theoretical Physics (NATO Advanced Research Workshop) Centro di Cultura Scientifica "Alessandro Volta" Villa Olmo, Como, Italy, 24-29 August 1987 / edited by K. Bleuler & M. Werner.
p. cm. -- (NATO Advanced Science Institutes series. Series C, Mathematical and physical sciences ; 250)
Includes index.
ISBN 9027728208
1. Geometry, Differential--Congresses. 2. Gauge fields (Physics)--Congresses. 3. String models--Congresses. 4. Mathematical physics--Congresses. I. Bleuler, Konrad, 1912- . II. Werner, M. III. Title. IV. Series: NATO ASI series. Series C, Mathematical and physical sciences ; no. 250.
QC20.7.D52I58 1987
530.1'5636--dc19 88-8262
CIP

ISBN 90-277-2820-8

Published by Kluwer Academic Publishers,
P.O. Box 17, 3300 AA Dordrecht, The Netherlands.

Kluwer Academic Publishers incorporates the publishing programmes of
D. Reidel, Martinus Nijhoff, Dr W. Junk, and MTP Press.

Sold and distributed in the U.S.A. and Canada
by Kluwer Academic Publishers,
101 Philip Drive, Norwell, MA 02061, U.S.A.

In all other countries, sold and distributed
by Kluwer Academic Publishers Group,
P.O. Box 322, 3300 AH Dordrecht, The Netherlands.

Printed in The Netherlands

TABLE OF CONTENTS

Organizers:

R.Cirelli, Milano, Italy
G.Casati, Milano, Italy
P.Cotta - Ramusino, Milano, Italy
K.Osterwalder, Zürich, Switzerland
K.Bleuler, Bonn, FRG

Advising Committee:

A.Lichnerowicz, Paris
A.Salam, Trieste
T.Regge, Torino
F.Hirzebruch, Bonn

INTRODUCTION AND SURVEY

The recent most impressive developments in theoretical physics, i.e. gauge and string theories with their far-reaching relations to basic concepts of present day mathematics brought a renewed and intense exchange between mathematicians and physicists. In a way the past few years might be compared to the great period of the mid-twenties with its really decisive mathematical contributions to the foundations of both quantum theory as well as general relativity. . Among others, this far-reaching exchange led H.Weyl in 1929 to the formulation of his (second) gauge principle (concerning the phase of the quantum mechanical wave function) which led, in turn (through the most natural generalization by C.N.Yang and R.Mills in 1954) to present day gauge theory which brought, first of all, a real breakthrough with respect to a theoretical interpretation of an enormous realm of data in modern particle physics. It led, in fact, to the well-known unification of electromagnetism with weak interactions (Weinberg-Salam theory) and, at the same time, to a most natural unification of particle and nuclear physics through QCD which deals with the basic quark-gluon structure of heavy matter,e.g. hadrons.

A very first and decisive step for this renewed exchange came, as emphasized by M.Atiyah (cp. his introductory article in this volume) through the insight that very much the same mathematical or, better, geometrical concepts (i.e. the most natural geometrical structure of gauge theory) represent a decisive ingredient in recent topological research (e.g. Donaldson's results for 4-manifolds). As an immediate consequence, however, of this new, apparently most satisfactory situation one of the greatest puzzles or, better, deepest quests of modern theory came up again (this time in an even more stringent form): It is the non-consistency between the two all-embracing basic theories, i.e. general relativity and quantum theory. Although gauge theory in its natural (i.e. pure) form (e.g. QCD) leads according to t'Hooft to a so far reasonable consistency (i.e. renormalizability), the theory of gravitation, (although at the origin of the very formulation of the gauge principle) still contradicts (in view of its more refined structure) the quantisation procedure (i.e. non-renormalizability in spite of all efforts; in a minor extent the same problem holds also for the combined electroweak theory). This most intriguing situation calls, of course, for a deeper understanding and naturally suggests - maybe in analogy to earlier situations[1] - an enlarged theoretical framework which then contains the conflicting theories just as special approximation schemes.

Such a solution was, so far, proposed in the form of the well-known string concept thus reviving, in a way, A.Einsteins 'historical vision' of a universal theory[2]. In simplified form the basic idea behind the string concept might be described by a relativistic theory of an extended object moving and interacting in a high-dimensional space (thus extending the Kaluza-Klein idea) whereby the 'quantized' supersymmetric inner degrees of freedom (i.e. the string variables) should represent - after projecting down to four-dimensional space - the physical particles. The special number of space-time dimensions (26 or 10) as well as the mutual interactions are due to certain consistency relations of the quantization scheme whereas the needed length scale - i.e. the space extension of the fundamental objects - is assumed to be the Planck length. Such an extreme numerical value is, in fact, needed in order to cover the three basic interactions (i.e. gravitational, electroweak and strong) which dimensionally are far apart from each other. In addition, the Planck length reminds us of the old quest for an elementary length or, better, of the basic and far-reaching idea of a fundamental departure from conventional Archimedean space-time to a new, maybe discrete structure.[3] In fact, this possibility was - in a most impressive way - already put forward by B.Riemann, later

---

[1] Special relativity contradicts conventional quantum theory for the N-particle system which was 'repaired' by the introduction of the then new notion of the quantized field according to Pauli and Heisenberg.

[2] In view of our present-day situation it appears - in contrast to earlier critics - most appealing that Einstein built his attempt entirely upon the concepts of 'classical' general relativity whereas Heisenberg's much later proposal introduced a basic but 'unphysical' 'quantized' field for generating all particles with their interactions. Thus both proposals exhibit - from opposite sides - some remote relations to our present attempt.

[3] The idea of a discrete geometry as basis for the explanation of the physical space (that means an elementary length) appeared already in Riemann's 'Habilitationsvortrag' in 1854
(cp. B.Riemann, Über die Hypothesen, welche der Geometrie zu Grunde liegen, ed. H.Weyl, 3rd edition, Springer 1923).

on (in a private discussion) by W.Pauli in relation with his deep dissatisfaction about the mathematical consistency of the (better, 'his') quantized fields and, eventually by E.Witten and Yu. Manin a.o. in connection with string theory[4].

Although the actual situation of string theories - i.e. the so-called 'heterotic' string theory as possible 'theory of everything' - is far from being satisfactory it brought an enormous stimulation of definitely needed research in mathematical physics which shed new light on these problems. In particular one has to mention the extended research in two-dimensional conformal quantum field theory (now representing the inner structure of strings) as well as in topological and extended group-theoretical methods which were applied and further developed in recent years. Overviewing the general situation it becomes evident that an enormous effort will be needed in order to achieve a deeper understanding of basic physical laws. On the other hand it is , however, realized - and this constitutes the idea and the real scope of this conference - that this attempt enforced, in fact, by empirical research led to an unexpected and most impressive extension of basic research in mathematical physics and to the above mentioned exchange with modern mathematics: Nearly all domains of present-day mathematical research which, in turn, were even (partly) suggested by physical viewpoints (e.g. supergroups, various infinite dimensional Lie algebras), play by now an important role in physics (e.g. for superstrings and conformal quantum field theory). It is always most impressive to realize that abstract but natural structures of modern mathematics play this decisive role in an interpretation and understanding of physical facts and that, on the other hand, empirical results lead to a better (i.e. intuitive) understanding of mathematical developments (cp. the introduction by M.Atiyah). One should, however, admit that the physicists who thought that they had approached the harbour of final understanding of empirical facts are again thrown out into the open sea of a most extended and far-reaching mathematical research. To quote a vision of W.Heisenberg (reported to one of us by C.Zuckmayer): The physicists enter a huge dome seeing, however, only the basis of large pillars bearing the closing vault which, however, appears still veiled. The lecture program (of strictly interdisciplinary character) thus contains a large number of only apparently

---

[4] P.G.O. Freund, E.Witten, Phys.Lett. B **199**,191(1987); Yu. Manin, Reflections on Arithmetical Physics, preprint, 1987; and references therein.

different (mathematical and physical) contributions serving, however, the same final goal.

In his introductory talk M. A t i y a h gave a general overview of the really striking relations between geometry (in modern mathematical sense) and theoretical physics in its far-reaching developments during the past years. Characteristic examples are the connections between Yang-Mills theory and the topology of 3- and 4-manifolds, strings and elliptic genera, as well as two-dimensional physical field theories and the invariants in mathematical knot theory.

The genuine mathematical contributions compiled in the first chapter yields an idea of the enormous extension of modern mathematical research which by now appears directly or indirectly related to physics, in particular to string theory. Catalyzed by the recent interest of physicists in nearly every aspect of group theory and the theory of infinite dimensional algebras caused by the appearance of such structures in string theory there is now also a renewed interest of mathematicians in these topics.
V. K a c et.al. thus discuss far-reaching structural properties of a special hyperbolic Lie algebra (E_{10}) which appears again in the contribution by H.Nicolai on the N = 16 supergravity. Also the genuine and most extended group and representation theoretical work of B. K o s t a n t has - especially in view of the discussion of special relations between important Lie groups - interesting connections to recent physical research. At the same time the topic of F. H i r z e b r u c h 's article (which represents a basic contribution to modern algebraic geometry), namely elliptic genera appears to be important in string theory, especially in the work on anomaly cancellations and calculations with respect to orbifolds (i.e manifolds with the exception of conical singularities) to be used for a special type of string compactification. An interesting example: A particlular operator whose index is the elliptic genus is obtained in physics as the supercharge of the supersymmetric nonlinear sigma model. Instead of (conventional) spin-manifolds Hirzebruch studies an alternative geometrical structure, namely complex manifolds with first Chern class divisible by N: The elliptic genus is in this case a modular form for $\Gamma_1(N) \subset SL_2(Z)$. In the equivariant case (occuring in physics) rigidity theorems according to Witten and Taubes are proved.
Within the framework of an introductory paper to this first chapter G. M a c k e y gives a review of H.Weyl's program of 1927 where Weyl tried to clear up the foundations of quantum mechanics with the help of group theory and relates Weyl's program to his own recent attempts to

use group theory again in order to clarify the foundations of modern particle theory thus presenting his deep insights into the fundamentals of present-day physics.

Our second chapter on strings contains - without expounding widely known string structures - special mathematical aspects of recent research in that domain.
L. Alvarez-Gaumé gives an overview on operator methods in string theory and shows the deep connections of these methods to the theory of soliton solutions of special non-linear PDE's such as the KP hierarchy which was referred to a.o. by the Kyoto school. The solutions of these equations can be obtained through any Riemann surface with a conformal field on it. Since the worldsheet of a propagating string is represented by a Riemann surface we can imagine that if there is progress in one field of research the other one will profit also. In fact, the work of Alvarez-Gaumé covers string pertubation theory as well as conformal field theories on Riemann surfaces of higher genus. Through these considerations it becomes clear that a huge amount of function theory, algebraic geometry (and even arithmetic algebraic geometry.i.e. the appearance of number theoretical results) is needed to go into the deeper structures of string theory. As in earlier times research in theoretical physics has brought together various branches of mathematics which, so far, did hardly take notice of one another. In very much the same context T. Regge deals in his short contribution with a classical (historical) mathematical problem (Plateau) which connects field theoretical methods with a special aspect of string theory. Also M. E. Mayer connects - in an unexpected way - most special (partly forgotten) mathematical concepts (i.e. groupoids and Lie bigebras) directly with modern string theory. Groupoids can, in fact, be used for a formulation of string field theory (cp. the work of E.Witten) whereas Lie bigebras - also known as Hopf algebras - play a decisive role in two-dimensional statistical physics and integrable systems.

Our third chapter then contains the two central topics of this conference, i.e. conformal invariance and integrable systems.
De Vega's overview about integrable models and quantum groups contained a general introduction from the physical viewpoint. Another definition of quantum group is the Hopf algebra (see M.Mayer's talk) which arose originally from the study of homology and cohomology of Lie groups and was introduced by H.Hopf already in 1941 in his investigations on the topology of group manifolds. Hopf algebras are used now as standard tools in algebraic topology. There are, at the same time, links with physics

via the "quantum inverse scattering method" (QISM) which, in turn, is a method for constructing and studying integrable quantum systems created already in 1978-1979 by Faddeev and collaborators. The QISM is (of course) closely related to the classical ISM and also to Baxter's work on exactly solvable models in classical statistical mechanics. The quantum version (QYBE) of the classical Yang-Baxter equation (CYBE) was also mentioned by de Vega in his talk. A general definition of a conformal field theory from a mathematical viewpoint in terms of an appropriate category and representation theory of groups was given in G. S e g a l ' s talk. As a special case M. K a r o w s k i examines conformal invariance for two-dimensional quantum field theories, in particular integrable models on the lattice which yield conformal QFT's with $c < 1$ in the scaling limit.
J. F r ö h l i c h examines in the first part of his contribution the statistics of fields in two-dimensional QFT in terms of representations of braid groups. The theory of braid groups was created already in 1925 by E.Artin as a tool for solving problems in knot theory. In this connection it is interesting to note that the representations of the braid groups are given by special solutions of the CYBE mentioned above. Finally B. S c h r o e r introduces a non-pertubative approach to QFT which is based on non-commutative algebraic structures such as "exchange algebras" and, again, braid groups to which these algebras exhibit, however, a great similarity and therefore are again related to the YBEs.

The <u>fourth chapter</u> then deals with a 'classical' domain of our conference series, namely the connection of symplectic structures with practically all domains of theoretical physics. Here S. S t e r n b e r g treats bosonic anomalies with the help of the pairing method which, in turn, was introduced in the seventies in the theory of geometrical quantization. The two shorter contributions of M. G o t a y and H. R ö m e r are, in a way, connected to S.Sternberg's far-reaching geometrical viewpoints.
A traditional domain of our earlier meetings is also general relativity: Y. N e ' e m a n discusses in an impressive way different gauge theoretical aspects of general relativity in relation with the problem of quantization. The diffeomorphism group which plays an important role in this research is treated in great detail in the paper by P. M i c h o r , whereas the general problem of defining quantized fields on a curved background is discussed in B. K a y ' s contribution.

The <u>fifth chapter</u> contains an overview on several directions in super-mathematics and super-physics:
S. F e r r a r a summarizes results from these domains

while H. N i c o l a i and O. P i g u e t et al. examine special models. M. S c h e u n e r t gives an introduction into the representation theory of Lie Super-algebras and U. B r u z z o presents an overview on 'superized' well-known mathematical structures.

Concluding remarks are due to M.F. A t i y a h who is deeply involved into the exchange between mathematics and physics and realizes better than anybody else the decisive importance of a mutual interaction and inspiration between the two sciences.

In the last additional (unconventional) paper the most interesting and enlighting personal memories of L. H a l p e r n who was an assistant of Erwin Schrödinger and of the late Paul Dirac.
As a final remark we would like to repeat Paul Dirac's (who was by no means religious in the common sense) beautiful words:
" Theory describing reality best ought to be the most beautiful - because God made it this way."

Acknowledgements

The organizers wish to express their thankfulness to the staff of the Centro di Cultura Scientifica "Alessandro Volta" at Villa Olmo for a really perfect organisation. In particular, we would like to thank the directors of "Centro Volta", G.Casati and F.Canobbio for their continous support, generosity and hospitality and Dott. Chiara de Santis for her enormous work without which the conference wouldn't have been a success. The help of J. Colchero, too, is greatly acknowledged. Much thanks also to the financing organizations:

- NATO Scientific Affairs Division
- Italian National Science Foundation
- University of Milano
- ETH Zürich

Bonn, June 1988

K.Bleuler
M.Werner

THE IMPACT OF PHYSICS ON GEOMETRY

Michael Atiyah
Mathematical Institute
University of Oxford
24-29 St. Giles
Oxford OX1 3LB
England.

ABSTRACT. I survey recent ideas in physics concentrating on their effect on mathematics. Yang-Mills theory has led to results by Donaldson on 4-manifolds and by Floer on 3-manifolds. Solvable models in statistical mechanics have been used by Vaughan Jones to construct polynomial invariants of knots. String theory has produced a new connection between manifolds and modular forms. In all cases it is quantum theory that ties in with topology.

1. INTRODUCTION

In the past decade the interactions between Physics and Geometry have been developing in a quite remarkable way. This is mainly due to the fact that physicists are now exploring complicated non-linear models of a geometrical character as possible explanations for fundamental physical processes. While differential geometry has been closely associated with physics ever since the introduction of Einsten's theory of General Relativity, the recent developments have some new features. On the one hand Quantum Theory is an essential ingredient of the relevant Physics (for elementary particles etc.). On the other hand the required Geometry involves global topological features. The first occurrence of topology in physics was Dirac's famous argument for the quantization of electric charge, and what we

K. Bleuler and M. Werner (eds.), Differential Geometrical Methods in Theoretical Physics, 1–9.

are now witnessing is essentially the non-abelian outgrowth of Dirac's initial idea.

This intimate relation between Geometry and Physics has had beneficial consequences in both directions. Physicists have been able to adopt and use sophisticated mathematical ideas and techniques without which the elaboration of the physical theories would be greatly impeded. Conversely mathematicians have used the insights derived from a physical interpretation to break new ground in Geometry.

As a mathematician I am particularly interested in this exploitation by Geometers of ideas coming in from Physics and, in this lecture, I will describe some of the striking consequences that have emerged. I will leave it to other speakers from the Physics community to describe the other side of the coin.

Perhaps I should begin by drawing attention to the important paper by Witten [12], written some five years ago, which has been very influential in mathematical circles. Many of the ideas I will be describing have their genesis in this paper of Witten, and I strongly recommend anyone interested in this field to begin by studying it.

In Quantum Field Theory Physicists have made an extensive study of two basic prototypes. These are Yang-Mills gauge theories in 4 dimensions and non-linear sigma models in 2 dimensions. The former is the basis for the standard Electro-week model and for quantum chromo dynamics while the latter, originally studied as a "toy" model, is now a basic first step in string theory. It is interesting that both types of theory have significant geometric applications as I shall explain.

In [12] Witten explained the basic relation between supersymmetric quantum field theories and topology. It is in fact through such supersymmetric models that the geometric applications arise. Roughly speaking the ground states of a supersymmetric theory give "homology groups" of an appropriate kind. These are topological invariants of the model in the sense that they are independent of such auxiliary data as Riemannian metrics. As Witten explains one should view this as an infinite-dimensional version, in a suitable sense, of Hodge theory.

In §2 I shall briefly review Yang-Mills theory as developed so spectacularly by S.K. Donaldson and more recently by A. Floer. Then in §3 I will say a few words about the "elliptic genus", a topic which will be developed in more detail in the lecture by F. Hirzebruch. Finally in §4 I shall mention the exciting new results of Vaughan Jones on polynomial invariants for knots.

2. YANG-MILLS THEORY

As is by now widely known Donaldson has used Yang-Mills instantons to prove many remarkable results in 4-dimensional geometry [5]. I recall that, if P is a pricipal $SU(2)$-bundle over a compact Riemannian 4-manifold X, a connection A on P defines a curvature F_A and the Yang-Mills Lagrangian is $\int_X |F_A|^2$. An instanton is a connection which gives the absolute minimum $8\pi^2 k$ for the Lagrangian where k is the second Chern class of P.

The instantons on X, for given k, are parametrized by a space $M_k(X)$. This depends on the background Riemannian metric on X, is a manifold with singularities and is in general non-compact.

Donaldson extracts information from $M_k(X)$ which is independent of the metric on X. In particular he defines, under appropriate restrictions on X, some numerical invariants of X, depending only on its differentiable structure. With these invariants he proves some striking theorems. For example a simply-connected complex algebraic surface cannot be decomposed (non-trivially) as a differentiable connected sum. A "trivial" decomposition is of the kind arising from a blowing-up (birational transformation). For details I refer to [5] and [6].

Instantons (on the 4-sphere $S^4 = R^4 \cup \infty$) were introduced by physicists because of their importance in computing the Euclidean Feynman integral for Yang-Mills theory. Their use by Donaldson came as a complete surprise and the physical significance of Donaldson's work was very unclear. Recently Floer [7] has developed a variant of Donaldson's ideas which is much closer to physics. Essentially Floer works with a Hamiltonian 3-space approach and closely follows the ideas in Witten's paper [12].

Floer starts with a compact oriented 3-manifold Y which is assumed to be a homology sphere, i.e. $H_1(Y,Z) = 0$. He then considers the non-trivial representations $\pi_1(Y) \to SU(2)$ of the fundamental group of Y. These give flat $SU(2)$-bundles over Y which we can interpret physically as Yang-Mills vacua. Quantum-tunnelling between such vacua is then used to construct (in an approximate sense) quantum ground states. As explained in [12] only tunnelling between "adjacent" vacua need be used in this approximation. Such tunnelling is computed by considering instantons on $Y \times R$ which connect the 2 vacua at $t \to \pm\infty$ $(t \in R)$. "Adjacent" means that only instantons with $k = 1$ are used.

Along these lines Floer defines homology groups HF(Y). These are invariants of Y and they are graded by the integers modulo 8. If X is a compact 4-manifold which is cut into 2 pieces along a 3-manifold Y then the Donaldson invariants of X are related to the Floer groups HF(Y).

For a more detailed and leisurely description of this theory I refer to [1].

Note added January 1988

Since my Como lecture Witten has found an appropriate super-symmetric extension of Yang-Mills in which the Donaldson invariants can be expressed directly as Feynman integrals. The relation with the Floer groups is also clear in this model. It is reasonable to hope that this new foundation of Donaldson's work will have some feed-back into physics.

3. MANIFOLDS AND MODULAR FORMS

The heat equation proof of the index theory for the Dirac operator [3] on a manifold M can be formulated in terms of integration on the loop space. As pointed out by Witten, and expounded in [2], the proof can be elegantly interpreted in terms of super-symmetric quantum mechanics. This point of view has been extensively followed up.

Similarly, moving from a particle to a string, one can look at the index of the Dirac-Ramond operator on loop space. This can be reformulated in terms of integration on the space of maps of a 2-dimensional torus into M. This implies an additional "symmetry" for the index of the Dirac-Ramond operator, namely that it is a modular function with respect to the group SL(2,Z).

In fact various anomalies require the preceding statements to be slightly qualified. For example for the loop space of M to be formally a "spin manifold", so that the Dirac-Ramond operators exists, requires the vanishing of the first Pontrjagin class of M.

There are a number of variants and generalizations of this basic idea. For details I refer to Witten [11] or Hirzebruch's lecture in this volume.

In addition to providing a natural explanation of modularity of this "elliptic genus" Witten's approach also leads to a proof of a vanishing theorem generalizing that of [4]. Under suitable assumptions Witten proves that the elliptic genus is constant for circle actions on M. This is a theorem effectively about characteristic class invariants of M of a classical finite-dimensional kind. However Witten's proof uses the infinite-dimensional loop space in an essential and transparent manner. The key observation is that the two circles (i.e. the circle acting geometrically on M and the "loop" circle) can be combined together in an appropriate way.

Taubes [10] has given a rigorous version of Witten's formal argument.

Note added January 1988

Bott and Taubes have further analyzed the proof in [10] and have produced a version similar in spirit to that of [4], with the essential difference that elliptic functions replace rational functions. Moreover Witten's trick of combining the two circles is still employed. Although quite elementary this proof would be somewhat mystifying without the physics background provided by Witten.

4. KNOT INVARIANTS

A few years ago Vaughan Jones [8] astounded topologists by producing a remarkable new polynomial invariant for knots and links in 3-space. It was superficially similar to the classical Alexander polynomial which had been known for over fifty years, but it was independent of it and could be used to distinguish certain knots from their mirror images.

Jones stumbled on his invariant in the course of a study of von Neumann algebras. It now appears that there are several important connections between the Jones polynomials and 2-dimensional physics. The original Jones approach, based on representations of the braid group, appears to be related to conformal field theory. A newer approach treats a plane projection of a knot as a statistical mechanical model. If this model satisfies the Yang-Baxter relations, so that the model is in a certain sense soluble, then the partition function of the model essentially gives a knot invariant. This is elaborated on in the lecture by Fröhlich.

The theory of the Yang-Baxter equations has been extensively devloped and using this it now seems that the original Jones polynomial is only the first of a whole series of knot invariants.

This subject remains to be more fully explored. Moreover it is an intriguing question whether this theory which gives information about

3-dimensional topology is related in any way to the Floer theory described in §3. Arguments in favour of such a connection are put forward in [1].

Note added January 1988

Some new ideas of Witten, motivated by cosmic strings, are at present being investigated for this purpose.

REFERENCES

1. M.F. Atiyah, New invariants of 3 and 4 dimensional manifolds, Amer. Math. Soc. Symposia in Pure Maths., Vol. 48 (1988).
2. M.F. Atiyah, Circular symmetry and stationary-phase approximation (Conference in honour of L. Schwartz), Soc. Math. France, Astérisque Vol. 131 (1985) 43-60.
3. M.F. Atiyah, R. Bott and V.K. Patodi, On the heat equation and the index theorem, Invent. Math. 19 (1973) 279-330.
4. M.F. Atiyah and F. Hirzebruch, Spin manifolds and group actions, Essays on topology and related topics (dedicated to G. de Rham), Springer Berlin (1970).
5. S.K. Donaldson, The Geometry of 4-manifolds, Proc International Congress of Mathematicians, Berkeley 1986.
6. S.K. Donaldson, Polynomial invariants for smooth 4-manifolds, Topology (to appear).
7. A. Floer, An instanton invariant for 3-manifolds (to appear).
8. V.F.R. Jones, A polynomial invariant for knots via von Neumann algebras, Bull. Amer. Math. Soc. 12 (1985) 103-111.

9. V.F.R. Jones, Hecke algebra representations of braid groups and link polynomials, Ann. of Math. 126 (1987) 335-388.

10. C.H. Taubes, S^1 actions and elliptic genera, preprint.

11. E. Witten, Quantum field theory, Grassmanians, and algebraic curves, Commun. in Math. Physics, Vol. 113, No. 4 (1988) 529-600.

12. E. Witten, Supersymmetry and Morse theory, J. Diff. Geom. 17 (1982) 661-692.

WEYL'S PROGRAM AND MODERN PHYSICS

George W. Mackey
Harvard University
Department of Mathematics
Science Center, One Oxford Street
Cambridge, Massachusetts 02138
USA

ABSTRACT. The author is presently engaged in trying to fit as much as possible of modern elementary particle physics into a group theoretical framework described at some length in the reference cited at the end of section IV. This article is in part an exposition of the relationship of this program to one formulated by Weyl in 1927, in part a review and revision of material from the reference mentioned above and in part a description of ideas and incomplete work relevant to carrying the program further.

I. Weyl's program.

Wigner and Weyl introduced group theoretical ideas into quantum physics in independent papers published in 1927 (Zeitschrift für Physik 43 and 46). While Wigner was concerned with simplifying calculations in the quantum theory of atomic spectra, Weyl used group theory in an attempt to clarify the foundations. In his introduction Weyl singles out two basic foundational questions: (1) How does one arrive at the self adjoint operators which correspond to various concrete physical observables? (2) What is the physical significance of these operators; i.e., how does one deduce physical statements? He goes on to say that question (2) has been more or less satisfactorily dealt with by von Neumann and that he proposes to concern himself chiefly with question (1). He asserts that question (1) is the deeper question and "Hier glaube ich mit Hilfe der Gruppentheorie zu eine tieferen Einsicht in der wahren Sachverhalt gelangt zu sein".

Weyl's answer to question (1) rested on his global reformulation of the Heisen-

K. Bleuler and M. Werner (eds.), Differential Geometrical Methods in Theoretical Physics, 11–36.

berg commutation relations in terms of associated unitary representations of the real line. If P and Q are self adjoint operators then $t \to e^{itQ}$ and $s \to e^{isP}$ are unitary representations of the additive group of the real line. Moreover it is easy to verify, at least on a formal level, that $QP - PQ = iI$ if and only if the associated group representations satisfy the commutation relation

$$e^{itQ}e^{isP} = e^{ist}e^{isP}e^{itQ}$$

for all s and t. More generally if $Q_1, Q_2 \cdots Q_n$ are mutually commuting self adjoint operators then $t_1, t_2, \cdots t_n \to e^{it_1 Q_1}e^{it_2 Q_2} \ldots e^{it_n Q_n} = A_{t_1, t_2, \cdots t_n}$ is a unitary representation of the additive group of an n dimensional real vector space. Let $s_1, s_2 \cdots s_n \to e^{is_1 P_1}e^{is_2 P_2} \ldots e^{is_n P_n} = B_{s_1, s_2, \cdots s_n}$ be a second such representation where $P_1, P_2 \cdots$ is another n tuple of mutually commuting self adjoint operators. Then it follows from the case $n = 1$ considered above that the Q_i and P_j satisfy the classical Heisenberg commutation relations if and only if $A_t B_s = e^{i(t \cdot s)} B_s\, A_t$ where $(t \cdot s) = t_1 s_1 + \cdots t_n s_n$, $s = s_1, s_2, \cdots s_n \quad t = t_1, t_2, \cdots t_n$. These commutation relations satisfied by the A_t and the B_s are sometimes referred to as the Heisenberg commutation relations in Weyl or global form.

Suppose now that $t \to A_t \quad s \to B_s$ are unitary representations of the above form and let $W_{t,s} = A_t B_s$. As Weyl observed (and as is easy to verify) to say that A and B satisfy the Heisenberg commutation relations in global form is equivalent to saying that

$$W_{t+t', s+s'} = e^{-i(t' \cdot s)} W_{t,s} W_{t', s'}.$$

But this is precisely the statement that the mapping $t, s \to A_t B_s$ is a so-called *projective unitary representative* of the commutative group of all $2n$ tuples of real numbers under addition with the function $\sigma(t, s; t', s') = e^{-i(t' \cdot s)}$ as "multiplier". Recall that, quite generally, one says that a mapping $x \to U_x$ of a group G into the unitary operators in some Hilbert space $\mathcal{H}(U)$ is a projective unitary representation if $U_{xy} = \sigma(x, y) U_x U_y$ for all x and y in G, where $\sigma(x, y)$ is a complex number of modulus one (which of course depends upon x and y). The function σ is called a projective multiplier.

As Weyl emphasized in his paper, the generalization of the concept of group representation to include projective representations is especially important for quantum mechanics. This is because unitary operators occur in quantum mechanics as automorphisms of the state space and two unitary operators U^1 and U^2 define the same automorphism whenever $U^2 = cU^1$ where c is a complex number of modulus one. This means that in a homomorphism of a group G into a group of automorphisms of a state space the unitary operators are determined only up to a constant factor of modulus one and if one chooses this factor at random one can only deduce that $U_{x_1 x_2}$ is a constant multiple of $U_{x_1} U_{x_2}$ *not* that $U_{x_1 x_2} = U_{x_1} U_{x_2}$. Of course one can attempt to choose the arbitrary constant in each U_x in such a manner that one does have $U_{x_1 x_2} = U_{x_1} U_{x_2}$ but this is not always possible. Indeed if $U_{x_1 x_2} \equiv \sigma(x_1, x_2) U_{x_1} U_{x_2}$ for some unitary representation of group G and we define $V_x = r(x) U_x$ where r is an arbitrary function from G to the complex numbers of modulus one then a simple calculation shows that $V_{x_1 x_2} \equiv \sigma'(x_1, x_2) V_{x_1} V_{x_2}$ where $\sigma'(x_1, x_2) \equiv \frac{\sigma(x_1, x_2) r(x_1 x_2)}{r(x_1) r(x_2)}$. Evidently one can choose r so that $\sigma'(x_1, x_2) \equiv 1$ if and only if $\sigma(x_1, x_2)$ can be written in the form $\frac{r(x_1) r(x_2)}{r(x_1 x_2)}$. It is easy to give examples in which one may prove that no such r exists. When it does exist one says that the multiplier is *trivial.* More generally one says that two multipliers are *equivalent* if their ratio is trivial. This is because if σ_1 and σ_2 are equivalent multipliers then every projective representation U with multiplier σ_1 may be made into one with multiplier σ_2 by simply multiplying each U_x by a suitable constant.

Consider now the multiplier $\sigma(t, s, t', s') = e^{-i(t' \cdot s)}$ introduced above. Weyl was much impressed by the fact that, to within equivalence, this is the *only* multiplier on the additive group R^{2n} of all $2n$ tuples of real numbers having a certain property of "non degeneracy". Moreover, although he did not give a rigorous proof, Weyl had reason to believe that every irreducible projective representation of R^{2n} with this particular multiplier is unitarily equivalent to every other. In other words there is essentially only one non degenerate projective multiplier σ for the commutative group R^{2n} and essentially only one irreducible unitary projective representation having σ as multiplier. Moreover this projective representation is just the global counterpart of a set of $2n$ self adjoint operators satisfying the Heisenberg commutation relations.

The fact that sets of operators satisfying the Heisenberg commutation relations may be characterized in such a simple and fundamental group theoretical manner was stated by Weyl as "The kinematical structure of a physical system is expressed by an irreducible Abelian group of unitary ray representations in system space". It seemed to him to be a fact of deep significance and was his answer to question (1).

Weyl also showed that the smallest and simplest commutative group having a non trivial projective multiplier σ is the four element group $Z_2 \times Z_2$. This group also admits an essentially unique irreducible projective representation. It is two dimensional and the three two by two matrices representing the non identity elements may be chosen to be $\begin{pmatrix} 1 & 0 \\ 0 & -1 \end{pmatrix}$, $\begin{pmatrix} 0 & 1 \\ 1 & 0 \end{pmatrix}$ and $\begin{pmatrix} 0 & i \\ -i & 0 \end{pmatrix}$. These of course are the well known Pauli spin matrices.

While Weyl's answer to question (1) is stimulating and suggestive it is far from conclusive or logically compelling. It is significant, however, as being perhaps the first step in the program of deriving the basic equations and relationships in quantum mechanics from group theoretical symmetry principles. It is perhaps therefore appropriate to refer to it as Weyl's program.

II. Stone's Paper on one Parameter Unitary Groups.

Weyl wrote in the spirit of a physicist and made no attempt to give rigorous mathematical proofs of all of the mathematical theorems that played a role in his analysis. These gaps were largely filled in 1930, in a note published in the Proceedings of the U.S. A. National Academy of Sciences, by Marshall Stone. In it he announced the proofs of two theorems about unitary groups in Hilbert space implicitly conjectured by Weyl. One of these states that every (strongly continuous) unitary representation of the additive group of the real line is of the form $t \to e^{itQ}$ where Q is a uniquely determined self adjoint operator and the other is the theorem that the Heisenberg commutation relations in Weyl form have (to within unitary equivalence) a unique irreducible solution. The first of these theorems is well known as "Stone's theorem" and the second as the theorem of "Stone and von Neumann". The reason for the terminology in the second case is that von Neumann published a proof shortly after Stone's paper appeared and Stone did not publish a proof.

Actually Stone formulated his theorem about unitary representations of the additive group of the real line as an analogue of Hilbert's spectral theorem and not as a theorem connecting such representations with self adjoint operators. Since this formulation will be important for what follows we give it in detail. Let S be any set in which there is given a Borel structure, that is, a distinguished family of subsets containing the whole space and closed under countable unions, countable intersections and the taking of complements. By a projection valued measure P on S, one means a mapping $E \to P_E$ of the distinguished subsets E (usually called the Borel subsets) of S into projection operators P_E in some separable Hilbert space $\mathcal{H}(P)$ such that the following conditions hold.

(1) $P_{\{0\}} = 0$ and $P_S = I$.

(2) $P_{E\cap F} = P_E P_F$ for all Borel sets E and F.

(3) $P_{E_1 \cup E_2 \cup \cdots} = P_{E_1} + P_{E_2} + \cdots$,

whenever $E_1, E_2 \cdots$ are Borel subsets of S such that $E_i \cap E_j = 0$ whenever $i \neq j$.

Since this concept is not as familiar to physicists as it could be – and perhaps ought to be, I pause to give some examples and make some remarks.

Example 1. Let H be a self adjoint operator with a discrete spectrum. Let $\lambda_1, \lambda_2, \cdots$ be its distinct eigenvalues and let M_{λ_j} be the closed subspace of the underlying Hilbert space spanned by all eigenvectors with eigenvalue λ_j. For each Borel subset E of the real line let P_E denote the projection operator whose range is the closed linear subspace spanned by the union of all M_{λ_j} with λ_j in E. One verifies easily that $E \to P_E$ is a projection valued measure on the real line. It is easy to see that given P one can reconstruct H and its decomposition into eigenspaces.

Example 2. Let μ be a measure in S which is a finite in the sense that $S = E_1 \cup E_2 \cdots$ where $\mu(E_j) < \infty$ for all j. Let $\mathcal{H}$ denote the Hilbert space $\mathcal{L}^2(S, \mu)$ of all complex valued functions on S which are measurable and square summable with respect to μ. For each Borel subset E of S let us define P^{μ}_{E} to be the operator which takes f in $\mathcal{L}^2(S,\mu)$ into $\phi_E f$ where ϕ_E is the function which is 1 or E and 0 on $S-E$. One checks at once that P^{μ}_{E} is a projection operator and that $E \to P^{\mu}_{E}$ is a projection valued measure.

The projection valued measures of example 2 come close to being universal. To state this precisely one needs the notion of the direct sum of two projection valued measures. Let P and P' be any two projection valued measures on the same Borel space S. Then their direct sum $P \oplus P'$ is the projection valued measure whose Hilbert space is the direct sum $\mathcal{H}(P) \oplus \mathcal{H}(P')$ and $(P \oplus P')_E$ is the operator taking f_1, f_2 into $P_E(f_1), P'_E(f_2)$. Similarly one defines the direct sum of any finite or countably infinite family $P, P', P'' \cdots$ of projection valued measures on S.

Example 2 is almost universal in the sense that *every* projection valued measure on S can be shown to be a finite or countably infinite direct sum of projection valued measures of the form P^{μ} where μ varies over different measures on S.

It may be helpful to remark that whenever S is countable and every one point set is a Borel set, then $\mathcal{H}(P)$ is simply the direct sum of the ranges of the $P_{\{s\}}$ and this direct sum decomposition completely determines P. A projection valued measure on S *is* a direct sum of Hilbert spaces parametrized by the points of S. More generally it is intuitively useful to think of a general projection valued measure as a positively continuous direct sum or direct integral of "infinitesimal" Hilbert spaces parametrized by the points of S.

In terms of projection valued measures Stone's theorem may be formulated as follows. For every strongly continuous unitary representation U of the additive group of the real line there exists a unique projection valued measure P defined on all Borel subsets E of the real line such that $\mathcal{H}(P) = \mathcal{H}(U)$ and for all elements f of this Hilbert space $(U_x(f)\cdot f) = \int_{-\infty}^{\infty} e^{ixt} d(P_t(f)\cdot f)$. Moreover every projection valued measure P on the real line occurs for some U and the correspondence $U \to P$ is one-to-one and onto between unitary equivalence classes. In understanding the meaning of the integral in the above formula one should note that for each f the mapping $E \to (P_E(f)\cdot f)$ is an ordinary measure on the line and the integral is just the integral of the function $t \to e^{ixt}$ with respect to this measure.

III. The Imprimitivity Theorem.

Using Stone's theorem it is easy to rewrite the Stone-von Neumann theorem in a form which admits far reaching generalizations. Observe first that there is no difficulty in generalizing Stone's theorem to apply to unitary representations of the additive group of all k tuples of real numbers. One simply replaces projection valued measures on the line by projection valued measures on k space and the fundamental equation by

$$(U_{x_1,x_2,\cdots x_k}(f)\cdot f) = \int e^{i(x_1 t_1 + x_2 t_2 + \cdots x_n t_n)} d(P_{t_1,t_2,\cdots t_n}(f)\cdot f).$$

Now consider the Weyl form of the Heisenberg commutation relations.

$$(*) \qquad A_{t_1,t_2,\cdots t_k} B_{s_1,s_2,\cdots s_k} = e^{i(t_1 s_1 + \cdots + s_k t_k} B_{s_1,s_2 \cdots s_k} A_{t_1,t_2 \cdots t_k}$$

where A and B can be regarded (in view of Stone's theorem) as otherwise unrestricted strongly continuous unitary representations of the additive group of all k tuples of real numbers.

The basic trick in putting the Stone-von Neumann theorem in a more generalizable form consists in applying the generalized Stone's theorem to one, but *not both*, of the two representations A and B. Let P denote the projection valued measure corresponding to B by this theorem. A straightforward calculation then shows that A and B satisfy $*$ if and only if A and P satisfy the commutation relations

$$(**) \qquad A_{t_1,t_2 \cdots t_n} P_E = P_{E-(t_1,t_2 \cdots t_n)} A_{t_1,t_2 \cdots t_n}$$

where $E - (t_1 \cdots t_n)$ is the set of all k-tuples of the form $x_1 - t_1, x_2 - t_2 \cdots x_n - t_n$ with $x_1, x_2, \cdots x_n \in E$. It follows that seeking all possible solutions of $(*)$ is completely equivalent to seeking all possible solutions of $(**)$.

The reason that $(**)$ is more generalizable than $(*)$ is that the factor $e^{i(t_1 s_1 + \cdots + t_k s_k)}$ has disappeared and only the group structure of the common domain of A and P plays a role. Let G be any group with a Borel structure and let A, P be a pair consisting of a unitary representation A of G and a projection valued measure P defined on the Borel subsets of G. Then one can write down the following commutation relation which A and P might satisfy

$$(***) \qquad A_x P_E = P_{[E]x^{-1}} A_x$$

where $[E]x^{-1}$ is the set of all yx^{-1} with y in E. Evidently $(* * *)$ reduces to $(**)$ when G is the additive group of all k tuples of real numbers.

The present author noticed these things some forty years ago, after reading Stone's paper and wondering how one might go about proving the uniqueness theorem. He was then pleased to discover that he could prove a corresponding uniqueness theorem when the additive group of k tuples of real number was replaced by an arbitrary separable locally compact group G. The results were published in 1949 in the Duke Mathematical Journal.

Finding a particular solution of $(* * *)$ for any separable locally compact group G is quite easy. By a celebrated theorem of Haar there is an essentially unique invariant σ finite measure μ defined on all Borel subsets E of G and one obtains a unitary representation U of G in the Hilbert space $\mathcal{L}^2(G, \mu)$ by setting

$U_x(f)(y) = f(yx)$. This is the so-called regular representation of G. Also using the Haar measure μ one obtains the projection valued measure P^μ defined in II. A trivial calculation now shows that

$$U_x P^\mu_E = P^\mu_{[E]x^{-1}} U_x \text{ for all } x \text{ and } E$$

so that U and P^μ satisfy $(***)$. One can also show that no non trivial closed subspace of $\mathcal{L}^2(\sigma, \mu)$ is invariant under all U_x and all P^μ_E so that the pair is irreducible. Our generalization of the Stone-von Neumann theorem now states that for every separable locally compact group G the solution of $(***)$ first constructed is (up to unitary equivalence) the only one that exists.

We are not finished however. The regular representation of a separable locally compact group G has an obvious generalization which also has an associated projection valued measure which satisfies $(***)$. It is natural to investigate uniqueness in this case as well. Let H be any closed subgroup of G and let G/H denote the set of all right H cosets; that is, the set of all closed subsets of G of the form Hx where $x \in G$. Two such cosets are either disjoint or identical and so define a partition of G. If Hx is a right H coset and y is an element of G then $(Hx)y$ is by definition $H(xy)$. Each y thus defines a permutation of G/H. G/H inherits a Borel structure from that in G. Let us suppose that G/H admits a σ-finite measure μ which is preserved by all the transformations $Hx \to Hxy$. Then taking $\mathcal{L}^2(G/H, \mu)$ instead of $\mathcal{L}^2(G, \mu)$ we produce a unitary representation U as before by setting $U_x(f)(y) = f(yx)$ where now y is a right coset instead of a group element. P^μ is defined just as before but now E is a Borel subset of G/H rather than G. The identity $(***)$ holds just as before where now $[E]x^{-1}$ is the set of all transforms by x^{-1} of a set E of right H cosets. To have $(***)$ make sense it is not necessary that the domain of P be subsets of G but only subsets of a space S on which G acts.

Our question now is this. Let U' be any unitary representation of G and let P' be any projection valued measure defined on G/H such that U' and P' are collectively irreducible and satisfy $(***)$. Is the pair U', P' unitarily equivalent to the U, P^μ defined above?

The answer is no! Given any unitary representation L of the closed subgroup H of G one can define a generalization of the construction of U, P^μ and all of these pairs satisfy $(***)$. Two such pairs are unitarily equivalent if and only if the corresponding representations L are unitarily equivalent and such a pair is irreducible if and only if L is irreducible. One no longer has a uniqueness theorem but a classification theorem. The solutions of $(***)$ for a given coset space G/H correspond one-to-one modulo unitary equivalence to the unitary representations of the subgroup H. One gets a uniqueness theorem only when H has only one equivalence clas of irreducible unitary representations, e.g., when H contains only the identity element e and $G/H = G$.

To motivate the generalized construction it is useful to note that $\mathcal{L}^2(G/H,\mu)$ can also be described as a space of functions on G which satisfy the identity

$$(\dagger) \qquad f(\xi x) = f(x) \text{ for all } \xi \in H \qquad x \in G.$$

It consists in fact of all measurable such functions such that $|f(x)|^2$ when regarded as a function on G/H has a finite integral with respect to μ. In our generalization we replace complex valued functions f by vector valued functions whose values are taken in the Hilbert space $\mathcal{H}(L)$ of the representation L of H and the identity ($\dagger$) is replaced by

$$(\dagger\dagger) \qquad f(\xi x) = L_\xi(f(x)).$$

For such functions the self scalar product $(f(x) \cdot f(x))$ satisfies ($\dagger$) and can be integrated with respect to μ as before.

Thus one is led to define a Hilbert space $\mathcal{H}'$ whose elements are measurable functions f from G to $\mathcal{H}(L)$ which satisfy the identity ($\dagger\dagger$) and are square summable in the sense that

$$\int_{G/H} (f(x) \cdot f(x)) d\mu(x) < \infty.$$

Of course one identifies two functions when they are equal almost everywhere with respect to μ. If f is any element in $\mathcal{H}'$ then for all x its translate $y \to f(yx)$ is also clearly in $\mathcal{H}'$ and one obtains a (strongly continuous) unitary representation of G in $\mathcal{H}'$ by setting $U_x^L(f)(y) = f(yx)$. One associates a projection valued measure P^L with U^L by simply noting that the definition given above applies equally well to vector valued functions. It is easy to check now that the pair U^L, P^L satisfies the commutation relations $(***)$.

We may now give a formal statement of the theorem alluded to in the title of this section.

THEOREM. *Let U be a unitary representation of the separable locally compact group G and let H be a closed subgroup of G. Let P be any projection valued measure in G/H whose values are projections in $\mathcal{H}(U)$ such that U and P satisfy the commutation relations $(***)$. Then there exists a unitary representation L of H such that the pair U^L, P^L is unitarily equivalent to the pair U, P. Moreover if L and M are two unitary representations of H then the pair U^L, P^L is unitarily equivalent to the pair U^M, P^M if and only if L and M are unitarily equivalent. Finally the pair U^L, P^L is irreducible if and only if L is irreducible and any direct sum decomposition of a reducible L implies an associated direct sum decomposition of U^L, P^L. $U^{M^1 \oplus M^2}, P^{M^1 \oplus M^2} \simeq U^{M^1}, P^{M^1} \oplus U^{M^2} \oplus P^{M^2}$.*

This theorem was first stated and its proof sketched by the present author in 1949 in a note in the U.S.A. National Academy of Sciences Proceedings. Strictly speaking, it only makes sense as stated when G/H has an invariant measure μ. However, using so-called quasi invariant measures, this limitation is easily removed. It is to be noticed that one obtains as a by-product a natural way of passing from an arbitrary irreducible unitary representation L of an arbitrary closed subgroup H of a separable locally compact group G to a unitary representation U^L of the whole group G. This representation U^L of G is called the representation L induced by L. It turns out that very many infinite dimensional unitary representations can be obtained as induced representations U^L where L is finite dimensional and even one dimensional.

In view of the role that projective representations play in quantum mechanics it is natural to ask whether the imprimitivity theorem may be extended so as to apply to projective unitary representations. The answer is yes and the formulation of the generalized theorem is obvious once we have extended the notion of inducing, $L \to U^L$, to apply when L is a projective unitary representation of H. Let σ be any multiplier for the whole group G and let L be any unitary projective representation of H whose multiplier is the restriction of σ to H. In the definition of U^L for ordinary representations, given above, replace the identity $f(\xi x) = L_\xi f(x)$ by $f(\xi x) = \sigma(\xi, x) L_\xi f(x)$ and the defining equation $U_x^L(f)(y) = f(yx)$ by $U_x^L(f)(y) = f(yx)/\sigma(y, x)$. With these changes U^L becomes a projective unitary representation with multiplier σ – in brief a unitary σ representation of G. Thus the inducing construction generalizes to allow us to construct a σ unitary representation of G for each σ unitary representation of H.

IV. The Imprimitivity Theorem and Weyl's Program.

In generalizing from the uniqueness theorem for solutions of the Heisenberg commutation relations to the imprimitivity theorem it may appear that we have been indulging in a purely mathematical digression which has taken us far away from any conceivable application to physics. We shall now show that this is definitely not the case and that in fact the imprimitivity theorem is just what is needed to give a much better answer to Weyl's question (1) than could possibly have been given at the time Weyl's paper was written.

We shall proceed in steps as follows.

(1) We shall recall the conventional quantum mechanical model for the unconstrained motion of a spinless particle of mass m.

(2) We shall reformulate this model in a simple and natural manner which is more intrinsic and coordinate free and which at the same time involves a pair U, P satisfying $(***)$ in an essential way.

(3) We shall observe that the existence of a pair U, P satisfying $(***)$ follows from simple and natural axioms involving little more than group theoretical invariance principles.

(4) Using the imprimitivity theorem we shall find the most general possible pair satisfying the axioms and observe that the Heisenberg commutation relations

are satisfied in all cases. Solutions of $(***)$ other than that provided by the conventional model will include the conventional model for particles with spin and other kinds of "internal degrees of freedom".

In the conventional model for a free spinless particle one chooses a rectangular coordinate system in space so that the points of space are in one-to-one correspondence with triples of real numbers x, y, z. The Hilbert space of quantum states is then $\mathcal{L}^2(x, y, z)$ where the relevant measure is the usual Lebesgue or volume measures $dx\, dy\, dz$. The self adjoint operators X, Y, Z corresponding to the observables defined by the three coordinates are then $X(\psi) = x\psi \quad Y(\psi) = y\psi \quad Z(\psi) = z\psi$ and those corresponding to the x, y, z components of the momentum P_X, P_Y, P_Z are

$$P_X(f) = \frac{1}{i}\frac{\partial f}{\partial x} \quad P_Y(f) = \frac{1}{i}\frac{\partial f}{\partial y} \quad P_Z(f) = \frac{1}{i}\frac{\partial f}{\partial z}.$$

Here we have chosen our units so that Planck's constant h is 2π.

In reformulating this model our main observation is that the unbounded operators X, Y and Z may be usefully replaced by a single projection valued measure on space. For each Borel subset E of $S(=$ set of all $x, y, z)$ let $Q_E\psi = \psi\phi_E$ where $\phi_E(x, y, z) = 1$ when $x, y, z \in E$ and $\phi_E(x, y, z) = 0$ when x, y, z is not in E. Using the spectral theorem for self adjoint operators one can easily construct the Q_E when X, Y and Z are given and vice versa. Actually, given the Q_E one can immediately describe the self adjoint operator corresponding to any real valued Borel function of x, y and z. As functions of E the Q_E evidently constitute a projection valued measure as defined in II. Moreover each Q_E has a simple physical interpretation. It is the self adjoint operator corresponding to the observable which is one when the particle is in the set E and zero otherwise.

One can describe the operators P_X, P_Y, P_Z in a coordinate free way using a rather different maneuver. Let $\mathcal{E}$ denote the group generated by the translations and rotations in space. Then for each α in $\mathcal{E}$. The mapping $x, y, z \rightarrow (x, y, z)\alpha$ is volume preserving and hence Lebesgue measure preserving. Hence the linear operator U_α such that $U_\alpha(f)(x, y, z) = f(x, y, z)\alpha)$ is a unitary operator and one can show that $\alpha \rightarrow U_\alpha$ is a unitary representation of $\mathcal{E}$. Indeed the fact that this is so is a special case of a much more general theorem stated in III. Consider now the restriction of U to the subgroup of all members of $\mathcal{E}$ of the form $x, y, z \rightarrow x+\lambda, y, z$ that is the subgroup of all translations in the x direction. By Stone's theorem this group may be written in the form $\lambda \rightarrow e^{iA\lambda}$ where A is a self adjoint operator and a simple calculation shows that $A = P_X$. We construct P_Y and P_Z analogously. Just as giving the projection valued measure Q gives all possible coordinate observables so giving the group representation U gives not only P_X, P_Y, and P_Z but the linear momentum in any direction. Actually giving U gives still more. In addition to the one parameter subgroups of translation $\mathcal{E}$ has other one parameter subgroups; in particular one parameter subgroups of rotations about the origin. Dealing with these in an analogous fashion leads to calculation of the self adjoint operators corresponding to the angular momentum observables. The connection with the imprimitivity theorem is now immediate. It follows by specializing a calculation already mentioned in III that U and Q satisfy the identity $(***)$.

With the foregoing considerations at our disposal let us now reconsider Weyl's question (1). Weyl wanted to know why the position and momentum operators take the special form that they do as described explicitly at the beginning of this section. We have now refined the question somewhat by replacing the position and moment operators by the pair U, Q consisting of a unitary representation of $\mathcal{E}$ and a projection valued measure Q on space. We repair an earlier omission here by pointing out the easily verified fact that space may be identified with $\mathcal{E}/R$ where R is the closed subgroup of $\mathcal{E}$ consisting of all rotations about $0,0,0$, i.e., all α in $\mathcal{E}$ which carry $0,0,0$ with itself.

Let us suppose that we have a quantum mechanical particle moving in a space S and let $\mathcal{H}$ be its Hilbert space of states. We no longer assume at the outset that $\mathcal{H} = \mathcal{L}^2(S)$ or that $\mathcal{H}$ has any other special representation. In addition we do not assume any particular form for the operators corresponding to the position and momentum observables. Instead we merely assume that for each Borel set E of S we are given a self adjoint operator Q_E in $\mathcal{H}$, that for each α in $\mathcal{E}$ the isometry group of space that we are given a unitary operator U_α in $\mathcal{H}$ and that these satisfy certain natural conditions which we now explain. Q_E is the self adjoint operator corresponding to the observable which is one when the particle is in E and zero when it is not. Since this observable takes on only the values 0 and 1, Q_E must be a projection operator. Moreover with a little reflection it is easy to convince oneself that, in its dependence on E, it is natural to assume that it is a projection valued measure. We do. In other words we assume that the position observables of or system are described by some (completely unspecified) projection valued measure on S. Our second assumption is that the physics of our system is invariant under all rigid motions in space and hence that every α in $\mathcal{E}$ defines an automorphism of $\mathcal{H}$ which we may assume to be described by a unitary operator U_α. The mapping $\alpha \to U_\alpha$ would be a representation of $\mathcal{E}$ if it were not for the fact that for every complex c with $|c| = 1$, U_α and cU_α define the same automorphism. We may assume only that $\alpha \to U_\alpha$ is a projective representation of $\mathcal{E}$ with some multiplier σ. Our only other assumption is that Q and U must be related in a way that is forced on us by the very definition of invariance. If T is the operator corresponding to some observable $\mathcal{O}$ and T' is the operator corresponding to the transform by α then T' must equal the transform of T by U_α. In particular we must have

$$U_\alpha Q_E U_\alpha^{-1} = Q_{[E]\alpha^{-1}}$$

for all α in $\mathcal{E}$ and all Borel subsets E of $S = \mathcal{E}/R$.

But this last identity is just $(***)$.

In other words, we have found an *a priori* argument leading to a description of the quantum mechanics of a free particle by a pair U, Q consisting of a projective unitary representation U of $\mathcal{E}$ and a projection valued measure Q on $S = \mathcal{E}/R$; this pair necessarily satisfying the crucial identity $(***)$ of the imprimitivity theorem. The imprimitivity theorem itself then allows us to survey *all* possibilities for U, Q. It tells us in particular that to within unitary equivalence there is only a discrete countable infinity of possibilties – one for each equivalence class of unitary projective representations of the compact subgroup R of $\mathcal{E}$.

The simplest cases are those in which the inducing projective representation L of R is irreducible. It turns out that $\mathcal{E}$ has just two equivalence classes of projective multipliers and that the one not containing the identity can be chosen so as to take on only the values 1 and -1. These multipliers remain inequivalent when restricted to R. Thus there are two fundamentally different kinds of particles depending upon whether L is an ordinary representation or a projective representation with a non trivial multiplier. In the first case there is precisely one equivalence class of irreducible representations whose dimension is d for $d = 1, 3 \cdots$ and in the second case there is just one equivalence clss of irreducible projective representations of dimension d for $d = 2, 4, 6 \cdots$. It is customary to denote the projection representation whose dimension is d by the symbol $D_{(\frac{d-1}{2})}$ whether its multiplier is trivial or not. The case in which L is irreducible and equal to $D_{\frac{d-1}{2}}$ coincides exactly with the model for a free particle of "spin $\frac{d-1}{2}$" which was introduced in the early days of quantum mechanics.

To summarize: Weyl based a suggestive, if not logically compelling, answer to his question (1) on the conjectured uniqueness of the solutions to the Heisenberg commutation rules and his elegant reformulation of this uniqueness theorem in terms of projective unitary representations of finite dimensional vector groups. We have just seen that a mathematically inspired generalization of the uniqueness theorem may be used to go much further in the direction Weyl wished to go.

It is possible to go further still. Just how far remains to be seen. Some steps in this direction were taken in the author's Oxford lectures during the academic year 1966-67. (See Chapters 17-24 in the published version of these lectures, "Unitary Group Representations in Physics, Probability and Number Theory", Benjamin Cummings 1978). The author is currently at work on extending the scope of the material in these chapters with the hope of making significant contact with a large part of present day elementary particle physics.

In the remaining sections we shall outline some of our ideas as well as relevant material from the book cited in the last paragraph.

V. The dynamics of a free particle and velocity operators.

Up to now we have only considered what might be called quantum statics. Nothing has been said about how the Hilbert space vectors describing the states of our particle change with time. It follows easily from an axiomatic approach to the Hilbert space formulation of von Neumann that there must be a unitary representation $t \to V_t$ of the additive group of the real line (thought of now as the group of translations in time) such that if ϕ defines the state of the system at some instant of time then $V_t(\phi)$ defines the state of the same system t times units later. In order to describe a "particle" completely we must not only be given Q and U but the unitary representation V. Weyl's question (1) must be amplified to include the question: (1′) Given Q and U what can we say about V? Are there principles with some degree of *a priori* justification which limit the possibilities in some useful way? In the case of a "free" particle it is easy to convince one's self that the mathematical counterpart of freeness is the assumption that U_α and V_t must commute for all α and t – at least up to a multiplicative factor. Thus the mapping $\alpha, t \to U_\alpha V_t = W_{\alpha,t}$ must be a projective unitary representation of the group $\mathcal{E} \times T$ where T denotes the time translation group. Actually an analysis of the possible projective multipliers on $\mathcal{E} \times T$ shows that we may assume without loss of generality that U_α and V_t actually commute.

By Stone's theorem V_t may be written uniquely in the form $V_t = e^{itH}$ where H is a self adjoint operator and one checks easily that $V_t U_\alpha = U_\alpha V_t$ for all t and α if and only if $U_\alpha H = H U_\alpha$ for all α. Finding the most general possible V is equivalent to finding the most general self adjoint operator H which commutes with all U_α. This is a problem in the theory of unitary representations which standard theory enables one to solve very completely by exploiting the fact that U is induced by a representation L of G. The answer is especially simple when L is irreducible and one-dimensional so that we are dealing with a free particle of zero spin. In that case one cam forget group representations and prove that H is necessarily of the form $\rho\left(-\left(\frac{\partial^2}{\partial x^2} + \frac{\partial^2}{\partial y^2} + \frac{\partial^2}{\partial z^2}\right)\right)$ where ρ is an arbitrary real valued Borel function defined on the positive real axis. (We assume of course that we have realized our Hilbert space as $\mathcal{L}^2(x,y,z)$ so that the operators corresponding to the coordinate and momentum observables have their conventional form.) To define ρ we observe that the Fourier transform of the operator $-\left(\frac{\partial^2}{\partial x^2} + \frac{\partial^2}{\partial y^2} + \frac{\partial^2}{\partial z^2}\right)$ is multiplication by $u^2 + v^2 + w^2$ so that $\rho\left(-\left(\frac{\partial^2}{\partial x^2} + \frac{\partial^2}{\partial y^2} + \frac{\partial^2}{\partial z^2}\right)\right)$ is the inverse Fourier transform of the operator of multiplication by $\rho(u^2 + v^2 + w^2)$.

Without making further assumptions one can conclude no more. *Any* choice of ρ will lead to a V satisfying all assumptions. One can select out the conventional choices for ρ by imposing space time invariance. The group $\mathcal{E} \times T$ occurs in a canonical way as a closed subgroup of the group $\mathcal{G}$ of all automorphisms of space time in the sense of Galileo and also as a closed subgroup of the Poincarè group $\mathcal{P}$ of all automorphisms of space time in the sense of Einstein. One can show that to have our particle space time invariant in the sense of Galileo (respectively Einstein) is equivalent to assuming that W has an extension to a (possibly projective) unitary representation $\widetilde{W}$ of $\mathcal{G}$ (respectively $\mathcal{P}$). Here the projective multiplier for $\widetilde{W}$ must restrict on $\mathcal{E} \times T$ to the given projective multiplier of $\mathcal{E} \times T$. It turns out that in the case of $\mathcal{G}$, $\widetilde{W}$ will have a non trivial projective multipier even when W is an ordinary representation of $\mathcal{E} \times T$.

Assuming Galilean relativity one can conclude that H must be multiplication by a constant λ which without loss of generality may be taken to be positive. Thus H is an operator of the form $-\lambda\left(\frac{\partial^2}{\partial x^2}+\frac{\partial^2}{\partial y^2}+\frac{\partial^2}{\partial z^2}\right)$ where λ turns out to depend upon the so-called mass m of the particle. $\lambda = \frac{1}{2m}$. On the other hand, one can deduce from the assumption of Einstein's relativity that ρ must be a function of the form $r \to \rho(r) = \sqrt{M + c^2 r}$ where c is a universal constant equal to the velocity of light and M is a constant which depends on the particle and can be shown to be equal to m^2c^4 where m is the so-called "rest mass" of the particle.

If we investigate the possibilities for V (or H) when L remains irreducible but is no longer one-dimensional we encounter a somewhat wider range of possibilities. Let ℓ denote the dimension of the space of L. The most general possible H is then specified by means of, not one, but ℓ independent functions ρ. To be more precise we note that the Hilbert space of our particle may be conveniently represented as the direct sum $\mathcal{H}_0 \oplus \mathcal{H}_0 + \cdots \oplus \mathcal{H}_0$ of ℓ replicas of $\mathcal{H}_0 = \mathcal{L}^2(x,y,z)$ in such a way that the position and linear momentum operators are simply direct sums of replicas of the corresponding operators Q^0_E and U^0_α (α a translation) in $\mathcal{H}_0$. It will be convenient to introduce some notation to describe operators formed in this way from operators in $\mathcal{L}^2(x,y,z)$. If T is any linear operator in $\mathcal{H}_0$ we shall denote the operator which takes $\phi_1, \phi_2, \cdots \phi_\ell$ in $\mathcal{H}_0 \oplus \mathcal{H}_0 + \cdots \mathcal{H}_0$ into $T(\phi_1), T(\phi_2) \cdots T(\phi_\ell)$ by $\tilde{T}$. Also if $A = \{a_{ij}\}$ is any $\ell \times \ell$ matrix of complex numbers we shall use $\hat{A}$ to denote the operator in $\mathcal{H}_0 \oplus \mathcal{H}_0 \oplus \cdots \mathcal{H}_0$ which takes $\phi_1, \phi_2, \cdots \phi_\ell$ into $\psi_1, \psi_2 \cdots \psi_\ell$ where $\psi_i = \sum_{j=1}^{\ell} a_{ij}\phi_j$. It is obvious that $\hat{A}$ and $\widetilde{T}$ commute for all A and T and easy to show that the set of all $\widetilde{T}$ is precisely the commutator of the set of all $\widetilde{A}$ and vice versa.

In these terms what one can prove is the following. There exist ℓ self adjoint operations $H_1, H_2, \cdots H_\ell$ in $\mathcal{H}_0 \oplus \mathcal{H}_0 \oplus \cdots \mathcal{H}_0$, each of which commutes with all U_α and hence is a possible H, such that the most *general* possible H is uniquely of the form

$$\widehat{T}_{\rho_1} H_1 + \widehat{T}_{\rho_2} H_2 + \cdots + \widehat{T}_{\rho_\ell} H_\ell$$

where $T_{\rho_j} = \rho_j(-(\frac{\partial^2}{\partial x^2} + \frac{\partial^2}{\partial y^2} + \frac{\partial^2}{\partial z^2})$ and each ρ_j is a Borel function.

To complete this description of the possible operators H one must specify the operators $H_1, H_2, \cdots H_\ell$ for each $\ell = 1, 2, 3, \cdots$. We shall content ourselves with a statement of the facts in the special case in which $\ell = 2$. Here H_1 is the identity operator and H_2 is the operator

$$\widetilde{\frac{1}{i}\frac{\partial}{\partial x}}\hat{S}_1 + \widetilde{\frac{1}{i}\frac{\partial}{\partial y}}\hat{S}_2 + \widetilde{\frac{1}{i}\frac{\partial}{\partial x}}\hat{S}_3$$

where S_1, S_2 and S_3 are the well known Pauli spin matrices $S_1 = \begin{pmatrix} 0 & 1 \\ 1 & 0 \end{pmatrix}$, $S_2 = \begin{pmatrix} 0 & -i \\ i & 0 \end{pmatrix}$ and $S_3 = \begin{pmatrix} 1 & 0 \\ 0 & -1 \end{pmatrix}$. Since the operators $\widetilde{T}_\rho$ are obvious generalizations of the possible H's for the spinless case, H_1 is the essentially new constituent in the spin 1/2 case. A straightforward computation shows that $H_2^2 = \widehat{T}_0$ where $T_0 = -(\frac{\partial^2}{\partial x^2} + \frac{\partial^2}{\partial y^2} + \frac{\partial^2}{\partial z^2})$. Thus H_2 is a square root of the Laplace operator of a quite different character from that obtained by Fourier transforms or the operational calculus.

Just as in the spin zero case one can sharply limit the possibilities for ρ_1 and ρ_2 and hence for H by imposing space time invariance. One finds that Galilean invariance holds if and only if $\rho_2 = 0$ and ρ_1 is multiplication by a constant and that Einsteinian invariance holds if and only if $\rho_1 = 0$ and $\rho_2(r)$ is of the form $\pm\sqrt{\frac{m^2 c^4}{r} + c^2}$ where c is once again the velocity of light and m is the rest mass of the particle.

There is a logical gap in the analysis of section IV in that we have tacitly identified the self adjoint operators associated with the restrictions of U to the one-parameter subgroups of $\mathcal{E}$ with those defining the momentum observables. The most obvious justification for the procedure is that the identification holds in the classical model of a spinless quantum particle. We recall however that our aim was to derive the classical model from axioms, so this justification is not acceptable. In order to repair this gap, as well as for later use in discussing particle interactions, we close this section with a discussion of velocity operators in quantum mechanics and their relationship to momentum operators.

In classical mechanics the linear momentum of a particle is defined as the mass of the particle multiplied by its velocity. Moreover the reason for the usefulness of the momentum concept is the invariance principle known as the law of conservation of momentum. One could approach the problem of defining linear momentum in quantum mechanics either by seeking an analogous invariance principle or by first seeking the operators which define the velocity observables and multiplying them by a suitably defined mass. It is gratifying to find that the two approaches lead to the same result.

Let H denote the dynamical operator discussed above and let X, Y, Z denote the self adjoint operators defining the x, y and z coordinates. Then as a function of time the expected value of the x coordinate on the trajectory $e^{iHt}(\phi)$ will be $(Xe^{iHt}(\phi) \cdot e^{iHt}(\phi))$ with analogous expressions for the expected values of the y and z coordinates. If these expected values have time derivatives at $t = 0$ their values will clearly be the expected values in the state ϕ of the self adjoint operators $\frac{1}{i}(HX - XH)$, $\frac{1}{i}(HY - YH)$ and $\frac{1}{i}(HZ - ZH)$. Whether these exist or not evidently depends upon H and when they do they must clearly be the operators corresponding to the x, y and z components of the particle velocity. Using what we have learned about the nature of H we may compute the velocity operators explicitly. In the case in which the particle has spin zero so that $H = \rho(-(\frac{\partial^2}{\partial x^2} + \frac{\partial^2}{\partial y^2} + \frac{\partial^2}{\partial z^2}))$ one computes that

$$\frac{1}{i}(HX - XH) = +\frac{2}{i}\rho'(-(\frac{\partial^2}{\partial x^2} + \frac{\partial^2}{\partial y^2} + \frac{\partial^2}{\partial z^2}))\frac{\partial}{\partial x}$$

where ρ' is the derivative of the function ρ. Of course the argument fails when ρ fails to have an almost everywhere defined derivative. Similarly

$$\frac{1}{i}(HY - YH) = \frac{2}{i}\rho'(-(\frac{\partial^2}{\partial x^2} + \frac{\partial^2}{\partial y^2} + \frac{\partial^2}{\partial z^2}))\frac{\partial}{\partial y}$$

$$\frac{1}{i}(HZ - ZH) = \frac{2}{i}\rho'(-(\frac{\partial^2}{\partial x^2} + \frac{\partial^2}{\partial y^2} + \frac{\partial^2}{\partial z^2}))\frac{\partial}{\partial z}.$$

If we denote the self adjoint operator $2\rho'(-(\frac{\partial^2}{\partial x^2} + \frac{\partial^2}{\partial y^2} + \frac{\partial^2}{\partial z^2}))$ by M we see that the operators defining the velocity observables may be written in the form $M(\frac{1}{i}\frac{\partial}{\partial x})$, $M(\frac{1}{i}\frac{\partial}{\partial y})$, $M(\frac{1}{i}\frac{\partial}{\partial z})$ where M is a self adjoint operator which commutes with all U_α and hence with H, $\frac{1}{i}\frac{\partial}{\partial x}$, $\frac{1}{i}\frac{\partial}{\partial y}$, $\frac{1}{i}\frac{\partial}{\partial z}$.

Note that in the special case in which our particle is Galilean invariant the operator M is a constant operator $\frac{1}{m}I$ where m is the classical mass of the particle. Hence in this case the abstract (group theoretical) definition of linear momentum agrees with the classical definition according to which the momentum is the mass times the velocity.

More generally the operator M will not be a constant operator but it will still be true that the group theoretically defined linear momentum operators $\frac{1}{i}\frac{\partial}{\partial x}$, $\frac{1}{i}\frac{\partial}{\partial y}$, $\frac{1}{i}\frac{\partial}{\partial z}$ may be written in the form

$$\frac{1}{M}(V_x),\ \frac{1}{M}(V_y),\ \frac{1}{M}V_z$$

where V_x, V_y, V_z are the velocity operators. What is, then, the meaning of the non constant self adjoint operator $\frac{1}{M}$? To answer this question let us consider the spectral resolution of $\frac{1}{M}$. Using it, the Hilbert space may be written as a discrete direct sum of time invariant subspaces in each of which $\frac{1}{M}$ has its spectrum confined to a very short interval. In each of these invariant subspaces $\frac{1}{M}$ will differ only slightly from a constant operator and the motion of our particle will differ only slightly from that of Galilean particle whose constant mass is in the very short interval just alluded to. *In other words our particle will move just like a Galilean particle whose mass is not constant but varies with its velocity in a manner indicated by the preceding discussion.*

It is interesting to compare this mass velocity relationship with the well known one predicted by applying Einstein's special theory of relativity to classical mechanics. First of all we see that some such variation is predicted not just by

Einsteinian relativity but by *any* denial of Galilean relativity. Secondly it is easy to check that the particular form of the variation predicted by the present method, in the case when ρ has the special form demanded by Einsteinian invariance, coincides with the one found by Einstein.

To keep this paper from getting too long we shall *not* at this time investigate what happens in the case of higher spin.

VI. Reducibility of L and connections with particle multiplets.

In section V and the latter part of section IV we have confined attention to the special case in which the pair U P is irreducible or equivalently (once the imprimitivity theorem has been invoked to write $U = U^L \quad Q = Q^L$) the case in which L is irreducible. It is time to explore what happens when this restriction is removed.

The simplest case is that in which L is the direct sum of two irreducible representations L^1 and L^2 so that U^L is $U^{L^1} \oplus U^{L^2}$ and in which the direct sum decomposition (it will not be unique if L^1 and L^2 are equivalent) can be so chosen that the V_t leave $\mathcal{H}(U^{L^1})$ and $\mathcal{H}(U^{L^2})$ separately invariant. In that case the system is simply the direct sum of two systems U^{L^1}, V^1 and U^{L^2}, V^2 in which L^1 and L^2 are both irreducible. Since L^1 and L^2 are irreducible the component systems describe particles with definite spins as explained above. How shall we interpret the direct sum of two systems each of which describes a particle? A little reflection makes it clear that the corresponding physical system can be thought of as a "particle" which has two "natures" or two "identities". Depending upon whether the state vector ϕ is in $\mathcal{H}(U^{L^1}), \mathcal{H}(U^{L^2})$ or is a sum $\phi_1 + \phi_2$ of two non zero vectors with ϕ_1 in $\mathcal{H}(U^{L^1})$ and ϕ_2 in $\mathcal{H}(U^{L^2})$ the particle will exhibit its first nature, exhibit its second nature or exhibit each nature with a certain probability. The extension to the case in which L is the direct sum of three or more irreducible representations should be obvious.

There is nothing very profound in all this. We have merely seen that it is possible to maintain a consistent particle interpretation for any system satisfying our axioms – even when L is reducible – provided that V happens to be compatible with the reduction of the system U^L, P^L in the manner described above. Actually it is sometimes both technically and conceptually convenient to group two or more particles together and consider them as a single entity in this way. In the case of the neutron and the proton this circumstance was first exploited by Heisenberg in his 1932 paper on nuclear structure. His point of view of course was quite different from ours. In the neutron-proton case the new joint entity has even been given a name. It is called the nucleon.

One can be led to such "groupings" of particles in a natural and suggestive way if one supposes *a priori* that the physical world is invariant under a somewhat larger group than the group $\mathcal{E}$ of all isometries; specifically a group of the form $\mathcal{E} \times K$ where K is some compact group which acts trivially on space. The analysis of sections four and five goes through without essential change, and leads to conclusions of the same sort, except that now projective representations of the rotation group R are replaced by projective representations of the product group $R \times K$. For simplicity let us assume that it is such that the only projective multipliers of $\mathcal{E} \times K$ are direct products of projective multipliers of $\mathcal{E}$ with the identity multiplier on K. Then the most general possible irreducible projective unitary representation of $R \times K$ will be of the former $L \times M$ where L is one of the irreducible unitary projective representations of R already discussed and M is some irreducible unitary representation of K. The unitary projective representation $U^{L \times M}$ describing the symmetry of our "particle" is easily shown to be equivalent to $U^L \times M$. From the point of view of an observer unaware of the symmetry group K this representation appears as its restriction to $\mathcal{E} \times e$ which is just U^L repeated as many times as the dimension of $\mathcal{H}(M)$. For those "particles" with $\mathcal{H}(M)$ one-dimensional there is nothing new, but for those for which $\mathcal{H}(M)$ has a dimension two or greater the unaware observer sees several different particles which are more properly regarded as the "same" particle in different identities or different "internal states" – related to symmetry under the group K in much the same manner as the different spin states are related to symmetry under the group R.

Although the technical description becomes more elaborate the same considerations apply when the symmetry is only "approximate". The neutron-proton pairing is a case in point. Here the group K is $SU(2)$ which is a double covering of R so that the internal degrees of freedom are quite similar to those associated with spin. Indeed when this symmetry was recognized the term "isotopic spin" was introduced to emphasize its similarity to spin. One says that the nucleon is a "particle" with isotopic spin 1/2. After the discovery of the pi mesons and their relationship to nucleon nucleon scattering it became clear that the three differently charged pi mesons fitted into the same scheme and could be regarded as components of an isotopic triplet. The pion is a "particle" with isotopic spin 1. Later still when the properties of the new so-called strange particles began to be known it was pointed out by Gell-Maun and Ne'eman that these could be best understood by assuming that the symmetry group $\mathcal{E} \times SU(2)$ is a subgroup of a (less exact) symmetry group $\mathcal{E} \times SU(3)$. Now if M is any irreducible unitary representation of $K = SU(3)$, and one looks at the "particle" associated with $L \times M$ for some irreducible projective unitary representation L of R, one can consider it not only from the point of view already described in which the observer knows only about $\mathcal{E}$ but also from the point of an observer who knows about $\mathcal{E} \times SU(2)$ but not about $SU(3)$. This observer will restrict $U^{L \times M} = U^L \times M$ to $\mathcal{E} \times SU(2)$ and get $U^L \times M\restriction$ where $M\restriction$ is the restriction of M to $SU(2)$ and may be reduced to a direct sum of irreducible representations of $SU(2)$. If $M\restriction = M^1 \oplus M^2 + \cdots M^n$ where $M^1, M^2, \cdots M^n$ are irreducible representations of $SU(2)$, then $U^L \times M\!\restriction = U^L \otimes M^1 \oplus U^L \otimes M^2 \oplus \cdots U^L \times M^n$. Thus our observer will see n isotopic multiplets having perhaps different isotopic spins. Taking $SU(3)$ into account these are all part of a single "supermultiplet" in the same sense that the individual particles are parts of multiplets.

Gell-Mann and Ne'eman (independently) observed that $SU(3)$ has an eight-dimensional irreducible representation whose restriction to $SU(2)$ is the direct sum $D_0 + 2D_{1/2} + D_1$ where D_j is the unique irreducible representation of dimension $2j + 1$. Taking this eight-dimensional representation for M one would have a supermultiplet made up of one isotopic singlet, two isotopic doublets and one isotopic triplet. This just fitted the then known particles of spin 1/2 and mass greater than or equal to that of the proton. In addition to the nucleon one had the Λ hyperon, the Ξ hyperon and the Σ hyperon which were respectively an isotopic singlet, an isotopic doublet and an isotopic triplet. Because of the dimension of the appropriate M one spoke of the theory as the "eightfold way". As more particles (or resonances) were found these too could be fitted into supermultiplets but with M sometimes having a dimension other than eight. The failure of a three-dimensional representation to occur led to the famous "quark" hypothesis of Gell-Mann and Zweig in 1964.

VII. Particle interactions and the gauge principle.

Up to this point we have said nothing about how quantum mechanics deals with interactions between particles. This of course is a serious omission since isolated particles do not exist in nature and, if they did, would not do anything very interesting. On the other hand, Weyl said nothing about extending his program to explain why interactions take the general form that they do. Presumably this is because once one knows the quantum mechanical formalism for free particles it is easy in many cases to guess the quantum mechanical counterpart of known classical interactions.

Nevertheless it is interesting and illuminating to put classical mechanics aside and attempt to understand interactions between quantum mechanical particles from first principles in the spirit of Weyl's original program. After all classical mechanics is presumably only a limiting form of the truer and richer quantum mechanics and important features can easily be blurred or eliminated in the classical limit. We shall describe here some incomplete but encouraging efforts in this direction.

The first remark to be made is that if U^{L^1}, V^1 and U^{L^2}, V^2 define two distinct particles with Hilbert spaces $\mathcal{H}^1$ and $\mathcal{H}^2$ and one constructs the tensor product $\mathcal{H}^1 \otimes \mathcal{H}^2$ and the associated tensor products of the operators $U_\alpha^{L^1}, V_t^1, U_\alpha^{L^2}, V_t^2$ one obtains a system which has an immediate and obvious interpretation as the quantum mechanics of a system of two particles each of which is moving as though the other were absent. The question before us then becomes: What changes must be made when the particles interact so that the motion of each is affected by the behavior of the other. It is natural to assume that the position and momentum observables as described by U^{L^1} and U^{L^2} remain unchanged but that $t \to V_t^1 \otimes V_t^2$ must be replaced by some other one-parameter unitary group $t \to V_t$; the exact nature of which depends upon the particular particles.

One plausible assumption, with interesting consequences, is that the interaction leaves unchanged, not only the position and momentum observables, but also the velocity observables as defined in section 5. Let H^1, H^2, and H respectively denote the self-adjoint generators of V^2, V^2 and V. Then $H^1 \times I + I \times H^2$ is the self-adjoint generator of $V_t^1 \times V_t^2$ and H is of course completely determined by H^1, H^2 and $J = H - (H^1 \times I + I \times H^2)$. But our hypotheses and the definition of velocity observable imply immediately that J, the so-called interaction operator, must commute with all $P_E^{L^1}$ and all $P_E^{L^2}$. In the simplest special case, that in which L^1 and L^2 are one-dimensional identity representations, so that the particles are spinless, one concludes without difficulty that the Hilbert space $\mathcal{H}^1 \otimes \mathcal{H}^2$ may be realized as $\mathcal{L}^2(x_1, y_1, z_1, x_2, y_2, z_2)$ so that J is a multiplication operator of the form $\psi \to V\psi$ where V is some real-valued function defined on six space. Using invariance under simultaneous translation and rotation of both particles by the same amount one shows that V depends only on the distance $\sqrt{(x_1 - x_2)^2 + (y_1 - y_2)^2 + (z_1 - z_2)^2}$ so that $V(x_1, y_1, z_1, x_2, y_2, z_2)$ $= v(\sqrt{(x_1 - x_2)^2 + (y_1 - y_2)^2 + (z_1 - z_2)^2})$ for some real function v defined on the positive real axis.

The Schrödinger equation which results in the special case in which the two constituent free particles are subject to Galilean invariance is just that customarily assigned to two interacting particles whose classical equations of motion are determined by the potential energy function

$$V(x_1, y_1, z_1, z_2, y_2, z_2) = v(\sqrt{(x_1 - x_2)^2 + (y_1 - y_2)^2 + (z_1 - z_2)}^2.$$

Now, however, we have a justification for this equation which is independent of classical mechanics. Indeed one can argue in reverse to show that certain features of classical mechanics that seem arbitrary are in fact consequences of the fact that classical mechanics is a limiting form of quantum mechanics – in a sense are "quantum effects".

When the particles have spin, a parallel analysis may be carried out to just as complete an extent but the results are more complicated and have no familiar classical analogues. The key difference is that a self-adjoint operator which commutes with all coordinate observables can no longer be proved to be multiplication by a real-valued function of the coordinates. Instead it is determined by an operator valued function of the coordinates in the following precise sense. Recall that one can realize the members of $\mathcal{H}^1$ and $\mathcal{H}^2$ respectively as vector valued functions on space having values in $\mathcal{H}(L^1)$ and $\mathcal{H}(L^2)$ respectively so that members of $\mathcal{H}^1 \otimes \mathcal{H}^2$ may be realized as functions of $x_1, y_1, z_1, x_2, y_2, z_2$ having values in $\mathcal{H}(L^1) \otimes \mathcal{H}(L^2)$. The interaction operation J will be defined by a function F of $x_1, y_1, z_1, x_2, y_2, z_2$

whose values are linear operators in the vector space $\mathcal{H}(L^1) \otimes \mathcal{H}(L^2)$. One can think of F as defining a "multiplication operator" as in the spinless case provided that one interprets the "product". A θ of an operator A with a vector θ is the transform $A(\theta)$ of θ by A. The analogue of the reduction from V to v in the spinless case can be carried out but involves a more elaborate discussion than seems appropriate in an article of this character. The final results can be interpreted as stating that in addition to an exact analogue of the "potential" found in the spinless case one must expect "potentials" corresponding to "forces" which depend upon the relative orientation of the spins of the particles. Such additional "forces" have been found experimentally and are especially important in the interactions of nucleons with one another.

While the immediately preceding considerations provide a certain amount of insight and illumination they can hardly be regarded as a final answer to the question. Our basic axiom about velocity observables is plausible but hardly logically compelling. Indeed it is clearly at most approximately true because it rules out interactions corresponding to so-called "velocity dependent forces" in classical mechanics. These occur in the interactions between charged particles and can only be neglected when the relevant velocities are small. It is thus gratifying to discover that a slight and plausible weakening of this axiom is also susceptible of a complete analysis and in the spinless case lets in just the missing velocity dependent "forces". Moreover when one examines the higher spin cases one finds the sort of new particle interactions implied by the introduction of "gauge fields" in the spirit of the work initiated in 1954 by Yang and Mills and continued so successfully by many others in the 1960's and after.

Our new assumption states not that the velocity operators are the *same* as they would be if the interaction were not there but only that the *commutation relations* between the velocity operators on the one hand and the position operators on the other are the same. In the case in which both particles are spinless this implies that each of the velocity operators for the two particles must differ from what it reduces to where there is no interaction by a multiplicative operator. In other words, there must exist six real-valued functions $A_1, B_1, C_1, A_2, B_2, C_2$ of the coordinates $x_1, y_1, z_1, x_2, y_2, z_2$ of the two particles such that the operator describing the x component of the velocity observable for the first particle is the sum of the operator $\frac{2}{i}\rho'(-(\frac{\partial^2}{\partial_1^2}+\frac{\partial^2}{\partial y_1^2}+\frac{\partial^2}{\partial x_1^2})\frac{\partial}{\partial x_1}$ and the operator which takes ψ into $A_1\psi$ and similarly for the other five velocity components. Once the six functions $A_1, B_1, C_1, A_2, B_2, C_2$ are known, the interaction operator is uniquely determined up to addition of an operator which commutes with all coordinate operators and hence is a multiplication operator in the above sense. In other words, under our new axiom the dynamics of two interacting spinless particles is determined by specifying not just one function V (the potential energy) but seven functions $A_1, B_1, C_1, A_2, B_2, C_2, V$. Of course invariance constraints put severe and useful constraints on the functions $A_1, B_1, C_1, A_2, B_2, C_2, V$ but time and space do not permit us to explore the matter here.

Passing on to the more difficult but important case in which the particles have nonzero spin one finds closely analogous results but in which the functions $A_1, B_1, \cdots$ are operator valued just as was the single function V in our earlier analysis. One is led into lengthy considerations involving linear algebra and group representations but finding all possible interactions is a doable problem. The author hopes to examine it in greater detail on a future occasion.

It is well known that in classical mechanics and in Galilean invariant quantum mechanics one can reduce the study of a pair of interacting particles to the study of a single particle moving in a "force field". It thus seems reasonable to attempt to gain insight into the above considerations by way of a parallel axiomitization which relates the velocity and position operators of a single free particle to those of a single particle which is not assumed to be free.

Consider a free spinless particle in the special case of Galilean invariance so that the operator H is of the form $-\frac{1}{2m}(\frac{\partial^2}{\partial x^2} + \frac{\partial^2}{\partial y^2} + \frac{\partial^2}{\partial z^2})$. The velocity operators are then $v_x = \frac{m}{i}\frac{\partial}{\partial x}$, $v_y = \frac{m}{i}\frac{\partial}{\partial y}$ and $v_z = \frac{m}{i}\frac{\partial}{\partial z}$ and one computes at once that the velocity operators and the position operators X, Y, Z all commute with one another except that

$$v_x X - X v_x = \frac{m}{i} I, \quad v_y Y - Y v_y = \frac{m}{i} I, \quad v_z Z - Z v_z = \frac{m}{i} I.$$

Conversely suppose that we ask for the most general self-adjoint operator H' having these properties; that is, such that if v_x^1, v_y^1 and v_z^1 are defined by the equations

$$v_x^1 = \frac{1}{i}(H'X - XH'), v_y^1 = \frac{1}{i}(H'Y - YH'), v_z^1 = \frac{1}{i}(H'Z - ZH')$$

then $v_x^1 X - X v_x^1 = v_y^1 Y - Y v_y^1 = v_z^1 Z - Z v_z^1 = \frac{m}{i} I$ and all other pairs commute. One deduces at once that $v_x^1 - v_x$, $v_y^1 - v_y$ and $v_z^1 - v_z$ all commute with X, Y and Z. Hence each of these three operators is a multiplication operator. Hence $v_x^1 = \frac{m}{i}\frac{\partial}{\partial x} + \widetilde{A}_1$, $v_y^1 = \frac{m}{i}\frac{\partial}{\partial y} + \widetilde{A}_2$, $v_z^1 = \frac{m}{i}\frac{\partial}{\partial z} + \widetilde{A}_3$ where A_1, A_2, A_3 denote real valued functions of x, y, z and $\widetilde{A}_j$ denotes the operator $\psi \to A_j \psi$.

Now consider the self-adjoint operator

$$H_0 = \frac{1}{2m}[(\frac{1}{i}\frac{\partial}{\partial x} + \widetilde{A}_1)^2 + (\frac{1}{i}\frac{\partial}{\partial y} + \widetilde{A}_2)^2 + (\frac{1}{i}\frac{\partial}{\partial z} + \widetilde{A}_3)^2].$$

One computes at once that H_0 and H' lead to exactly the same velocity operators and hence that $H' - H_0$ commutes with all position observables and so is a multiplication operator. In short the most general possible H' is $H_0 + \widetilde{V}$ where $\widetilde{V}$ is multiplication by the real-valued function V. But the operator $H_0 + \widetilde{V}$ is precisely the one that occurs in the quantum mechanics of a charged spinless particle moving in an electric field of potential V and a magnetic field of vector potential $A = A_1, A_2, A_3$.

Notice that if we had made the stronger assumption that the velocity operators are the *same* for free and unfree particles then the magnetic field would have been forced out. Notice also that the physics is completely determined by the quadruple consisting of H', X, Y and Z. This means that if H'', X, Y, Z and H', X, Y, Z are unitarily equivalent quadruples H'' and H' yield the same physical predictions. But with a spinless particle the most general unitary operator which carries X into X, Y into Y and Z into Z is a unitary operator which commutes with X, Y, Z; that is, a multiplication operator $\widetilde{U}$ in which U is a function of the form $x, y, z \rightarrow e^{i\sigma(x,y,z)}$ where σ is real valued. One computes at once that for such a U

$$U^{-1}\left(\frac{1}{i}\frac{\partial}{\partial x} + \widetilde{A}_1\right)U = U^{-1}\left(\frac{1}{i}\frac{\partial}{\partial y}\right)U + \widetilde{A}_1 = \frac{1}{i}\frac{\partial}{\partial x} + \widetilde{\frac{\partial \sigma}{\partial x}} + \widetilde{A}_1$$

with the obvious analogous formulas for $\frac{1}{i}\frac{\partial}{\partial y} + \widetilde{A}_2$ and $\frac{1}{i}\frac{\partial}{\partial z} + \widetilde{A}_3$. Thus $U^{-1}H'U$ will have the same form as H' and will differ from it only in that the vector potential A_1, A_2, A_3 will be replaced by $A_1 + \frac{\partial \sigma}{\partial x}$, $A_2 + \frac{\partial \sigma}{\partial y}$, $A_3 + \frac{\partial \sigma}{\partial z}$. This is interesting because in the classical theory of electricity and magnetism the components of the magnetic field defined by a given vector potential A_1, A_2, A_3 have the components $\frac{\partial A_3}{\partial x_2} - \frac{\partial A_2}{\partial x_3}$, $\frac{\partial A_1}{\partial x_3} - \frac{\partial A_3}{\partial x_1}$, $\frac{\partial A_2}{\partial x_1} - \frac{\partial A_1}{\partial x_2}$ so that A_1, A_2, A_3 and $A_1 + \frac{\partial \sigma}{\partial x}$, $A_2 + \frac{\partial \sigma}{\partial y}$, $A_3 + \frac{\partial \sigma}{\partial z}$ define the same magnetic field. What is interesting about it is the emergence of the result from abstract axioms. It is the embryo of the principle of gauge invariance and the multiplicative unitary operators U are called *gauge* transformations. The principle of gauge invariance as introduced by Weyl in his celebrated book of 1928, "Gruppentheorie und Quantummechanik" (page 100 of the English translation) differs from the above chiefly in allowing U as well as the functions A_1, A_2, A_3 and V to be time dependent.

Now consider the same problem with the restriction removed that the inducing representation L of the free particle be one-dimensional. In the simplest case L is just the two-dimensional identity representation and the free H is just as before, except that it now acts on functions which have two vectors as values instead of complex numbers. The velocity operators will change in the same way and obey the same commutation relations with X, Y and Z as before. Now, however, when one seeks the velocity operators associated with the most general H' having velocity operators satisfying these same commutation relations one finds that they are defined not by a triple A_1, A_2, A_3 of real-valued functions but by a triple A_1, A_2, A_3 of 2×2 self-adjoint matrix valued functions and that V is similarly replaced by a self-adjoint matrix-valued function. This suggests a generalization of an electro-magnetic field in which once again one has analogues of vector and scalar potentials, velocity dependent forces and a principle of gauge invariance.

However the gauge transformations will be defined by operator valued functions rather than scalar valued functions and will not commute with one another.

The author has not yet done the detailed calculations required to check this, but would be very surprised if pursuing this line of thought did not lead to something very close to, and possibly more general than, the modern theory of non Abelian gauge fields which originated in the celebrated work of Yang and Lee. If so, it is interesting that one could have been led to it by systematically seeking to extend Weyl's program in the manner indicated in this section. On page 216 of the author's book (mentioned at the end of section 4) one will find the words "Although more general cases are interesting and important, we shall consider here only the case in which A'_j, B'_j and C'_j are all zero so that there are no 'velocity dependent forces"'. This sentence was written down in the fall of 1965 but unfortunately the author was not moved to look further in the direction indicated until quite recently.

VIII. Super symmetry.

Let us return to the theory of a single free particle and consider the case in which the particle has spin 1/2. We saw in section VI that the most general possible dynamical operator is of the form

$$\rho_1\left(-\left(\frac{\partial^2}{\partial x^2}+\frac{\partial^2}{\partial y^2}+\frac{\partial^2}{\partial z^2}\right)\right)I+\rho_2\left(-\left(\frac{\partial^2}{\partial x^2}+\frac{\partial^2}{\partial y^2}+\frac{\partial^2}{\partial z^2}\right)\right)H_2$$

where H_2 is the operator

$$\widetilde{\frac{1}{i}\frac{\partial}{\partial x}}\hat{S}_1+\widetilde{\frac{1}{i}\frac{\partial}{\partial y}}\hat{S}_2+\widetilde{\frac{1}{i}\frac{\partial}{\partial z}}\hat{S}_3 \quad \text{and} \quad S_1, S_2, S_3$$

are the Pauli spin matrices. Let us choose $\rho_1(r) = r$ and $\rho_2(r) \equiv 0$ so that our dynamical operator is $-(\frac{\partial^2}{\partial x^2}+\frac{\partial^2}{\partial y^2}+\frac{\partial^2}{\partial z^2}) = (P^2+Q^2+R^2) = T_0$ where for convenience we have denoted $\widetilde{\frac{1}{i}\frac{\partial}{\partial x}}$, $\widetilde{\frac{1}{i}\frac{\partial}{\partial y}}$, and $\widetilde{\frac{1}{i}\frac{\partial}{\partial z}}$ by P, Q and R respectively. Recall that $H_2^2 = F_0$ so that H_2 commutes with T_0. Also it is obvious that $\hat{S}_1, \hat{S}_2, \hat{S}_3$ commute with P, Q, R and hence with T_0.

We now focus our attention on the four dimensional vector space spanned by $\hat{S}_1, \hat{S}_2, \hat{S}_3, H_2$. Its self-adjoint members are all "symmetries" of the system in the sense that they commute with T_0. However this vector space has only zero in common with the usual Lie algebra of symmetries generated by $P, Q, R, T, 0$ and the angular momentum operators. Moreover it is not a Lie algebra in the sense of being closed under the usual bracket operation. Instead it is closed under a slightly different operation; namely the operation $A, B \to AB^2 - B^2A$. Given that $(B+C)^2 = B^2+BC+CB+C^2$ one sees at once that closure under $A, B \to AB^2-B^2A$ is equivalent to closure under the ternary operation $A, B, C \to A(BC+$

$CB) - (BC + CB)A$. This ternary operation may be regarded in turn as a sort of composite of the commutator $AB - BA$ and the anti-commutator $AB + BA$.

These extra symmetries are called super symmetries and the vector space generated by them and the usual Lie algebra of symmetries is called a Lie superalgebra. The two vector spaces, whose direct sum it is, are called the odd and even parts respectively. They are related in an interesting way which it is instructive to explore in the abstract.

Let V be any vector space of linear operators which is such that $AB^2 - B^2A$ is in V whenever A and B are in V. Let $\tilde{V}$ denote the vector space of all linear combinations of squares of elements of V. Then the following mildly surprising lemma has a very short and simple proof.

LEMMA 1. *$\tilde{V}$ is a Lie algebra in the sense that $AB - BA$ is in $\tilde{V}$ whenever A and B are in $\tilde{V}$.*

PROOF: It will clearly suffice to show that $A^2B^2 - B^2A^2$ is in $\tilde{V}$ whenever A and B are in V. But $A^2B^2 - B^2A^2 = A^2B^2 - AB^2A + AB^2A - B^2A^2 = A(AB^2 - B^2A) + (AB^2 - B^2A)A$. By hypothesis $AB^2 - B^2A = C$ is in V so $A^2B^2 - B^2A = AC + CA = (A + C)^2 - A^2 - C^2$ where C is in V. But since A and C are in V, $(A + C)^2 - A^2 - C^2$ is in V^2 and the proof is complete.

Since $\tilde{V}$ is in an obvious sense the *square* of V it is natural to call V a *square root* of $\tilde{V}$ and to define a Lie algebra square root to be any V closed under $A, B \to AB^2 - B^2A$.

In the example with which this section started V is the four-dimensional vector space generated by $\hat{S}_1, \hat{S}_2, \hat{S}_3$ and $H_2 = (\hat{S}_1P + \hat{S}_2Q + \hat{S}_3R)$. One checks easily that $\tilde{V}$ is the five-dimensional vector space generated by P Q R T_0 and I and is an invariant subalgebra of the Lie algebra generated by these and the angular momentum operators. Actually one can verify that V itself is invariant under commutation with the angular momentum operators and then the fact that $\tilde{V}$ is an invariant subalgebra is a consequence of a simple extension of Lemma 1.

LEMMA 2. *Let V be a Lie algebra square root and let $\mathcal{L}$ be a Lie algebra of operators in the same vector space such that $AB - BA$ is in V whenever A is in V and B is in $\mathcal{L}$. Then $\tilde{V} + \mathcal{L}$ the linear sum of $\tilde{V}$ and $\mathcal{L}$ is a Lie algebra and contains $\tilde{V}$ as an invariant subalgebra.*

The proof, which we omit, is very similar to that of Lemma 1.

The example of a Lie algebra square root presented above is quite special in that its square is a commutative Lie algebra. The first example to appear in the physical literature is of quite a different character in that it is two-dimensional and has a three-dimensional square which is not only non commutative but simple. In fact, it is possible to give a complete analysis of all possible two-dimensional Lie algebra square roots V and show that if $\tilde{V}$ is three-dimensional it is either commutative or simple. Obviously $\tilde{V}$ cannot be more than three-dimensional when V is two-dimensional.

The two-dimensional example mentioned above appeared in a paper of Neveu and Schwartz on string theory which appeared in 1971. It was constructed there in a rather complicated way. A much simpler construction was shown to the author by H. Bacry in the spring of 1986. Let P and Q be two self-adjoint operators which satisfy the Heisenberg commutation relations. Then the two-dimensional vector space of linear combinations of P and Q is readily checked to be a Lie algebra square root whose square is three-dimensional and simple.

Because of the fundamental role played by the Heisenberg commutation relations in the early stages of the implementation of Weil's program it is interesting to notice that Weyl's integrated or global form of the commutation relations has an analogue – or better an extension – which gives a global form to the Lie super algebra generated by the linear span of P and Q and its square. Consider the semi direct product $N \odot G$ of a two-dimensional real vector space N and the group G of all automorphisms of N which leave fixed a non degenerate projective multiplier σ on N. Then σ can be extended to a projective multiplier $\widetilde{\sigma}$ on $N \odot G$ by setting $\widetilde{\sigma}(n_1, x_1, n_2, x_2) = \sigma(n_1, n_2)$ for all n_1 and n_2 in N and all x_1 and x_2 in G. From the general theory of induced projective representations one can then show that there exists a unique irreducible $\widetilde{\sigma}$ representation U of $N \odot G$ whose restriction to N is irreducible. The self adjoint generators of U restricted to N then generate a Lie algebra square root whose square is generated by the self adjoint generators of G.

IX. Concluding Remarks..

The author had originally hoped to discuss the elements of quantum field theory from a point of view consistent with the foregoing. However there turned out to be neither enough time nor enough space for this to be possible. The author is very conscious of many other sins of omission and over sketchiness and asks the reader's indulgence on the grounds that the later sections must be regarded as a premature report of vague ideas and work in progress.

ACKNOWLEDGEMENTS: The material in section VIII was found while the author was a guest at the Max Planck Institut für Mathematik in Bonn in 1985-86 while being supported by a prize from the Alexander von Humboldt foundation. More details will be found in "Some remarks on Lie superalgebras", Czechoslovak Journal of Physics, Vol. B 37, 1987, pp. 373-386.

ELLIPTIC GENERA OF LEVEL N FOR COMPLEX MANIFOLDS

Friedrich Hirzebruch
Max-Planck-Institut für Mathematik
Gottfried-Claren-Str. 26
D-5300 Bonn 3
Federal Republic of Germany

My lecture at the Como Conference was a survey on the theory of elliptic genera as developed by Ochanine, Landweber, Stong and Witten. A good global reference are the Proceedings of the 1986 Princeton Conference [1]. In this contribution to the Proceedings of the Como Conference I shall not reproduce my lecture, but rather sketch a theory of elliptic genera of level N for compact complex manifolds which I presented in the last part of my course at the University of Bonn during the Winter-semester 1987/88. For a natural number $N > 1$ the elliptic genus of level N of a compact complex manifold M of dimension d is a modular form of weight d for the group $\Gamma_1(N)$. In the cusps of $\Gamma_1(N)$ the genus degenerates either to $\chi_y(M)/(1+y)^d$ where $-y$ is an Nth-root of unity different from 1 or to $\chi(M,K^{k/N})$ where K is the canonical line bundle and $0 < k < N$. Here $\chi_y(M) = \sum_{p=0}^{d} \chi^p(M)y^p$ with $\chi^p(M) = \chi(M,\Omega^p) = \sum_{q=0}^{d} (-1)^q h^{p,q}$ is the χ_y-genus introduced in [13] and $\chi(M,K^{k/N})$ is the genus with respect to the characteristic power series

$$\frac{x}{1-e^{-x}} \cdot e^{-(k/N)\cdot x}$$

which equals the holomorphic Euler number of M with coefficients in the line bundle L^k provided $K = L^N$ (see [13]).

K. Bleuler and M. Werner (eds.), Differential Geometrical Methods in Theoretical Physics, 37–63.

For $N = 2$ the genus is expressible in Pontrjagin numbers and hence defined for an oriented differentiable manifold M. The only possible value of $-y$ is -1 and the genus degenerates in the two cusps to

$$\chi_1(M)/2^d = \mathrm{sign}(M)/2^d \qquad (\dim_{\mathbb{R}} M = 2d)$$

or to

$$\chi(M,K^{1/2}) = \hat{A}(M) .$$

Only recently I realized that Witten in [19] studied also complex manifolds. His discussion includes the genus studied here, at least if one restricts attention to the cusps with specialization $\chi(M,K^{k/N})$.

In this report I shall also try to give an account of the rigidity theorems for complex manifolds with circle actions which for $N = 2$ are due to Taubes [18] and Witten with a new exposition by Bott [9]. These rigidity theorems hold if the first Chern class of M is divisible by N, i.e. if a holomorphic line bundle L with $L^N = K$ exists.

The results in this paper hold also for manifolds with a stable almost complex structure and for circle actions which preserve this structure. For simplicity we have formulated the results for complex manifolds only.

I would like to thank the students of my course Thomas Berger and Rainer Jung for writing notes. Many thanks to Nils-Peter Skoruppa who lectured several times in my course when I was away and with whom I had helpful discussion on modular forms. After my course I had intensive discussions with Michael Atiyah on the rigidity theorem in Oxford and also with Don Zagier at the Max-Planck-Institut.

1. In the following N is a fixed natural number >1, the "variable" x runs through the complex numbers, $\mathbb{H}$ is the upper half plane, $\tau \in \mathbb{H}$ and $q = e^{2\pi i\tau}$. For a lattice L in $\mathbb{C}$ we consider the elliptic function $g(x)$ with divisor $N\cdot 0 - N\cdot\alpha$ where $\alpha \in \mathbb{C}$ is an N-division point ($\alpha \notin L$, $N\alpha \in L$). The function g is uniquely determined by L and α (regarded as element of $\mathbb{C}/L$) if we demand that the power series of g in the origin begins with x^N. The function $f(x) = g(x)^{1/N}$ is uniquely defined if we request $f(x) = x$ + higher terms. The function f is elliptic with respect to a sublattice L' of L whose index in L equals the order of α as element of $\mathbb{C}/L$. For $\omega \in L$ the function $f(x+\omega)/f(x)$ is constant and equals an N-th root of unity. After multiplying L with a non-vanishing complex number we can assume that

$$L = 2\pi i(\mathbb{Z}\tau + \mathbb{Z}) \quad \text{and} \tag{1}$$

$$0 \neq \alpha = 2\pi i\left[\frac{k}{N}\tau + \frac{\ell}{N}\right] \quad \text{with} \quad 0 \leq k < N \quad \text{and} \quad 0 \leq \ell < N\ .$$

To write down a product development for $f(x)$ in the case that L and α are as in (1) we introduce the entire function

$$\Phi(x) = (1-e^{-x}) \prod_{n=1}^{\infty} (1-q^n e^{-x})(1-q^n e^x)/(1-q^n)^2 \tag{2}$$

which has zeros of order 1 in the points of L. The function $\Phi(x)$ equals the Weierstrass sigma-function for L up to a factor of the form $\exp(b_1x + b_2x^2)$. It can be proved easily that

$$f(x) = e^{\frac{k}{N}x}\Phi(x)\Phi(-\alpha)/\Phi(x-\alpha)\ . \tag{3}$$

Namely, it suffices to check

$$\Phi(x+2\pi i\tau) = -e^{-x}e^{-2\pi i\tau}\Phi(x).$$

For this replace in (2) the exponential e^x by λ and then substitute λ by λq to see that the factor $-\lambda^{-1}q^{-1}$ comes out. In fact, we have (for $\zeta = e^{2\pi i/N}$)

(4)
$$f(x+2\pi i) = \zeta^{k} f(x)$$
$$f(x+2\pi i\tau) = \zeta^{-\ell} f(x)$$

The function $f(x)$ as belonging to L and α (see (1)) degenerates for $q \longrightarrow 0$ to a function $f_\infty(x)$.

We have

(5)
$$f_\infty(x) = e^{(k/N)x} \cdot (1-e^{-x}) \quad \text{for} \quad k > 0$$
$$f_\infty(x) = (1-e^{-x})(1-e^{\alpha})/(1-e^{\alpha-x}) \quad \text{for} \quad k = 0.$$

For reasons which are apparent from the introduction we put $e^{\alpha} = -y$ for $k = 0$ and have in this case

$$f_\infty(x) = (1-e^{-x})(1+y)/(1+ye^{-x}) \quad \text{with} \quad -y = \zeta^{\ell} \neq 1.$$

The involution $x \longrightarrow -x+\alpha$ interchanges the zeros and poles of $f(x)$. Therefore,

$$f(x)f(-x+\alpha),$$

which is elliptic for L, is in fact a constant $\neq 0$. We write the constant as c^{-2}. Then c^{2N} depends only on L and the chosen N-division point as point of $\mathbb{C}/L$. If the lattice and α are normalized as in (1), then

$$f(x)f(-x+\alpha) = e^{(k/N)\alpha-\alpha}\Phi(-\alpha)^2 = c^{-2}$$

and

(6)
$$c^{2N} = \Phi(-\alpha)^{-2N} q^{\frac{k(N-k)}{N}} \cdot \zeta^{-k\ell}.$$

The coefficients of the power series developments of $f(x)/x$, $x/f(x)$ and $x\frac{f'(x)}{f(x)}$ determine each other. If one replaces in such a series x by λx, one obtains the corresponding function for the lattice $\lambda^{-1}L$ and the n-division point $\lambda^{-1}\alpha$. Therefore the coefficient of x^r in any of these series as function of the pair L, α with $\alpha \in \mathbb{C}/L$ is homogeneous of degree $-r$. Also c^{2N} is such a function of L and α. It is homogeneous of degree $-2N$ and is related to Dedekind's η-function.

If the pair L,α is chosen as in (1), then the coefficients of $f(x)/x$ are indeed modular forms of weight r for the subgroup consisting of the matrices $\begin{pmatrix} a & b \\ c & d \end{pmatrix}$ of $SL_2(\mathbb{Z})$ which satisfy

$$\begin{pmatrix} a & c \\ b & d \end{pmatrix}\begin{pmatrix} k \\ \ell \end{pmatrix} = \begin{pmatrix} k \\ \ell \end{pmatrix} \mod N.$$

Also c^{2N} is a modular form of weight $2N$ for this group. (It still has to be shown that these forms are holomorphic in the cusps. See the next section.)

2. The classification of pairs L,α where L is a lattice in $\mathbb{C}$ and $\alpha \in \mathbb{C}/L$ with $N\alpha = 0$ (but $\alpha \neq 0$), up to multiplication by some complex number $\lambda \neq 0$, leads to the introduction of the modular group

$$\Gamma_1(N) = \left\{ \begin{pmatrix} a & b \\ c & d \end{pmatrix} \in SL_2(\mathbb{Z}) \mid c \equiv 0 \mod N, \ a \equiv d \equiv 1 \mod N \right\}.$$

If we assume that the N-divison point has order N in $\mathbb{C}/L$, then the classes of pairs L,α are in one-to-one correspondence with the points of the modular curve $\mathbb{H}/\Gamma_1(N)$ where $\begin{pmatrix} a & b \\ c & d \end{pmatrix}$ acts on $\mathbb{H}$ by

$$\tau \longmapsto \frac{a\tau+b}{c\tau+d}.$$

This follows from the fact that each pair L,α is equivalent to a pair of type (1) with $\alpha = 2\pi i/N$ (i.e. $k = 0$, $\ell = 1$). The coefficients of x^r in $\frac{x}{f(x)}$, $\frac{f(x)}{x}$, $x\frac{f'(x)}{f(x)}$ are modular forms of weight r for $\Gamma_1(N)$. It remains to show that such a coefficient is holomorphic in each cusp of $\mathbb{H}/\Gamma_1(N)$. Transforming $f(x)$ (taken for the lattice (1) with $\alpha = 2\pi i/N$) to a cusp gives a function $f(x)$ for some $\alpha = 2\pi i(\frac{k}{N}\tau + \frac{\ell}{N})$ and the same lattice. The formulas (2) and (3) show immediately that the q development of each coefficient of $f(x)$ has only non-negative powers of q, fractional if $k > 0$, but then a suitable root of q is the local uniformizing variable at the cusp.

The coefficients e_r of $x\frac{f'(x)}{f(x)}$ are the Eisenstein series. Their q-developments (for $k = 0$ and $\ell = 1$) for example can be read off from

$$(7) \quad x\frac{f'(x)}{f(x)} = \sum_{n=0}^{\infty} \frac{xq^n e^{-x}}{1-q^n e^{-x}} - \sum_{n=1}^{\infty} \frac{xq^n e^x}{1-q^n e^x}$$

$$- \sum_{n=0}^{\infty} \frac{x\zeta q^n e^{-x}}{1-\zeta q^n e^{-x}} + \sum_{n=1}^{\infty} \frac{x\zeta^{-1} q^n e^x}{1-\zeta^{-1} q^n e^x} .$$

Furthermore, c^{2N} (see (6)) is a modular form of weight $2N$. For more detailed formulae concerning these q-developments in the case $N = 2$ see [20]. For $N > 2$ and $N \neq 4$ the number of cusps of $\Gamma_1(N)$ equals

$$\frac{1}{2}\sum_{d|N} \varphi(d)\varphi(\tfrac{N}{d}),$$

where φ is Euler's function. Each cusp can be represented by several division points α as in (1).

3. Let M_d be a compact complex manifold. The Chern classes c_i of M_d are elements of the 2i-dimensional cohomology group $H^{2i}(M_d,\mathbb{Z})$. Let c be the total Chern class of M_d split up formally

(8) $$c = \sum_{i=0}^{d} c_i = (1+x_1)(1+x_2)\ldots(1+x_d)$$

where $x_1, x_2, \ldots, x_d$ can be regarded as 2-dimensional cohomology classes in some manifold fibred over M_d (see [13], § 13.3). Let $Q(x)$ be a fixed power series in the indeterminate x starting with 1 whose coefficients are in some commutative ring containing $\mathbb{Z}$. Then

(9) $$\varphi_Q(M_d) = Q(x_1)Q(x_2)\ldots Q(x_d)[M_d]$$

is the genus of M_d with respect to the power series Q where in (9) the symmetric expression $Q(x_1)Q(x_2)\ldots Q(x_d)$ is written in terms of the Chern classes in view of (8) and the 2d-dimensional component of this expression is evaluated on M_d (compare [13], § 10.2). We define the elliptic genus $\varphi_N(M_d)$ by using the power series

(10) $$Q(x) = \frac{x}{f(x)} = \frac{x\Phi(x-\alpha)}{\Phi(x)\Phi(-\alpha)}$$

where $\alpha = 2\pi i/N$ and $f(x)$ is taken for the pair L,α with $L = 2\pi i(\mathbb{Z}\tau+\mathbb{Z})$. We put again $\zeta = e^{2\pi i/N}$.

Theorem. The elliptic genus $\varphi_N(M_d)$ _is a modular form of weight_ d _for the group_ $\Gamma_1(N)$. _If one represents a cusp of_ $\Gamma_1(N)$ _by_ $2\pi i(\frac{k}{N}\tau + \frac{\ell}{N})$ _with_ $0 \leq k < N$ _and_ $0 \leq \ell < N$, _then the value of_ $\varphi_N(M_d)$ _in this cusp equals_

$$\chi(M_d, K^{k/N}) \quad \text{if} \quad k > 0$$

and $\chi_y(M_d)/(1+y)^d$ if $k = 0$ and $-y = \zeta^\ell$.

The theorem follows from the remarks in section 2 and from (5) by recalling that $\chi(M_d, K^{k/N})$ is the genus for the power series

$$\frac{x}{1-e^{-x}} \cdot e^{-(k/N)x}$$

and $\chi_y(M_d)/(1+y)^d$ is the genus for the power series

$$\frac{x}{1-e^{-x}}(1+ye^{-x})/(1+y) ,$$

see [13].

A genus can be defined also by a power series $Q(x)$ not beginning with 1 (we assume $Q(0) = a_0 \neq 0$). The definition is done by equation (9) again. Then $a_0^{-1}Q(a_0x)$ gives the same genus with a normalized power series (i.e. the constant term equals 1). We now define $\tilde{\varphi}_N(M_d)$ using the power series

$$\tilde{Q}(x) = \frac{x\Phi(x-\alpha)}{\Phi(x)}, \qquad \alpha = 2\pi i/N. \tag{11}$$

<u>Theorem.</u> <u>The elliptic genus</u> $\tilde{\varphi}_N(M_d)$ <u>is a modular function for</u> $\Gamma_1(N)$ <u>if</u> $d \equiv 0 \bmod 2N$. <u>We have</u>

$$\varphi_N(M_d) = \tilde{\varphi}_N(M_d)(\Phi(-\alpha))^{-d} = \tilde{\varphi}_N(M_d)\cdot c^d .$$

The result follows from the preceding theorem and the considerations in section 1 and 2 which show that $\Phi(-\alpha)^{-d} = c^d$ is a modular form for $\Gamma_1(N)$ of weight d.

If d is not divisible by $2N$, then

$$\tilde{\varphi}_N(M_d)^{2N/(d,2N)}$$

is a modular function (here $(d,2N)$ is the greatest common divisor of d and $2N$).One simply applies the theorem to the $2N/(d,2N)$-th power of M_d.

The function $\tilde{\varphi}_N$ has poles in the cusps represented by (1) with $k > 0$. The order of the pole is given by (6). Let us restrict to the case that N is a prime. For $N = 2$ we have 2 cusps represented by $(k,\ell) = (0,1)$

and $(k,\ell) = (1,0)$. For N odd, we have $2 \cdot \frac{N-1}{2}$ cusps. There are $\frac{N-1}{2}$ cusps represented by $(k,\ell) = (0,\ell)$ and $1 \leq \ell \leq \frac{N-1}{2}$ and $\frac{N-1}{2}$ cusps represented by $(k,\ell) = (k,0)$ and $1 \leq k \leq \frac{N-1}{2}$. In the first kind of cusps the q-development of $\tilde{\varphi}_N(M_d)$ begins with the constant term $\chi_y(M_d)$ with $y = -\zeta^\ell$, in the latter case it starts with

$$\chi(M_d, K^{k/N}) \cdot \tilde{q}^{-k(N-k)d/2N}$$

where $\tilde{q}$ is a local uniformizing parameter for this cusp of $\mathbb{H}/\Gamma_1(N)$. (We have $\tilde{q}^N = q$ in (6)).

4. For a complex vector bundle W of dimension r the exterior powers $\Lambda^i W$ and the symmetric powers $S^i W$ are well-defined vector bundles. Their Chern classes can be calculated from those of W. If $c_1, \ldots, c_r$ are the Chern classes of W (where c_i is in the 2i-dimensional cohomology of the base space) and if we write formally

$$c = 1 + c_1 + c_2 + \ldots + c_r = (1+x_1)(1+x_2)\ldots(1+x_r)$$

then the Chern character (in the rational cohomology of the base space) is given by

$$ch(W) = e^{x_1} + e^{x_2} + \ldots + e^{x_r} .$$

Over the rationals c and ch determine each other. For the exterior powers we write with some indeterminate t

$$\Lambda_t(W) = \sum_{i=0}^{r} \Lambda^i W \cdot t^i$$

and for the symmetric powers

$$S_t(W) = \sum_{i=0}^{\infty} S^i W \cdot t^i .$$

Then we have for the Chern character

(12) $$\mathrm{ch}(\Lambda_t W) = \prod_{i=1}^{r} (1 + te^{x_i}) ,$$

(13) $$\mathrm{ch}(S_t W) = \prod_{i=1}^{r} (1 - te^{x_i})^{-1} .$$

Formula (12) was often used in [13]. Formula (13) is, of course, a special case of the general method to calculate the Chern classes associated to given vector bundles by representations [7].

Following Witten's idea (see [19]) we write the elliptic genus $\tilde{\varphi}_N(M_d)$, or rather its q-development in the standard cusp, in the form

(14) $$\tilde{\varphi}_N(M_d) = \sum_{n=0}^{\infty} \chi_y(M_d, R_n) q^n .$$

Here, as before, $-y = \zeta = e^{2\pi i/N}$. Furthermore R_n is a virtual vector bundle associated to the complex tangent bundle of M_d by a virtual representation of $GL(d,\mathbb{C})$ (with coefficients in $\mathbb{Z}[\zeta]$).

For a vector bundle W the polynomial $\chi_y(M_d, W)$ is defined in [13]. We have, if T is the tangent bundle of M_d,

$$\chi_y(M_d, W) = \sum_{p=0}^{d} \chi(M_d, \Lambda^p T^* \otimes W) y^p .$$

We now can specify the R_n in (14). Let us recall that $\tilde{\varphi}_N(M_d)$ is the genus belonging to the power series (11)

(15) $$\tilde{Q}(x) = \frac{x}{1-e^{-x}}(1+ye^{-x})\prod_{n=1}^{\infty}\frac{1+yq^ne^{-x}}{1-q^ne^{-x}} \cdot \frac{1+y^{-1}q^ne^{x}}{1-q^ne^{x}}$$

Therefore (by (12) and (13))

(16) $$\sum_{n=0}^{\infty} R_n q^n = \prod_{n=1}^{\infty} \Lambda_{yq^n} T^* \cdot \prod_{n=1}^{\infty} \Lambda_{y^{-1}q^n} T \cdot \prod_{n=1}^{\infty} S_{q^n}(T+T^*)$$

(with $-y = \zeta = e^{2\pi i/N}$).

We have

$$R_0 = 1, \quad R_1 = (1-\zeta)T^* + (1-\zeta^{-1})T$$

Modulo the ideal $(1-\zeta)$ of $\mathbb{Z}[\zeta]$, the elliptic genus $\tilde{\varphi}_N(M_d)$ equals the Euler-Poincaré number $e(M_d)$.

According to Witten's philosophy (compare also [2] and [3]) if we had a χ_y-operator on the loop space $\mathcal{L}M_d$ of M_d, we could try to calculate (or define) its equivariant χ_y-genus for the natural S^1-action on $\mathcal{L}M_d$ with $q \in S^1$ $(q = e^{2\pi i\tau},\ \tau \in \mathbb{R})$ by the Atiyah-Bott-Singer ([4], [6]) fixed point theorem (fixed point set M_d (constant loops) in $\mathcal{L}M_d$). The result for the equivariant χ_y-genus $\chi_y(\mathcal{L}M_d,q)$ would be that it is the genus with respect to the power series

$$x\prod_{n=-\infty}^{\infty}\frac{1+yq^ne^{-x}}{1-q^ne^{-x}}$$

This does not make sense as a power series in q, but formal manipulations bring $\chi_y(\mathcal{L}M_d,q)$ to the genus $\tilde{\varphi}_N(M_d)$ belonging to $\tilde{Q}(x)$ (see (15)) provided $(-y)^d = 1$. Observe that (15) is a meromorphic function in the two variables x and q where $(x,q) \in \mathbb{C}^2$ and $|q| < 1$.

5. The genus $\tilde{\varphi}_N(M_d)$ has in the standard cusp a development whose coefficients are integral. They are elements of $\mathbb{Z}[\zeta]$. See formula (14). In the cusps (represented by (1) with $0 < k < N$) this is not so. The coefficients are of the form

$$\chi(M_d, K^{k/N} \otimes W_n) ,$$

where W_n is a virtual vector bundle associated to the tangent bundle by a virtual representation of $GL(d,\mathbb{C})$ with coefficients in $\mathbb{Z}[\zeta]$. The W_n can be calculated using (3). These coefficients are in general not integral. If, however, the first Chern class c_1 of M_d is divisible by N they are integral. This divisibility condition is equivalent to the existence of a holomorphic complex line bundle with $L^N = K$ and the coefficients

$$\chi(M_d, L^k \otimes W_n)$$

become "Riemann-Roch numbers" [13] which are integral.

<u>Theorem.</u> <u>If the first Chern class</u> c_1 <u>of the complex manifold</u> M_d <u>is divisible by</u> N, <u>then the coefficients of the</u> q-<u>developments of the genus</u> $\tilde{\varphi}_N(M_d)$ <u>in all cusps (given by (1)) are integral</u> ($\in \mathbb{Z}[\zeta]$); <u>for the elliptic genus</u> $\varphi_N(M_d)$ <u>the coefficients are integral in a cusp with</u> $k > 0$, <u>in the cusps with</u> $k = 0$ <u>they become integral after multiplication with</u> $(1-\zeta)^d$.

6. Let M_d be a compact complex manifold together with an action of the circle S^1 on M_d by holomorphic maps. We write elements of the circle as $\lambda = e^{2\pi iz}$ where $z \in \mathbb{R}/\mathbb{Z}$. The group S^1 acts on the virtual bundles R_n (see (16)). It also acts on the "cohomology group"

$$H^q(M_d, \Lambda^p T^* \otimes R_n) \tag{17}$$

which is in fact a formal direct sum of cohomology groups $H^q(M_d,\Lambda^p T^* \otimes W)$ with coefficients in $\mathbb{Z}(\zeta)$. Since S^1 acts, we get from (17) (considered equivariantly) a character of S^1, i.e. a finite Laurent series in λ. Taking alternating sums over q in (17) gives us a character

$$\chi(M_d,\Lambda^p T^* \otimes R_n,\lambda)$$

and finally (with $-y = e^{2\pi i/N}$),

$$\chi_y(M_d,R_n,\lambda) = \sum_{p=0}^{d} \chi(M_d,\Lambda^p T^* \otimes R_n,\lambda)y^p$$

and

(18) $$\tilde{\varphi}_N(M_d,\lambda) = \sum_{n=0}^{\infty} \chi_y(M_d,R_n,\lambda)q^n$$

It may be more convenient to return to our elliptic genus φ_N with characteristic power series $Q(x)$ (see (10)) and consider it equivariantly

(19) $$\varphi_N(M_d,\lambda) = \tilde{\varphi}_N(M_d,\lambda)\cdot\Phi(-\alpha)^{-d}$$

$$= \sum_{n=0}^{\infty} \chi_y(M_d,S_n,\lambda)q^n$$

where the S_n are virtual bundles (coefficients in $\mathbb{Q}(\zeta)$). We can calculate $\varphi_N(M_d,\lambda)$ using the Atiyah-Bott-Singer fixed point theorem (holomorphic Lefschetz theorem [6], p. 566). Before doing this some remarks concerning the fixed point set $M_d^{S^1}$ of the action are necessary.

The set $M_d^{S^1}$ is a smooth submanifold of M_d being a disjoint union of connected submanifolds of various dimensions. For each fixed point p, the circle acts in the tangent space T_p, hence integers $m_1,\dots,m_d$ are defined such that $\lambda \in S^1$ acts by the diagonal matrix $(\lambda^{m_1},\lambda^{m_2},\dots,\lambda^{m_d})$. For each $r \in \mathbb{Z}$ we consider those m_i which are equal to r. This leads to the eigenspace E_r of T_p. Of course, E_0 is the tangent space in p of the connected component of $M_d^{S_1}$ to which p belongs. The numbers $m_1,\dots,m_d$ (well defined up to order) depend only on the component of $M_d^{S_1}$. Over each component we have eigenspace bundles, also denoted by E_r.

The characteristic power series of the elliptic genus φ_N is given in (10) in the form $Q(x) = x/f(x)$. For the fixed point theorem we need $1/f(x)$. We put

$$F(x) = 1/f(x) = \frac{\Phi(x-\alpha)}{\Phi(x)\Phi(-\alpha)} \tag{20}$$

We shall now give a formula for $\varphi_N(M_d,\lambda)$ using the holomorphic Lefschetz theorem writing it down in short hand form which will need some explanation:

$$\varphi_N(M_d,\lambda) = (e_0\cdot F(x_1+2\pi i m_1 z)\dots F(x_d+2\pi i m_d z))[M_d^{S^1}] \tag{21}$$

where e_0 is the product of those x_i for which $m_i = 0$. Recall $\lambda = e^{2\pi i z}$. Formula (21) has the following meaning. For each component of the fixed point set, e_0 is the Euler class (highest Chern class) of its tangent bundle E_0, the formal roots of the total Chern class of E_0 are the x_i with $m_i = 0$. The x_i with $m_i = r \neq 0$ are the formal roots of the total Chern class of the eigenspace bundle E_r over the component. Thus for E_0 one uses in the above product $xF(x) = Q(x)$ and for E_r $(r \neq 0)$ the function $F(x+2\pi i r z)$ which for $rz \notin \mathbb{Z}\tau+\mathbb{Z}$ has no pole for $x = 0$ (and we use a general z) and hence is a power series in x. Then one evaluates the expression in (21) on the component and adds over all components. Observe that the rotation numbers $(m_1,\dots,m_d)$ depend on the component and also the meaning of the x_i which are the formal roots of the total Chern class restricted to the component. According to (4) our function $F(x)$ has the property

(22) $$F(x+2\pi i) = F(x), \quad F(x+2\pi i\tau) = \zeta F(x)$$

where $\zeta = e^{2\pi i/N}$.

It follows immediately that $\varphi_N(M_d,\lambda)$ can be extended to an elliptic function in z (with $\lambda = e^{2\pi i z}$) for the lattice $\mathbb{Z}\cdot N\tau+\mathbb{Z}$. More precisely:

Let υ be an index for the connected components $(M_d^{S^1})_\upsilon$ of the fixed point set $M_d^{S^1}$. Then according to (21)

(23) $$\varphi_N(M_d,\lambda) = \sum_\upsilon \varphi_N(M_d,\lambda)_\upsilon$$

where $\varphi_N(M_d,\lambda)_\upsilon$ is an elliptic function for the lattice $\mathbb{Z}N\tau+\mathbb{Z}$ associated to $(M_d^{S^1})_\upsilon$. Indeed,

(24) $$\begin{aligned}\varphi_N(M_d,\lambda q)_\upsilon &= \varphi_N(M_d,e^{2\pi i(z+\tau)})_\upsilon \\ &= \zeta^{m_1+\dots+m_d}\varphi_N(M_d,\lambda)_\upsilon\end{aligned}$$

The exponent $m_1+\dots+m_d$ depends on υ, even the residue class of the exponent mod N depends on υ in general.

<u>Definition:</u> The S^1-action on M_d is called N-balanced if for the components $(M_d^{S^1})_\upsilon$ of the fixed point set the residue class of $m_1+\dots+m_d$ modulo N does not depend on υ. If the action is N-balanced, the common residue class of $m_1+\dots+m_d$ is called the type of the action and denoted by t.

We have proved

<u>Theorem.</u> <u>For an</u> N-<u>balanced</u> S^1-<u>action of type</u> t <u>on the complex manifold</u> M_d, <u>the equivariant elliptic genus</u> $\varphi_N(M_d,\lambda)$ <u>with</u> $\lambda = e^{2\pi iz}$ <u>is an elliptic function for the lattice</u> $\mathbb{Z}\cdot N\tau + \mathbb{Z}$ <u>which satisfies</u>

$$(25) \qquad \varphi_N(M_d,\lambda q) = \varphi_N(M_d,e^{2\pi i(z+\tau)}) = \zeta^t \varphi_N(M_d,\lambda)$$

<u>Remark</u> Of course, φ_N can be regarded as a function of τ and z. In τ it is a modular form of weight d. In fact, φ_N is a meromorphic Jacobi form on $\Gamma_1(N)$ of weight d and index 0 (see [11]).

7. We now shall approach the rigidity theorems which under certain conditions state that the finite Laurent series $\chi_y(M_d,R_n,\lambda)$ (see (16) and (18)) do not depend on λ. (Recall $-y = e^{2\pi i/N}$). This rigidity means that the elliptic function $\varphi_N(M_d,\lambda)$ of the preceeding theorem is a constant (see (19)), i.e. we have to show that it has no poles. The rigidity results were not included in my course at the University of Bonn. When Michael Atiyah came to Bonn in February 1988 he explained to me Bott's approach [9] and that it is rather close to our old paper [5] and we discussed it in Oxford in March. I did not study Taubes' paper [18] in detail, but rather looked in Bott's report [9]. Then I carried out the proof for the level N case during my visit in Cambridge (England) in March 1988 as a guest of Robinson college.

Let us consider $\varphi_N(M_d,\lambda)$ as a function of λ and q. It is meromorphic for $\lambda \in \mathbb{C}^*$ and $|q| < 1$. According to (21) it can have poles only for $mz \in \mathbb{Z}\tau+\mathbb{Z}$ where m is a rotation number $\neq 0$ occuring for one of the components of $M_d^{S^1}$. Of course, $mz \in \mathbb{Z}\tau + \mathbb{Z}$ means $\lambda^m = q^n$ where $n \in \mathbb{Z}$. We have

(26) $$\varphi_N(M_d,\lambda) = \sum_{n=0}^{\infty} c_n(\lambda)q^n$$

with $c_n(\lambda)$ a finite Laurent series (compare (18),(19)). The meromorphic function $\varphi_N(M_d,\lambda)$ can have poles only on the curves $\lambda^m = q^n$. If (λ,q) does not lie on such a curve, then the series (22) converges. If $\lambda_0 = e^{2\pi ik/r}$ is a primitive r-th root of unity, then (λ_0,q) lies precisely on the curves $\lambda^m = q^n$ with $m \equiv 0 \bmod r$ and $n = 0$. But still

$$\varphi_N(M_d,\lambda_0) = \sum_{n=0}^{\infty} c_n(\lambda_0)q^n$$

converges for $|q| < 1$, because this q-development can be calculated from the fixed point set $M_d^{\lambda_0}$ of λ_0, namely

$$\varphi_N(M_d,\lambda_0) = (e_0F(x_1+2\pi im_1k/r)\dots(F(x_d+2\pi im_dk/r))[M_d^{\lambda_0}]$$

where e_0 is now the product over those x_j for which $m_j \equiv 0 \bmod r$ and where we interpret the formula in a similar way as in (21). If $|q| \leq R < 1$, then there is a neighborhood U of λ_0 in $\mathbb{C}^*$ such that no point (λ,q) with $\lambda \in U - \{\lambda_0\}$ and $|q| \leq R$ lies on one of the curves $\lambda^m = q^n$. We know that $\sum_{n=0}^{\infty} c_n(\lambda)q^n$ converges for $\lambda \in U$ and $|q| < R$.

It is not immediately clear that our elliptic function $\varphi_N(M_d,\lambda)$ is holomorphic in λ_0 and has there the value $\varphi_N(M_d,\lambda_0)$ calculated from the fixed point set of λ_0. But in fact $\varphi_N(M_d,\lambda)$ is holomorphic for $\lambda \in U$ and $|q| < R$, and (26) (for this range of λ and q) is the power series development of a holomorphic function in two variables with respect to one of the variables. <u>In particular</u>, <u>there is no pole for $\lambda = \lambda_0$</u>. We conclude this from the following lemma.

<u>Lemma.</u> <u>Suppose</u> $g(\lambda,q)$ <u>is a meromorphic function in the two complex variables</u> λ,q <u>where</u> $\lambda \in U \subset \mathbb{C}^*$ <u>and</u> $q \in D_R$ <u>where</u> D_R <u>is the open disc around the origin of positive radius</u> R. <u>Assume that</u> $\lambda_0 \in U$ <u>and</u> $h(\lambda,q) = (\lambda-\lambda_0)^m g(\lambda,q)$ (<u>with</u> $m \geq 0$, $m \in \mathbb{Z}$) <u>is holomorphic in</u> $U \times D_R$. <u>If</u>

$$(27) \qquad g(\lambda,q) = \sum_{n=0}^{\infty} c_n(\lambda_0)q^n$$

<u>in</u> $(U - \{\lambda_0\}) \times D_R$ <u>where</u> $c_n(\lambda)$ <u>is a finite Laurent series, then</u> $g(\lambda,q)$ <u>is holomorphic in</u> $U \times D_R$ <u>and</u> (27) <u>holds in</u> $U \times D_R$.

<u>Proof:</u> The holomorphic function $g(\lambda,q)$ has a development in $U \times D_R$ of the form

$$h(\lambda,q) = \sum_{n=0}^{\infty} d_n(\lambda)q^n$$

where the functions $d_n(\lambda)$ are holomorphic in U. Since $c_n(\lambda) = (\lambda - \lambda_0)^{-m} d_n(\lambda)$ for $\lambda \neq \lambda_0$ and $\lim_{\lambda\to\lambda_0} c_n(\lambda) = c_n(\lambda_0)$ we see that the $d_n(\lambda)$ are divisible by $(\lambda - \lambda_0)^m$.

Actually, we did not need the above discussion of the fixed point set of λ_0 and the convergence of (26) for $\lambda = \lambda_0$.

An S^1-action is semi-free (i.e. the fixed point set of any $\lambda \in S^1$, $\lambda \neq 1$, equals $M_d^{S^1}$) if and only if all non-vanishing rotation numbers m equal ± 1. Therefore, for a semi-free action, $M_d^{S^1}$ can have poles only for $\lambda = q^n$ with $n \in \mathbb{Z}$.

<u>Theorem.</u> <u>For an N-balanced semi-free</u> S^1<u>-action of type</u> t <u>on the complex manifold</u> M_d, <u>the equivariant elliptic genus</u> $\varphi_N(M_d,\lambda)$ <u>does not</u>

<u>depend on</u> λ. <u>It equals the elliptic genus</u> $\varphi_N(M_d)$. <u>If</u> $t \not\equiv 0 \bmod N$, <u>then</u> $\varphi_N(M_d) = 0$. (Compare [14] and [16].)

<u>Proof</u>: By the lemma, there is no pole for $\lambda = 1$. Because of (25) there are no poles for $\lambda = q^n$. The vanishing of $\varphi_N(M_d)$ follows also from (25).

8. Let M_d be a complex manifold with first Chern class $c_1 \in H^2(M_d,\mathbb{Z})$ divisible by N. The importance of this condition was already apparent in section 5. We choose a holomorphic line bundle L with $L^N = K$. Now suppose we have an S^1-action on M_d. Consider the N-fold covering $S^1 \longrightarrow S^1$ with $\mu \longmapsto \lambda = \mu^N$. Then μ acts on M_d and K through λ. This action can be lifted to L. If p is a fixed point of the given S^1-action with rotation numbers $m_1, m_2, \dots, m_d$, then μ acts in the fibre L_p by $\mu^{-(m_1+\dots+m_d)}$. However, if $\mu = \zeta = e^{2\pi i/N}$, then it operates trivially on M_d. Therefore the action of ζ in each fibre of L is by multiplication with ζ^{-t}, where t is a residue class mod N which does not depend on the base point of the fibre. (Assume that M_d is connected.) It follows that the action is N-balanced of type t (see the definition in section 6).

The condition $c_1 \equiv 0 \bmod N$ implies a stronger property than N-balanced. Let $G_m \subset S^1$ be the group of mth roots of unity. The fixed point set of G_m is a submanifold of M_d which includes $M_d^{S^1}$ and is strictly larger if and only if there is a rotation number divisible by m. We denote the fixed point set of G_m by M_d^m. There is the map $S^1 \longrightarrow S^1$ with $\mu \longmapsto \lambda = \mu^N$ which we considered before. Hence any $\mu \in S^1$ with $\mu^{mN} = 1$ operates trivially on M_d^m, however it operates on every fibre L_p $(p \in M_d^m)$ by multiplication with some mN-th root of unity which only depends on the connected component of M_d^m which contains p. Since

μ acts on L_p (for $p \in M_d^{S^1}$) by $\mu^{-(m_1+\ldots+m_1)}$ where the m_j are the rotation numbers of the action in p, it follows that the residue class of $m_1+\ldots+m_d$ mod mN depends only on the connected components of M_d^m and not on the components of $M_d^{S^1}$ contained in them.

Let X be a connected component of M_d^m and $(M_d^{S^1})_\upsilon$ a component of $M_d^{S^1}$ contained in X with rotation numbers $m_1,\ldots,m_d$. Over X the tangent bundle T of M_d splits into vector bundles $\tilde{E}_k$ where $k = 0,1,\ldots,m-1$ and the action of G_m in $\tilde{E}_k$ is by multiplication with λ^k if $\lambda \in G_m$. Of course, $\tilde{E}_0$ is the tangent bundle of X. Over $(M_d^{S^1})_\upsilon$ we have

(28) $$\tilde{E}_k = \sum_{r \equiv k \bmod m} E_r \qquad \text{(see section 6)} .$$

We write the rotation numbers in the following form

(29) $$m_i = r_i m + k_i \quad \text{where} \quad k_i = 0,1,\ldots,m-1 .$$

Since the integer $\sum k_i = \sum_{k=0}^{m-1} k \dim \tilde{E}_k$ depends only on X, we see that $m \cdot \sum r_i$ mod mN depends only on X. Hence, $\sum r_i$ mod N depends only on X and not on the components $(M_d^{S^1})_\upsilon$ contained in it. We put

(30) $$\sum r_i = t(m,X) \bmod N .$$

Of course, $t(1,M_d)$ is the type t of the action (for connected M_d).

9. Let M_d be a compact complex manifold with $c_1 \equiv 0 \bmod N$. We assume that we have an S^1-action and wish to show that the elliptic function $\varphi_N(M_d,\lambda)$ has no poles. Let X be a connected component of M_d^m (see section 8). We define

$$\varphi_N(X,\lambda) = \sum_{\upsilon} \varphi_N(M_d,\lambda)_{\upsilon} \tag{31}$$

where the summation is over those connected components $(M_d^{S^1})_{\upsilon}$ which are contained in X (see (23)). This is a shorthand notation. Do not confuse (31) with the elliptic genus of X. Let $(M_d^{S^1})_{\upsilon}$ have the rotation numbers $m_1,\ldots,m_d$. According to (21) we have

$$\varphi_N(M_d,\lambda)_{\upsilon} = (e_0 F(x_1+2\pi i m_1 z)\ldots F(x_d+2\pi i m_d z))[(M_d^{S^1})_{\upsilon}] \ . \tag{32}$$

Let s be an integer and replace in (32) the variable z by $z + \frac{s}{m}\tau$ (in other words, replace λ by $\lambda\cdot q^{s/m}$), then $\varphi_N(M_d,\lambda q^{s/m})_{\upsilon}$ is again an elliptic function in z for the lattice $\mathbb{Z}\cdot N\tau+\mathbb{Z}$. It follows from (22), (29) and (30) that

$$\varphi_N(M_d,\lambda q^{s/m})_{\upsilon} = \tag{33}$$

$$\zeta^{st(m,X)}\cdot(e_0 \prod_{j=1}^{d} F(x_j+2\pi i m_j z + 2\pi i \frac{sk_j}{m}\tau)[(M_d^{S^1})_{\upsilon}] \ .$$

If we write down the q-development of the right hand side of (33) (with fractional powers of q) we see that $\varphi_N(X,\lambda q^{s/m})$ is of the form

$$\varphi_N(X,\lambda q^{s/m}) = \sum_{n=0}^{\infty} \chi_y(X,S_n,\lambda) q^{n/m} \tag{34}$$

where the S_n are virtual equivariant bundles constructed from the bundles $\tilde{E}_k$ over X. For $m = 1$ we come back to (19).

The elliptic function (34) has no poles for $|\lambda| = 1$. We use again the lemma in section 7.

10. We are now able to prove the rigidity theorem.

Theorem. Let M_d be a compact complex manifold with first Chern class $c_1 \in H^2(M_d, \mathbb{Z})$ divisible by N. Suppose an S^1-action on M_d is given. Then the equivariant elliptic genus $\varphi_N(M_d, \lambda)$ does not depend on $\lambda \in S^1$. It equals the elliptic genus $\varphi_N(M_d) = \varphi_N(M_d, 1)$. If the type t of the action is $\not\equiv 0 \bmod N$, then $\varphi_N(M_d) \equiv 0$.

Proof. Let m be a natural number ≥ 1 and $\lambda^{\pm m} = q^n$. Then λ is of the form $\lambda = \lambda_0 q^{s/m}$ where $\lambda_0^m = 1$ and $s = \pm n$. We have

$$\varphi_N(M_d, \lambda_0 q^{s/m}) = \sum_X \varphi_N(X, \lambda_0 q^{s/m})$$

where the summation is over all the connected components of M_d^m. Since the elliptic function (34) has no poles for λ_0, the result follows. The vanishing $\varphi_N(M_d) = 0$ for $t \not\equiv 0 \bmod N$ follows again from (25).

11. We want to point out some applications of the rigidity theorem.

If we develop in a cusp (1) with $k > 0$, we get a different version of the rigidity theorem (compare [19]). In particular, we get that $\chi(M_d, L^k, \lambda)$ does not depend on λ for $k = 1, \ldots, N-1$, in fact $\chi(M_d, L^k) = 0$, if the action is non-trivial. This is a well-known result ([12],[15]). For $N = 2$ and $k = 1$ it corresponds to the theorem in [5] on the $\hat{A}$-genus.

The elliptic genus of level N is strictly multiplicative in fibre bundles with a manifold M_d with $c_1(M_d) \equiv 0 \bmod N$ as fibre and a compact connected Lie group G of automorphisms of M_d as structure group (compare [14] and [16]).

This we wish to apply, for example, to the compact irreducible hermitian symmetric spaces G/U studied in [7] § 16. There we gave a formula for the coefficient $\lambda(G/U)$ in

$$c_1(G/U) = \lambda(G/U) \cdot g$$

where g is a positive generator of the infinite cyclic group $H^2(G/U)$.

Take a system $w_1,\ldots,w_d$ of positive complementary roots for G/U (see [7]). Here d is the complex dimension of G/U. The roots $w_1,\ldots,w_d$ are linear forms in $x_1,\ldots,x_\ell$ where $\ell = \mathrm{rank}(U) = \mathrm{rank}(G)$, the $x_1,\ldots,x_\ell$ can be identified with a base of $H^1(T,\mathbb{Z})$ where T is the maximal torus of U. Without proof we state the following result which is equivalent to the strict multiplicativity of the elliptic genus for G/U-bundles.

<u>Theorem.</u> <u>Let</u> $F(x) = f(x)^{-1}$ <u>be the elliptic function introduced for level</u> N (see section 1). <u>Let</u> $w_1,\ldots,w_d$ <u>be positive complementary roots for the irreducible hermitian symmetric space</u> G/U. <u>Suppose</u> $\lambda(G/U) \equiv 0 \bmod N$. <u>Then</u>

$$(35) \qquad \sum_{\sigma\in W(G)/W(U)} F(\sigma(w_1))F(\sigma(w_2))\ldots F(\sigma(w_d)) = \varphi_N(G/U).$$

Here $W(G)$, $W(U)$ are the Weyl groups. (An element $\sigma \in W(U)$ permutes $w_1,\ldots,w_d$. Therefore the sum over the $W(U)$-cosets is well-defined.)

The formula (35) is an identity in the ℓ variables $x_1,\ldots,x_\ell$. The sum is a constant, i.e. does not depend on these variables anymore.

The rigidity theorem in section 10 also gave a vanishing result. We give an example: Consider the Grassmannian

$$W(m,n) = U(m+n)/(U(m) \times U(n)) .$$

We use the notation of [7]. As a system of positive roots of $U(m+n)$, we take

$$\{-x_i + x_j \mid 1 \leq i < j \leq m + n\}.$$

The complementary roots w_r are given by $1 \leq i \leq m$ and $m + 1 \leq j \leq m + n$. Their sum equals

$$(36) \qquad \sum w_r = -n \sum_{i=1}^{m} x_i + m \sum_{j=m+1}^{m+n} x_j =$$

$$-(m + n) \sum_{i=1}^{m} x_i + m \sum_{j=1}^{m+n} x_j$$

We put $-\sum_{i=1}^{m} x_i = g$ and $\sum_{j=1}^{m+n} x_j = \sigma_1$. Then g becomes the positive generator of $H^2(W(m,n),\mathbb{Z})$ whereas σ_1 vanishes if regarded as element of this cohomology group. Therefore,

$$\lambda(W(m,n)) = m + n.$$

We also see from (36) that $W(m,n)$ admits an N-balanced circle action of type m if $m + n \equiv 0 \bmod N$. We obtain

Proposition: The elliptic genus $\varphi_N(W(m,n))$ vanishes if $m + n \equiv 0 \bmod N$ and $m \not\equiv 0 \bmod N$. For the complex projective spaces $P_n(\mathbb{C}) = W(n,1)$ we have

$$\varphi_N(P_n(\mathbb{C})) = 0 \quad \text{if} \quad n + 1 \equiv 0 \bmod N.$$

12. The elliptic function f defined in section 1 satisfies a differential equation

$$(37) \qquad (\frac{f'}{f})^N + a_1(\frac{f'}{f})^{N-1} + \ldots + a_{N-1}(\frac{f'}{f}) + a_N = \frac{1}{f^N} + a_{2N}f^N$$

$$\text{with } a_{2N} = c^{2N} \quad \text{(see section 1)}$$

where the a_j are modular forms of weight j for $\Gamma_1(N)$. (if $\alpha = 2\pi i/N$ in (1)). The polynomial

$$P(\xi) = \xi^N + a_1\xi^{N-1} + \ldots + a_{N-1}\xi + a_N$$

has the following properties:

1) $a_{N-1} = 0$,

2) If $P'(\xi) = 0$, but $\xi \neq 0$, then $P(\xi)^2 = 4a_{2N}$.

The property 2) implies that the values at the critical points ξ with $\xi \neq 0$ are all equal up to sign. In this case, the polynomial might be called almost - Čebyšev. Theodore J. Rivlin wrote to me that polynomials with essentially such properties occur in the literature under the name Zolotarev-polynomials. Also their relation to elliptic functions is known (see for example [10]). I plan to write a separate paper on these matters. For $N = 2$ the differential equation is of the form

$$(f')^2 = 1 - a_2f^2 + a_4f^4,$$

very well known for the elliptic genus of level 2 (see [14]).

REFERENCES

[1] Proceedings of the conference on elliptic curves and modular forms in algebraic topology, Princeton, September 1986, to appear: Springer Lecture Notes in Mathematics 1326 (1988).

[2] O. ALVAREZ, T.-P. KILLINGBACK, M. MANGANO, P. WINDEY, String theory and loop space index theorems, Comm. Math. Phys. 111 (1987), 1-10.

[3] O. ALVAREZ, T.-P. KILLINGBACK, M. MANGANO, P. WINDEY, The Dirac-Ramond operator in string theory and loop space index theorems, Nuclear Physics B (Proc. Suppl.) 1A (1987), 189-216.

[4] M.F. ATIYAH and R. BOTT, A Lefschetz fixed-point formula for elliptic complexes II: Applications, Ann. of Math. 88 (1968), 451-491.

[5] M.F. ATIYAH and F. HIRZEBRUCH, Spin-manifolds and group actions. In: Memoires dédiés à Georges de Rham, p. 18-28, Berlin-Heidelberg-New York: Springer 1970.

[6] M.F. AITYAH and I.M. SINGER, The index of elliptic operators: III, Ann. of Math. 87 (1986), 546-604.

[7] A. BOREL and F. HIRZEBRUCH, Characteristic classes and homogeneous spaces I, Am. J. Math. 80 (1958), 458-538.

[8] A. BOREL and F. HIRZEBRUCH, Characteristic classes and homogeneous spaces II, Am. J. Math. 81 (1959), 315-382.

[9] R. BOTT, On the fixed point formula and the rigidity theorems of Witten, Lectures at Cargèse 1987, to appear.

[10] B.C. CARLSON and John TODD, Zolotarev's first problem- the best approximation by polynomials of degree $\leq n-2$ to $x^n - n\sigma x^{n-1}$ in $[-1,1]$, Aequationes mathematicae 26(1983), 1-33.

[11] M. EICHLER and D. ZAGIER, The theory of Jacobi forms, Boston-Basel-Stuttgart, Birkhäuser 1985.

[12] A. HATTORI, $Spin^c$-structures and S^1-actions, Inventiones math. 48 (1978), 7-31.

[13] F. HIRZEBRUCH, Topological methods in algebraic geometry (3. edition, with appendices by R.L.E. Schwarzenberger and A. Borel), Grundlehren der mathematischen Wissenschaften, Bd. 131, Berlin-Heidelberg-New York: Springer 1966.

[14] P.S. LANDWEBER, Elliptic genera: An introductory overview, in [1].

[15] K.H. MAYER, A remark on Lefschetz formulae for modest vector bundles, Math. Ann. 216 (1975), 143-147.

[16] S. OCHANINE, Genres elliptiques équivariants, in [1].

[17] G. SEGAL, Elliptic cohomology, Séminaire Bourbaki, 40e année, 1987-88, no. 695 (Février 1988).

[18] C.H. TAUBES, S^1 actions and elliptic genera, Harvard preprint 1987.

[19] E. WITTEN, The index of the Dirac operator in loop space, in [1].

[20] D. ZAGIER, Note on the Landweber-Stong elliptic genus, in [1].

The Principle of Triality and A Distinguished Unitary Representation of SO(4,4)

by Bertram Kostant*

0. Introduction. The group $SO(4,2)$ is a familiar symmetry group in physics since among other things it is the conformal group of compactified Minkowski space $S^3 \times S^1$. The double covering $SU(2,2)$ of its identity component $SO(4,2)_e$ appears as a subgroup of the symmetry group $Sp(8,\mathbf{R})$. In fact, if one considers the real and imaginary part of the Hermitian form in $\mathbf{C}^4$ with signature $++--$ one sees immediately that

$$U(2,2) = Sp(8,\mathbf{R}) \cap SO(4,4) \tag{A}$$

The occurrence of $SU(2,2)$ as a subgroup of $Sp(8,\mathbf{R})$ also has physical significance since one knows that if one goes again to coverings ($\sim$) the restriction of the metaplectic (or oscillator) representation

$$\pi' : \widetilde{Sp(8,\mathbf{R})} \to \text{ Unit } L_2(\mathbf{R}^4)$$

to $\widetilde{SU(2,2)}$ decomposes into a direct sum of all mass zero, arbitrary spin representations of $\widetilde{SU(2,2)}$. (I thank M. Flato for informing me sometime ago of this fact).

From the point of view of representation theory the metaplectic representation has an analogue π for many other simple Lie groups G. If $\mathfrak{g} =$ Lie G one characterization of π is that as a non-trivial unitary representation its Gelfand-Kirillov dimension is equal to $\frac{1}{2}\dim_{\mathbf{C}} O$ where $O \subseteq \mathfrak{g}_{\mathbf{C}}$ is the Ad $\mathfrak{g}_{\mathbf{C}}$ orbit of the highest root vector, and hence is as small as possible. In addition if $J =$ Ker π in the universal enveloping algebra $U(\mathfrak{g})$ of $\mathfrak{g}$ then J is the Joseph ideal and hence by [G] is quadratically generated. As in the case of π' one might very well expect the theory of Howe pairs to be valid for π in general – although this has not been proved.

Returning to (A), we have been motivated for a number of reasons to replace $Sp(8,\mathbf{R})$ by $SO(4,4)$ and to see what sort of physical or geometric information is involved in a construction of π for the case where $G = SO(4,4)_e$ and to consider as in the case of π', the restriction $\pi \mid U(2,2)$. (Representation theoretically, the latter has been solved independently by B. Orsted and D. Vogan and indeed the subgroups $U(1)$ and $SU(2,2)$ behave again as Howe pairs). The overriding motivation for us is that we see $SO(4,4)$ as the point of departure from classical groups to the exceptional Lie groups. Indeed, we recall that the Dynkin diagram

* This paper was supported in part by the NSF Grant No. DMS8403203

K. Bleuler and M. Werner (eds.), Differential Geometrical Methods in Theoretical Physics, 65–108.

of D_4 is

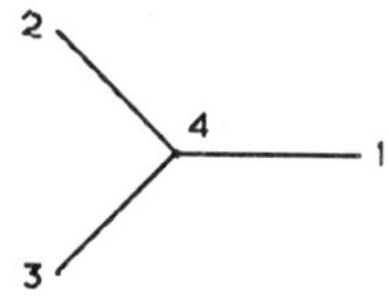

and hence the outer automorphism group of D_4 is the symmetric group Σ_3 of (1,2,3). In particular, there exists an outer automorphism $\tau' \in \Sigma_3$ of D_4 of order 3. (We have been informed that the discovery of τ' goes back to an observation of E. Study at the turn of the century). The automorphism τ' is called triality in [C]. D_4 is unique having such an automorphism. The relation to the exceptional Lie groups is (1) τ' can be defined so that the exceptional Lie group $G_2 \subseteq D_4$ is the fixed subalgebra. Also, (2) τ' defines 3 inequivalent 8-dimensional representations $\mathbb{C}_i^8$, $i = 1, 2, 3$, of D_4 and induces a Lie algebra structure on

$$D_4 \oplus \mathbb{C}_1^8 \oplus \mathbb{C}_2^8 \oplus \mathbb{C}_3^8 = F_4 \tag{B}$$

which is the exceptional Lie algebra F_4. Replacing $\mathbb{C}_i^8$ by End $\mathbb{C}_i^8$ it induces a Lie algebra structure on

$$(D_4 \oplus D_4) \oplus \text{ End } \mathbb{C}_1^8 \oplus \text{ End } \mathbb{C}_2^8 \oplus \text{ End } \mathbb{C}_1^8 = E_8 \tag{C}$$

which is the exceptional Lie algebra E_8.

Among the non-compact real forms of D_4 only $\mathfrak{g} = \text{Lie } SO(4,4)$ admits the action of Σ_3 so that Σ_3 operates on the simply-connected covering group $\tilde{G}$ or on Proj $G = G/\text{Cent } G$.

It should be pointed out to twister theorists that τ' carries $(SO(4,\widetilde{2)\times SO}(2))$ $\subseteq \tilde{G}$ to $U(\tilde{2},2) \subseteq \tilde{G}$ performing a Penrose transform inside of $\tilde{G}$.

0.1. The unitary representation

$$\pi \colon SO(4,4)_e \to \text{ Unit } \mathcal{H} \tag{D}$$

is in fact a representation of Proj G and from abstract considerations one can show that it extends to the outer automorphism group Σ_3. In fact, Vogan has shown that the exceptional Lie groups G_2 and Σ_3 are Howe pairs with respect to π.

One knows about π from abstract representational considerations. However, the real significance of π (it seems to us) would only come about (as with π') from a concrete realization of the Hilbert space $\mathcal{H}$, together with a concrete geometric realization of the action of G. This paper is the first in a series devoted to the construction of π and $\mathcal{H}$, the action of Σ_3 and physical applications. The group $SO(4,4)$ is the conformal group of $S^3 \times S^3$ for Riemannian metrics with signature $+++---$ (space and time are 3-dimensional). Compactified Minkowski space

$$S^3 \times S^1 \subseteq S^3 \times S^3 \tag{E}$$

is the unique closed orbit of $SU(2,2)$ on $S^3 \times S^3$ and such metrics on $S^3 \times S^3$ induce a Lorentzian metric on $S^3 \times S^1$. One of the results we obtain (although not presented in this first paper) is that $\mathcal{H}$ contains a G-invariant cone $\mathcal{C} \subseteq \mathcal{H}$ whose elements are in fact metrics g on $S^3 \times S^3$ in the conformal class of a flat metric with the additional property that

$$\text{Scalar Curvature } g = 0 \tag{F}$$

That is, **in this case the vanishing of scalar curvature is a conformal invariant.** Each such metric $g \in \mathcal{C}$ induces a Lorentzian metric in $S^3 \times S^1$ together (by considering the normal bundle in $S^3 \times S^3$) with an electromagnetic field ω on $S^3 \times S^1$. Thus, each $g \in \mathcal{C}$ defines a sort of Kaluza-Klein theory. One may go out on a long limb here and suggest that $(\pi, \mathcal{H})$ has the rudiments of a quantized Kaluza-Klein theory.

In this paper we geometrically construct $\mathcal{H}$, giving an explicit (and remarkably simple) formula for its unitary structure, (Theorem 6.28), its beautiful K-type decomposition (Theorem 5.12) as well as an explicit formula for the action of a critical element $\tau \in \Sigma_3$ on $\mathcal{H}$ (Theorem 7.4).

One knows (going back to E. Study) that there exists a twisted G-isomorphism of the space of all isotropic lines in $\mathbb{R}^8$ (with ultra-hyperbolic metric $(++++----)$) with either one, Y, or the two equivalence classes of isotropic 4-planes in $\mathbb{R}^8$. In the next paper we will show that the unitary structure on $\mathcal{H}$ arises also from a composite of this isomorphism

$$X \to Y \tag{G}$$

together with a Radon transform involving integration over isotropic 4-planes. This result came about in a conversation with V. Guillemin and E. Grinberg. We wish to thank V. Guillemin for other conversations involving line bundles over X.

1.1 Definition. Let $\mathfrak{g}$ be a Lie algebra over $\mathbb{R}$ which is isomorphic to the split real form of the complex simple Lie algebra D_4. Let G be a simply-connected Lie group having $\mathfrak{g}$ as its Lie algebra. Let $\mathfrak{a} \subseteq \mathfrak{g}$ be a split Cartan subalgebra. Thus $\dim \mathfrak{g} = 28$ and $\dim \mathfrak{a} = 4$.

Let M be an $\mathbb{R}$-vector space of dimension 8 on which a symmetric nonsingular bilinear B of index 4 (i.e. having signature (4,4)) is defined. Let $O(M)$ (resp. $SO(M)$) be the orthogonal group (resp. special orthogonal group) of M with respect to B. Thus, using standard notation one has an isomorphism

$$SO(M) \cong SO(4,4).$$

Furthermore, if $SO(M)_e$ is the identity component of $SO(M)$ then, as one knows, one has an epimorphism

$$\eta: G \to SO(M)_e$$

defining a representation of G on M. We will use η to also denote the corresponding representation of $\mathfrak{g}$ on M.

Let $\mathfrak{a}'$ be the real dual space to $\mathfrak{a}$. Recalling well known facts from representation theory we may find a weight basis $v_1, ..., v_4, u_1, ..., u_4$ of M with respect to $\mathfrak{a}$ such that if $\lambda_i \subset \mathfrak{a}'$ is the weight corresponding to v_i then $-\lambda_i$ is the weight corresponding to $u_i, i = 1, ..., 4$. One knows that the λ_i are a basis of $\mathfrak{a}'$ and the set $\Delta \subseteq M'$ of roots (card $\Delta = 24$) is given by

$$\Delta = \{\pm\lambda_i \pm \lambda_j\}, \quad 1 \leq i < j \leq 4. \tag{1}$$

We may choose a set of positive roots Δ_+ so that

$$\Delta_+ = \{\lambda_i \pm \lambda_j\}, \quad 1 \leq i < j \leq 4$$

The corresponding set $\Pi = \{\alpha_1, ..., \alpha_4\}$ of simple positive roots may then be labelled so that

$$\alpha_1 = \lambda_1 - \lambda_2, \quad \alpha_2 = \lambda_3 + \lambda_4, \quad \text{and}$$

$$\alpha_3 = \lambda_3 - \lambda_4, \quad \alpha_4 = \lambda_2 - \lambda_3. \tag{2}$$

The corresponding Dynkin diagram then appears as

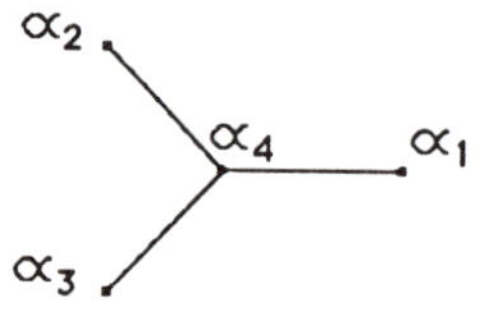

(3)

The following is immediate. With respect to the above choice of Δ_+

Proposition 1.2. *v_1 is a highest weight vector and u_1 is a lowest weight vector of η.*

1.3. The v_k and u_k, $k = 1, ..., 4$ can clearly be normalized so that

$$B(v_i, v_j) = B(u_i, u_j) = 0$$

$$B(v_i, u_j) = \delta_{ij} \tag{4}$$

Now put $\epsilon_i = v_i + u_i$ and $\epsilon_{i+4} = v_i - u_i, i = 1, ..., 4$. It is therefore clear that $\epsilon_j, j = 1, ..., 8$, is an orthogonal basis of M and if $Q(w) = 1/2 \cdot B(w, w)$ is the quadratic form corresponding to B one has

$$Q(\epsilon_i) = \begin{cases} 1 & \text{for } 1 \le i \le 4; \\ -1 & \text{for } 5 \le i \le 8; \end{cases} \tag{5}$$

If E is the span of ϵ_i for $1 \le i \le 4$ and F is the span of ϵ_i for $5 \le i \le 8$ then

$$M = E \oplus F \tag{6}$$

is an orthogonal direct sum with B positive definite on E and negative definite on F. We may regard $SO(E)$ and $SO(F)$ as subgroups of $SO(M)$. The following is well known.

Proposition 1.4. *$SO(E) \times SO(F)$ is a maximal compact subgroup of $SO(M)_e$.*

As an invariant bilinear form on $\mathfrak{g}$ we will define $(x, y) = 1/2 \cdot \operatorname{tr} \eta(x)\eta(y)$ for $x, y \in \mathfrak{g}$. Since $\mathfrak{g}$ is simple it is a positive multiple of the Killing form.

We now note

Lemma 1.5. *There exists an element $\kappa \in G$ such that $\eta(\kappa)v_i = u_i$ and $\eta(\kappa)u_i = v_i, \quad i = 1, ..., 4$.*

Proof. Let $b \in \operatorname{Aut} M$ be the operator such that $bv_i = u_i$ and $bu_i = v_i$. Then $b = 1$ on E and -1 on F. Thus $b \in SO(F) \subseteq SO(M)_e$. Since η is surjective such an element $\kappa \in G$ exists. QED.

Let $\theta = \operatorname{Ad} \kappa \in \operatorname{Aut} \mathfrak{g}$. Since $\eta(\kappa)$ is involutory it follows that θ is an involutory automorphism of $\mathfrak{g}$. Let $\mathfrak{k} = \{x \in \mathfrak{g} \mid \theta x = x\}$ and $\mathfrak{q} = \{y \in \mathfrak{g} \mid \theta y = -y\}$.

Proposition 1.6. *$\mathfrak{g} = \mathfrak{k} + \mathfrak{q}$ is a Cartan decomposition of $\mathfrak{g}$ with $\mathfrak{k}$ the Lie algebra of the maximal compact subgroup of G and $\mathfrak{q}$ its orthogonal complement.*

Proof. It is clear that $x \in \mathfrak{k}$ if and only if $\eta(x)$ commutes with $\eta(\kappa)$. That is, if and only if $\eta(x)$ stabilizes both E and F. Again, that is, if and only if $\eta(x)$ is in the Lie algebra of $SO(E) \times SO(F)$. The proof then follows from Proposition 1.4. QED.

Now since $\eta(x)$ permutes the weight basis v_i, u_i it follows that θ stabilizes $\mathfrak{a}$. Since $\eta(\kappa)$ interchanges v_i and u_i it follows that $\theta = -1$ on $\mathfrak{a}$ since the weight for v_i and u_i are respectively λ_i and $-\lambda_i$. One therefore has

Proposition 1.7. $\mathfrak{a} \subseteq \mathfrak{q}$.

We next note that if $0 \neq e \in \mathfrak{g}$ is a root vector for $(\mathfrak{a}, \mathfrak{g})$ then $(\theta e, e) < 0$. This is clear since $e + \theta e \in \mathfrak{k}$ is not zero and hence $(e + \theta e, e + \theta e) < 0$. It follows therefore that we can choose root vectors $e_\varphi \in \mathfrak{g}, \varphi \in \Delta$, uniquely up to sign, by the condition that

$$(\theta e_\varphi, e_\varphi) = -1. \tag{6.5}$$

The choice of e_φ determines that of $e_{-\varphi}$ by requiring

$$\theta e_\varphi = -e_{-\varphi} \tag{6.6}$$

or

$$(e_\varphi, e_{-\varphi}) = 1. \tag{7}$$

Now for any $\varphi \in \Delta$ let $n_i(\varphi) \in \mathbb{Z}$, $i = 1, ..., 4$ be defined so that $\varphi = \sum_{i=1}^4 n_i(\varphi)\alpha_i$. We wish to consider 3 reductive subalgebras $\mathfrak{g}_i \subseteq \mathfrak{g}, i = 1, 2, 3$, all isomorphic to $\mathfrak{sl}(4, \mathbb{R})$. These subalgebras are defined by putting

$$\mathfrak{g}_i = \mathfrak{a} + \sum_{n_i(\varphi)=0} \mathbb{R} e_\varphi$$

where the sum is over all $\varphi \in \Delta$ such that $n_i(\varphi) = 0$.

Let $\mathfrak{n} = \sum_{\varphi>0} \mathbb{R} e_\varphi$ so that

$$\mathfrak{p}_i = \mathfrak{g}_i + \mathfrak{n}$$

i=1,2,3, is a maximal parabolic subalgebra of $\mathfrak{g}$. Let $P_i \subseteq G$ be the normalizer under Ad of $\mathfrak{p}_i$. As one knows $\mathfrak{p}_i$ is the subalgebra of $\mathfrak{g}$ corresponding to P_i.

1.8. Let Σ_3 be the permutation group of the set $\{1, 2, 3\}$. From the automorphism theory of Lie algebras, one knows that to each $\sigma \in \Sigma_3$ there exists a unique (which by abuse of notation we represent with the same symbol) automorphism σ of $\mathfrak{g}$ which stabilizes $\mathfrak{a}$ and is such that $\sigma e_{\alpha_4} = e_{\alpha_4}$ and

$$\sigma e_{\alpha_i} = e_{\alpha_{\sigma(i)}}, \quad i = 1, 2, 3 \tag{8}$$

Since G is simply connected σ induces a corresponding automorphism of G which we continue to denote by σ. Clearly $\sigma \mathfrak{p}_i = \mathfrak{p}_{\sigma(i)}$ and hence $\sigma P_i = P_{\sigma(i)}$. Thus, if for any $g \in G$ we let g_i be its left coset representative in G/P_i, one has a diffeomorphism

$$\sigma: G/P_i \to G/P_{\sigma(i)} \tag{9}$$

given by $\sigma(g_i) = (\sigma(g))_{\sigma(i)}$ which is such that for any $a \in G$

$$\sigma(a \cdot g_i) = \sigma(a) \cdot \sigma(g_i) \tag{10}$$

where dot here denotes the action of G on G/P_j, $j = 1,2,3$.

2. The varieties X and Y. Now let $C \subseteq M$ be the 7-dimensional cone defined by putting $C = \{w \in M \mid Q(w) = 0\}$. It is clear from (4) that $u_i, v_j \in C$. Let M^* and C^* be the set of all non-zero elements in M and C respectively. Now let Proj M be the real projective variety of all 1-dimensional subspaces in M so that one has a map

$$M^* \to \text{Proj } M, \quad v \mapsto \overline{v} \tag{11}$$

Let $X \subseteq$ Proj M be the image of C^* under (11) so that X is a 6-dimensional subvariety of Proj M.

Now the action of G on M induces an action of G on Proj M. Obviously C is stable under the action of G on M and consequently X is stable under the action of G on Proj M. One knows (and easily proves) that G is transitive on C^* and hence G is transitive on X.

The significance of this as far as the parabolic subgroup $P_i \subseteq G$ is concerned is given in

Proposition 2.1. *The map $G \to X$ given by $g \mapsto g \cdot \overline{v}_1$ defines an isomorphism*

$$G/P_1 \to X \tag{12}$$

Proof. We have only to show that P_1 is the isotropy group at $\overline{v}_1$. But now if $\varphi \in \Delta$ then one notes that $n_1(\varphi) = 0$ if and only if φ is of the form

$$\varphi = \pm\lambda_i \pm \lambda_j \quad \text{for} \quad 2 \leq i < j \leq 4. \tag{13}$$

But clearly $\eta(e_\varphi)v_1 = 0$ in such a case.

Since v_1 is a highest weight vector this easily implies

$$\mathfrak{p}_1 = \{x \in \mathfrak{g} \mid \eta(x)v_1 \in \mathbf{R}v_1\} \tag{14}$$

Since η is irreducible v_1, up to a scalar, is the only vector satisfying (14). Thus $\overline{v}_1$ is fixed by P_1. By maximality P_1 is necessarily the isotropy group at $\overline{v}_1$. QED

A subspace $W \subseteq M$ is called isotropic if $B(u,v) = 0$ for all $u, v \in W$. Since B has index 4 one knows an isotropic subspace W is maximal if and only if dim $W = 4$. Let J be the set of all maximally isotropic subspaces of M. If U and V are respectively the span of the u_i and v_i, then clearly $U, V \in J$.

Now let

$$Y = \{W \in J \mid \dim W \cap V \text{ is even }\}$$

and

$$Y' = \{W \in J \mid \dim W \cap V \text{ is odd}\}$$

so that

$$J = Y \cup Y'$$

is a disjoint union.

The action of G on M induces an action of G on J. One knows

Proposition 2.2. *There are exactly two orbits for the action of G on J. In fact if $W_1, W_2 \in J$ the condition*

$$\dim W_1 \cap W_2 \text{ is even}$$

is an equivalence relation on J. There are two equivalence classes and these are the two orbits of G on J. In particular, in the notation above the two orbits are Y and Y'.

Proof. See [C]. QED.

It follows from Proposition 2.2 that Y and Y' have the structure of G-homogeneous spaces and they will be so regarded.

Now the action of G on M extends to an action of G on the exterior algebra $\wedge M$. In particular the 70-dimensional subspace $\wedge^4 M$ has the structure of a G-module.

For any $W \in J$ let $[W]$ be the one dimensional subspace $\mathbf{R} \cdot w_1 \wedge \cdots \wedge w_4$ where $W = \sum_{i=1}^{4} \mathbf{R} w_i$. Let N_2 be the span of all $[W]$ for $W \in Y'$ and N_3 the span of all $[W]$ for $W \in Y'$. It is clear from Proposition 2.2 that N_2 and N_3 are G-submodules of $\wedge^4 M$. One knows

Proposition 2.2.3. *One has*

$$\wedge^4 M = N_2 \oplus N_3 \tag{15}$$

Furthermore N_2 and N_3 are inequivalent, irreducible (even over $\mathbf{C}$) submodules of $\wedge^4 M$ so that (15) is the unique decomposition of $\wedge^4 M$ into a sum of simple submodules.

Proof. See [C]. QED.

It is clear that the subspace $V \subseteq M$ is stable under the action of $\mathfrak{n}$. But $v_1 \wedge \cdots \wedge v_4 \in N_2$. Hence one has

Proposition 2.4. *$v_1 \wedge \cdots \wedge v_4$ is a highest weight vector for the irreducible G-module N_2.*

Recalling (12) one notes that if $\varphi \in \Delta$ then $n_2(\varphi) = 0$ if and only if φ is of the form

$$\varphi = \lambda_i - \lambda_j, \quad 1 \le i, j \le 4, i \ne j. \tag{16}$$

It follows immediately that V is stable under $\mathfrak{g}_2$. Thus V is stable under $\mathfrak{p}_2 = \mathfrak{g}_2 + \mathfrak{n}$.

Proposition 2.5. *The correspondence $g \mapsto g \cdot V$ induces a G–isomorphism*

$$G/P_2 \to Y \tag{17}$$

Proof. It suffices only to prove that $g \cdot V = V$ if and only if $g \in P_2$. Since $p_2 \cdot V \subseteq V$ it follows that $(P_2)_e \cdot v_1 \wedge \cdots \wedge v_4 \subseteq [V]$ where $(P_2)_e$ is the identity component of P_2. But if $a \in P_2$ and $a \cdot v_i = w_i, a \cdot V = W$, then one would have by normality $(P_2)_e \cdot w_1 \wedge \cdots \wedge w_4 \subseteq [W]$. But $[W] \subseteq N_2$. By the uniqueness up to scalar, of the highest weight vector, one has $W = V$ and $P_2 \cdot V = V$. The result then follows from the maximality of P_2. QED.

Definition 2.5.5. The automorphism of G defined by the transposition $\upsilon = (2,3) \in \Sigma_3$ is common to all even dimensional rotation groups. If τ is the transposition $(1,2)$ then the automorphism $\upsilon\tau = \tau'$ of order 3 is unique to the case of 8 dimensions and is referred to as triality in [C]. But in fact the automorphism τ of order 2 is also unique to 8 dimensions and readily defines triality. In this paper we find it convenient to deal with τ instead of τ', and even though it has order 2 we will, by abuse of notation, refer to it as triality.

Recalling (9) and Propositions 2.1 and 2.5 we have recovered the following classical result.

Theorem 2.6. *Let C be the cone of all $w \in M^{\cdot}$ such that $Q(w) = 0$ and let X be the projectivization of C. Let Y be the variety of all maximal isotropic (i.e. dimension $= 4$) subspaces $W \subseteq M$ which are in the equivalence class of V. (See Proposition 2.2). Let $\tau \in \Sigma_3$ be the transposition (1,2). Then there exists a unique diffeomorphism*

$$\tau_X : X \to Y \tag{18}$$

such that (1) $\tau_X \overline{v}_1 = V$ and (2) for any $w \in C*$ and $g \in G$

$$\tau_X(g \cdot \overline{w}) = \tau(g) \cdot \tau_X \overline{w} \tag{19}$$

where $\overline{w} \in X$ is the image of w in X and we use the notation of (10).

3. Triality and the maximal compact subgroup K.

To proceed further we will need a much more detailed understanding of the map τ_X. This will come about by understanding how τ operates on a maximal compact subgroup of G. Let $K \subseteq G$ be the subgroup of G corresponding to $\mathfrak{k}$. From the structure theory of semi-simple groups one knows that K is simply connected and by definition of $\mathfrak{k}$ the restriction of η to K defines a map

$$K \to SO(E) \times SO(F) \tag{20}$$

and that (20) is a covering of Lie groups. Since the fundamental group of the right side of (20) is $\mathbf{Z}_2 \times \mathbf{Z}_2$ it follows that (20) is a 4-fold covering and hence K is a maximal compact subgroup of G.

To begin with we note

Proposition 3.1. *The group K is stable under $\sigma: G \to G$ for any $\sigma \in \Sigma_3$. In particular K is stable under τ.*

Proof. We have only to observe that $\sigma: \mathfrak{g} \to \mathfrak{g}$ commutes with θ. For this it suffices to see that they commute on the simple root vectors $e_{\alpha_i}, i = 1, ..., 4$. But this is immediate from (8) since $\theta e_{\alpha_i} = -e_{\alpha_i}$. QED.

It will be convenient to give a second label to the basis $u_j, j = 1, ..., 4$, of $U \subseteq M$. Let $v_j = u_{9-j}$ and put $\lambda_j = -\lambda_{9-j}$ for $j = 5, ..., 8$. Then $v_1, ..., v_8$ is a basis of M and the corresponding weights are $\lambda_1, ..., \lambda_8$. Further, the 24 roots of $\varphi \in \Delta$ may now be labelled $\varphi = \varphi(i,j)$ for unordered pairs (i,j) where

$$\varphi(i,j) = \lambda_i + \lambda_j, \quad 1 \leq i,j \leq 8, \quad i \neq j, \quad i+j \neq 9 \tag{21}$$

We can now describe the action of $\eta(e_\varphi)$ on M for any $\varphi \in \Delta$.

Lemma 3.2. *Let $\varphi \in \Delta$. If $\varphi = \varphi(i,j)$ then $\eta(e_\varphi)$ vanishes on v_k for all $1 \leq k \leq 8$ unless $k = 9 - i$ or $9 - j$. Furthermore, in such a case one either has the map*

$$\eta(e_\varphi) \begin{cases} v_{9-i} \mapsto v_j \\ v_{9-j} \mapsto -v_i \end{cases} \tag{22}$$

or its negative

$$\eta(e_\varphi) \begin{cases} v_{9-i} \mapsto -v_j \\ v_{9-j} \mapsto v_i \end{cases} \tag{23}$$

Furthermore, in any case

$$\eta(e_{-\varphi})\eta(e_\varphi)v_k = \begin{cases} 0 & \text{if} \quad k \neq 9-i, 9-j \\ v_k & \text{if} \quad k = 9-i, 9-j \end{cases} \tag{24}$$

Proof. Let $\mathfrak{g}_\varphi$ be the TDS generated by e_φ and $e_{-\varphi}$. Since $\varphi = \lambda_i + \lambda_j$ from the representation theory of a TDS one has $\eta(e_\varphi)v_k = 0$ unless $k = 9 - i, 9 - j$ and $\eta(e_\varphi)v_{9-i} = cv_j, \eta(e_\varphi)v_{9-j} = -cv_i$ for some non-zero scalar c. We are also using the fact that $\eta(e_\varphi)$ annihilates the bilinear form B. It follows that v_{9-i} and v_{9-j} are, with respect to $\eta|\mathfrak{g}_\varphi$, lowest weight vectors for two distinct irreducible 2-dimensional submodules. Thus $\eta(e_{-\varphi})\eta(e_\varphi)v_{9-i} = dv_{9-i}$ and $\eta(e_{-\varphi})\eta(e_\varphi)v_{9-j} = dv_{9-j}$ for some scalar d. However, since $(e_\varphi, e_{-\varphi}) = 1$ (see (7)) one has that $d = 1$ establishing (24). Thus $\eta(e_{-\varphi})v_j = 1/c \cdot v_{9-i}$ and $\eta(e_{-\varphi})v_i = -1/c \cdot v_{9-j}$.

Let $z = e_\varphi - e_{-\varphi}$ since $i, j, 9-i, 9-j$ are 4 distinct numbers $\eta(z)$ maps $v_{9-i} \mapsto cv_j, v_{9-j} \mapsto -cv_i, v_i \mapsto 1/c \cdot v_{9-j}$ and $v_j \mapsto -1/c \cdot v_{9-i}$. Also $\eta(z)v_k = 0$ for k distinct from these four numbers. However, clearly $\theta z = z$ so that $\eta(z)$ commutes with $\eta(\kappa)$. Since $\eta(\kappa)$ interchanges v_k and v_{9-k} it follows that $c = \pm 1$.

QED.

For any $\varphi \in \Delta$ let

$$z_\varphi = e_\varphi - e_{-\varphi} \tag{25}$$

Proposition 3.3. *One has $z_\varphi \in \mathfrak{k}$ and the element $z_\varphi, \varphi \in \Delta_+$, are a basis of $\mathfrak{k}$. Furthermore if $\varphi = \varphi(i,j)$ then $i, j, 9-i, 9-j$ are distinct and $\eta(z_\varphi)$ maps*

$$\begin{array}{ccc} v_{9-i} & \to & cv_j \\ v_{9-j} & \to & -cv_i \\ v_i & \to & cv_{9-j} \\ v_j & \to & -cv_{9-i} \end{array} \tag{26}$$

where $c = \pm 1$. Moreover, $\eta(z_\varphi)$ annihilates $v_k, 1 \le k \le 8$, if $k \notin \{i, j, 9-i, 9-j\}$.

Proof. This is immediate from the proof of Lemma 3.2 and the definition of $\mathfrak{k}$.

QED.

3.4. Let $\mathbf{H}$ be the skew field of quaternions over $\mathbf{R}$. We adopt the usual notation so that $1, i, j, k$ is an $\mathbf{R}$-basis with $ij = k$. We will now define on both E and F the structure of a two-sided $\mathbf{H}$-module. Let $h_E: \mathbf{H} \to E$ be the linear isomorphism so that h_E maps $1 \mapsto \epsilon_1$, $i \mapsto \epsilon_2, j \mapsto \epsilon_3$ and $k \mapsto \epsilon_4$. One defines a two-sided $\mathbf{H}$ structure on E so that if $w \in E$, $\alpha, \beta \in \mathbf{H}$ then $\alpha w \beta = h_E(\alpha(h_E^{-1}(w))\beta)$. Similarly one defines a two-sided $\mathbf{H}$-module structure on F where $h_F: \mathbf{H} \to F$ is such that h_F maps $1 \to \epsilon_5, i \to \epsilon_6, j \to \epsilon_7$ and $k \to \epsilon_8$.

Now let S (which is isomorphic to $SU(2)$) be the unit sphere in $\mathbf{H}$. Put

$$\tilde{K} = S \times S \times S \times S \tag{27}$$

so that $\tilde{K}$ is a simply-connected compact Lie group. The elements of $\tilde{K}$ are thus of the form $(\alpha, \beta, \gamma, \delta)$ where $\alpha, \beta, \gamma, \delta \in S$. Since $M = E \oplus F$ any element $w \in M$ can be uniquely represented in the form (e, f) where $e \in E, f \in F$. The following is an immediate consequence of the structure of $SO(4)$.

Proposition 3.5. *One defines the structure of a $\tilde{K}$-module on M by putting, for $\alpha, \beta, \gamma, \delta \in S, e \in E$ and $f \in F$,*

$$(\alpha, \beta, \gamma, \delta) \cdot (e, f) = (\alpha e \beta^{-1}, \gamma f \delta^{-1}). \tag{28}$$

Moreover this action of $\tilde{K}$ defines a covering of Lie groups

$$\tilde{K} \to SO(E) \times SO(F) \tag{29}$$

Now recalling the covering (20) one has a unique isomorphism

$$\tilde{K} \to K \tag{30}$$

which makes the diagram with the maps (20), (29), and (30) commutative. Henceforth we will identify $\tilde{K}$ with K by the map (30).

For any $\varphi \in \Delta$ the choice of e_φ was unique up to sign. By Lemma 3.2 one has that $\eta(e_{\alpha_1})v_2$ is either v_1 or $-v_1$. We now fix the choice of e_{α_1} as well as e_{α_2} and e_{α_3} by requiring $\eta(e_{\alpha_1})v_2 = v_1, \eta(e_{\alpha_2})u_4 = v_3$, $\eta(e_{\alpha_3})v_4 = v_3$. Recall that this fixes the action of σ for any $\sigma \in \Sigma_3$. We can now determine the action of τ on K. In effect it just interchanges the second and third components.

Theorem 3.6. *Let $\tau \in \Sigma_3$ be the transposition* $(1,2)$. *Let $\alpha, \beta, \gamma, \delta \in S$. Then*

$$\tau(\alpha, \beta, \gamma, \delta) = (\alpha, \gamma, \beta, \delta) \tag{31}$$

In order to prove Theorem 3.6 we will need to make some easy computations. First let $\alpha_0 = \alpha_1 + \alpha_2 + \alpha_3 + 2\alpha_4$ be the highest root. One sees easily that

$$\alpha_0 = \lambda_1 + \lambda_2 \tag{32}$$

By Lemma 3.2 we may then normalize the choice of e_{α_0} so that

$$\eta(e_{\alpha_0})u_2 = v_1 \tag{32.5}$$

Note that $\alpha_i, i = 0, ..., 3$, is a set of 4-orthogonal roots so that $z_{\alpha_i}, i = 0, ..., 3$, spans the Lie algebra of a maximal torus of K. Recall $\epsilon_i = v_i + u_i$ and $\epsilon_{i+4} = v_i - u_i, i = 1, ..., 4$.

Lemma 3.7. *The effect of $\eta(z_{\alpha_i}), i = 0, 1, 2, 3$, on the basis $\epsilon_j, 1 \le j \le 8$, of M is given by*

$$\eta(z_{\alpha_0}): \begin{cases} \epsilon_1 \to -\epsilon_2, & \epsilon_2 \to \epsilon_1 \\ \epsilon_5 \to \ \ \epsilon_6, & \epsilon_6 \to -\epsilon_5 \end{cases}$$

$$\eta(z_{\alpha_1}): \begin{cases} \epsilon_1 \to -\epsilon_2, & \epsilon_2 \to \epsilon_1 \\ \epsilon_5 \to -\epsilon_6, & \epsilon_6 \to \epsilon_5 \end{cases}$$

$$\eta(z_{\alpha_2}): \begin{cases} \epsilon_3 \to -\epsilon_4, & \epsilon_4 \to \epsilon_3 \\ \epsilon_7 \to \epsilon_8, & \epsilon_8 \to -\epsilon_7 \end{cases}$$

$$\eta(z_{\alpha_3}): \begin{cases} \epsilon_3 \to -\epsilon_4, & \epsilon_4 \to \epsilon_3 \\ \epsilon_7 \to -\epsilon_8, & \epsilon_8 \to \epsilon_7 \end{cases}$$

Furthermore $\eta(z_{\alpha_i})\epsilon_j = 0$ *for all* i, j *not indicated above.*

Proof. This is a direct computation using Proposition 3.3 and the various values of c which have now been fixed. QED.

Lemma 3.8. *Let* $\varphi \in \Delta_+ \subseteq \mathfrak{a}'$ *be fixed under the action of* τ *on* $\mathfrak{a}'$. *Then* $\tau e_\varphi = e_\varphi$ *and hence* $\tau e_{-\varphi} = e_{-\varphi}$.

Proof. Note that φ is fixed under τ if and only if $n_1(\varphi) = n_2(\varphi) = 0$ or 1. Furthermore, even if $n_1(\varphi) = n_2(\varphi) = 1$ we may write

$$e_\varphi = c \cdot \text{ad } e_{\psi_1} \cdots \text{ ad } e_{\psi_k} \text{ ad } e_{\alpha_1} \text{ ad } e_{\alpha_2}(e_{\alpha_4})$$

for some constant c where $\psi_1, ..., \psi_k$ are certain simple roots distinct from α_1, α_2. But then τe_φ has the same expression except that ad e_{α_1} and ad e_{α_2} are exchanged. But then $\tau e_\varphi = e_\varphi$ since these two operators commute. QED.

As a consequence one clearly has $\tau z_{\alpha_0} = z_{\alpha_0}$. In fact one notes

Lemma 3.9. $\tau z_{\alpha_i} = z_{\alpha_i}$, $i = 0, 3$. *But* $\tau z_{\alpha_1} = z_{\alpha_2}$ *and* $\tau z_{\alpha_2} = z_{\alpha_1}$.

Let $S_i \subseteq K$, $i = 1, ..., 4$, be the (normal) subgroup defined, using the expression (27), by retaining S in the i-th component but replacing the other components by 1. Thus

$$K = \prod_{i=1}^{4} S_i \tag{33}$$

Let $\mathfrak{s} \subseteq \mathbf{H}$ be the 3-dimensional space of all imaginary quaternions. Then as a Lie algebra $\mathfrak{s}$ is the Lie subalgebra of $\mathbf{H}$ and as usual we may regard $\mathfrak{s}$ as the Lie algebra of S. Note that i and j generate $\mathfrak{s}$ as a Lie algebra.

Remark. As in the case of $Gl(n, \mathbb{C}) \subseteq M(n, \mathbb{C})$, an element $a \in S \cap \mathfrak{s}$ can be regarded as a Lie algebra or a group element. In the paper it should be clear from the context which of the two possibilities is being understood.

Let $\mathfrak{s}_i \subseteq \mathfrak{k}$, $i = 1, ..., 4$, be the Lie algebra of S_i. The $\mathfrak{s}_i$ are simple ideals of $\mathfrak{k}$ and

$$\mathfrak{k} = \sum_{i=1}^{4} \mathfrak{s}_i. \tag{34}$$

Also if $t \in \mathfrak{k}$ we may represent t by a 4-tuple (u, v, x, y) where $u, v, x, y \in \mathfrak{s}$. The $\mathfrak{k}$-module structure on M is clearly given by

$$(u, v, x, y) \cdot (e, f) = (ue - ev, xf - fy) \tag{35}$$

for $e \in E, f \in F$.

We now have

Lemma 3.10. *The elements* $(i, 0, 0, 0)$ *and* $(0, 0, 0, i)$ *of* $\mathfrak{k}$ *are fixed by* τ *whereas* $(0, i, 0, 0)$ *and* $(0, 0, i, 0)$ *are interchanged by* τ.

Proof. By Lemma 3.9 it suffices to observe that

$$\begin{aligned}
\frac{-1}{2}(z_{\alpha_0} + z_{\alpha_1} + z_{\alpha_2} + z_{\alpha_3}) &= (i, 0, 0, 0) \\
\frac{1}{2}(-z_{\alpha_0} + z_{\alpha_1} + z_{\alpha_2} - z_{\alpha_3}) &= (0, 0, 0, i) \\
\frac{1}{2}(z_{\alpha_0} + z_{\alpha_1} - z_{\alpha_2} - z_{\alpha_3}) &= (0, i, 0, 0) \\
\frac{1}{2}(z_{\alpha_0} - z_{\alpha_1} + z_{\alpha_2} - z_{\alpha_3}) &= (0, 0, i, 0)
\end{aligned} \tag{36}$$

But by Lemma 3.7 the left side of the first equation annihilates F and maps $\epsilon_1 \to -\epsilon_2, \epsilon_2 \to \epsilon_1, \epsilon_3 \to -\epsilon_4$ and $\epsilon_4 \to \epsilon_3$. But this is just left multiplication by i on E. The left side of the second equation annihilates E and maps $\epsilon_5 \to -\epsilon_6, \epsilon_6 \to \epsilon_5, \epsilon_7 \to \epsilon_8, \epsilon_8 \to -\epsilon_7$. This, however, is accomplished by right multiplication by $-i$ on F. The left side of the third equation annihilates F and maps $\epsilon_1 \to -\epsilon_2, \epsilon_2 \to \epsilon_1, \epsilon_3 \to \epsilon_4, \epsilon_4 \to -\epsilon_3$. But right multiplication by $-i$ on E yields this map.

Finally, the left side of the fourth equation annihilates E and maps $\epsilon_5 \to \epsilon_6, \epsilon_6 \to -\epsilon_5, \epsilon_7 \to \epsilon_8, \epsilon_8 \to -\epsilon_7$. However, the left multiplication by i produces this map. The four equations then follow from (35). QED.

Now let $\beta_0 = \alpha_1 + \alpha_2 + \alpha_3 + \alpha_4 = \lambda_1 + \lambda_3$, $\beta_1 = \alpha_1 + \alpha_4$, $\beta_2 = \alpha_2 + \alpha_4$, $\beta_3 = \alpha_3 + \alpha_4$. Then β_i, $i = 0, ..., 3$, is a set of 4 orthogonal positive roots. Recalling Lemma 3.2 we may normalize e_{β_0} so that $\eta(e_{\beta_0})u_3 = v_1$. By (6.5) we may define e_{β_1},e_{β_2}, and e_{β_3} so that $e_{\beta_1} = [e_{\alpha_1}, e_{\alpha_4}]$, $e_{\beta_2} = [e_{\alpha_2}, e_{\alpha_4}]$ and $e_{\beta_3} = [e_{\alpha_3}, e_{\alpha_4}]$. Now by Lemma 3.2 we may normalize e_{α_4} so that $\eta(e_{\alpha_4})v_3 = v_2$. Then one easily has $\eta(e_{\beta_1})v_3 = v_1$, $\eta(e_{\beta_2})u_2 = v_4$ and $\eta(e_{\beta_3})u_2 = u_4$.

One has, analogously with Lemma 3.9

Lemma 3.11. $\tau(z_{\beta_i}) = z_{\beta_i}$ *for* $i = 0, 3$, *and* τ *interchanges* z_{β_1} *and* z_{β_2}.

Next, corresponding to Lemma 3.7, one has

Lemma 3.12. *The effect of $\eta(z_{\beta_i})$, $i = 0, \ldots, 3$, on the basis ϵ_j, $1 \le j \le 8$, of M is given by*

$$\eta(z_{\beta_0})\colon \begin{cases} \epsilon_1 \to -\epsilon_3, & \epsilon_3 \to \epsilon_1 \\ \epsilon_5 \to \quad \epsilon_7, & \epsilon_7 \to -\epsilon_5 \end{cases}$$

$$\eta(z_{\beta_1})\colon \begin{cases} \epsilon_1 \to -\epsilon_3, & \epsilon_3 \to \epsilon_1 \\ \epsilon_5 \to -\epsilon_7, & \epsilon_7 \to \epsilon_5 \end{cases}$$

$$\eta(z_{\beta_2})\colon \begin{cases} \epsilon_2 \to \quad \epsilon_4, & \epsilon_4 \to -\epsilon_2 \\ \epsilon_6 \to -\epsilon_8, & \epsilon_8 \to \epsilon_6 \end{cases} \tag{37}$$

$$\eta(z_{\beta_3})\colon \begin{cases} \epsilon_2 \to \epsilon_4, & \epsilon_4 \to -\epsilon_2 \\ \epsilon_6 \to \epsilon_8, & \epsilon_8 \to -\epsilon_6 \end{cases}$$

Also $\eta(z_{\beta_i})\epsilon_j = 0$ for i, j not indicated above.

Proof. Same argument as in the proof of Lemma 3.7. QED.

We now have

Lemma 3.13. *The elements $(j,0,0,0)$ and $(0,0,0,j)$ of $\mathfrak{k}$ are fixed by τ whereas the elements $(0,j,0,0)$ and $(0,0,j,0)$ of $\mathfrak{k}$ are interchanged by τ.*

Proof. Using Lemma 3.12 we note

$$-\frac{1}{2}(z_{\beta_0} + z_{\beta_1} + z_{\beta_2} + z_{\beta_3})\colon \begin{cases} \epsilon_1 \to \epsilon_3 \\ \epsilon_2 \to -\epsilon_4 \\ \epsilon_3 \to -\epsilon_1 \\ \epsilon_4 \to \epsilon_2 \end{cases}$$

and vanishes on F. Thus by

$$-\frac{1}{2}(z_{\beta_0} + z_{\beta_1} + z\beta_2 + z_{\beta_3}) = (j,0,0,0)$$

Similarly

$$\frac{1}{2}(-z_{\beta_0} + z_{\beta_1} + z_{\beta_2} - z_{\beta_3})\colon \begin{cases} \epsilon_5 \to -\epsilon_7 \\ \epsilon_6 \to -\epsilon_8 \\ \epsilon_7 \to \epsilon_5 \\ \epsilon_8 \to \epsilon_6 \end{cases}$$

and vanishes on E so that

$$\frac{1}{2}(-z_{\beta_0} + z_{\beta_1} + z_{\beta_2} - z_{\beta_3}) = (0,0,0,j)$$

Again

$$\frac{1}{2}(z_{\beta_0} + z_{\beta_1} - z_{\beta_2} - z_{\beta_3})\colon \begin{cases} \epsilon_1 \to -\epsilon_3 \\ \epsilon_2 \to -\epsilon_4 \\ \epsilon_3 \to \epsilon_1 \\ \epsilon_4 \to \epsilon_2 \end{cases}$$

and vanishes on F so that

$$\frac{1}{2}(z_{\beta_0} + z_{\beta_1} - z_{\beta_2} - z_{\beta_3}) = (0, j, 0, 0)$$

and finally

$$\frac{1}{2}(z_{\beta_0} - z_{\beta_1} + z_{\beta_2} - z_{\beta_3})\colon \begin{cases} \epsilon_5 \longrightarrow \epsilon_7 \\ \epsilon_6 \longrightarrow -\epsilon_8 \\ \epsilon_7 \longrightarrow -\epsilon_5 \\ \epsilon_8 \longrightarrow \epsilon_6 \end{cases}$$

and vanishes on E so that

$$\frac{1}{2}(z_{\beta_0} - z_{\beta_1} + z_{\beta_2} - z_{\beta_3}) = (0, 0, j, 0)$$

The result then follows as in the proof of Lemma 3.10. QED.

Proof of Theorem 3.6. It suffices to prove the corresponding statement for the Lie algebra $\mathfrak{k}$ of K. But this follows immediately from Lemmas 3.10 and 3.13 since i and j generate $\mathfrak{s}$. QED.

3.14. Let S_E and S_F be, respectively, the unit spheres in E and F (i.e. $Q(w) = 1$ characterizes S_E and $Q(w) = -1$ characterizes S_F). Recalling the cone $C \subseteq M$ one notes, regarding $M = E \times F$, that $S_E \times S_F \subseteq C$.

Now it is clear that $S_E \times S_F$ is stable under the action of K. In fact the map

$$S_1 \times S_3 \longrightarrow S_E \times S_F, \ (\alpha, 1, \gamma, 1) \longmapsto (\alpha\epsilon_1, \gamma\epsilon_5) \tag{38}$$

is clearly a diffeomorphism so that $S_E \times S_F$ is a principal homogeneous space for the subgroup $S_1 \times S_3 \subseteq K$. Recall the projection

$$C^* \longrightarrow X, \qquad w \longrightarrow \overline{w} \tag{39}$$

From the general principle one knows that K operates transitively on X. The following asserts that it is even true for the subgroup $S_1 \times S_3$.

Proposition 3.15. *The restriction of (39) to $S_E \times S_F$ induces a two to one simply-connected covering*

$$S_E \times S_F \longrightarrow X \tag{40}$$

Furthermore (40) is a map of homogeneous K-(also $S_1 \times S_3$) spaces.

Proof. It is only necessary to observe that if $e_i \in E$, $f_i \in F$, $i = 1, 2$, then $\overline{(e_1, f_1)} = \overline{(e_2, f_2)}$ if and only if $e_1 = e_1$ and $f_1 = f_2$ or $e_1 = -e_2$ and $f_1 = -f_2$. QED.

3.16. Now let $O(E,F)$ be the space of all linear isomorphisms $s\colon E \to F$ such that $Q(s(e)) = -Q(e)$. If $W \in J$ (the space of all maximal isotropic subspaces of M) upon considering the components of any $w \in W$ in E and F it is clear that there is a unique element $s_W \in O(E,F)$ such that

$$W = \{(e, s_W e) \mid e \in E\} \tag{41}$$

Furthermore, one clearly has

Proposition 3.17. *The map*

$$J \to O(E,F), \quad W \to s_W \tag{42}$$

is a bijection.

Now since $2v_i = \epsilon_i + \epsilon_{i+4}$, $i = 1, ..., 4$, one notes that $s_V(\epsilon_i) = \epsilon_{i+4}$. From the definition of the $\mathbf{H}$ structure on E and F one has

Proposition 3.18. *The map $s_V : E \to F$ is an isomorphism of two-sided $\mathbf{H}$-modules.*

Now $O(M)$ operates on J. In particular the $O(E) \subseteq O(M)$ operates on J.

Proposition 3.19. *For any $h \in O(E)$ and $W \in J$ one has*

$$s_W \circ h^{-1} = s_{hW} \tag{43}$$

Furthermore the map

$$O(E) \to J, \quad h \mapsto hV \tag{44}$$

is a bijection so that J is a principal $O(E)$-homogeneous space.

Proof. The proof is immediate. One needs only to observe that any $s \in O(E,F)$ is uniquely of the form $s_V \circ h^{-1}$ for some $h \in O(E)$.

With regard to equivalence relations in J (see Proposition 2.2) one has

Proposition 3.20. *Let $W_1, W_2 \in J$. Then W_1 and W_2 are equivalent if and only if $hW_1 = W_2$ where $h \in SO(E)$.*

Proof. If $a \in O(4)$ one notes that the eigenvalue 1 of a has even multiplicity if and only if $a \in SO(4)$. The result then follows from Proposition 2.2. QED.

It follows from Proposition 3.20 that Y is stable under the action of $SO(E)$. In fact one easily has

Proposition 3.21. *The map*

$$SO(E) \to Y, \quad h \to hV \tag{45}$$

is a bijection so that Y *is a principal* $SO(E)$*-homogeneous space.*

Remark. We have noted before that X is a homogeneous $S_1 \times S_3$ space. We now note that Y on the other hand is a homogeneous $S_1 \times S_2$ space. This follows from (28) which implies $\eta(S_1 \times S_2) = SO(E)$.

For any $a, b \in S$ let $g(a,b) \in S_1 \times S_2$ be defined by putting

$$g(a,b) = (a, b, 1, 1) \tag{46}$$

With respect to the action of $S_1 \times S_2$ on Y one has

$$\begin{aligned} g(a,b)V &= \{(aeb^{-1}, s_V e) \mid e \in E\} \\ &= \{(e, a^{-1} s_V e\, b) \mid e \in E\} \end{aligned} \tag{47}$$

The second equality follows from the first using Proposition 3.18. One thus has

Proposition 3.22. $g(a,b)V$ *is the span of* $(e_i, a^{-1} e_{i+4}\, b), i = 1, ..., 4.$

We can now put the map τ_X of Theorem 2.6 in a very simple form.

Theorem 3.23. *Let the notation be as in Theorem 2.6. Then for any* $a, b \in S$ *(group of unit quaternions) one has*

$$\tau_X((\overline{a\epsilon_1, b\epsilon_5})) = g(a,b)V \tag{48}$$

where $g(a,b)V \in Y$ *is explicitly given by (47).*

Proof. Since $2v_1 = \epsilon_1 + \epsilon_5$ this follows immediately from (47), Proposition 3.22 and Theorems 2.6 and 3.6. Theorem 3.6 implies that

$$\tau(S_1 \times S_3) = S_1 \times S_2 \tag{49}$$

and more to the point,

$$\tau(a, 1, b, 1) = (a, b, 1, 1) \tag{50}$$

QED.

4. Line bundles over X. Let $L = L(X)$ be the complex line bundle over X, defined over $\mathbf{R}$, whose dual L^{-1} satisfies $(L^{-1}_{\overline{w}})_{\mathbf{R}} = \mathbf{R}w$ for any $w \in C^* = C - \{0\}$. By taking tensor products one defines the line bundle L^k for any $k \in \mathbf{Z}$.

If $D \subseteq M$ is a subset which is stable under multiplication by $\mathbf{R}^*$ and such that $D^* = D - \{0\}$ is a submanifold of M let $\Gamma^k(D)$ be the space of all $\mathbf{C}$-valued C^∞ functions f on D^* which admits the homogeneity

$$f(tw) = t^k f(w) \tag{51}$$

for any $w \in D^*$ and $t \in \mathbf{R}^*$.

If D is stable under the action of G then clearly $\Gamma^k(D)$ has the structure of a G (and hence $\mathfrak{g}$) module where if $f \in \Gamma^k(D), g \in G$ and $w \in D^*$ one has $(g \cdot f)(w) = f(g^{-1}w)$. If $x \in \mathfrak{g}$ then $(x \cdot f)(w) = d/dt \cdot f(exp(-tx) \cdot w)$ at $t = 0$.

If $w \in C^*$ let $w^{-k} \in L^k_{\overline{w}}$ be the $|k|$-fold tensor product of w if k is negative and the k-fold tensor product of w^{-1} if k is positive, where $w^{-1} \in L_{\overline{w}}$ is defined so that $\langle w, w^{-1}\rangle = 1$. If $u \in L^k_{\overline{w}}$ let $\tilde{u}(w) \in \mathbf{C}$ be such that

$$\tilde{u}(w)w^{-k} = u \tag{52}$$

Clearly,

$$\tilde{u}(tw) = t^k \tilde{u}(w) \tag{53}$$

(Note that if $k = 0$ then $L_{\overline{w}} = \mathbf{C}$ and $w^0 = 1$).

Now let $\Gamma(L^k)$ be the space of smooth sections $s: X \to L^k$ of L^k. For any $s \in \Gamma(L^k)$ let $\tilde{s}$ be the function on C^* defined so that $\tilde{s}(w) = \widetilde{s\overline{w}}(w) \in \mathbf{C}$, recalling (52). Using (53) one obviously has $\tilde{s}(tw) = t^k \tilde{s}(w)$.

Now the action of G on C induces an action of G on L^k for any k, and $\Gamma(L^k)$ has the structure of a G-module where if $g \in G$, $s \in \Gamma(L^k)$ and $p \in M$ one has $(g \cdot s)(p) = g \cdot (s(g^{-1} \cdot p))$. The following is straight forward.

Proposition 4.1. *The map*

$$\Gamma(L^k) \to \Gamma^k(C) \qquad s \to \tilde{s} \tag{54}$$

is an isomorphism of G-modules.

4.2. Henceforth we will identify $\Gamma(L^k)$ with $\Gamma^k(C)$ so that we are always dealing with scalar-valued functions f on C^*. For convenience we write Γ^k for $\Gamma^k(C)$. Clearly one has a pairing

$$\Gamma^i \otimes \Gamma^j \to \Gamma^{i+j} \tag{55}$$

for any $i, j \in \mathbf{Z}$ and in addition one notes that (55) is a map of G-modules. Also one has the obvious identification

$$\Gamma^0 = C^\infty(X)$$

Our next concern is with Γ^{-6}.

Let r_i, $i = 1, ..., 8$, be the coordinate functions on M relative to the basis ϵ_i so that if $w \in M$ then $w = \sum_{i=1}^{\infty} r_i(w)\epsilon_i$. The vector field $I = \sum_{i=1}^{\infty} r_i \dfrac{\partial}{\partial r_i}$ (the Euler operator is tangent to rays and hence to D^*, where D is as above, and one notes that

$$If = kf \quad \text{for any } f \in \Gamma^k(D) \tag{56}$$

Henceforth we will let $D = \{(e,f) \mid e \in E,\ f \in F,\ (e,f) = (0,0) \text{ or } e \neq 0 \text{ and } f \neq 0\}$. One clearly has $C \subseteq D$ and D^* is open in M. Let $R_E = \sum_{i=1}^{4} r_i^2$ and $R_F = \sum_{j=5}^{8} r_j^2$. Let R be the restriction of $(R_E\, R_F)^{1/4}$ to D^*. If $k \in \mathbf{Z}$ one has $R^k \in \Gamma^k(D)$ so that

$$IR^k = kR^k \tag{57}$$

Also,

$$R^k \mid S_E \times S_F = 1. \tag{58}$$

Now for some $c \in \mathbf{R}^*$ let Ω be the volume element on M defined by $\Omega = c dr_1 \wedge \cdots \wedge dr_8$. If θ_I denotes Lie differentiation one clearly has

$$\theta_I \Omega = 8\Omega \tag{59}$$

Now let $\sigma_i = 1$, for $i = 1, ..., 4$, and $\sigma_i = -1$ for $i = 5, ..., 8$. Then clearly $Q = \sum_{i=1}^{\infty} \sigma_i r_i^2$ and $dQ = \sum_{i=1}^{8} 2\sigma_i r_i dr_i$. But then $(dQ)_p$ is non-zero for any $p \in C^*$ and hence there exists a 7-form $\tilde{\omega}_7$ of M defined along C^* such that $\Omega = dQ \wedge \tilde{\omega}_7$ on C^*. Furthermore, since C is defined by the equation $Q = 0$ one knows that the volume element ω_7 on C^*, defined by pulling back $\tilde{\omega}_7$, is independent of $\tilde{\omega}_7$ and one has

Proposition 4.2. *The volume element ω_7 on C^* is invariant under the action of G. Moreover, one has*

$$\theta_I \omega_7 = 6\omega_7 \tag{60}$$

More explicitly, if C_i is the open subset of C defined by the equation $r_i \neq 0$ then the restriction $\omega_7 \mid C_i$ is just the pull-back of $(-1)^{i+1} \dfrac{c\sigma_i}{2r_i} dr_1 \wedge \cdots \wedge \widehat{dr_i} \wedge \cdots \wedge dr_8$ to C_i.

Proof. The first statement is immediate since Ω and dQ are G-invariant. But now if $\tilde{\omega}_7 = (-1)^{i+1} \dfrac{c\sigma_i}{2r_i} dr_1 \wedge \cdots \wedge \widehat{dr_i} \wedge \cdots \wedge dr_8$ then clearly $\Omega = dQ \wedge \tilde{\omega}_7$ on C_i. This establishes the last statement of the proposition. One obtains (60) using the explicit form of $\tilde{\omega}_7$ together with the relation

$$Ir_i^k = kr_i^k \tag{61}$$

for any $k \in \mathbf{Z}$. QED.

Let $\Omega_E = c_E\, dr_1 \wedge \cdots \wedge dr_4$ for some constant c_E. Now S_E is defined by the equation $R_E - 1 = 0$. But $d(R_E - 1) = dR_E = \sum_{i=1}^4 2r_i dr_i$ vanishes nowhere on S_E. Thus $\Omega_E = dR_E \wedge \tilde{\mu}_E$ for a 3-form $\tilde{\mu}_E$ on E defined along S_E. The pull-back μ_E on $\tilde{\mu}_E$ is a volume element on S_E which is independent of $\tilde{\mu}_E$.

Proposition 4.3. *The volume element μ_E on the 3-sphere S_E vanishes nowhere and is $SO(E)$-invariant. More explicitly, if $(S_E)_i$ is an open subset of S_E defined by the equation $r_i \neq 0$, $i = 1, ..., 4$, then the restriction $\mu_E \mid (S_E)_i$ is just the pull-back of $(-1)^{i+1}\dfrac{c\sigma_i}{2r_i} dr_1 \wedge \cdots \wedge \widehat{dr_i} \wedge \cdots \wedge dr_4$ to $(S_E)_i$.*

Proof. The argument is similar to the proof of Proposition 4.2. QED.

4.4. Henceforth we define an orientation of S_E such that μ_E is positive for positive c_E and we normalize c_E such that $\int_{S_E} \mu_E = 1$. We define μ_F similarly. (No account is taken of the fact that $Q = -1$ on S_F). Now let μ be the volume element on $S_E \times S_F$ defined by putting $\mu = \mu_E \wedge \mu_F$. Thus, in the neighborhood of $(\epsilon_1, \epsilon_5) \in S_E \times S_F$ where $r_1 r_5 \neq 0$, one has, for some positive constant c^1,

$$\mu = P\left(\frac{c^1}{r_1 r_5} dr_2 \wedge dr_3 \wedge dr_4 \wedge dr_6 \wedge dr_7 \wedge dr_8\right) \tag{62}$$

where P denotes the pull-back to $S_E \times S_F$, and

$$\int_{S_E \times S_F} \mu = 1. \tag{63}$$

Now let ι_I denote the operator of the interior product by I on differential forms. Let $\omega = \iota_I \omega_7$.

Proposition 4.5. *ω is a G-invariant 6-form on C^*. Moreover, one has $\theta_I \omega = 6\omega$ and $\iota_I \omega = 0$. Furthermore the constant c in the definition of Ω may be chosen so that*

$$2\mu = P\omega \tag{64}$$

where P is the pull-back to $S_E \times S_F$.

Proof. Obviously I is invariant under the action of G so that ω is G-invariant by Proposition 4.2. Also θ_I commutes with ι_I so that $\theta_I \omega = 6\omega$ using the same reference. One has $\iota_I \omega = 0$ since $(\iota_I)^2 = 0$. The final statement follows from invariance under $SO(E) \times SO(F)$. However, it is illuminating to make the computation.

Let $\bar{r}_i = r_i \mid S_E \times S_F$ so that $\Sigma_{i=1}^4 \bar{r}_i^2 = 1$. Thus $\sum_{i=1}^4 \bar{r}_i d\bar{r}_i = 0$ and hence where $\bar{r}_1 \neq 0, (-1)^{j+1} \bar{r}_j d\bar{r}_1 \wedge \cdots \wedge \widehat{d\bar{r}_j} \wedge \cdots \wedge d\bar{r}_4 = \dfrac{\bar{r}_j^2}{\bar{r}_1} d\bar{r}_2 \wedge d\bar{r}_3 \wedge d\bar{r}_4$, since

$d\overline{r}_1 = \frac{-1}{\overline{r}_1}(\sum_{j=2}^4 \overline{r}_j d\overline{r}_j)$. Consequently,

$$\Sigma_{j=1}^4 (-1)^{j+1} \overline{r}_j d\overline{r}_1 \wedge \cdots \wedge \widehat{d\overline{r}_j} \wedge \cdots \wedge d\overline{r}_4 = \frac{1}{\overline{r}_1} d\overline{r}_2 \wedge d\overline{r}_3 \wedge d\overline{r}_4 \tag{65}$$

But since $d\overline{r}_1 \wedge \cdots \wedge d\overline{r}_4 = 0$ it follows, taking $i = 5$ in Proposition 4.2 that, where $\overline{r}_1 \overline{r}_5 \neq 0$,

$$P\omega = \frac{-c}{\overline{r}_1 \overline{r}_5} d\overline{r}_2 \wedge d\overline{r}_3 \wedge d\overline{r}_4 \wedge d\overline{r}_6 \wedge d\overline{r}_7 \wedge d\overline{r}_8 \tag{66}$$

The result then also follows from (62) QED.

4.6. Let $h \in \Gamma^{-6}$. Now consider $h\omega$. Clearly $h\omega$ is invariant under the map $C^* \to C^*$, $w \mapsto -w$. But also by Proposition 4.5 one has that

$$\theta_I h\omega = \iota_I h\omega = 0. \tag{67}$$

Thus $h\omega$ descends to a volume element $\widetilde{h\omega}$ on X. Moreover, $h \mapsto \widetilde{h\omega}$ is clearly a G-isomorphism of Γ^{-6} with the space of all smooth volume elements on X. It is clear then that we may identify L^{-6} with the canonical bundle on X so that, recalling (54), h itself is identified with $\widetilde{h\omega}$. We can (using (57)) and do choose the orientation on X such that R^{-6} is positive. Hence one has, by (64)

$$\int_{S_E \times S_F} h\mu = \int_X h \tag{68}$$

for any $h \in \Gamma^{-6}$, since (40) is a two-to-one covering map.

By (55) one has a pairing of Γ^{-k} and Γ^{k-6} into Γ^{-6}. But then one has a pairing of Γ^{-k} and Γ^{k-6} defined by putting

$$\langle \varphi, \psi \rangle = \int_X \varphi\psi \tag{69}$$

for $\varphi \in \Gamma^{-k}$, $\psi \in \Gamma^{k-6}$. Furthermore, one immediately has

Proposition 4.7. *The pairing (69) is G-invariant. That is, for any* $g \in G$, $\varphi \in \Gamma^{-k}$ *and* $\psi \in \Gamma^{k-6}$

$$\langle g \cdot \varphi, g \cdot \psi \rangle = \langle \varphi, \psi \rangle \tag{70}$$

5. The Laplacian map $\Delta_X: \Gamma^{-2} \to \Gamma^{-4}$. Now since $S \cong SU(2)$ one knows that for any $n \in \mathbf{Z}_+$ there exists an irreducible complex representation

$$\rho_n: S \to \operatorname{Aut} Z_n \tag{70.5}$$

where dim $Z_n = n+1$. Moreover ρ_n is self-contragredient so that there is a complex linear operation End $Z_n \to$ End Z_n, $A \to A^t$, satisfying $(AB)^t = B^tA^t$, tr $A =$ tr A^t and $(\rho_n(a))^t = \rho_n(a)^{-1}$ for any $a \in S$.

Recall the diffeomorphism $S \to S_E$, $a \mapsto a \cdot \epsilon_1$. Now for any $n \in \mathbb{Z}_+$, and $A \in$ End Z_n let $\varphi_A \in C^\infty(S_E)$ be defined by putting

$$\varphi_A(a \cdot \epsilon_1) = (n+1) \text{ tr } A\rho_n(a^{-1}). \tag{71}$$

But $C^\infty(S_E)$ is a $S_1 \times S_2$-module and one notes that for $(a,b,1,1) \in S_1 \times S_2$, $a, b \in S$ one has

$$(a,b,1,1) \cdot \varphi_A = \varphi_{\rho_n(a)A\rho_n(b^{-1})} \tag{72}$$

On the other hand by Proposition 4.3 μ_E corresponds to normalized Haar measure on S_E so that by the Peter-Weyl theory one has for $A \in$ End Z_n, $B \in$ End Z_m

$$\int_{S_E} \varphi_A \varphi_B \mu_E = \begin{cases} 0 & \text{if } \quad n \neq m \\ n+1 \quad \text{tr} \quad AB^t & \text{for } \quad n = m \end{cases} \tag{73}$$

Note that we are using the readily established fact that $\varphi_B(a \cdot \epsilon_1) = \varphi_{B^t}(a^{-1} \cdot \epsilon_1)$.

Recalling (33) the unitary dual $\widehat{K}$ of K may be parameterized by the set of all 4-tuples $(m,n,p,q) = \ell$ where $m,n,p,q \in \mathbb{Z}_+$. The corresponding representation is just $\rho_m \otimes \rho_n \otimes \rho_p \otimes \rho_q = \rho_\ell$. If we restrict ourselves to the subgroup $S_1 \times S_2$ then $\widehat{(S_1 \times S_2)}$ is parameterized by the subset where $p = q = 0$.

Now let $C_K(S_E)$ be the space of all $S_1 \times S_2$ finite elements in $C^\infty(S_E)$. One defines a pre-Hilbert space structure on $C_K(S_E)$ by putting

$$\{\varphi, \psi\}_K = \int_{S_E} \varphi \bar{\psi} \mu_E \tag{74}$$

Note that for any $n \in \mathbb{Z}_+$ there exists a conjugate linear operation End $Z_n \to$ End Z_n, $A \to A^*$, where $\varphi_{A^*} = \bar{\varphi}_{A^t}$. Let $C_n(S_E) = \{\varphi_A \mid A \in \text{End } Z_n\}$. One of course has

$$\dim C_n(S_E) = (n+1)^2 \tag{75}$$

The following is part of the Peter-Weyl theorem for $SU(2)$.

Proposition 5.1. *One has an orthogonal direct sum*

$$C_K(S_E) = \oplus_{n=0}^\infty C_n(S_E) \tag{76}$$

Moreover $C_n(S_E)$ is an irreducible $S_1 \times S_2$ submodule and $(n,n,0,0)$ is the corresponding parameter.

Proof. The last statement follows from (72). QED.

Recall that R_E is the quadratic form on E defined by the restriction $Q|E$. That is, $R_E = r_1^2 + \cdots + r_4^2$. This induces a Riemannian structure on E and the latter induces a Riemannian structure on S_E. Let $\Delta(S_E)$ be the corresponding Laplace-Beltrami operator. One of course has

$$\Delta(E) = \left(\frac{\partial}{\partial r_1}\right)^2 + \cdots + \left(\frac{\partial}{\partial r_4}\right)^2 \tag{77}$$

and one knows

Proposition 5.2. *There exists a constant c_0 such that $\Delta(S_E)$ reduces to the scalar $c_0(n+2)n$ on $C_n(S_E)$ for any $n \in \mathbb{Z}_+$.*

Proof. This follows from Proposition 5.1 since one knows that $\Delta(S_E)$ may be given by a fixed multiple of a Casimir element for S_1 operating on $C_K(S_E)$. QED.

Now we may write $E^* = E - \{0\}$ as a direct product $E^* = \mathbb{R}_+ \times S_E$ and hence if $\varphi \in C^\infty(S_E)$ and $j \in \mathbb{R}$ one defines a function $\varphi^{(j)} \in C^\infty(E^*)$ by putting for any $t > 0$, $e \in S_E$

$$\varphi^{(j)}(te) = t^j \varphi(e) \tag{78}$$

Lemma 5.3. *Let $j \in \mathbb{R}$. Also let $n \in \mathbb{Z}_+$ and let $A \in$ End Z_n. Then if c_0 is as in Proposition 5.2 one has*

$$\Delta(E)((\varphi_A)^{(j)}) = (j(j+2) + c_0 n(n+2))(\varphi_A)^{(j-2)} \tag{79}$$

Proof. In the notation of (77) the radial part of $\Delta(E)$ is $\left(\frac{d}{dr}\right)^2 + \frac{3}{r}\frac{d}{dr}$. It then follows from the polar coordinate decomposition of $\Delta(E)$ that the left side, ψ, of (79), restricted to S_E, is just $(j(j+2) + c_0 n(n+2))\varphi_A$. This of course uses Proposition 5.2. However, one has $\psi(te) = t^{j-2}\psi(e)$ since if $I(E)$ is the Euler operator on E then clearly one has $[I(E), \Delta(E)] = -2\Delta(E)$. This proves the proposition. QED.

5.4. Now for any $n \in \mathbb{Z}_+$ and $B \in$ End Z_n let $\psi_B \in C^\infty(S_F)$ be defined so that $\psi_B(a \cdot \epsilon_5) = (n+1)$ tr $B\rho_n(a^{-1})$ for any $a \in S$. Analogously, one defines $C_K(S_F)$ and $C_n(S_F)$. Also, analogously

$$C_K(S_F) = \oplus_{n=0}^\infty C_n(S_F) \tag{80}$$

is an orthogonal direct sum where $C_n(S_F)$ is an irreducible $S_3 \times S_4$ submodule and $(0, 0, n, n)$ is the corresponding representation.

Let $C_K(S_E \times S_F)$ be the space of all K-finite elements in $C^\infty(S_E \times S_F)$. One easily has

$$C_K(S_E \times S_F) = C_K(S_E) \otimes C_K(S_F) \tag{81}$$

where, of course, $(\varphi \otimes \psi)(e,f) = \varphi(e)\psi(f)$ for $\varphi \in C_K(S_E), \psi \in C_K(S_F)$, $e \in S_E, f \in S_F$. For $m,n \in \mathbf{Z}_+$ let

$$C_{m,n}(S_E \times S_F) = C_m(S_E) \otimes C_n(S_F) \tag{82}$$

Also $C_K(S_E \times S_F)$ has the K-invariant structure of a pre-Hilbert space where $(\gamma_1, \gamma_2) = \int\limits_{S_E \times S_F} \gamma_1 \overline{\gamma}_2 \mu$ for $\gamma_i \in C_K(S_E \times S_F),\quad i = 1,2$.

Proposition 5.5. *One has an orthogonal direct sum*

$$C_K(S_E \times S_F) = \oplus_{m,n=0}^{\infty} C_{m,n}(S_E \times S_F) \tag{83}$$

Moreover, $C_{m,n}(S_E \times S_F)$ is an irreducible K-submodule and (m,m,n,n) is the corresponding parameter.

Now let Δ be the Laplace-Beltrami operator on M defined with respect to the pseudo-Riemannian structure corresponding to Q. Obviously Δ is G-invariant and with respect to the notation of section 4.2, one has $\Delta = \sum_{i=1}^{8} \sigma_i \left(\frac{\partial}{\partial r_i}\right)^2$. That is, if $\Delta(E)$ is, as in (77) (except that it is now regarded as an operator on $C^\infty(M)$) and $\Delta(F) = \sum_{i=5}^{8} \left(\frac{\partial}{\partial r_i}\right)^2$ then Δ is the ultra-hyperbolic operator given by

$$\Delta = \Delta(E) - \Delta(F) \tag{84}$$

For notational convenience let Q denote the operator of multiplication by the function $Q = \sum_{i=1}^{8} \sigma_i r_i^2$. Then where I, as usual, is the Euler operator, one has the familiar relation

Proposition 5.6. $[I,\Delta] = -2\Delta, [I,Q] = 2Q$ *and* $[\Delta, Q] = 4(I+4)$ *where 4, inside the parenthesis, denotes the operator of multiplication by the constant 4.*

Let $C \subseteq D' \subseteq M$ where D' is stable under multiplication by $\mathbf{R}$ and $D' - \{0\}$ is open in M. Recall the notation of section 4. For the following we thank V. Guillemin.

Lemma 5.7. *One has*

$$\Delta : \Gamma^{-2}(D') \to \Gamma^{-4}(D') \tag{85}$$

Furthermore, if $\varphi \in \Gamma^{-2}(D')$ vanishes on C^*, then the same is true for $\Delta\varphi$.

Proof. One has $I\Delta\varphi = -4\Delta\varphi$ since $[I, \Delta] = -2\Delta$. On the other hand since $\varphi(w) = \varphi(-w)$ for all $w \in D' - \{0\}$ the same is true for $\Delta\varphi$. This establishes (85). But now since dQ does not vanish on C^* it follows that φ vanishes on C^* if and only if we can write $\varphi = Q\psi$ on $D' - \{0\}$ for some $\psi \in \Gamma^{-4}(D')$. This is clear since $Q \mid D' - \{0\} \in \Gamma^2(D')$. But then $\Delta\varphi = 4(I+4)\psi + Q\Delta\psi$. However, $(I+4)\psi = 0$. Thus, $\Delta\varphi = Q\Delta\psi$ which implies that $\Delta\varphi$ vanishes on C^*. QED.

If $\varphi \in \Gamma^{-2} = \Gamma^{-2}(C)$ then φ clearly admits an extension $\varphi_e \in \Gamma^{-2}(D')$ where D' is as in Lemma 5.7. One may then apply Δ and note that $\Delta\varphi_e \mid C^*$, by Lemma 5.7, is independent of the extension φ_e. Hence, one has

Proposition 5.8. *The Laplacian Δ induces a map*

$$\Delta_X : \Gamma^{-2} \to \Gamma^{-4} \tag{86}$$

which intertwines for the actions of G on Γ^{-2} and Γ^{-4}.

Proof. This is immediate since Δ is G-invariant. QED.

Lemma 5.9. *Let $k \in \mathbb{Z}$. Then the map*

$$\Gamma^k \to C^\infty(S_E \times S_F). \tag{87}$$

defined by restriction is a K-isomorphism onto the space of all $\varphi \in C^\infty(S_E \times S_F)$ such that $\varphi(w) = (-1)^k\varphi(-w)$ for all $w \in (S_E \times S_F)$.

Proof. We have only to observe that if $\varphi(w) = (-1)^k\varphi(-w)$ then $\varphi^{(k)} \in \Gamma^k$ where $\varphi^{(k)}(tw) = t^k\varphi(w)$ for $t \in \mathbb{R}_+$ and $w \in (S_E \times S_F)$. QED.

Let Γ_K^k be the space of all K-finite elements in Γ^k. Also, for $m, n \in \mathbb{Z}_+$ let $\Gamma_{m,n}^k$ be the ρ_ℓ primary component where $\ell = (m, m, n, n)$. As a consequence of (83) one clearly has

Proposition 5.10. *$\Gamma_{m,n}^k$ is non-zero if and only if $m + n = k$ mod 2. Furthermore, in such a case $\Gamma_{m,n}^k$ is K-irreducible and summing over all such pairs (m, n) one has the direct sum*

$$\Gamma_K^k = \oplus\Gamma_{m,n}^k \tag{88}$$

Proof. This follows immediately from the fact that $\rho_n(-1) = (-1)^n$ for any $n \subset \mathbb{Z}_+$. QED.

5.11. We now consider the harmonic component H^∞ of Γ^{-2}. By Proposition 5.8 it is clear that H^∞ is a G and $\mathfrak{g}$ submodule of Γ^{-2}. Let $H \subseteq H^\infty$ be the space of K-finite elements in H^∞ for any $n \in \mathbb{Z}_+$. Let H_n be the ρ_ℓ primary component where $\ell = (n,n,n,n)$. Clearly H is a $\mathfrak{g}$ submodule of H^∞.

Theorem 5.12. *The ρ_ℓ primary component of H vanishes unless $\ell = (n,n,n,n)$. Furthermore, H_n is non-zero and irreducible so that*

$$\dim H_n = (n+1)^4 \tag{89}$$

and one has the direct sum

$$H = \oplus_{n=0}^{\infty} H_n \tag{90}$$

so that it is a Harish-Chandra module for $\mathfrak{g}$.

Proof. Let $m,n \in \mathbb{Z}_+$ and let $A \in \mathrm{End}\, Z_m$, $B \in \mathrm{End}\, Z_n$. Then $\varphi_A \otimes \psi_B \in C_{m,n}(S_E \times S_F)$ by (82) and $C_{m,n}(S_E \times S_F)$ is spanned by such elements. Now let D be as in section 4.2 so that $D^* = E^* \times F^*$. Let ${\varphi_A}^{(-1)} \in C^\infty(E^*)$ be defined as in (78) and let ${\psi_B}^{(-1)} \in C^\infty(F^*$ be defined analogously. Furthermore, if $\gamma_e \in C^\infty(D^*)$ is defined by putting $\gamma_e = {\varphi_A}^{(-1)} \otimes {\psi_B}^{(-1)}$ and $\gamma = \gamma_e \mid C^*$ then choosing m,n so that $m+n = 0 \bmod 2$ one has that $\gamma_e \in \Gamma^{-2}_{m,n}(D)$ is an extension of $\gamma \in \Gamma^{-2}_{m,n}$. Furthermore, $\Gamma^{-2}_{m,n}$ is clearly spanned by all such γ. But then $\Delta_X \gamma = \Delta \gamma_e \mid C^*$. However, by (84) one has

$$\Delta\gamma_e = \Delta(E){\varphi_A}^{(-1)} \otimes {\psi_B}^{(-1)} + {\varphi_A}^{(-1)} \otimes \Delta(F){\psi_B}^{(-1)} \tag{91}$$

But then by (79) and the analogous statement for F one has

$$\Delta\gamma_e \mid S_E \times S_F = c_0(m(m+2) - n(n+2))\varphi_A \otimes \psi_B \tag{92}$$

But then the right side of (92) is just $\Delta_X \gamma \mid S_E \times S_F$. One thus has

$$\Gamma^{-2}_{n,n} \subseteq H \tag{93}$$

On the other hand, using Proposition 5.5, Δ_X clearly induces an isomorphism

$$\Gamma^{-2}_{m,n} \to \Gamma^{-4}_{m,n} \tag{94}$$

in case $m \neq n$. The result then follows from Proposition 5.10. QED.

We have noted in the proof above that

$$H_n = \Gamma^{-2}_{n,n} \tag{95}$$

so that

$$H = \oplus_{n=0}^{\infty} \Gamma^{-2}_{n,n} \tag{96}$$

6. The Harish-Chandra Module H. Recalling the notation of Section 1, for any root $\varphi \in \Delta$ let $h_\varphi \in \mathfrak{a}$ be defined by putting $h_\varphi = [e_\varphi, e_{-\varphi}]$. It then follows from (24) that the non-zero eigenvalues of $\eta(h_\varphi)$ are $1, 1, -1, -1$. But then from the representation theory of a TDS it follows that $(h_\varphi, e_\varphi, e_{-\varphi})$ is an S-triple. That is, in addition to $[e_\varphi, e_{-\varphi}] = h_\varphi$ one has

$$[h_\varphi, e_\varphi] = 2e_\varphi \text{ and } [h_\varphi, e_{-\varphi}] = -2e_{-\varphi} \tag{97}$$

Thus $\varphi(h_\varphi) = 2$ for any $\varphi \in \Delta$. But then from the explicit form of the $\alpha_i, i = 1, ..., 4$, one has that if $a_{ij} = \lambda_i(h_{\alpha_j})$ then the matrix (a_{ij}) is just

$$\begin{pmatrix} 1 & -1 & 0 & 0 \\ 0 & 0 & 1 & 1 \\ 0 & 0 & 1 & -1 \\ 0 & 1 & -1 & 0 \end{pmatrix} \tag{98}$$

Now let $\nu_i \in \mathfrak{a}'$ be defined so that $\nu_i(h_{\alpha_j}) = \delta_{ij}$. That is, the ν_i are the highest weights of the fundamental representation of $\mathfrak{g}$. Inverting the matrix (98) yields

Proposition 6.1. *One has* $\nu_1 = \lambda_1$, $\nu_2 = \frac{1}{2}(\lambda_1 + \lambda_2 + \lambda_3 + \lambda_4)$, $\nu_3 = \frac{1}{2}(\lambda_1 + \lambda_2 + \lambda_3 - \lambda_4)$ *and* $\nu_4 = \lambda_1 + \lambda_2$.

Remark. Of course $\nu_1 = \lambda_1$ is the highest weight of η and $\nu_4 = \lambda_1 + \lambda_2$ is the highest weight (i.e. highest root) of the adjoint representation. In the notation of (32) we have also $\nu_4 = \alpha_0 = \lambda_1 + \lambda_2$.

Now if (h, e, e_-) is an S-triple spanning, over $\mathbb{R}$, a TDS, then from the structure of $\mathfrak{sl}(2, \mathbb{C})$ one notes that (iz, x, x_-) is again an S-triple spanning over $\mathbb{C}$, the complexification, $\mathfrak{u}_\mathbb{C}$ where $z = e - e_-$, $x = \left(\frac{e + e_- + ih}{2}\right)$ and $x_- = \left(\frac{e + e_- ih}{2}\right)$.

Thus if subscript $\mathbb{C}$ denotes complexification, $(i, z_\varphi, x_\varphi, x_{-\varphi})$ is an S-triple in $\mathfrak{g}_\mathbb{C}$, for any $\varphi \in \Delta$ where $z_\varphi = e_\varphi - e_{-\varphi}$, $x_\varphi = \frac{e_\varphi + e_{-\varphi} + ih_\varphi}{2}$ and $x_{-\varphi} = \frac{e_\varphi + e_{-\varphi} - ih_\varphi}{2}$. Let $U_\varphi \subseteq \text{Ad}\, \mathfrak{g}_\mathbb{C}$ be the complex three dimensional subgroup corresponding to the Lie algebra $\mathfrak{u}_\varphi$ over $\mathbb{C}$ spanned by h_φ, e_φ and $e_{-\varphi}$. The first statement in the following lemma asserts that in the terminology of [KR] the new S-triple is normal.

Lemma 6.2. *For any* $\varphi \in \Delta$ *one has* $iz_\varphi \in \mathfrak{k}_\mathbb{C}$ *and* $x_\varphi, x_{-\varphi} \in \mathfrak{q}_\mathbb{C}$. *Furthermore, there exists* $c_\varphi \in U_\varphi$ *such that* $c_\varphi h_\varphi = iz_\varphi$, $c_\varphi e_\varphi = x_\varphi$ *and* $c_\varphi e_{-\varphi} = x_{-\varphi}$.

Proof. We recall (see Proposition 3.3) that $z_\varphi \in \mathfrak{k}$. Also, by linearly extending the action of θ to $\mathfrak{g}_\mathbb{C}$ one has $\theta x_\varphi = \frac{-e_{-\varphi} - e_\varphi - ih_\varphi}{2} = -x_\varphi$ by (6.6).

Thus $x_\varphi \in q_{\mathbf{C}}$. Similarly, $x_{-\varphi} \in \mathfrak{q}_{\mathbf{C}}$. The second statement of the proposition follows since any 2 S-triples are conjugate. That is, all automorphisms of $\mathfrak{sl}(2,\mathbf{C})$ are inner. QED.

For any $\varphi \in \Delta$ let $c_\varphi \in U_\varphi$ be chosen satisfying the conditions of Lemma 6.2. Since the roots $\alpha_i, i = 0, ..., 3$, are (strongly) orthogonal, the elements $c_{\alpha_i} \in \text{Ad}\ \mathfrak{g}_{\mathbf{C}}$ necessarily commute with each other. Let $c_* = c_{\alpha_0} \cdots c_{\alpha_3}$.

Now extend the action of any element $\sigma \in \Sigma_3$ from $\mathfrak{g}$ to $\mathfrak{g}_{\mathbf{C}}$. Hence it defines an action on Ad $\mathfrak{g}_{\mathbf{C}}$.

The element c_φ is unique up to a central element of U_φ. By (8) and Lemma 3.8 we may normalize the choice so that

$$\begin{aligned} \tau c_{\alpha_1} &= c_{\alpha_2},\ \tau c_{\alpha_2} = c_{\alpha_1} \quad \text{and} \\ \tau c_{\alpha_j} &= c_{\alpha_j} \quad \text{for} \quad j = 0, ..., 3 \end{aligned} \tag{99}$$

Lemma 6.3. *One has $\tau c_* = c_*$ so that with respect to their action on $\mathfrak{g}_{\mathbf{C}}$, the operators, defined by τ and c_*, commute with each other.*

Proof. The proof is immediate from (99). QED.

Let $\mathfrak{t}$ be the span of $z_{\alpha_j}, j = 0, ..., 3$, so that $\mathfrak{t} \subseteq \mathfrak{k}$ is a Lie algebra of a maximal torus of K. By the orthogonality of the α_i and Lemma 6.2 one has $c_* h_{\alpha_j} = i z_{\alpha_j}$ so that

$$c_* \mathfrak{a} = i\mathfrak{t} \tag{100}$$

Remark. An automorphism satisfying (100) is often referred to as a Cayley transform.

Let $\mathfrak{b}_0 = \mathfrak{a}_{\mathbf{C}} + \mathfrak{n}_{\mathbf{C}}$ so that $\mathfrak{b}_0$ is a Borel subalgebra of $\mathfrak{g}_{\mathbf{C}}$. Now put $\mathfrak{b} = \sigma_* \mathfrak{b}_0$.

Proposition 6.4. *$\mathfrak{b}$ is a θ-stable Borel subalgebra of $\mathfrak{g}_{\mathbf{C}}$ so that*

$$\mathfrak{b} = \mathfrak{b}_{\mathfrak{k}} + \mathfrak{b}_{\mathfrak{q}} \tag{101}$$

where $\mathfrak{b}_{\mathfrak{k}} = \mathfrak{b} \cap \mathfrak{k}_{\mathbf{C}}$ and $\mathfrak{b}_{\mathfrak{q}} = \mathfrak{q}_{\mathbf{C}}$. Furthermore, $\mathfrak{b}_{\mathfrak{k}}$ is a Borel subalgebra of $\mathfrak{k}_{\mathbf{C}}$. In addition, $\mathfrak{b}$ and, in fact, $\mathfrak{b}_{\mathfrak{k}}$ and $\mathfrak{b}_{\mathfrak{q}}$, are stable under τ.

Proof. By (100) one has $\mathfrak{t}_{\mathbf{C}} \subseteq \mathfrak{b}$. But $\mathfrak{t}_{\mathbf{C}} \subseteq \mathfrak{k}_{\mathbf{C}}$ so that the first two statements follow from general structure theory. The final statement follows from Lemma 6.3 since $\mathfrak{b}_0$, $\mathfrak{k}$ and $\mathfrak{q}$ are τ-stable. (See Proposition 3.1). QED.

We use $\mathfrak{b}$ to define a positive system of roots relative to $(\mathfrak{t}_{\mathbf{C}}, \mathfrak{g}_{\mathbf{C}})$ and $\mathfrak{b}_{\mathfrak{k}}$ is used similarly relative to $(\mathfrak{t}_{\mathbf{C}}, \mathfrak{k}_{\mathbf{C}})$.

Proposition 6.5. *The elements x_{α_0} and $x_{-\alpha_0}$ in $\mathfrak{g}_{\mathbf{C}}$ are, respectively, highest and lowest root vectors relative to $(\mathfrak{t}_{\mathbf{C}}, \mathfrak{g}_{\mathbf{C}})$. Also they are, respectively, highest and lowest weight vectors with respect to the representation of $\mathfrak{k}_{\mathbf{C}}$ on $\mathfrak{q}_{\mathbf{C}}$ relative to $(\mathfrak{t}_{\mathbf{C}}, \mathfrak{k}_{\mathbf{C}})$. Furthermore, both are fixed by τ.*

Proof. One has $x_{\alpha_0}, x_{-\alpha_0} \in \mathfrak{q}_{\mathbf{C}}$ by Lemma 6.2. The remaining statements are immediate, using Lemma 6.3, since $c_* e_{\alpha_0} = x_{\alpha_0}$ and $c_* e_{-\alpha_0} = x_{-\alpha_0}$ by definition of c_* and since α_0 is the highest root relative to $(\mathfrak{a}_{\mathbf{C}}, \mathfrak{g}_{\mathbf{C}})$. QED.

Now let $\tilde{\eta}$ be the representation of G and $\mathfrak{g}$ on $C^\infty(M)$ induced by the action of G and $\mathfrak{g}$ on M. By complex linearity we extend $\tilde{\eta}$ so it is defined on $\mathfrak{g}_{\mathbf{C}}$. Thus $\tilde{\eta}(\mathfrak{g}_{\mathbf{C}})$ is a Lie algebra of complex vector fields on M.

It will be convenient to introduce holomorphic coordinates $s_1, \ldots, s_4$ and their conjugates $\bar{s}_1, \ldots, \bar{s}_4$ on M. We define

$$s_1 = r_1 - ir_2,\ s_2 = r_3 - ir_4,\ s_3 = r_5 + ir_6,\ s_4 = r_7 + ir_8 \tag{101.5}$$

One recalls that $\dfrac{\partial}{\partial s_1} = \dfrac{1}{2}\left(\dfrac{\partial}{\partial r_1} + i\dfrac{\partial}{\partial r_2}\right), \ldots, \dfrac{\partial}{\partial s_4} = \dfrac{1}{2}\left(\dfrac{\partial}{\partial r_7} - i\dfrac{\partial}{\partial r_8}\right)$.

Let

$$\begin{aligned} x_o &= ix_{\alpha_0} \quad \text{and} \\ y_o &= ix_{-\alpha_0} \end{aligned} \tag{101.6}$$

Proposition 6.6. *One has*

$$\begin{aligned} \tilde{\eta}(x_o) &= s_1\frac{\partial}{\partial \bar{s}_3} + s_3\frac{\partial}{\partial \bar{s}_1} \quad \textit{and} \\ \tilde{\eta}(y_o) &= -\left(\bar{s}_1\frac{\partial}{\partial s_3} + \bar{s}_3\frac{\partial}{\partial s_1}\right) \end{aligned} \tag{102}$$

Proof. By the choice in (32.5) we recall $\eta(e_{\alpha_0})$ maps u_2 to v_1 and hence, by Lemma 3.2, u_1 to $-v_2$. But then $\eta(e_{-\alpha_0})$ maps v_1 to u_2 and v_2 to $-u_1$. Thus v_1, v_2, u_1, u_2 are eigenvectors for $\eta(h_{\alpha_0}) = [\eta(e_{\alpha_0}), \eta(e_{-\alpha_0})]$ corresponding, respectively, to the eigenvalues $1, 1, -1$, and -1. The operators $\eta(e_{\alpha_0})$, $\eta(e_{-\alpha_0})$ and $\eta(h_{\alpha_0})$ vanish on all u_1, v_j not specifically mentioned above. Now

$$\begin{aligned} x_{\alpha_0} &= \frac{e_{\alpha_0} + e_{-\alpha_0} + ih_{\alpha_0}}{2} \quad \text{and} \\ x_{-\alpha_0} &= \frac{e_{\alpha_0} + e_{-\alpha_0} - ih_{\alpha_0}}{2} \end{aligned} \tag{102.5}$$

But then $\eta(x_{\alpha_0})$ maps u_1 to $\frac{1}{2}(-v_2 - iu_1)$, v_1 to $\frac{1}{2}(u_2 + iv_1)$, u_2 to $\frac{1}{2}(v_1 - iu_2)$ and v_2 to $\frac{1}{2}(-u_1 + iv_2)$. But then in terms of the basis ϵ_i one has that $\eta(x_{\alpha_0})$ maps ϵ_1

to $\frac{1}{2}(-\epsilon_6+i\epsilon_5)$, ϵ_2 to $\frac{1}{2}(\epsilon_5+i\epsilon_6)$, ϵ_5 to $\frac{1}{2}(\epsilon_2+i\epsilon_1)$ and ϵ_6 to $\frac{1}{2}(-\epsilon_1+i\epsilon_2)$. It vanishes on all other ϵ_j. Furthermore, for any i one clearly has $\eta(x_{-\alpha_0}\epsilon_i) = \overline{\eta(x_{\alpha_0}\epsilon_i)}$. But if $M' \subseteq C^\infty(M)$ is the dual space to M then the map $M \to M'$ where $\epsilon_i \to r_i$, $i = 1,...,4$, and $\epsilon_j \to -r_j$, $j = 5,...,8$, clearly intertwines for the representations η and $\tilde{\eta}$. Thus the vector field $\tilde{\eta}(x_{\alpha_0})$ maps r_1 to $r_6 - ir_5$, r_2 to $\frac{1}{2}(-r_5 - ir_6)$, r_5 to $\frac{1}{2}(-r_2 - ir_1)$ and r_6 to $r_1 - ir_2$ and vanish on all other r_j. Hence

$$i\tilde{\eta}(x_{\alpha_0}) = \frac{1}{2}\left((r_5 + ir_6)\left(\frac{\partial}{\partial r_1} - i\frac{\partial}{\partial r_2}\right) + (r_1 - ir_2)\left(\frac{\partial}{\partial r_5} + i\frac{\partial}{\partial r_6}\right)\right)$$

But this is just $s_1\frac{\partial}{\partial s_3} + s_3\frac{\partial}{\partial s_1}$. This proves the proposition since $\tilde{\eta}(x_{-\alpha_0})$ is clearly just the conjugate of the vector field $\tilde{\eta}(x_{\alpha_0})$. QED.

6.7. Now let $U(\mathfrak{g})$ be the universal enveloping algebra of $\mathfrak{g}_{\mathbf{C}}$. We are mainly concerned now with the $U(\mathfrak{g})$-module structure on $H \subseteq \Gamma^{-2}$ induced by the action of $\mathfrak{g}$ on H. (See section 5.13). If $w \in U(\mathfrak{g}), \psi \in H$ let $w \cdot \psi \in H$ denote the action of w on ψ. To describe the module we first note that the representation $\tilde{\eta}$ of $\mathfrak{g}_{\mathbf{C}}$ extends to a representation (also denoted by $\tilde{\eta}$) of $U(\mathfrak{g})$ as differential operators on M. Thus if $w \in U(\mathfrak{g}), \mathfrak{k} \in \mathbf{Z}$, $\varphi \in \Gamma^k(D)$ where D is defined as in section 4.2, then $\tilde{\eta}(w)\varphi \in \Gamma^k(D)$ is well defined. Furthermore, if $\psi \in H$ and $\varphi \in \Gamma^{-2}(D)$ is such that $\varphi \mid C^*= \psi$ then since C is G-stable,

$$w \cdot \psi = \tilde{\eta}(w)\varphi \mid C^* \tag{103}$$

Now, recalling the notation of (57) one has that $R^{-2} \in \Gamma^{-2}(D)$ where $R = (R_E\ R_F)^{\frac{1}{4}} \mid D^*$. Restricting further to C^* let $\xi = R^{-2} \mid C^*$ so that $\xi \in \Gamma^{-2}$. Recall that H_0, by Theorem 5.12, is the space of K-invariants in H and that it is 1-dimensional.

Lemma 6.8. *One has $H_0 = R\xi$. Furthermore, $\xi \mid S_E \times S_F$ is the constant function 1.*

Proof. By (89) and (93.5) $\Gamma_{0,0}^{-2}$ is 1-dimensional and $\Gamma_{0,0}^{-2} = H_0$. But obviously $\xi \in \Gamma^{-2}$ is K-invariant and $\xi \mid S_E \times S_F = 1$. Thus $\Gamma_{0,0}^{-2} = \mathbf{R}\xi$. QED.

Now, recalling (101.5) note that $R_E = s_1\bar{s}_1 + s_2\bar{s}_2$ and $R_F = s_3\bar{s}_3 + s_4\bar{s}_4$. But then by (102)

$$\tilde{\eta}(x_o)R_E = \tilde{\eta}(x_o)R_F = s_1 s_3$$

and

$$\tilde{\eta}(y_o)R_E = \tilde{\eta}(y)R_F = -\bar{s}_1\bar{s}_3 \tag{104}$$

On the other hand

$$\tilde{\eta}(x_o)s_i = \tilde{\eta}(y_o)\bar{s}_i = 0 \tag{105}$$

for $i = 1, ..., 4$.

We also note that if $a, b \in \mathbb{R}$ and $k = 2(a+b)$ then since R_E and R_F both reduce to the constant 1 on $S_E \times S_F$ one has

$$R_E^a R_F^b = R^k \text{ on } C^* \tag{106}$$

Lemma 6.9. *Let $k \in \mathbb{R}$ and $n \in Z_+$ Then if* $d_{k,n} = \dfrac{k(k-2)\cdots(k-2n+2)}{2^n}$ *one has*

$$\tilde{\eta}(x_o^n)R^k = d_{k,n}(s_1 s_3)^n R^{k-2n} \tag{107}$$

and

$$\tilde{\eta}(y_o^n)R^k = (-1)^n d_{k,n}(\bar{s}_1\bar{s}_3)^n R^{k-2n} \tag{108}$$

on C^*. In particular $(k = -2)$ so that one has

$$x_o^n \xi = (-1)^n \; n!(s_1 s_3)^n \tag{109}$$

and

$$y_o^n \xi = -n!(\bar{s}_1\bar{s}_3)^n \tag{110}$$

on $S_E \times S_F$.

Proof. By (106) one has $R_E^{k/2} = R^k$ on C^*. Let $m = k/2$. But then by (104) one has $\tilde{\eta}(x_o^n)R_E^m = m(m-1)\cdots(m-n+1)(s_1 s_3)^n R_E^{m-n}$. But this then yields (107) and (109). A similar argument yields (108) and (110). QED.

Consider the adjoint representation of $\mathfrak{k}_{\mathbb{C}}$ (or K) on $U(\mathfrak{g})$. By Proposition 6.5 x_o is a highest weight vector. But then $x_o^n \in U(\mathfrak{g})$ is also a highest weight vector for any $n \in \mathbb{Z}_+$. Let $P_n \subseteq U(\mathfrak{g})$ be the irreducible $\mathfrak{k}_{\mathbb{C}}$-module generated by x_o^n and let $P = \sum_{n=0}^{\infty} P_n$. Also let $\ell(n) \in \mathbb{Z}_+^4$ be the parameter for the irreducible representation $\rho_{\ell(n)} \in \widehat{K}$ corresponding to P_n.

Proposition 6.11. *One has $\ell(n) = (n,n,n,n)$ so that* $\dim P_n = (n+1)^4$. Also $P_1 = \mathfrak{q}_{\mathbb{C}}$.

Proof. If $\varphi \in \Delta$ then from the root structure of D_4 one has that among eigenvalues of ad h_φ the eigenvalue 1 and the eigenvalue -1 both have multiplicity 8. The same then is true for the elements $iz_{\alpha_j} = c_* h_{\alpha_j} \in \mathfrak{t}_{\mathbb{C}}$, $j = 0, ..., 3$. But from the structure of $\mathfrak{k}_{\mathbb{C}}$ these eigenvalues correspond to non-compact roots with respect to $(\mathfrak{t}_{\mathbb{C}}, \mathfrak{g}_{\mathbb{C}})$. That is, the eigenvalues of ad iz_{α_j} on $\mathfrak{q}_{\mathbb{C}}$ are either 1 or -1. It follows immediately then from Proposition 6.5 that $\ell(1) = (1,1,1,1)$

so that $\mathfrak{q}_{\mathbf{C}} = P_1$ since dim $\mathfrak{q}_{\mathbf{C}} = 16$. But then since the highest weight corresponding to x_o^n is n times the highest weight corresponding to x_o, one has $\ell(n) = (n,n,n,n)$. QED.

We now have

Lemma 6.12. *H is a cyclic $U(\mathfrak{g})$-module with the spherical vector ξ as cyclic generator. In fact the map*

$$P \to H, \quad w \to w \cdot \xi \tag{111}$$

is a linear isomorphism mapping P_n onto H_n.

Proof. By Proposition 6.11 one has $x_o^n \cdot \xi \in H_n$. But $x_o^n \cdot \xi \neq 0$ by (109) so that

$$x_o^n \cdot \xi \in H_n \tag{111.5}$$

is a highest weight vector in H_n. It follows then that the map induces a linear isomorphism $P_n \to H_n$. The result follows from Theorem 5.12. QED.

Let $z = -y_o$. Then $\eta(z)\bar{s}_i = 0,\ i = 1, ..., 4$, and

$$\tilde{\eta}(z)s_1 = \bar{s}_3, \quad \tilde{\eta}(z)\bar{s}_3 = \bar{s}_1 \text{ and } \tilde{\eta}(z)R_E = \bar{s}_1\bar{s}_3 \tag{112}$$

If $k \in \mathbf{R}$ and $i \in \mathbf{Z}_+$ let $(k;i) = k(k-1) \cdots (k-i+1)$.

Lemma 6.13. *One has, for $n, m \in \mathbf{Z}_+$ and $m \in \mathbf{R}$*

$$\tilde{\eta}(z^n)((s_1s_3)^n R_E^m) = \sum b_{ij}|s_1|^{2(n-i)}|s_3|^{2(n-j)}R_E^{m-n+i+j} \tag{113}$$

where the sum is over all pair $(i,j) \in \mathbf{Z}_+^2$ such that $i + j \leq n$ and $b_{ij} = (n;i)(n;j)(m;n-i-j)e_{ij}$ where $e_{ij} = n!i!j!(n-i-j)!$

Proof. Let $(k,\ell,p,q,r) \in \mathbf{Z}^4 \times \mathbf{R}$ and let $\alpha(k,\ell,p,q,r) = s_1^k\, \bar{s}_1^\ell\, s_3^p\, \bar{s}_3^q R_E^r$. Then, by (112) one notes that $\tilde{\eta}(z)\alpha(k,\ell,p,q,r)$ is the sum of (3 terms) $k\alpha(k-1,\ell,p,q+1,r) + p\alpha(k,\ell+1,p-1,q,r) + r\alpha(k,\ell+1,p,q+1,r-1)$. But then where the sum is over all $(i,j) \in \mathbf{Z}_+^2$ such that $i + j \leq n$ one has

$$\tilde{\eta}(z^n)\alpha(k,\ell,p,q,r) = \sum c_{ij}\alpha(k-i,\ell+n-i,p-j,q+n-j,r-n+i+j) \tag{114}$$

where $c_{ij} = (k;i)(p;j)(r;n-i-j)e_{ij}$. Indeed, e_{ij} is the number of sequences of length n whose entries are either s_1, s_3, or R_E and where s_1 occurs i times, s_3 occurs j times and R_E occurs $n-i-j$ times. That is, by making a choice of one of the three terms above at each stage, there are c_{ij} ways of obtaining the component $(k;i)(p;j)(r;n-i-j)\alpha(k-i,\ell+n-i,p-j,q+n-j,r-n+i+j)$ in the computation of $\tilde{\eta}(z^n)\alpha(k,\ell,p,q,r)$. Applying (114) to the case where $(k,\ell,p,q,r) = (n,0,n,0,m)$ proves the lemma. QED.

Recalling the product measure μ in $S_E \times S_F$ one has

Lemma 6.14.

$$\int_{S_E\times S_F} |s_1|^{2(n-i)}|s_3|^{2(n-j)}\mu = \frac{1}{(n-i+1)(n-j+1)} \tag{115}$$

Proof. By symmetry it clearly suffices to prove

$$\int_{S_E} |s_1|^{2k}\mu_E = \frac{1}{k+1} \tag{116}$$

for any $k \in \mathbb{Z}_+$.

Let $k \in \mathbb{Z}_+$. and let $S^{2k}(E)$ denote the space of all polynomial functions on E of degree $2k$ on E and let $P^{2k}(E)$ be the unique $SO(E)$ stable complement of $\mathbf{R}R_E^k$ in $S^{2k}(E)$. Then, if λ is the left side of (116) one has

$$|s_1|^{2k} = \lambda R_E^k + p \tag{117}$$

where $p \in P^{2k}(E)$. This is clear since the trivial representation of $SO(E)$ does not occur in $P^{2k}(E)$. Let $\Delta_0 = \frac{1}{4}\Delta(E)$ where, we recall, $\Delta(E)$ is the Laplacian on E. One has $\Delta_0 R_E = 2^k$ and more generally one has the classic formula $\Delta_0 R_E^k = k(k+1)R_E^{k-1}$. Thus, $\Delta_0 R_E^k = k!(k+1)!$ But $\Delta_0^k p = 0$ for the same reason as the one establishing (117). Thus $\Delta_0^k|s_1|^{2k} = \lambda k!(k+1)!$ But $\Delta_0 = \dfrac{\partial}{\partial s_1}\dfrac{\partial}{\partial \bar{s}_1} + \dfrac{\partial}{\partial s_2}\dfrac{\partial}{\partial \bar{s}_2}$. Thus, $\Delta_0^k|s_1|^{2k} = \left(\dfrac{\partial}{\partial s_1}\dfrac{\partial}{\partial \bar{s}_1}\right)^{1k}|s_1|^{2k}$. But $\left(\dfrac{\partial}{\partial s_1}\dfrac{\partial}{\partial \bar{s}_1}\right)|s_1|^{2k} = k^2|s_1|^{2(k-1)}$. Thus, $\Delta_0^k|s_1|^{2k} = (k!)^2$. Therefore, $\lambda k!(k+1)! = (k!)^2$ and hence $\lambda = \dfrac{1}{k+1}$. QED.

Lemma 6.15. *Let the notation be as in* (113). *Then*

$$\int_{S_E\times S_F} \eta_0(z^n)((s_1s_3)^n R_E^m)\mu = \frac{1}{(n+1)^2}(m+2n+2;n) \tag{118}$$

Proof. Let α be the left side of (118). Then since R_E reduces to 1 on $S_E \times S_F$ one has, by Lemma 6.14

$$\alpha = \sum b_{ij}\frac{1}{(n-i+1)}\frac{1}{(n-j+1)}$$

where the sum is over all $(i,j) \in \mathbf{Z}_+^2$ such that $i+j \leq n$. But $(n;i) = \frac{n!}{(n-i)!}$ and $(n;j) = \frac{n!}{(n-j)!}$. Thus, if $a_{ij} = b_{ij}\frac{1}{(n-i+1)}\frac{1}{(n-j+1)}$ then $a_{ij} = n!(\frac{n!}{(n-i+1)!i!}\frac{n!}{(n-j+1)!j!}\frac{(m;n-i-j)}{(n-i-j)!}$. That is,

$$a_{ij} = \frac{n!}{(n+1)^2}\binom{n+1}{i}\binom{n+1}{j}\frac{(m;n-i-j)}{(n-i-j)!}.$$

Hence, if m is a natural number one has

$$a_{ij} = \frac{n!}{(n+1)^2}\binom{n+1}{i}\binom{n+1}{j}\binom{m}{n-i-j}$$

But then in such a case

$$\alpha = \frac{n!}{(n+1)^2}\sum_{p=0}^{n}\left(\sum_{i=0}^{p}\binom{n+1}{i}\binom{n+1}{p-i}\right)\binom{m}{n-p}$$

Thus, from a classical formula on binomial sums one has

$$\begin{aligned}\alpha &= \frac{n!}{(n+1)^2}\sum_{p=0}^{n}\binom{2n+2}{p}\binom{m}{n-p}\\ &= \frac{n!}{(n+1)^2}\binom{m+2n+2}{n}\end{aligned}$$

That is, $\alpha = \frac{1}{(n+1)^2}(m+2n+2;n)$ establishes the lemma in case m is a natural number. But both α and the right side of (118) are polynomials in m for any fixed n and hence are necessarily equal for all values of m. QED.

6.16. Let $k \in \mathbf{R}$ and, recalling (88), let

$$\Gamma_K^k \to \Gamma_{0,0}^k, \quad v \to v_0 \tag{119}$$

be the projection defined relative to the sum. Clearly, Γ_K^k has the structure of a $U(\mathfrak{g})$-module and one notes that by the Chinese remainder theorem

$$v_0 \in U(\mathfrak{g})\cdot v \tag{120}$$

for any $v \in \Gamma_K^k$. Let

$$1_k \in \Gamma_{0,0}^k \tag{120.5}$$

be defined by putting $1_k = R^k \mid C^*$ so that $\Gamma_{0,0}^k = \mathbf{R}1_k$. Note $\xi = 1_{-2}$.

Theorem 6.15. *Let $k \in \mathbf{R}$ and let $x_o, y_o \in \mathfrak{g}$ be defined as in* (101.6). *Then for any $n \in \mathbb{Z}_+$ one has*

$$\left(\frac{y_o^n}{n!}\frac{x_o^n}{n!} \cdot 1_k\right)_0 = \alpha_{k,n} 1_k \tag{121}$$

for $\alpha_{k,n} = \dfrac{(-1)^n}{2^{2n}((n+1)!)^2}\beta_{k,n}\gamma_{k,n}$ *where*

$$\beta_{k,n} = k(k-2)\cdots(k-2n+2)$$
$$\gamma_{k,n} = (k+6)(k+8)\cdots(k+2n+4)$$

Proof. By (106) and (107) one has $\frac{x_o^n}{n!} \cdot 1_k$ is the restriction to C^* of $d_{k,n}\frac{(s_1 s_3)^n}{n!}R_E^{\frac{k}{2}-n}$. But recalling the notation of (112) one has $y_o^n = (-1)^n z^n$ and hence $\frac{y_o^n}{n!}\frac{x^n}{n!}1_k$ is the similar restriction of $\varphi = (-1)^n \frac{d_{k,n}}{(n!)^2}\tilde{\eta}(z^n)(s_1 s_3)^n R_E^{\frac{k}{2}-n}$. But if $\left(\frac{y^n}{n!}\frac{x^n}{n!}1_k\right)_0 = \lambda 1_k$ it is immediate that $\lambda = \int_{S_E \times S_F} \varphi\mu$ since any $\psi \in \Gamma_{p,q}^k$ integrates to 0 unless $(p,q) = (0,0)$. But then if $m = \frac{k}{2} - n$ one has $\lambda = \frac{(-1)^n}{(n!)^2} d_{k,n} \cdot \frac{1}{(n+1)^2}(\frac{k}{2}+n+2; n)$ by (118). But clearly $d_{k,n} = \frac{\beta_{k,n}}{2^n}$ and $\left(\frac{k}{2}+n+2; n\right) = \frac{\gamma_{k,n}}{2^n}$. This proves the theorem. QED.

One important observation for us is

Corollary 6.16. *Let the notation be as in Theorem 6.15. Then if $-6 < k < 0$ one has $\alpha_{k,n} > 0$ for all $n \in \mathbb{Z}_+$.*

Proof. Clearly $\gamma_{k,n} > 0$ if $-6 < k$. On the other hand $(-1)^n\beta_{k,n} > 0$ if $k < 0$.
QED.

One notes the symmetry of $\alpha_{k,n}$ about $k = -3$. That is, clearly $\beta_{k,n} = (-1)^n\gamma_{-6-k,n}$ so that

$$\alpha_{k,n} = \alpha_{-6-k,n} \tag{122}$$

Our main concern is for the case where $k = -2$ since for that case one has the harmonic $U(\mathfrak{g})$-submodule $H \subseteq \Gamma_K^{-2}$. Recall that $\xi = 1_{-2} \in H_0$.

Theorem 6.17. *One has*

$$\left(\frac{y_o^n}{n!}\frac{x_o^n}{n!} \cdot \xi\right)_0 = \frac{1}{n+1}\xi \tag{123}$$

for any $n \in \mathbb{Z}_+$.

Proof. Recalling Theorem 6.15 one notes that $\beta_{-2,n} = (-1)^n 2^n n!$ and $\gamma_{-2,n} = 2^n (n+1)!$. Thus $\alpha_{-2,n} = \frac{1}{n+1}$. QED.

Now as we have noted H is a Harish-Chandra module for $\mathfrak{g}$. Let

$$\pi: U(\mathfrak{g}) \to \text{End } H \tag{124}$$

denote the corresponding representation of $U(\mathfrak{g})$.

Theorem 6.18. *The representation π is irreducible.*

Proof. By Lemma 6.12 the spherical vector $\xi \in H_0$ is a cyclic vector for π. We have therefore only to show if $0 \neq h \in H$ is arbitrary there exists $u \in U(\mathfrak{g})$ such that $u \cdot h = \xi$. By the Chinese remainder theorem we may assume $h \in H_n$ for some n. However, since H_n is $\mathfrak{k}$-irreducible it suffices by (111.5) to take $h = x_o^n \cdot \xi$. But then $y_o^n \cdot h$, by Theorem 6.17, has a non-zero component in H_0 relative to the $\mathfrak{k}$-type decomposition. The result then follows from the Chinese remainder theorem. QED.

6.19. Now let $k \in \mathbb{Z}$ and $\gamma \in \Gamma^k$. One then defines a smooth function f_γ on G by putting

$$f_\gamma(g) = \gamma(g \cdot v_1) \tag{125}$$

Let χ_{λ_1} be the character on P_1 defined so that if $b \in P_1$ then $\eta(b)v_1 = \chi_{\lambda_1}(b)v_1$. We will also write $b^{\lambda_1} = \chi_{\lambda_1}(b)$. If $k \in \mathbb{Z}$ put $\chi^k_{\lambda_1}(b) = b^{k\lambda_1} = (\chi_{\lambda_1}(b))^k$. Clearly, from (125)

$$f_\gamma(gb) = b^{k\lambda_1} f_\gamma(g) \qquad \text{for } g \in G,\ b \in P_1 \tag{126}$$

If we let $\text{Ind}^G_{P_1}\chi_\lambda^{-k}$ be the space of all smooth functions f on G satisfying (126) then $\text{Ind}^G_{P_1}\chi_\lambda^{-k}$ is the (smooth) module for the induced representation $\text{ind}^G_{P_1}\chi_\lambda^{-k}$ where if $a \in G$ and $f \in \text{Ind}^G_{P_1}\chi_\lambda^{-k}$ and we write $(\text{ind}^G_{P_1}\chi_\lambda^{-k})(a)f = a \cdot f$ then $(a \cdot f)(g) = f(a^{-1}g)$. The following is an immediate consequence of Proposition 2.1.

Proposition 6.20. *The map*

$$\Gamma^k \to \text{Ind}^G_{P_1}\chi_{\lambda_1}^{-k}, \qquad \gamma \to f_\gamma \tag{127}$$

is an isomorphism of G-modules.

Now let $(\Gamma^k)'$ be the algebraic dual to Γ^k. Then $(\Gamma^k)'$ is a G (and also $\mathfrak{g}$) module where if $\langle\gamma,\gamma'\rangle$ is the pairing then $\langle g\cdot\gamma, g\cdot\gamma'\rangle = \langle\gamma,\gamma'\rangle$ for $g\in G$, $\gamma\in\Gamma^k, \gamma'\in(\Gamma^k)'$.

Now let $\delta\in(\Gamma^k)'$ be defined by putting

$$\langle\gamma,\delta\rangle = f_\gamma(e) = \gamma(v_1) \tag{128}$$

where $e\in G$ is the identity.

Lemma 6.21. *One has*

$$b\cdot\delta = \chi^k_{\lambda_1}(b)\delta \tag{129}$$

for any $b\in P_1$.

Proof. For $\gamma\in\Gamma^k$ then $\langle\gamma, b\delta\rangle = \langle b^{-1}\gamma,\delta\rangle = f_\gamma(b) = b^{k\lambda_1}f_\gamma(e) = b^{k\lambda_1}\langle\gamma,\delta\rangle$.
QED.

6.22. Now from the Iwasawa decomposition

$$\mathfrak{g} = \mathfrak{k}+\mathfrak{a}+\mathfrak{n} \tag{130}$$

One has, using the Birkhoff-Witt theorem, the direct sum

$$U(\mathfrak{g}) = U(\mathfrak{a})\oplus(\mathfrak{k}U(\mathfrak{g})+U(\mathfrak{g})\mathfrak{n}) \tag{131}$$

The second direct sum is the sum of the left ideal generated by $\mathfrak{k}$ and the right ideal generated by $\mathfrak{n}$. For any $u\in U(\mathfrak{g})$, let $p_u\in U(\mathfrak{a})$ be the component of u in $U(\mathfrak{a})$ relative to (131). We may regard $U(\mathfrak{a})$ as the ring of polynomial functions on $\mathfrak{a}'_{\mathbf{C}}$. Recall (see (120.5)) the definition of the spherical vector $1_k\in\Gamma^k_{0,0}$.

Proposition 6.23. *For any $u\in U(\mathfrak{g})$*

$$\langle 1_k, u\cdot\delta\rangle = p_u(k\lambda_1)\langle 1_k,\delta\rangle \tag{132}$$

Proof. It is obvious from (129) that $\mathfrak{n}\cdot\delta = 0$. Thus the left side of (132) vanishes if $u\in(\mathfrak{k}U(\mathfrak{g})+U(\mathfrak{g})\mathfrak{n})$. Hence

$$\langle 1_k, u\cdot\delta\rangle = \langle 1_k, p_u(k\lambda_1)\delta\rangle \tag{133}$$

by (129).
QED.

Now let $U(\mathfrak{g})^{\mathfrak{k}}$ be the centralizer of $\mathfrak{k}$ in $U(\mathfrak{g})$. Then by reductivity one has the direct sum

$$U(\mathfrak{g}) = U(\mathfrak{g})^{\mathfrak{k}} \oplus [\mathfrak{k}, U(\mathfrak{g})] \tag{134}$$

For any $u \in U(\mathfrak{g})$ let $u^\sharp$ be the projection of u in $U(\mathfrak{g})^{\mathfrak{k}}$ relative to (134). For any $\lambda \in \mathfrak{a}'_{\mathbb{C}}$ let

$$I_\lambda = \{u \in U(\mathfrak{g}) \mid p_{(vu)^\sharp}(\lambda) = 0 \quad \text{for all} \quad v \in U(\mathfrak{g})\} \tag{135}$$

and let $V_\lambda = U(\mathfrak{g})/I_\lambda$.

Let $\rho = \nu_1 + \nu_2 + \nu_3 + \nu_4 \in \mathfrak{a}'_{\mathbb{C}}$ using the notation of Proposition 6.1. Thus by Proposition 6.1

$$\rho = 3\lambda_1 + 2\lambda_2 + \lambda_3 \tag{136}$$

Let W be the Weyl group of $(\mathfrak{a}_{\mathbb{C}}, \mathfrak{g}_{\mathbb{C}})$. One knows that W is isomorphic to a semi-direct product

$$W \cong \Sigma_4 \times \mathbb{Z}_2^3 \tag{136.5}$$

where Σ_4 operates as the group of all permutations of $\{\lambda_1, ..., \lambda_4\}$ and $\mathbb{Z}_2^3$ is the group of sign changes $\lambda_i \to -\lambda_i$, $i \in Q \subseteq \{1,2,3,4\}$ where card Q is even. This clearly implies

$$-1 \in W \tag{136.6}$$

We recall

Theorem 6.24. *I_λ is a maximal left ideal in $U(\mathfrak{g})$ for any $\lambda \in \mathfrak{a}'_{\mathbb{C}}$ so that V_λ has the structure of an irreducible $U(\mathfrak{g})$ module. Furthermore, V_λ is spherical and the spherical vector is unique up to scalar multiple. Moreover, I_λ is the annihilator of the spherical vector and*

$$I_{\lambda_1} = I_{\lambda_2} \quad \textit{if and only if} \quad \lambda_1 + \rho \sim \lambda_2 + \rho \tag{137}$$

where $\sim$ denotes the conjugate with respect to W. Conversely, any irreducible spherical $U(\mathfrak{g})$ module is of the form V_λ for some $\lambda \in \mathfrak{a}'_{\mathbb{C}}$.

In addition, V_λ has a non-trivial and hence non-singular invariant (necessarily unique up to scalar multiple) Hermitian structure if and only if

$$-\overline{(\lambda + \rho)} \sim \lambda + \rho \tag{138}$$

Proof. See [K], Theorem 1.3.5 and Theorem 1.6.5. QED.

In [K] we defined an operation $\lambda \mapsto \lambda'$ in $\mathfrak{a}'_{\mathbb{C}}$. Note that in the present case

$$\lambda = \lambda' \tag{138.5}$$

because of (136.6).

Theorem 6.24 sets up a bijection (Harish-Chandra) between the set of equivalence classes of irreducible spherical $U(\mathfrak{g})$ modules and the Weyl group orbits in $\mathfrak{a}'_{\mathbf{C}}$. The orbit is determined as soon as one knows the character on $U(\mathfrak{g})^{\mathfrak{k}}$ defined by its action on the spherical vector. See [K] Theorem 1.3.1. For the case of V_λ one knows (see [K], Corollary 1.3.4) that the character is given by

$$u \to p_u(\lambda) \tag{139}$$

Let

$$U(\mathfrak{g}) \to U(\mathfrak{g}) \qquad u \to u^t \tag{139.5}$$

be the anti-automorphism of $U(\mathfrak{g})$ such that $x^t = -x$ for $x \in \mathfrak{g}$. By (138.5) all finite dimensional, and in particular spherical, irreducible representations, are self-contragredient. One therefore has

$$p_u = p_{u^t} \tag{139.6}$$

for $u \in U^{\mathfrak{k}}$ by (139). On the other hand, in the notation of (133) note that

$$\begin{aligned} \langle u \cdot 1_k, \delta\rangle &= \langle 1_k, u^t \cdot \delta\rangle \\ &= p_{u^t}(k\lambda_1)\langle 1_k, \delta\rangle \end{aligned} \tag{139.7}$$

by (133).

Returning to our spherical representation H, we have

Theorem 6.25. *The annihilator ξ in $U(\mathfrak{g})$ is $I_{-2\lambda_1}$. Moreover, H admits a unique $\mathfrak{g}$-invariant pre-Hilbert space structure $\langle\gamma_1|\gamma_2\rangle$ such that $\langle\xi|\xi\rangle = 1$.*

Proof. If $u \in U^{\mathfrak{k}}$ then, since $\langle\xi, \delta\rangle$ is clearly not zero,

$$u\xi = p_u(-2\lambda_1)\xi \tag{140}$$

by (139.6-7) since H_0 is clearly stable under $U^{\mathfrak{k}}$ and we are considering the case where $k = -2$. On the other hand by Theorem 6.24 and Theorem 6.18 the annihilator of ξ is I_λ for some $\lambda \in \mathfrak{a}'_{\mathbf{C}}$. But then by (139) one has $u\xi = p_u(\lambda)\xi$. Thus one has that $I_\lambda = I_{-2\lambda_1}$ and hence

$$H \cong V_{-2\lambda_1} \quad \text{as } U(\mathfrak{g}) - \text{modules} \tag{141}$$

But $-2\lambda_1 + \rho = \overline{-2\lambda_1 + \rho}$ and

$$-2\lambda_1 + \rho \sim 2\lambda_1 - \rho \tag{142}$$

by (136.6). Thus H has a non-singular $\mathfrak{g}$ invariant Hermitian inner product $\langle\gamma_1|\gamma_2\rangle$ which may be normalized so that $\langle\xi|\xi\rangle = 1$. Since the K-types H_n are

$\mathfrak{k}$-irreducible and orthogonal to each other, to prove that the inner product is positive definite, it suffices to exhibit just one vector $v_n \in H_n$ for each $n \in \mathbf{Z}_+$ such that $\langle v_n | v_n \rangle > 0$. But by definition of x_o, y_o one clearly has (see (101.5) and (102.5))

$$\langle x_o \cdot \gamma_1 | \gamma_2 \rangle = \langle \gamma_1 | y_o \cdot \gamma_2 \rangle \tag{143}$$

But then if $v_n = \dfrac{x_o^n \cdot}{n!} \cdot \xi$ one has $v_n \in H_n$ by (111.5) and

$$\begin{aligned}\left\langle \frac{x_o^n}{n!} \cdot \xi \,\middle|\, \frac{x_o^n}{n!} \cdot \xi \right\rangle &= \left\langle \xi \,\middle|\, \frac{y_o^n}{n!} \frac{x_o^n}{n!} \cdot \xi \right\rangle \\ &= \frac{1}{n+1}\end{aligned} \tag{144}$$

by (123). QED.

Let $\mathcal{H}$ be the Hilbert space completion of H relative to $\langle \gamma_1 | \gamma_2 \rangle$. By Harish-Chandra's well known result the representation π defines an irreducible unitary representation

$$\pi: G \to \text{ Unit } \mathcal{H} \tag{145}$$

Remark 6.26. Since $\mathcal{H}$ has a spherical vector and the center of G lies in K the representation π annihilates the center of G. In particular, π defines a unitary representation of $SO(4,4)_e$.

6.27. Let $A, B \in$ End Z_n (see (70.5)) and let $\alpha_{A,B} \in C^\infty(S_E \times S_F)$ be defined by putting $\alpha_{A,B}(a\epsilon_1, b\epsilon_5) =$ tr $A\rho_n(a^{-1})$ tr $B\rho_n(b^{-1})$ for any $a, b \in S$. (We are dropping the $n+1$ in (71)). Clearly, $\alpha_{A,B}(a\epsilon_1, b\epsilon_5) = \alpha_{A,B}(-a\epsilon_1, -b\epsilon_1)$ and hence by Lemma 5.9 there exists a unique element $\gamma_{A,B} \in H_n$ such that

$$\gamma_{A,B}(a\epsilon_1, b\epsilon_5) = \text{ tr } A\rho_n(a^{-1}) \text{ tr } B\rho_n(b^{-1}) \tag{146}$$

for all $a, b \in S$. It is also clear that H_n is spanned by all such $\gamma_{A,B}$.

Since Z_n has an S-invariant Hermitian structure one has the operation $A \to A^*$ on End Z_n. By Peter-Weyl the relation (73) becomes for $A, B \in$ End Z_n

$$\int_{S_E} \varphi_A \overline{\varphi}_B \mu_E = (n+1) \text{ tr } AB^* \tag{146.5}$$

The existence of the unitary structure in $\mathcal{H}$ on H, given by Theorem 6.25, may be established by the general theory of the unitary dual of $SO(4,4)_e$. However, using results obtained here we can now be very explicit about that structure.

Theorem 6.28 *Let $n \in \mathbb{Z}_+$ and let $\gamma_1, \gamma_2 \in H_n$. Then*

$$\langle \gamma_1 \mid \gamma_2 \rangle = (n+1) \int\limits_{S_E \times S_F} \gamma_1 \overline{\gamma}_2 \mu \tag{147}$$

In particular, if $A_1, A_2, B_1, B_2 \in$ End Z_n then

$$\langle \gamma_{A_1,B_1} \mid \gamma_{A_2,B_2} \rangle = \frac{1}{n+1} \text{tr } A_1 A_2{}^* \text{ tr } B_1 B_2{}^* \tag{148}$$

Proof. Since H_n is K-irreducible one has

$$\langle \gamma_1 \mid \gamma_2 \rangle = c \int\limits_{S_E \times S_F} \gamma_1 \overline{\gamma}_2 \mu \tag{149}$$

for some scalar c. To determine c for $\gamma_1, \gamma_2 \in H_n$ we have only to determine c for $\gamma_1 = \gamma_2 = \frac{x_o^n}{n!} \cdot \xi$. But now $\frac{x_o^n}{n!} \cdot \xi = (-1)^n (s_1 s_3)^n$ on $S_E \times S_F$ by (109). Thus the right side of (147) becomes $\frac{1}{n+1}$ by (115). However, the left side of (147) is $\frac{1}{n+1}$ by (144). This proves $c = (n+1)$. But now by (146.5) the right side of (147) for $\gamma_1 = \gamma_{A_1,B_1}$ and $\gamma_2 = \gamma_{A_2,B_2}$ is $\frac{1}{n+1}$ tr $A_1 A_2{}^*$ tr $B_1 B_2{}^*$. Hence (148) follows from (147). QED.

7. The effect of triality on H. Now the action of Σ_3 on $\mathfrak{g}$ extends to an action of Σ_3 on $U(\mathfrak{g})$. We note that $\mathfrak{n}$, by (8) is stable under Σ_3. On the other hand $\mathfrak{k}$ is stable under Σ_3 by Proposition 3.1 Of course, $\mathfrak{a}$ is stable under Σ_3 by definition.

Proposition 7.1. *For any $u \in U(\mathfrak{g})$ and $\sigma \in \Sigma_3$ one has*

$$\sigma p_u = p_{\sigma u} \tag{150}$$

Proof. This is immediate since Σ_3 stabilizes the summands of (131). QED.

Lemma 7.2. *The maximal left ideal $I_{-2\lambda_1}$ is stable under Σ_3.*

Proof. Assume $u \in I_{-2\lambda}$. Then $p_{(vu)^\sharp}(-2\lambda_1) = 0$ for all $v \in U(\mathfrak{g})$. But then $(\sigma v \sigma u)^\sharp = \sigma (vu)^\sharp$ since the decomposition (134) is clearly Σ_3 stable. Thus by (150)

$$\begin{aligned} p_{(\sigma v \sigma u)^\sharp}(-2\lambda_1) &= (\sigma(p_{(vu)^\sharp}))(-2\lambda_1) \\ &= p_{(vu)^\sharp}(-\sigma^{-1} 2\lambda_1) \end{aligned} \tag{151}$$

But now by Proposition 6.1 one has $\lambda_1 = \nu_1$ and $\sigma\nu_1 = \nu_1$, ν_2 or ν_3 by (8). To prove the lemma we must show that $p_{(vu)^\sharp}(-2\nu_i) = 0$, for $i = 1, 2, 3$. That is, by (137) it suffices to prove that

$$\rho - 2\nu_1 \sim \rho - 2\nu_2 \sim \rho - 2\nu_3 \tag{152}$$

But now $\nu_2 = \frac{1}{2}(\lambda_1 + \lambda_2 + \lambda_3 + \lambda_4)$ and $\nu_3 = \frac{1}{2}(\lambda_1 + \lambda_2 + \lambda_3 - \lambda_4)$. Hence by (136) $\rho - 2\nu_2 = 2\lambda_1 + \lambda_2 - \lambda_4$ and $\rho - 2\nu_3 = 2\lambda_1 + \lambda_2 + \lambda_4$. However, $\rho - 2\nu_1 = \rho - 2\lambda_1 = \lambda_1 + 2\lambda_2 + \lambda_3$. But clearly

$$\lambda_1 + 2\lambda_2 + \lambda_3 \sim 2\lambda_1 + \lambda_2 - \lambda_4 \sim 2\lambda_1 + \lambda_2 + \lambda_4 \tag{153}$$

upon recalling the action of (136.5). QED.

As a consequence one has

Theorem 7.3. *There exists a unique action of Σ_3 on H such that* (1) $\sigma\xi = \xi$ *for any* $\sigma \in \Sigma_3$ *and* (2)

$$\sigma(u \cdot \gamma) = \sigma u \cdot \sigma\gamma \tag{154}$$

for $u \in U(\mathfrak{g})$ and $\gamma \in H$.

Proof. This is immediate from Lemma 7.2 and (141). QED.

We can be very explicit about the action of Σ_3 on H. However, we restrict the explicitness to the element $\tau \in \Sigma_3$. We note that it carries the product of traces (146) to a trace of a product.

Theorem 7.4. *The action Σ_3 on H is unitary. Furthermore, each K-type H_n is stable under Σ_3. Moreover, for $A, B \in$ End Z_n one has, for $x, y \in S$*

$$(\tau \cdot \gamma_{A,B})(x \cdot \epsilon_1, y \cdot \epsilon_5) = \text{tr } A^t \rho_n(x) B \rho_n(y^{-1}) \tag{155}$$

Proof. It is immediate from (154) that $\langle \gamma_1 | \gamma_2 \rangle_\sigma = \langle \sigma\gamma_1 | \sigma\gamma_2 \rangle$ for $\gamma_1, \gamma_2 \in H$ and $\sigma \in \Sigma_3$ is a $\mathfrak{g}$-invariant Hermitian inner product on H. By uniqueness, since $\sigma\xi = \xi$, it follows that $\langle \gamma_1 | \gamma_2 \rangle_\sigma = \langle \gamma_1, \gamma_2 \rangle$ so that the action of Σ_3 on H is unitary.

Now the action of Σ_3 on K induces an action of Σ_3 on $\widehat{K}$. But an element $\sigma \in \Sigma_3$ must permute the factors of (27). Hence $(n, n, n, n) \in \widehat{K}$ is necessarily fixed by σ. Thus H_n is stabilized by σ.

Now since Z_n has an S-invariant bilinear form (w_1, w_2) we may identify End Z_n with $Z_n \otimes Z_n$ where $w_1 \otimes w_2 \in$ End Z_n is such that if $w \in Z_n$ then $(w_1 \otimes w_2)(w) = (w, w_2)w_1$. One has

$$\text{tr } w_1 \otimes w_2 = (w_1, w_2) \tag{155.4}$$

and

$$(w_1 \otimes w_2)^t = w_2 \otimes w_1 \tag{155.5}$$

Also, if $a, b \in S$ then

$$\rho_n(a) \cdot w_1 \otimes w_2 \cdot \rho_n(b^{-1}) = aw_1 \otimes bw_2 \tag{156}$$

But now if $a, b, c, d \in S$ so that $(a, b, c, d) \in K$ and $A, B \in \text{End } Z_n$ then clearly

$$(a, b, c, d) \cdot \gamma_{A,B} = \gamma_{\rho_n(a)A\rho_n(b^{-1}),\, \rho_n(c)B\rho_n(d^{-1})} \tag{157}$$

using (28). In particular, if $w_1, w_2, w_3, w_4 \in Z_n$

$$(a, b, c, d) \cdot \gamma_{w_1 \otimes w_2, w_3 \otimes w_4} = \gamma_{aw_1 \otimes bw_2, cw_3 \otimes dw_4} \tag{158}$$

It is clear from (158) that there exists non-zero $w_1, w_2, w_3, w_4 \in Z_n$ so that $\gamma_{w_1 \otimes w_2, w_3 \otimes w_4} \in H_n$ is a highest weight vector. In fact by (31) and Proposition 6.4 we may assume that $w_2 = w_3 = w$ for some $w \in Z_n$.

But then by (111.5) there exists a scalar c so that

$$\gamma_{w_1 \otimes w, w \otimes w_4} = cx_o^n \cdot \xi \tag{159}$$

But then x_o and ξ are τ-invariant (see Proposition 6.5). Thus $\gamma_{w_1 \otimes w, w \otimes w_4}$ is τ-invariant and hence clearly by (154), (31) and (159) one has

$$\tau \cdot \gamma_{aw_1 \otimes bw, cw \otimes dw_4} = \gamma_{aw_1 \otimes cw, bw \otimes dw_4}$$

Thus if $A = aw_1 \otimes bw$ and $B = cw \otimes dw_4$ one has

$$\begin{aligned} (\tau \cdot \gamma_{A,B})(x\epsilon_1, y\epsilon_2) &= \langle aw_1, xcw \rangle \langle bw, ydw_4 \rangle \\ &= \langle x^{-1}aw_1, cw \rangle \langle bw, ydw_4 \rangle \end{aligned} \tag{160}$$

by (146) and (156).

But the right side of (160) is, by (155.4-5), tr $A^t \rho_n(x) B \rho_n(y^{-1})$. But decomposable elements of the form A and B independently span End Z_n so that one has (155) for all $A, B \in$ End Z_n. QED.

References

[C], C. Chevalley, *The Algebraic Theory of Spinors,* Columbia University Press, 1954.

[G], D. Garfinkle, MIT Thesis.

[K], B. Kostant, "On the Existence and Irreducibility of Certain Series of Representations", *Lie Groups and Their Representations,* Edited by I.M. Gelfand, Halsted Press, Wiley and Sons, 1975, 231-329.

[KR], B. Kostant and S. Rallis, "Orbits and Representations Associated with Symmetric Spaces", Amer. J. Math., vol. 93, 1971, 753-809.

On E_{10}

V.G. Kac

Massachusetts Institute of Technology[1]

R.V. Moody

University of Saskatchewan [2]

M. Wakimoto

Mie University[3]

1. Introduction. The simply-laced exceptional simple Lie algebras form part of an infinite series of Kac-Moody Lie algebras

$$A_2 \times A_1, \quad A_4, \quad D_5, \quad E_6, \quad E_7, \quad E_8, \quad E_9, \quad E_{10}, \ldots$$

with Coxeter-Dynkin diagrams $E_k, k \geq 3$, of the form

One sees by induction that the determinant of the corresponding generalized Cartan matrix, and hence also of the corresponding root lattice, is $9 - k$. In particular there are

[1] Work supported by NSF grant DMS 8508953

[2] Work supported by a research grant from the National Sciences and Engineering Research Council of Canada and by the Tata Institute for Fundamental Research in Bombay

[3] Work supported by the Massechusetts Institute of Technology

K. Bleuler and M. Werner (eds.), Differential Geometrical Methods in Theoretical Physics, 109–128.

precisely two unimodular root lattices: E_8, which is positive definite, and E_{10}, which is Minkowskian (signature $(+^9, -1)$ *), and indeed they are the unique even unimodular lattices of these types with these dimensions [Se]. The Lie algebras E_8, E_9, E_{10} are thus singled out, E_8 and E_{10} on account of their root lattices and E_9 because it is the affine Lie algebra $E_8^{(1)}$ and occurs as a subalgebra of E_{10} containing E_8.

Our present interest in E_{10} is in trying to understand its root multiplicities. We recall the rootspace decomposition

$$\underline{\hat{g}} = \underline{g}(E_{10}) = \underline{\hat{h}} \oplus \bigoplus_{\alpha \in \hat{\Delta}} \underline{\hat{g}}^{\alpha}$$

where $\hat{\Delta}$ is the set of roots of $\underline{\hat{g}}$ relative to $\underline{\hat{h}}$. Thanks to the conjugacy theorem [P-K] any two such decompositions are conjugate by the adjoint group $\hat{G}_{ad}$ of $\underline{\hat{g}}$, and hence the root multiplicities, $mult(\alpha) = dim\underline{\hat{g}}^{\alpha}$, are invariants of $\underline{\hat{g}}$. It is remarkable how little we know about root multiplicities except in the finite and affine cases. There are two formulas which give root multiplicities as complicated sums, and hence provide algorithms for computing root multiplicities, but these are otherwise very unrevealing [B-M] and [P] (the proof of the latter may be found in [Ka]).

In [F-F] Feingold and Frenkel considered the hyperbolic Lie algebra given by $\circ \Longleftrightarrow \circ - \circ$ and gave explicit formulas for the root multiplicities for roots of level $0, 1$, and 2 (see §2 for definitions). Since levels 0 and 1 were already understood their contribution was in understanding the structure of the level 2 module of level 2 root spaces relative to the affine subalgebra of type $A_1^{(1)}$ under the adjoint action.

In this paper we give the root multiplicities for levels $0, 1$, and 2 in E_{10}. The level 2 module is isomorphic to $L(\Lambda_0) \wedge L(\Lambda_0)/L(\Lambda_1)$, exactly analogous to the case considered in [F-F]. We establish this fact as an immediate consequence of the Gabber-Kac theorem [G-K].

The task is then to compute the characters of $L(\Lambda_0) \wedge L(\Lambda_0)$ and $L(\Lambda_1)$ (their ratio

* E_k is Minkowskian for all $k > 9$ since $detE_k < 0$ and $A_{k-1} \subset E_k$.

is already known [K-W1]). In the end this reduces to computing the three level 2 string functions of $\underline{g} = \underline{g}(E_9)$. It is remarkable that two of these are equal. This fact ultimately shows that for level 2 roots $dim\hat{\underline{g}}^\alpha$ depends only on the square length $(\alpha|\alpha)$ of α. Precisely, we prove the following. Define $p^{(8)}(n)$ and $\xi(n)$, by

$$\frac{1}{\phi(q)^8} = \sum_{n\geq 0} p^{(8)}(n) q^n$$

$$(1.1) \qquad \frac{1}{\phi(q)^8}[1 - \frac{\phi(q^2)}{\phi(q^4)}] = \sum_{n\geq 0} \xi(n) q^n \ .$$

Then for $\alpha \in \hat{\Delta}$

$$(1.2) \qquad dim\hat{\underline{g}}^\alpha = \begin{cases} p^{(8)}\left(1 - \frac{(\alpha|\alpha)}{2}\right) & \text{if } \alpha \text{ is of level 0 or 1} \\ \xi\left(3 - \frac{(\alpha|\alpha)}{2}\right) & \text{if } \alpha \text{ is of level 2 .} \end{cases}$$

Since the second series expands as

$$q^2\{\sum p^{(8)}(n)q^n + q^4 \sum p^{(8)}(n)q^n - q^6 \sum p^{(8)}(n)q^n + \ldots\}$$

we see that $\xi(6) = p^{(8)}(4) + 1 > p^{(8)}(4)$ which disproves the conjecture in [Fr] that for hyperbolic Lie algebras

$$dim\hat{\underline{g}}^\alpha \leq p^{(rank-2)}(1 - \frac{(\alpha|\alpha)}{2}).$$

In fact, in all cases that we know (see the Table at the end of the paper) the reverse inequality occurs.

One of the interesting features of the string functions of level 2 of E_9 is the high degree of modular invariance that appears within them. We use this to prove the key identities which allow us to compute the characters of $L(\Lambda_1)$ and $L(\Lambda_0) \wedge L(\Lambda_0)$. One of the authors [RVM] would like to thank D.-N. Verma for some helpful discussions regarding modular forms.

2. Notation. In general we use the notation of [Ka]. Since the Cartan matrix of $\hat{\underline{g}} = \underline{g}(E_{10})$ is nonsingular its Cartan subalgebra $\hat{\underline{h}}$ and dual space $\hat{\underline{h}}^*$ afford a realization for the Cartan matrix of E_9. Thus denoting the Lie algebras for E_8, E_9, E_{10} by $\dot{\underline{g}}$, $\underline{g}$, and $\hat{\underline{g}}$ we have

$$\dot{\underline{g}} \subset \underline{g} \subset \hat{\underline{g}}$$

with Cartan subalgebras

$$\dot{\underline{h}} \subset \underline{h} = \hat{\underline{h}}.$$

We denote the corresponding root lattices by $\dot{Q}$, Q, and $\hat{Q}$ respectively. We adopt the following indexing convention for E_{10}:

$$\begin{matrix} & & & & & & 8 & & \\ -1 & 0 & 1 & 2 & 3 & 4 & 5 & 6 & 7 \end{matrix}$$

so that $\hat{\underline{g}}$ has simple roots α_i, $i = -1, \ldots, 8$.

Let $A = (a_{ij})_{i,j=-1}^{8}$ be the Cartan matrix of E_{10}. Introduce a bilinear form $(\cdot|\cdot)$ on $\hat{\underline{h}}^*$ by

$$(\alpha_i|\alpha_j) = a_{ij} \qquad i, j, = 1, 0, \ldots, 8 \ ,$$

and identify $\hat{\underline{h}}$ and $\hat{\underline{h}}^*$ via this form.

Let

$$\gamma_1 : \ = \sum_{i=1}^{8} n_i \alpha_i \tag{2.1}$$

be the highest root of E_8 and set $n_0 = 1$. Then the null root $\delta \in Q$ of E_9 is

$$\delta = \sum_{i=0}^{8} n_i \alpha_i$$

and the center of $\underline{g}$ is $\mathbf{C}\delta$. Explicitly the n_i are given by

$$\begin{matrix} & & & & & 3 & & \\ 1 & 2 & 3 & 4 & 5 & 6 & 4 & 2 \end{matrix} \tag{2.2}$$

If $\alpha \in \hat{Q}$, $\quad \alpha = a_{-1}\alpha_{-1} + a_0\alpha_0 + \ldots$, then the level of α is defined as $(\alpha|\delta) = -a_{-1}$. We define

$$\hat{\Delta}_{[k]} := \{\alpha \epsilon \hat{\Delta} | (\alpha|\delta) = -k\}$$

and

$$\hat{\underline{g}}_{[k]} := \bigoplus_{\alpha \epsilon \hat{\Delta}_{[k]}} \hat{\underline{g}}^{\alpha} \tag{2.3}$$

so that we obtain a $\mathbb{Z}$-gradation

$$\hat{\underline{g}} = \bigoplus_{k \epsilon \mathbb{Z}} \hat{\underline{g}}_{[k]} \ . \tag{2.4}$$

The fundamental weights Λ_i, $i \geq -1$, of $\hat{\underline{g}}$ are given by the equations

$$(\Lambda_i | \alpha_j) = \delta_{ij} \qquad i, j = -1, 0, \ldots, 8 \ .$$

Let $A^{-1} = -(\bar{a}_{ij})^{8}_{i,j=-1}$, so we have

$$\Lambda_i = -\sum_j \bar{a}_{ij} \alpha_j, \quad (\Lambda_i | \Lambda_j) = -\bar{a}_{ij} \ .$$

A calculation gives

$$(\bar{a}_{ij}) = \begin{pmatrix} 0 & 1 & 2 & 3 & 4 & 5 & 6 & 4 & 2 & 3 \\ 1 & 2 & 4 & 6 & 8 & 10 & 12 & 8 & 4 & 6 \\ 2 & 4 & 6 & 9 & 12 & 15 & 18 & 12 & 6 & 9 \\ 3 & 6 & 9 & 12 & 16 & 20 & 24 & 16 & 8 & 12 \\ 4 & 8 & 12 & 16 & 20 & 25 & 30 & 20 & 10 & 15 \\ 5 & 10 & 15 & 20 & 25 & 30 & 36 & 24 & 12 & 18 \\ 6 & 12 & 18 & 24 & 30 & 36 & 42 & 28 & 14 & 21 \\ 4 & 8 & 12 & 16 & 20 & 24 & 28 & 18 & 9 & 14 \\ 2 & 4 & 6 & 8 & 10 & 12 & 14 & 9 & 4 & 7 \\ 3 & 6 & 9 & 12 & 15 & 18 & 21 & 14 & 7 & 10 \end{pmatrix} \tag{2.5}$$

Note that

$$\begin{aligned} \Lambda_{-1} &= -\delta \\ \Lambda_0 &= -\alpha_{-1} - 2\delta \\ \Lambda_1 &= -2\alpha_{-1} - \alpha_0 - 3\delta = 2\Lambda_0 + \delta - \alpha_0 = 2\Lambda_0 + \gamma_1 \\ \Lambda_7 &= 2\Lambda_0 + \gamma_7 \end{aligned} \tag{2.6}$$

where

$$\gamma_1 = \begin{pmatrix} & & & & 3 & & \\ 2 & 3 & 4 & 5 & 6 & 4 & 2 \end{pmatrix} \in \dot{Q} \qquad (see(2.1))$$

$$\gamma_7 = \begin{pmatrix} & & & & 5 & & \\ 2 & 4 & 6 & 8 & 10 & 7 & 4 \end{pmatrix} \in \dot{Q}$$

Using transformations of the Weyl group (which preserve root multiplicities), we can bring every negative imaginary root into the set

$$C = \left\{ \sum_{i=-1}^{8} k_i \Lambda_i \;\middle|\; k_i \in \mathbb{Z}_+ \right\} .$$

Thus, it suffices to determine multiplicities of the roots from the set C. (Recall that multiplicities of all real roots are 1.)

Note that

(2.7) $$\hat{\Delta}_{[0]} = \Delta, \quad \text{the set of roots of} \quad \hat{\underline{g}}_{[0]} = \underline{g} = E_9 ,$$

(2.8) $$\hat{\Delta}_{[1]} \cap C = \{\Lambda_0 + k_{-1}\Lambda_{-1} = -\alpha_{-1} - (k_{-1}+2)\delta | k_{-1} \in \mathbb{Z}_+\} ,$$

(2.9) $$\hat{\Delta}_{[2]} \cap C = \{2\Lambda_0 + k_{-1}\Lambda_{-1}, \quad \Lambda_1 + k_{-1}\Lambda_{-1}, \quad \Lambda_7 + k_{-1}\Lambda_{-1} | k_{-1} \in \mathbb{Z}_+\} ,$$

(2.10)
$$|2\Lambda_0 + k_{-1}\Lambda_{-1}|^2 = -4k_{-1} - 8,$$
$$|\Lambda_1 + k_{-1}\Lambda_{-1}|^2 = -4k_{-1} - 6,$$
$$|\Lambda_7 + k_{-1}\Lambda_{-1}|^2 = -4k_{-1} - 4 ,$$
$$|\gamma_1|^2 = 2, \quad |\gamma_7|^2 = 4 .$$

The root multiplicities of $\hat{\underline{g}}_{[0]} = \underline{g}$ are, of course, well-known: all imaginary roots are non-zero integral multiples of δ and their multiplicities are equal to 8. Since $mult(\alpha) = mult(-\alpha)$, we may restrict our attention to $\hat{\Delta}_{[k]}$ with $k > 0$.

To proceed further we look at $\hat{\underline{g}}$ as a $\underline{g}$-module under the adjoint representation. It is clear that $\hat{\underline{g}}_{[k]}$ is a $\underline{g}$-submodule of $\hat{\underline{g}}$. For $k > 0$, $\hat{\underline{g}}_{[k]}$ is an integrable $\underline{g}$-module in the category O (since $-k\alpha_{-1} + \sum_{i=0}^{8} a_i\alpha_i$ cannot be a root if any $a_i > 0$) and hence completely reducible into highest weight modules for $\underline{g}$ [Ka, Chapter 10]. For example $\hat{\underline{g}}_{[1]}$ is irreducible since it is generated by f_{-1}, which is a highest weight vector relative to $\underline{g}$. Indeed any commutator $[f_{i_k}, [\ldots, [f_{i_2}, f_{i_1}]\ldots]]$ in which exactly one index is -1 can be rewritten as a linear combination of commutators

$$[f_{j_{k-1}}, [\ldots, [f_{j_1}, f_{-1}]\ldots]]$$

where $j_1, \ldots, j_{k-1} \in \{0, \ldots, 8\}$.

Now since $-\alpha_{-1} = \Lambda_0 + 2\delta$, $\hat{\underline{g}}_{[1]}$ affords the basic representation $L(\Lambda_0 + 2\delta)$. But according to [Ka, Chapter 12] we have

$$(2.11) \qquad \sum_{k \in \mathbb{Z}_+} mult(\Lambda_0 + 2\delta - k\delta) q^k = \phi(q)^{-8}$$

where

$$(2.12) \qquad \phi(q) := \prod_{j=1}^{\infty} (1 - q^j) \; .$$

Denoting by $p^{(8)}(n)$ the coefficient of q^n in $\phi(q)^{-8}$ and using (2.7) and (2.8) we obtain

$$(2.13) \qquad mult(\alpha) = p^{(8)}\big(1 - \frac{1}{2}(\alpha|\alpha)\big) \qquad \text{for all} \quad \alpha \in \Delta_{[0]} \cup \Delta_{[1]} \; .$$

A simple consequence of string function identity (2.11) is the character formula

$$(2.14) \qquad chL(\Lambda_0 + 2\delta) = \sum_{\lambda \in \Lambda_0 + Q} p^{(8)}\Big(1 - \frac{(\lambda|\lambda)}{2}\Big) e^{\lambda} \; .$$

Recall that if a weight λ lies in the weight system $P(\Lambda)$ of an irreducible hightest weight module $L(\Lambda)$ then so do all the weights $\lambda - k\delta$, $k \in \mathbb{Z}_+$, and λ is called *maximal* if $\lambda+\delta\notin P(\Lambda)$. The weight string $\lambda,\lambda-\delta,\lambda-2\delta,\ldots$ headed by a maximal weight contributes the terms $e^{\lambda}b_{\lambda}^{\Lambda}$, where

$$b_{\lambda}^{\Lambda}:=\sum_{k\geq 0} mult(\lambda-k\delta)e^{-k\delta}\ ,$$

to the character of $L(\Lambda)$. Replacing $e^{-\delta}$ by q we obtain the *string function:*

$$b_{\lambda}^{\Lambda}(q)=\sum_{k\geq 0} mult(\lambda-k\delta)q^{k}\ . \tag{2.15}$$

These become genuine functions later on when we set $q=e^{\pi i\tau}, \tau\in\mathcal{H}=$ upper half plane in $\mathbf{C}$. (Often string functions are multiplied by some power of q to render them as modular forms).

Since weight multiplicities are W-invariant, we need only determine string functions b_{λ}^{Λ} for λ dominant (and maximal). There are only finitely many such λ, a fact which is easily seen once one realizes that all weights in $P(\Lambda)$ have the same level, i.e. $(\lambda|\delta)=(\Lambda|\delta)$ for all $\lambda\in P(\Lambda)$. In the case of E_9 and $\Lambda=\Lambda_0+2\delta$, $\Lambda_0+2\delta$ itself is the only dominant maximal weight and the string function $b_{\Lambda_0+2\delta}^{\Lambda_0+2\delta}(q)$ is given by (2.11), whence (2.14) (it becomes a modular form when multiplied by $q^{-1/3}$).

3. The structure of $\hat{\underline{g}}_{[2]}$. Recall that $\hat{\underline{g}}=\oplus_{k\in\mathbb{Z}}\hat{\underline{g}}_{[k]}$ is a $\mathbb{Z}$-graded Lie algebra. Put $\hat{\underline{g}}_{[0]}=\underline{g}$, $\hat{\underline{g}}_{[1]}=V$ and $\hat{\underline{g}}_{[-1]}=\overline{V}$ for short. It is clear that $\hat{\underline{g}}$ is generated by its *local part* $\overline{V}\oplus\underline{g}\oplus V$. Let $\tilde{\underline{g}}=\oplus_{j\in\mathbb{Z}}\tilde{\underline{g}}_j$ be the maximal $\mathbb{Z}$-graded Lie algebra with this local part [K], so that

$$\tilde{\underline{g}}_{-1}\oplus\tilde{\underline{g}}_0\oplus\tilde{\underline{g}}_1\simeq\overline{V}\oplus\underline{g}\oplus V$$

and $\tilde{\underline{g}}_{\pm}:=\oplus_{j>0}\tilde{\underline{g}}_{\pm j}$ is a free Lie algebra on a basis of $\tilde{\underline{g}}_{\pm 1}$ [K]. Then $\hat{\underline{g}}$ is a quotient of $\tilde{\underline{g}}$ by a sum of two $\mathbb{Z}$-graded ideals $I_{\pm}=\oplus_{j\geq 2}I_{\pm j}$ of $\tilde{\underline{g}}_{\pm}$, where $I_{\pm j}\subset\tilde{\underline{g}}_{\pm j}$.

We shall identify $\tilde{\underline{g}}_0$ with $\underline{g}$ and $\tilde{\underline{g}}_1$ with V, so that all $\tilde{\underline{g}}_j$ are $\underline{g}$-modules under the adjoint representation. Note that the element

$$v^{+}:=[f_{-1},[f_{-1},f_0]]$$

is a highest weight vector of the $\underline{g}$-module $\tilde{\underline{g}}_2$ with highest weight (see §2.2) $\Lambda_1 + 3\delta$. Since V is an integrable $\underline{g}$-module so is $[V, V]$ and hence v^+ generates an irreducible highest weight $\underline{g}$-module $U_+ = L(\Lambda_1 + 3\delta)$ in $\tilde{\underline{g}}_2$. Similarly, we have $v^- = [e_{-1}, [e_{-1}, e_0]]$, which is a lowest weight vector of the submodule $U_- = L^*(\Lambda_1 + 3\delta)$ of $\tilde{\underline{g}}_{-2}$. We can now easily prove the following

Proposition 3.1 The ideal $I_\pm$ of the (free) Lie algebra $\tilde{\underline{g}}_\pm$ is generated by the subspace $U_\pm$. In particular, $I_2 = U_+, \quad I_3 = [V, U_+], \quad I_4 = [V, [V, U_+]] + [U_+, [V, V]]$, etc.

Proof. According to the Gabber-Kac theorem, $\hat{\underline{g}}$ is a quotient of a free Lie algebra L_{30} on generators $e_i,\ f_i,\ \alpha_i, \quad i = -1, 0, \ldots, 8$, by the unique *maximal* ideal J generated by certain well-known elements. The kernel of the canonical map $L_{30} \to \tilde{\underline{g}}$ contains all these elements except for v^+ and v^-. Let $\tilde{I}_\pm$ be the ideals of $\tilde{\underline{g}}_\pm$ generated by $U_\pm$. Then $\tilde{I}_+ + \tilde{I}_-$ is an ideal of $\tilde{\underline{g}}$, which contains v^+ and v^-. Thus, considering $L_{30} \to \tilde{\underline{g}} \to \tilde{\underline{g}}/(\tilde{I}_+ + \tilde{I}_-)$, the kernel of the composite homomorphism $L_{30} \to \tilde{\underline{g}}/(\tilde{I}_+ + \tilde{I}_-)$ contains J which is maximal. We conclude that $\tilde{\underline{g}}/(\tilde{I}_+ + \tilde{I}_-) = \hat{\underline{g}}$ and $\tilde{I}_\pm = I_\pm$. ∎

Corollary 3.2. We have the following isomorphism of $\underline{g}$-modules:

$$\hat{\underline{g}}_{[2]} \simeq (L(\Lambda_0 + 2\delta) \wedge L(\Lambda_0 + 2\delta))/L(\Lambda_1 + 3\delta)\ .$$

Remarks.

1. The Proposition holds equally well for $\underline{g}(A)$ and $\underline{g}(\hat{A})$ given by any pair of generalized Cartan matrices

$$A = (A_{ij})_{0 \le i,j \le \ell}, \quad \hat{A} = (A_{ij})_{-1 \le i,j \le \ell}$$

where

$$A_{-1,i} = 0 \quad \text{for} \quad i > 0 \qquad \text{and} \qquad A_{-1,0} = -1\ .$$

2. The complete structure of $\hat{\underline{g}}_{[k]}$ for $k \ge 3$ as $\underline{g}$-modules seems to be unknown.

3. Let $\hat{\underline{g}}_+ = \oplus_{k>0}\hat{\underline{g}}_{[k]}$. Then $U(\hat{\underline{g}}_+)$ is a quadratic algebra, i.e. a quotient of the tensor algebra $T(V)$ by a 2-sided ideal generated by $U_+ \subset T^2(V)$. Its dual quadratic algebra, the algebra $T(\overline{V})$ divided by the 2-sided ideal generated by $U_+^\perp \subset T^2(\overline{V})$, is isomorphic to $\wedge(\overline{V})$ divided by the ideal generated by the complement to the highest component in $\wedge^2\overline{V}$. Note the similarity of the latter algebra with the coordinate ring of the orbit of the highest weight vector of $L(\Lambda_0)$ (= the space of based loops on the compact Lie group of type E_8), which is the quotient of $S(\overline{V})$ by the ideal generated by the complement to the highest component in $S^2\overline{V}$ (cf. Kac-Peterson, Progress in Math. 36, Birkhäuser, pp 141–166). A natural question then arises: is $U(\hat{\underline{g}}_+)$ a Koszul algebra?

4. The string functions of $L(\Lambda_1)$. The dominant weights of level 2 for E_9 are, modulo δ, $2\Lambda_0$, Λ_1, and Λ_7. Using (2.6) we find that

$$\Lambda_1 - \delta = 2\Lambda_0 - \alpha_0 \qquad \text{is a maximal weight of} \quad L(\Lambda_0) \wedge L(\Lambda_0)$$

and

$$2\Lambda_0 \quad \text{and} \quad \Lambda_7 - \delta \qquad \text{are maximal weights of} \quad L(\Lambda_1)$$

(for this last pair use the explicit description of weights given in [Ka, Prop. 12.5]).

The W-orbit of Λ_0 is $\{\Lambda_0 - \frac{1}{2}|\alpha|^2\delta + \alpha \mid \alpha \in \dot{Q}\}$ [Ka] from which we have the well-known factorization

$$chL(\Lambda_0) = \frac{1}{\phi(q)^8} \sum_{\alpha\in\dot{Q}} e^{\Lambda_0 - \frac{1}{2}|\alpha|^2\delta + \alpha} \qquad (q = e^{-\delta})\ . \tag{4.1}$$

Since

$$ch(L(\Lambda_0) \wedge L(\Lambda_0)) = \frac{1}{2}\Big\{[chL(\Lambda_0)(e^{\lambda})]^2 - chL(\Lambda_0)(e^{2\lambda})\Big\}$$

we have

$$\begin{aligned} ch(L(\Lambda_0) \wedge L(\Lambda_0)) = \frac{1}{2}\Big\{ & \frac{1}{\phi(q)^{16}} \sum_{\beta,\gamma\in\dot{Q}} e^{2\Lambda_0 - \frac{1}{2}(|\beta|^2 + |\gamma|^2)\delta + \beta + \gamma} \\ & - \frac{1}{\phi(q^2)^8} \sum_{\alpha\in\dot{Q}} e^{2\Lambda_0 - |\alpha|^2\delta + 2\alpha}\Big\}\ . \end{aligned} \tag{4.2}$$

Now one knows [K-W 1] that

$$L(\Lambda_0)\wedge L(\Lambda_0)\simeq L(\Lambda_1)\otimes V\Big(\frac{1}{2},\frac{1}{16}\Big)$$

where $V(\frac{1}{2},\frac{1}{16})$ refers to the irreducible representation of the Virasoro algebra with character $\phi(q^2)/\phi(q)$. Thus we arrive at

$$ch(L(\Lambda_0)\wedge L(\Lambda_0)) = q\frac{\phi(q^2)}{\phi(q)}chL(\Lambda_1) \tag{4.3}$$

(the extra factor $q=e^{-\delta}$ occurs because $\Lambda_1-\delta$ is the highest weight of $L(\Lambda_0)\wedge L(\Lambda_0)$).

Setting $\beta+\gamma=\alpha$ in (4.2) and using (4.3) we obtain

$$chL(\Lambda_1) = \frac{\phi(q)}{2q\phi(q^2)}\Big\{\frac{1}{\phi(q)^{16}}\sum_{\alpha\in\dot{Q}} S_\alpha e^{2\Lambda_0-\frac{1}{2}|\alpha|^2\delta+\alpha} - \frac{1}{\phi(q^2)^8}T\Big\} \tag{4.4}$$

where

$$S_\alpha := \sum_{\beta\in\dot{Q}} e^{-[|\beta|^2-(\alpha|\beta)]\delta}$$

$$T := \sum_{\alpha\in\dot{Q}} e^{2\Lambda_0-|\alpha|^2\delta+2\alpha}$$

From this one reads off the string functions directly:

$$b^{\Lambda_1}_{2\Lambda_0}(q) = \frac{\phi(q)}{2q\phi(q^2)}\Big\{\frac{1}{\phi(q)^{16}}\sum_{\beta\in\dot{Q}} q^{|\beta|^2} - \frac{1}{\phi(q^2)^8}\Big\} \tag{4.5}$$

$$b^{\Lambda_1}_{\Lambda_1}(q) = \frac{\phi(q)}{2\phi(q^2)}\cdot\frac{1}{\phi(q)^{16}}\sum_{\beta\in\dot{Q}} q^{|\beta|^2-(\gamma_1|\beta)} \tag{4.6}$$

$$b^{\Lambda_1}_{\Lambda_7-\delta}(q) = \frac{\phi(q)}{2\phi(q^2)}\cdot\frac{1}{\phi(q)^{16}}\sum_{\beta\in\dot{Q}} q^{|\beta|^2-(\gamma_7|\beta)} \tag{4.7}$$

For example, terms of the form $e^{2\Lambda_0-k\delta}$ appear in the two parts of (4.4) only when $\alpha=0$. This gives (4.5).

In order to simplify these we need to introduce a number of lattices related to $\dot{Q}$. The root lattice $\dot{Q}$ of type E_8 is the even part of the lattice

$$D_8^\circ := \Big\{y=(y_0,\dots,y_7)\big|y\in\mathbb{Z}^8\cup\Big(\frac{1}{2}+\mathbb{Z}\Big)^8\Big\}$$

in $\mathbb{R}^8$ with the standard inner product. That is

(4.8) $$E_8 = \{y \in D_8^\circ | \sum y_i \equiv 0 \mod 2\}$$ [*Se*].

D_8° is the $\mathbb{Z}$-dual of the D_8 root lattice

$$D_8 = \{y \in \mathbb{Z}^8 | \sum y_i \equiv 0 \mod 2\} .$$

We have

$$\begin{array}{ccccc} & & D_8^\circ & & \\ & / & & \backslash & \\ E_8 & & & & \mathbb{Z}^8 \\ & \backslash & & / & \\ & & D_8 & & \end{array}$$

In some of the notation below we drop the subscripts 8. For any lattice L we denote by ϑ_L the corresponding theta function:

$$\vartheta_L(\tau) = \sum_{\alpha \in L} e^{\pi i \tau (\alpha|\alpha)} = \sum_{\alpha \in L} q^{(\alpha|\alpha)}$$

where $q = e^{\pi i \tau}$ and $\tau \in \mathcal{H}$. We will also have need of the classical one dimensional theta functions

(4.9)
$$\begin{aligned} \vartheta_2(\tau) &:= \sum_{n \in \mathbb{Z}} q^{(n+\frac{1}{2})^2} \\ \vartheta_3(\tau) &:= \sum_{n \in \mathbb{Z}} q^{n^2} \\ \vartheta_4(\tau) &:= \sum_{n \in \mathbb{Z}} (-1)^n q^{n^2} \end{aligned}$$

The reader is referred to [Ra] where all the properties of these functions that we need are proved. In particular ϑ_2, ϑ_3, and ϑ_4 are entire modular forms of weight $\frac{1}{2}$ for $\Gamma(2)$ with certain nontrivial multipliers involving 8th roots of 1. Finally we recall Dedekind's η-function

$$\eta(\tau) = q^{\frac{1}{12}} \prod_{n=1}^{\infty} (1 - q^{2n}) = q^{\frac{1}{12}} \phi(q^2)$$

which is an entire modular form of weight $\frac{1}{2}$ for $\Gamma(1)$ with transformation formula

(4.10) $$\eta\left(\frac{a\tau + b}{c\tau + d}\right) = \varepsilon(a, b, c, d)\{-i(c\tau + d)\}^{\frac{1}{2}} \eta(\tau)$$

where ε is a certain 24th root of 1 whose explicit description can be found in [Ap].

We now set about rewriting (4.5) – (4.7) in terms of these theta functions and η. Set

$$\beta_i := \alpha_i \qquad 1 \le i \le 5$$
$$\beta_6 := \alpha_8$$
$$\beta_7 := \alpha_6$$

Then $\{\beta_0, \beta_1, \ldots, \beta_7\}$ (where $-\beta_0 = \gamma_1$ is the highest root of E_8) forms the basis of a D_8 root lattice:

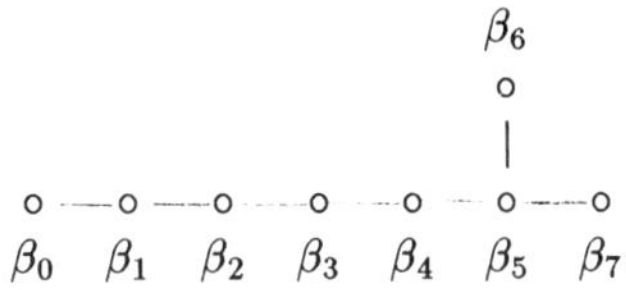

and

$$-\alpha_7 = \frac{1}{2}\beta_0 + \beta_1 + \frac{3}{2}\beta_2 + 2\beta_3 + \frac{5}{2}\beta_4 + 3\beta_5 + \frac{3}{2}\beta_6 + 2\beta_7 \ .$$

Thus through $\beta = \sum x_i\beta_i \longleftrightarrow (x_0 \ldots, x_7)$ we have

$$\dot{Q} = \{x = (x_0, \ldots, x_7) | (x_1, x_3, x_5, x_7) \in \mathbb{Z}^4, \quad (x_0, x_2, x_4, x_6) \in \mathbb{Z}^4 \cup (\frac{1}{2} + \mathbb{Z})^4\} \ .$$

Making the coordinate transformation

$$y_0 := x_0$$
$$y_i := x_i - x_{i-1} \qquad 1 \le i \le 5$$
$$y_6 := x_6 + x_7 - x_5$$
$$y_7 := x_7 - x_6$$

we obtain the description (4.8) of the lattice E_8. Note that for $\beta = \sum_{i=0}^{7} x_i\beta_i = (y_0, \ldots, y_7)$ we have from (2.6)

(4.11)
$$-(\gamma_1|\beta) = (\alpha_0|\beta) = 2x_0 - x_1 = y_0 - y_1$$
$$-(\gamma_7|\beta) = -(\gamma_7 - 2\delta|\beta)$$
$$= \left(\begin{matrix} & & & & & 1 & \\ 2 & 2 & 2 & 2 & 2 & 2 & 1 \end{matrix} \;\middle|\; \beta \right) = 2x_0 = 2y_0 \ .$$

Set

$$F(q) = \frac{\phi(q)}{2\phi(q^2)\phi(q)^{16}} . \tag{4.12}$$

Proposition 4.1

(i) $b^{\Lambda_1}_{2\Lambda_0}(q) = q^{-1}F(q)\{\vartheta_E(\tau) - \eta(\tau/2)^{16}\Big/\eta(\tau)^8\} = q^{-1}F(q)\{\vartheta_E(\tau) - \vartheta_4^8(\tau)\}$

(ii) $b^{\Lambda_1}_{\Lambda_1}(q) = \frac{1}{2}q^{-\frac{1}{2}}F(q)\{\vartheta_2^2(\tau)\vartheta_3^6(\tau) + \vartheta_2^6(\tau)\vartheta_3^2(\tau)\}$

(iii) $b^{\Lambda_1}_{\Lambda_7-\delta}(q) = q^{-1}F(q)\{\vartheta_{D^\circ}(\tau) - \vartheta_E(\tau)\}$.

Proof: (i) The first equality is clear from (4.5) and the second from $\eta(\tau/2)^{16}\Big/\eta(\tau)^8 = \vartheta_4^8(\tau)$ which is seen from the product expansions of these functions.

(ii)
$$\begin{aligned} F(q)^{-1}b^{\Lambda_1}_{\Lambda_1}(q) &= \sum_{y\in E} q^{|y|^2+y_0-y_1} \\ &= \sum_{y\in D} q^{|y|^2+y_0+y_1} + \sum_{y\in E\setminus D} q^{|y|^2+y_0-y_1} . \end{aligned}$$

Denote these two sums by Σ_D and $\Sigma_{E\setminus D}$ respectively. Replacing y_0 by $-(y_0+1)$ in Σ_D makes no difference to the terms appearing in it but replaces D by $\mathbb{Z}^8\setminus D$. Thus

$$\begin{aligned} \Sigma_D &= \frac{1}{2}\sum_{y\in\mathbb{Z}^8} q^{|y|^2+y_0+y_1} \\ &= \frac{1}{2}q^{-\frac{1}{2}}\Big(\sum_{n\in\mathbb{Z}} q^{n^2}\Big)^6\Big(\sum_{n\in\mathbb{Z}} q^{(n+\frac{1}{2})^2}\Big)^2 \\ &= \frac{1}{2}q^{-\frac{1}{2}}\vartheta_3^6(\tau)\vartheta_2^2(\tau) . \end{aligned}$$

Replacing y_7 by $-y_7$ in $\Sigma_{E\setminus D}$ changes the parity of the sum of the coordinates and hence

$$\Sigma_{E\setminus D} = \sum_{y\in(\frac{1}{2}+\mathbb{Z})^8,\Sigma y_i \text{ odd}} q^{|y|^2+y_0-y_1} .$$

Thus

$$\begin{aligned}\Sigma_{E\backslash D} &= \frac{1}{2}\sum_{y\in(\frac{1}{2}+\mathbb{Z})^8} q^{|y|^2+y_0-y_1}\\ &= \frac{1}{2}\sum_{y\in(\frac{1}{2}+\mathbb{Z})^8} q^{|y|^2+y_0+y_1}\\ &= \frac{1}{2}q^{-1/2}\vartheta_2^6(\tau)\vartheta_3^2(\tau)\ .\end{aligned}$$

(iii) $$\begin{aligned}F(q)^{-1}b^{\Lambda_1}_{\Lambda_7-\delta}(q) &= \sum_{y\in E} q^{|y|^2+2y_0}\\ &= \frac{1}{q}\sum_{y\in D^0\backslash E} q^{|y|^2}\\ &= \frac{1}{q}\{\vartheta_{D^0}(\tau)-\vartheta_E(\tau)\}\ .\end{aligned}$$

Lemma 4.2

(i) $2\vartheta_E(\tau)-\vartheta_{D^\circ}(\tau)=\eta(\frac{\tau}{2})^{16}/\eta(\tau)^8=\vartheta_4^8(\tau)$

(ii) $\vartheta_2^6(2\tau)\vartheta_3^2(2\tau)+\vartheta_2^2(2\tau)\vartheta_3^6(2\tau)+2\{\vartheta_{D^\circ}(2\tau)-\vartheta_E(2\tau)\}=4\eta(\tau)^{16}/\eta(\frac{\tau}{2})^8$.

Proof: We prove that each side of each equation is an entire modular form of weight 4 for $\Gamma(2)$ with trivial multiplier. Since the space M of such forms is of dimension 3 [Ra, Table 7] we are reduced to checking that the series agree for a few terms.

(i) Since $\vartheta_{D^\circ}(\tau)=\vartheta_2^8(\tau)+\vartheta_3^8(\tau)$ (this is clear from the definitions), $\vartheta_{D^\circ}(\tau)\in M$. Similarly $\vartheta_4^8(\tau)\in M$. Now $\vartheta_E(\tau)$ is known to be modular with trivial multiplier for $\Gamma(1)$ [Se].

A basis for M is the set of three functions

$$\begin{aligned}\vartheta_2^8(\tau) &= 2^8(q^{\frac{1}{4}}+q^{\frac{9}{4}}+\ldots)^8=256q^2+\ldots\\ \vartheta_3^8(\tau) &= (1+2q+2q^4+\ldots)^8=1+16q+112q^2+\ldots\\ \vartheta_E(\tau) &= 1+240q^2+\ldots\ .\end{aligned}$$

Since these are independent, even modulo q^3, we have to check (i) only for the first 3 terms on each side, which begin

$$1 - 16q + 112q^2 + \ldots .$$

(ii) Among the remarkable transformation properties of ϑ_2, ϑ_3, ϑ_4 we find [Ra, 7.1.8]

$$\vartheta_2^2(2\tau) = \frac{1}{2}[\vartheta_3^2(\tau) - \vartheta_4^2(\tau)]$$
$$\vartheta_3^2(2\tau) = \frac{1}{2}[\vartheta_3^2(\tau) + \vartheta_4^2(\tau)] .$$

Thus

$$\vartheta_2^6(2\tau)\vartheta_3^2(2\tau) + \vartheta_2^2(2\tau)\vartheta_3^6(2\tau) = \frac{1}{8}[\vartheta_3^8(\tau) - \vartheta_4^8(\tau)]$$

which is evidently in M. Similarly

$$\vartheta_{D^\circ}(2\tau) = \vartheta_2^8(2\tau) + \vartheta_3^8(2\tau) = \frac{1}{8}[\vartheta_3^8(\tau) + 6\vartheta_3^4(\tau)\vartheta_4^4(\tau) + \vartheta_4^8(\tau)] .$$

Furthermore $\vartheta_3^4(\tau)$ and $\vartheta_4^4(\tau)$ have the same multiplier (consisting of square roots of 1) for $\Gamma(2)$ [Ra, §7.2]. Thus $\vartheta_{D^\circ}(2\tau) \in M$. For $\eta(\tau)^{16}/\eta(\frac{\tau}{2})^8$ one checks that the multiplier is trivial for $\left(\begin{smallmatrix}1&2\\0&1\end{smallmatrix}\right)$ (obvious) and $\left(\begin{smallmatrix}1&0\\2&1\end{smallmatrix}\right)$ using (4.10). Thus $\eta(\tau)^{16}/\eta(\frac{\tau}{2})^8 \in M$. Computing the first few terms of both sides of (ii) we obtain $4q(1 + 8q + 28q^2 + \ldots)$. ∎

Combining with Prop. 4.1 we obtain

Proposition 4.3

(i) $b^{\Lambda_1}_{2\Lambda_0}(q) = b^{\Lambda_1}_{\Lambda_7-\delta}(q)$

(ii) $b^{\Lambda_1}_{\Lambda_1}(q^2) + q b^{\Lambda_1}_{\Lambda_7-\delta}(q^2) = \phi(q^2)\big/\phi(q^4)\phi(q)^8$. ∎

Remark: The coincidence (i) and other coincidences of string functions for level 2 representations of E_9 are explained in [K-W 2].

5. The dimension formulas. Set

$$\phi(q^2)\big/\phi(q^4)\phi(q)^8 = \sum_{k\geq 0} \tau(k)q^k .$$

Proposition 5.1

$$chL(\Lambda_1 + 3\delta) = \sum_{\lambda \in 2\Lambda_0 + Q} \tau\Big(3 - \frac{(\lambda|\lambda)}{2}\Big) e^{\lambda} .$$

Proof: Consider the multiplicity of a weight λ of $L(\Lambda_1 + 3\delta)$. We can assume that λ is dominant and hence $\lambda = \Lambda - j\delta$ for some $j \geq 0$ where Λ is one of $\Lambda_1 + 3\delta$, $2\Lambda_0 + 3\delta$, $\Lambda_7 + 2\delta$. But

$$j = \Big\lfloor \frac{1}{2}\Big(3 - \frac{(\lambda|\lambda)}{2}\Big)\Big\rfloor$$

since

$$3 - \frac{1}{2}|\Lambda_1 + 3\delta|^2 = 0$$

$$3 - \frac{1}{2}|\Lambda_7 + 2\delta|^2 = 1 = 3 - \frac{1}{2}|2\Lambda_0 + 3\delta|^2$$

(see (2.10)) and the quantity $3 - \frac{1}{2}(\lambda|\lambda)$ increases by 2 when λ (which is of level 2) is replaced by $\lambda - \delta$. The result follows from Prop. 4.3 and the definition of τ . ∎

Let

$$\frac{1}{\phi(q)^8}\Big\{1 - \frac{\phi(q^2)}{\phi(q^4)}\Big\} = \sum_{k \geq 0} \xi(k) q^k .$$

Proposition 5.2

(i) $ch(L(\Lambda_0 + 2\delta) \wedge L(\Lambda_0 + 2\delta)) = \sum_{\lambda \in 2\Lambda_0 + Q} p^{(8)}\Big(3 - \frac{(\lambda|\lambda)}{2}\Big) e^{\lambda}$

(ii) $ch\Big\{L(\Lambda_0 + 2\delta) \wedge L(\Lambda_0 + 2\delta)/L(\Lambda_1 + 3\delta)\Big\}$

$$= \sum_{\lambda \in 2\Lambda_0 + Q} \xi\Big(3 - \frac{(\lambda|\lambda)}{2}\Big) e^{\lambda} .$$

Proof: Use $ch(L(\Lambda_0) \wedge L(\Lambda_0)) = q\frac{\phi(q^2)}{\phi(q)} chL(\Lambda_1)$ (see (4.3)) and Prop. 5.1. ∎

From Corollary 3.2 we have

Proposition 5.3.

$$ch\hat{\underline{g}}_{[2]} = \sum_{\alpha \in \hat{\Delta}_{[2]}} \xi\Big(3 - \frac{(\alpha|\alpha)}{2}\Big) e^{\alpha} .$$

∎

n	$p^{(8)}(n)$	$\xi(n)$
0	1	
1	8	
2	44	1
3	192	8
4	726	44
5	2464	192
6	7704	727
7	22528	2472
8	62337	7747
9	164560	22712
10	417140	63020
11	1020416	166840
12	2418710	424161
13	5573568	1040664
14	12520744	2474026
15	27484160	5717880
16	59068372	12882567
17	124505880	28360256
18	257770964	61125215
19	524871424	129203160

Table of Values of $p^{(8)}(n)$ and $\xi(n)$

References.

[Ap] T. Apostol, Modular functions and Dirichlet series in number theory, Springer-Verlag, 1976.

[BM] S.Berman, R.V. Moody, Lie algebra multiplicities, Proc. Am. Math. Soc. **76**(1979), 223–228.

[FF] A. Feingold, I. Frenkel, A hyperbolic Kac-Moody algebra and the theory of Siegel modular forms of genus 2, Math. Ann. **263**(1983), 87–144.

[Fr] I. Frenkel, Representations of Kac-Moody algebras and dual resonance models, Applications of Group Theory in Physics and Mathematical Physics **21**(1985), 325–354.

[GK] O.Gabber, V.G. Kac, On defining relations of certain infinite dimensional Lie algebras, Bull. Am. Math. Soc. **5**(1981), 185–189.

[K] V.G. Kac, Simple irreducible graded Lie algebras of finite growth, Math. USSR Izvestija **2**(1968), 1271–1311.

[Ka] V.G. Kac, Infinite dimensional Lie algebras, 2nd ed., Cambridge University Press, 1985.

[KW1] V.G. Kac, M. Wakimoto, Unitarizable highest weight representations of the Virasoro, Neveu-Schwarz and Ramond algebras, Lecture Notes in Physics, **261**(1986), 345-371.

[KW2] V.G. Kac, M. Wakimoto, Modular and conformal invariance constraints in representation theory of affine algebras, Adv. in Math. (1988).

[P] D. Peterson, Affine Lie algebras and theta functions, Lecture Notes in Math 933, Springer Verlag, 1982, 166–175.

[PK] D. Peterson, V.G. Kac, Infinite flag varieties and conjugacy theorems, Proc. Nat. Acad. Sci. USA **80**(1983), 1778–1782.

[Ra] R.A. Rankin, Modular forms and functions, Cambridge University Press, 1977.

[Se] J.-P. Serre, A course in arithmetic, Springer-Verlag, 1973.

OPERATOR METHODS IN STRING THEORY

Luis Alvarez-Gaumé * **

Physics Department

Boston University

590 Commonwealth Avenue

Boston, MA 02215, USA

Lectures Delivered at the Cargése Summer School

on Mathematical Physics, July 1987

and

at the Varenna School

in Mathematical Physics, Lake Come, August 1987

* Supported in part by DOE contract and by Alfred P. Sloan Foundation Fellowship

** Address after Sept. 1, 1988: Theory Division, CERN, CH-1211, Geneva 23 Switzerland

K. Bleuler and M. Werner (eds.), Differential Geometrical Methods in Theoretical Physics, 129–144.

OPERATOR METHODS IN STRING THEORY

I. Introduction

These lectures are based on joint work with C. Gomez, C. Reina, G. Moore and C. Vafa [1], [2], [3]. The motivation is two-fold. On the one hand we would like to develop an operator formalism describing high orders of string perturbation theory, as well as conformal field theories on Riemann Surfaces of genus bigger than one. On the other hand, developments in the theory of soliton solutions of the K-P hierarchy (Kadomtsev-Petviashvili equations); for a detailed geometrical account of this theory and references to the literature see for example [4], [5]) made it clear that many of the geometrical features of Riemann Surfaces and their móduli spaces can be formulated in terms of the properties of certain two dimensional quantum field theories [6], so that the geometrical complexity of a Riemann Surface with a field on it can be coded into a state in the standard Fork space of the field theory [1], [7], [8], [9], [10], [11], [12]. This approach gives an important understanding of the action of the Virasoro algebra on the moduli space of surfaces [13], [14]. The space of solutions of the K.P. equation can be described in terms of an infinite dimensional grassmannian Gr. To any algebraic curve X with a point P selected on it, a local coordinate around P, and a line bundle L over X, we can associate a point in Gr Krichever construction [4], [5]). Moreover, the collection of all those points in Gr is a dense set. It is thus plausible to expect that some subspace of Gr provides an explicit model for the Universal Moduli Space of Friedan and Shenker [15, 16] which plays a central role in their non-perturbative approach to string theory. This possibility still remains largely unexplored, mainly because it seems difficult to find integration measures (the non-perturbative analog of the ghost of the bosonic string). After all the progress achieved in string theory over the last few years, we are still scratching the surface of a deep structure whose building block seems far from being unravelled. One of the most important clues we seem to be getting from string theory is that the space of classical solutions to a quantum theory of strings is the space of two dimensional conformal theory with a specific value of the central term in the Virasoro algebra.

Conformal theories have been developed along two different lines. One approach is to use path integrals to define the theory, and the other is to use operator techniques. The operator formulation is simplest for the sphere and the torus, which has been discussed in detail in the literature (see for example [16], [17]), in fact, most of the substantial work on conformal field theories has been carried out using operator methods for theories formulated in the twice pictured sphere (conformally equivalent to the cylinder). For higher genus surfaces, many results have been obtained using a functional integral formulation, because there is no simple way of constructing a hamiltanian formulation.

It is well known that the energy momentum tensor can be viewed as providing a connection on moduli space. This is quite important in the operator formulation of string theory. One of the main aims in our work was to better understand the Virasoro algebra and moduli space [13]. We can use the Virasoro algebra to move the states in moduli space. Below we describe these issues for the simplest conformal theories, even though the set up applies in principle to more general situations. One of the spin-offs of formulation we present is an alternative proof of chiral bosonization completely in the operator language. The interested reader is referred to [3] for details. We will not present these results here.

These notes are organized as follows: In section II we review the general ideas of the operator formalism of conformal field theories in arbitrary surfaces, and work out some examples in detail. In section III we turn to the action of the Virasoro algebra and the Polyakov measure for the bosonic string. Section IV contains an outline of the extension of this work to superstrings, together with the conclusions.

Many authors have developed independently an operator formalism for string perturbation theory with a motivation based on string field theory. An incomplete (but representative) list of references can be found in [18-20]. There is obviously a good deal of overlap between these approaches and the one presented here; and in many respects they can be considered complementary.

II. OVERVIEW AND EXAMPLES

Consider a Riemann Surface X with parametrized boundaries, and a quantum field theory on it, whose fields will be collectively denoted by ϕ. If $\mathcal{H}$ is the Hilbert space of the theory in the standard conformal plane, we can construct a state in $\mathcal{H}^{\otimes n}$ (one copy per parametrized boundary) associated to the surface and the field theory by performing the Feynman path integral with fixed boundary conditions $\phi_i(t_i) i = 1, 2.., n$ at each of the boundaries. This procedure yields a wave functional $\Psi[\phi_1(t_1), ... \phi_n(t_n)]$ in $\mathcal{H}^{\otimes n}$. If $\mathcal{P}(g,n)$ denotes the moduli space of Riemann surfaces of genus g with n points on it $P..., P_n \epsilon X$ and local coordinates about them $t_i(P_i) = 0 \ 1 \leq i \leq n$; and we fix the field theory, we get a mapping $\Psi = \mathcal{P}(g,n) \rightarrow \mathcal{H}^{\otimes n}$. When the theory is not conformally invariant, we have to specify a metric on each surface together its moduli. For conformal theories however, Ψ only depends on the information parametrized by $\mathcal{P}(g,n)$. Punctured surfaces are of interest in string theory, because a surface with semi-infinite cylinders attached to the boundaries describes string scattering amplitudes, and these semi-infinite cylinders are conformally equivalent to punctured discs. Mathematically, the string scattering amplitudes can be understood as measured on the moduli space of punctured surfaces. To illustrate the previous ideas, we consider a surface X divided into two pieces X_1, X_2 by a circle S^1, and a b-c system [17] of spin j on it $(j \epsilon \frac{1}{2} Z)$. We would like to represent the result of functionally integrating over the whole surface as an overlap integral on the Fork space of the circle of two states $\Psi_1[f], \Psi_2[f]$, where $f(z)$ is the field configuration of $c(z)$ on S^1. To obtain Ψ_1 we functionally integrate the fields over X_1 subject of the boundary condition $C|_{S^1} = f$, and free boundary conditions on $b(z)$. Standard arguments in functional integration give the following representation for Ψ_1:

$$\psi[f] = \int_{c|_{s^1}=f} dbdc \; exp(-\int_{X_1} (b\bar{\partial}c + c\bar{\partial}b) + \oint_{S^1} cb) \tag{2.1}$$

One of the basic properties of $\Psi[f]$ is its invariant under shifts of f by section of c holomorphic in X_1:

$$\psi[f + w_n] = \psi[f] \tag{22}$$

if for simplicity we choose X_2 to be a disc centered upon the point P chosen on X, and with a local parameter $t, t(P) = 0$; the section $w_n(t)$ may have arbitrary order poles at $t = 0$, but extend S holomorphically to $X - p$. Since the translation of f by w_n is implemented by the operator $\oint_{s^1} w_n(t)b(t) \equiv Q_n$, a more invariant way of expressing (2.2) is $Q_n|\psi>= 0$. A similar argument shows that $|\psi>$ is invariant under shifts of b by sections $\tilde{w}_n$ which extend holomorphically off P. Hence the ray $|\psi>$ is completely defined by

$$\begin{gathered} Q_n|\psi> = \tilde{Q}_n|\psi>= 0 \\ Q_n = \oint_{s^1} w_n(\pm)b(t) \quad \tilde{Q}_n = \oint_{s^1} \tilde{w}_n(t)b(t) \end{gathered} \tag{2.3}$$

It is a simple consequence of the $b-c$ canonical (anti)-commutation relations that $[Q_n, Q_m] = [\tilde{Q}_n, \tilde{Q}_m] = 0$. the Q_n's and $\tilde{Q}_m$'s also commute:

$$[Q_n, \tilde{Q}_m]+ = \oint_{s^1} w_n(t)\tilde{w}_m(t) = 0 \tag{2.4}$$

since $w_n(t)$ (resp. $\tilde{w}_m(t)$) is a $1 - j$(resp. j) differential, the product $w_n(t)\tilde{w}_m(t)$ is a meromorphic differential on X whose only poles are located at P. Then the residue (2.4) vanishes identically. The condition (2.3) provide the largest set of compatible conditions which can be imposed on a state $|\psi>$; and in most known cases, they uniquely determine the ray represented by $|\psi>$. The exceptions known always involve superconformal ghosts (the $\beta - \gamma$ system in [17]).

a) As a first example, we consider an anticommuting b.c. system of spin j on the sphere with one puncture P. By the Riemann-Roch Theorem, there are $2j - 1$ global holomorphic $(1-j)$ differentials: $1, t, t^2, ..., t^{2j-2}$. We can also find $1 - j$ differentials with poles of arbitrary order at P. The w_n's generate a set $W = \{t^{2j-2}, ..., t, 1, t^{-1}, t^{-2}, ...\}$. The j-differentials holomorphic off P are generated by $\tilde{W} = \{t^{-2j}, t^{-2j-1}, ...\}$. The conserved charges are:

$$Q_n = \oint_0 b(t)t^n = \{b_n, n > -j\} \tag{2.5a}$$

$$\tilde{Q}_n = \oint_0 c(t)t^n = \{c_n, n > j-1\} \tag{2.5b}$$

with

$$b(t) = \sum_n b_n t^{n-j} \quad , \quad c(t) = \sum_n c_n t^{n+j-1} \tag{2.6}$$

$n\epsilon Z$ if j is integral, and $n\epsilon Z + 1/2$ if j is half-integral. The state $|\psi>$ is then the $SL_2(C)$ invariant vacuum of the $b-c$ system. We introduce the notation $|\phi_g^n, j>$ for the state obtained on a surface of genus g with n punctures and an anticommuting $b-c$ system of spin j. Thus, the example we have just discussed gives $|\phi_0^1, j>$. When j is an integer, we also have the zero mode algebra $\{b_0, c_0\} = 1$, and the highest weight states of the oscillator algebra are the states $|\pm>$ satisfying.

$$c_n|+>= b_n|->= 0 \quad n \geq 0 \quad b_0|+>= |->, \quad c_o|->= |+> \tag{2.7}$$

The relation between $|\pm>$ and $|\phi_0^1>$ is:

$$\begin{aligned} |+> &= c_0 c_1 ... c_{j-1} |\phi_0^1> \\ |\phi_0^1> &= b_{-j+1} ... b_{-1} b_0 |+> \end{aligned} \tag{2.8}$$

SL_2 invariance of $|\phi_0^1>$ implies

$$L_0|\pm>= -\frac{j(j-1)}{2}|\pm> \tag{2.9}$$

and the natural ghost charge assignment follows from the Atiyah-Patochi-Singer index theorem:

$$Q_{gh}|\phi_0^1>= -(j-1/2)|\phi_0^1> \tag{2.10}$$

b) For our second example, consider a spin 1/2 system on a surface X with a single puncture $P\epsilon X$ and genus g. If the spin structure is even and non-singular (the kernel of the Dirac operator is equal to zero), the w_n sections are obtained from the Szegö kernel [21]:

$$S(z,w) = \frac{\theta\left[\begin{smallmatrix}\alpha\\ \beta\end{smallmatrix}\right]\left(\int_w^z w|_\tau\right)}{\theta\left[\begin{smallmatrix}\alpha\\ \beta\end{smallmatrix}\right](o|\tau)} E(z,w) \tag{2.11}$$

where E is the prime form on the surface, and $[\alpha, \beta]$ represent an even spin structure. For z, w in a neighborhood of P, (2.11) can be written as:

$$S(z,w) = \frac{1}{z-w} + \sum_{n,m=1}^{\infty} B_{nm} z^{n-1} w^{m-1} \tag{2.12a}$$

$$B_{nm} = \frac{1}{(n-1)!(m-1)!} \frac{\partial^{n-1}}{\partial z^{n-1}} \frac{\partial^{m-1}}{\partial w^{m-1}} (S(z,w) - \frac{1}{z-w}) \tag{2.12b}$$

Then the w_n's are obtained by differentiating $S(z,w)$ with respect to w and setting $w = \text{P}$:

$$w_n(t) = t^{-n} + \sum_{m=1}^{\infty} B_{nm} t^{m-1} \tag{2.13}$$

thus giving conserved charges:

$$Q_n = \oint_0 w_n(t) c(t) = (c_{n-1/2} + \sum_m B_{nm} c_{-m+1/2}) \tag{2.14}$$

since $\{c_n \; n > 0\}$ are annihilation operators, we can represent c_n by $\partial/\partial b_{-n}$. We obtain a differential equation for $|\Phi_g^1, 1/2>$:

$$(\frac{\partial}{\partial b_{-n+1/2}} + \sum_m B_{nm} c_{-m+1/2}) |\Phi_g^1, 1/2> = 0 \tag{2.15}$$

with solution:

$$|\Phi_g^1 1/2> = C \; exp - \sum_{n,m=1}^{\infty} B_{nm} c_{-n+1/2} b_{-m+1/2} |\Phi_0^1, 1/2> \tag{2.16}$$

If the spin structure is singular or if it is odd, there will be holomorphic 1/2 differential on the surface: $h(t)$. Then the associated conserved charge $Q(h) = b(h) \equiv \oint b(t)h(t)$ only involves creation operators. Thus we have to impose the conservation of $Q(h)$ and $\tilde{Q}(h)$; apart from Bogolinbov transformation obtained from the $b-c$ two point function with zero modes removed. If $g(B)$ denotes the Bogolinbov transformation, the state obtained becomes:

$$|\Phi_g^1, 1/2> = g(B) Q(h) \tilde{Q}(h) |\Phi_0^1, 1/2> \tag{2.16}$$

Finally, if we wish to study a Weyl-Majorana fermion, there is a single fermionic field $\psi(t)$ with action $S[\psi] = \int \psi\bar{\partial}\psi$, with commutation relations $\{\psi_n, \psi_m\} = \delta_{m+n}, 0; n, m\epsilon\mathcal{Z} + 1/2$, and the state obtained in this case is:

$$|\Phi_g^1, Maj >= C \exp -\frac{1}{2} \sum_{n,m=1}^{\infty} B_{nm}\psi_{-n+1/2}\psi_{-m+1/2}|\Phi_0^1 Maj > \tag{2.17}$$

(for an even non-singular spin structure). The 2_n-point functions are easily proved to be expressed in terms of pfaffians:

$$\frac{\langle o|\psi(t_1)\psi(t_2)..\psi(t_{2n})|\Phi_g^1, M_{aj}\rangle}{\langle 0|\Phi_g^1 M_{aj}\rangle} = Pfaff_{ij}S(t_i, t_j) \tag{2.18}$$

and the partition functions involve Prym theta functions.

c) Next we explicitly treat the ghost system of the bosonic string. This is a fermionic $b-c$ system with $j = 2$. On any surface X with $g > 1$, there are $3g-3$ holomorphic quadratic differentials; and we can always find quadratic differentials with poles of arbitrary order at a point P. These statements need extra qualification when P is a Weierstrass point on X, but we will not insist on this issue here. A convenient basis for the meromorphic 2-differentials is (see [3] for details).

$$S_N = (t^N + \sum_{m \geq K} B_{Nm}^{(j)} t^m)(dt)^2, -\infty < N \leq k-1 \quad k = 3g-3 \tag{2.19}$$

and as a consequence of the Riemann-Roch theorem, the vector fields holomorphic off P̲ must have poles of order larger than k:

$$V_M = (t^M + \sum_{m \geq -k} B_{Mm}^{(1-j)} t^m)(dt)^{-1} \quad M \leq -(k+1) \tag{2.20}$$

since the charges:

$$\oint c(t)S_N(t) \quad -1 \leq N \leq k-1 \tag{2.21}$$

only involve creation operators, the state $|\phi_g^1, 2>$ takes the form:

$$|\phi_g^1, 2> =$$

$$= c(S_0)c(S_1)..c(S_{3g-4})c(S_{-1}) \; exp\,(-\sum_{n,m} B_{n+2,m}b_{-n}c_{-m-2}).|+> \tag{2.22}$$

$$c(Si) = \oint_0 c(t)S_i(t)$$

so that the ghost number of $|\phi_g, 2>$ is $3g-3+3/2$, after we assign ghost number 1 (-1) to $c(t)(b(t))$. The extension of (2.22) to surface with more punctures is straightforward. We only point out that the ghost number charge of $|\phi^n_{g,j}>$ is $(2j-1)(g-1)+(j-1/2)n$.

d) Finally we consider a free scalar on X. This case is important in string theory because the embedding coordinates of the flat bosonic string are free fields from the two dimensional point of view. The state associate to the surface X with a parametrized boundary S^1 and a free single-valued scalar field ϕ can be represented by mean of a functional integral

$$\Psi[f] = \int_{\phi|_{s^1}=f} d\phi \; exp(-\frac{1}{8n}\int_{X_1} d\phi \wedge *d\phi + \frac{1}{8n}\oint_{S^1} \phi * d\phi) \tag{2.23}$$

where $*$ is the duality operation on forms $*dz = -i\; dZ$. The state (2.23) is invariant under shifts of f by functions on S^1 which are boundary values of harmonic functions on X_1 (using the same notation as in (2.1)). As first pointed out by Vafa [9], the explicit construction of $\Psi[f]$ is more complicated than in previous cases. The origin of the difficulty is that for generic points PϵX, the Weierstrass gap theorem implies that functions holomorphic off P must have a pole at P of order at least $g+1$. We can however construct multivalued meromorphic functions with poles of order less that $g+1$. Since the shifts involve harmonic function, we can compensate the multivaluedness by adding some antiholomorphic piece. Explicitly, if $w_i, \bar{w}_i$ are the abelian differentials on X, and we introduce the differentials:

$$\begin{aligned} &\eta_n(t) - A_n(Im\tau)^{-1}(w-\bar{w}) \\ &w_i(t) = \sum_{n=1}^{\infty} A_{in} t^{n-1} dt \\ &\eta_n(t) = \frac{1}{(n-1)!}\frac{\partial}{\partial t}\frac{\partial^n}{\partial y^n}|_n E(t,|y)|_{y=0} \end{aligned} \tag{2.24}$$

the harmonic functions of interest are:

$$\begin{aligned} \phi_n(t) &= \int^t (\eta_n(t) - A_n(Im\tau)^{-1}(w-\bar{w})) \\ &= \phi_n^{hol}(t) + \phi_n^{ah}(\bar{t}) \end{aligned} \tag{2.25}$$

Expanding $\phi(t)$ in terms of oscillators,

$$\phi(t) = q + iplnt + ip|n\bar{t}\bar{t} + i\sum_{n\neq 0}(\frac{t^n}{n}j_n + c.c.)$$
$$[q,p] = +i \quad [j_n, j_m] = n\delta_{n+m,0} \tag{2.26}$$

The conserved charges are:

$$Q_n = \oint(\partial\phi\phi_n^{hol}(t) - \bar{\partial}\phi\phi_n^{ah}(\bar{t})) =$$
$$= j_n - \sum_{n=1}^{\infty}(2Q_{nm} + \pi A_n(Im\tau)^{-1}A_m)\frac{j_{-m}}{m} + \pi A_n(Im\tau)^{-1}\bar{A}_m\frac{\bar{j}_{-m}}{m} \tag{2.27}$$

plus their complex conjugates. Representing j_n as nx_n, and $j_n = \partial/\partial x_n, \quad n > 0$, the differential equations $Q_n\psi >= \bar{Q}_n|\psi >= 0$ yield:

$$|\psi >= c\ exp((x,\bar{x})M(\frac{x}{x}) \tag{2.28}$$

$$M = \begin{pmatrix} Q_{nm} + \frac{\pi}{2}A_n(Im\tau)^{-1}A_m & -\frac{\pi}{2}A_n(Im\tau)^{-1}\bar{A}_m \\ -\frac{\pi}{2}\bar{A}_n(Im\tau)^{-1}A_m & \bar{Q}_{nm} + \frac{\pi}{2}\bar{A}_n(Im\tau)^{-1}\bar{A}_m \end{pmatrix}$$

The overall constants appearing in front of the states depend on the moduli parameters of the surface. Notice that (2.28) is the first example where the dependence of $|\psi >$ on the moduli parameters is not holomorphic. In the operator language this plays an important role in identifying the correct bosonic string measure.

The previous arguments are easily extended to surfaces with more punctures. Many more details and examples can be found in [3].

III. VIRASORO ALGEBRA AND THE BOSONIC STRING MEASURE

It was shown in [13] that there is a natural action of the Virasoro algebra on the moduli space of punctured surfaces with a local parameter at the puncture. To understand this, let $P\epsilon X$, and D be the disc centered upon P where the local parameter t vanishing at P is defined. A local meromorphic vector fields with poles only at P, $v(t)\partial/\partial t$ can be written in terms of a Laurent expansion:

$$v(t) = \sum a_{n-1}t^n \tag{3.1}$$

these vector fields can be divided into three subsets

i) Vector fields holomorphic in $D : C||t||\partial/\partial t$. Among these, those vectors vanishing at P generate local coordinate changes. Those satisfying $v(P) \neq 0$ will also produce infinitesiaml motions in the position of the puncture.

ii) Vector fields which extend holomorphically off P. These generate infinitesimal conformal automorphisms of the punctured surface.

iii) Vector fields holomorphic on D-P which do not extend holomorphically to either D or $X - P$. The Riemann-Roch theorem shows that this space has dimension 3g-3; and it is in fact the same as the tangent space to moduli space at the curve X. If we think in terms of the moduli space $P(g,1)$ of genus g Riemann surfaces with a point and a local parameter, we find that its tangent space at (X,P,t) is isomorphic to i) + iii). Using the Kodaira-Spencer construction we can generate a new surface S' close to S. Only when v is either in i) or iii) will we get an inequivalent point on $\mathcal{P}(g,1)$. We can go even further than that, and exponentiate locally these vector fields to generate a local neighborhood of S in $\mathcal{P}(g,1)$. One of the basic properties of the Virasoro algebra in a conformal field theory is that the corresponding variation of $|\Psi>$ under the variation of the data caused by the vector field $v(t)$, is given by the energy-momentum tensor:

$$\begin{aligned} T(v) &= \oint T(t)v(t), T(t) = \sum_{n\epsilon\mathbb{Z}} L_n t^{n-2} \\ \delta|\psi> &= (T(v) + \bar{T}(\bar{v}))|\psi> \end{aligned} \tag{3.2}$$

it is easy to show explicitly that (3.2) holds for all the examples considered in the previous section; and it also gives an equation determining the moduli dependence of the constants appearing in front of the states. For instance, if $|\psi>$ is the state associate to a spinor with an even spin structure on the torus, then there is only one vector field in class iii): $\frac{1}{t}\frac{d}{dt} = v$, and $T(v) = L_2$. Then

$$\delta(\mathcal{C}g(B)|0>) = \delta\mathcal{C}g(B)|0> + \mathcal{C}g(B)\delta B|0> = T(v)\mathcal{C}g(B)|0> \tag{3.3}$$

where $g(B)$ is the Bogolinbov transformation (2.16). From (3.3) we obtain a differential equation

$$\frac{\partial}{\partial\tau} \log C \propto \frac{\partial}{\partial\tau} \log \frac{\theta[^{\alpha}_{\beta}](0|\tau)}{\eta(\tau)}$$

a well known result for the fermion determinant. More generally, the vector fields in D-P which do not extend to either D or $X-$P have the form $V_i(t) = t^{-i}\frac{d}{dt}$ $i = 1, 2, ..., 3g-3$. Hence L_{-n} $n \geq 0$ changes the local coordinate, L_1 moves the puncture, and L_k, $k = 2, 3..., 3g-3+1$ change the moduli. If $v(t)$ is holomorphic in $X-p$, then the ray associated to S should not change under the action of $T(v)$. However it is not consistent to require $T(v)|\psi >= 0$, unless the central charge of the Virasoro algebra vanishes in the theory considered, otherwise $T(t)$ does not behave as a quadratic differential under changes of variables. Only when we cover the surface X with patches with $SL_2(C)$ transition functions can we consistently impose $T(v)|\psi >= 0$. To construct the bosonic string measure, let $|W>$ be the state associated to $S\epsilon\mathcal{P}(g,1)$, and the string coordinates $X|^{\mu}$ and ghost fields $b, c, \bar{b}, \bar{c}$. The state $|W>$ has ghost charge $3g-3+3/2$. Let $v_i, i = 1, 3g-3$ be the vector fields representing the tangent vectors to the moduli space $M_g : V_1, ..., V_g$. If $|0>$ is the SL_2 invariant vacuum on the disc, then an obvious candidate for the measure is:

$$\mu(V,\bar{V}) =< 0|\ b(v_1)\bar{b}(\bar{v}_1)...b(V_{3g-3})\bar{b}(\bar{v}_{3g-3})|W> \tag{3.4}$$

To prove that this is indeed a measure on M_g, we have to exhibit the independence of $\mu(V,\bar{V})$ with respect to the data in $\mathcal{P}(g,1)$ not contained in M_g:

a) We can replace v_i by $v_i + v_0$, where v_0 extend holomorphically to $X-p$.

b) Independence of the local parameter and the position of the puncture.

The first requirement follows from the definition of $|W>$ in terms of conserved charges, and the second is a consequence of the properties of the SL_2 invariant vacuum $<0|$. Since $<0|L_{-n} = 0$ $n = -1, 0, 1, 2, ...$ The measure (3.4) is well defined. To finally make sure that (3.4) agrees with the measure obtained using path integral techniques, we will show that $\mu(v,\bar{v})$ is of the form:

$$\mu = (\rho \wedge \bar{\rho})\ (det\ I_m \tau)^{-13} \tag{3.5}$$

where ρ is a holomorphic $3g-3$ form on moduli space. To prove this, we only need to show that

$$\partial\bar{\partial}\ log\ \mu(..V_i..) = -13\partial\bar{\partial} tr\ log\ I_{m\ \tau} \tag{3.6}$$

where $\partial, \bar{\partial}$ are derivatives with respect to the moduli space coordinates. From (3.2) and (3.4), we obtain:

$$\partial\bar{\partial}\ log\ \mu(V_i,\bar{V}_i) =$$

$$= \frac{1}{(<0|...|W>)^2}(<0|..|W><0|...T()\bar{T}()|W> - <0|...\bar{T}()|W>) <0|...\bar{T}()|W>) \tag{3.7}$$

where the dots refer to the $b(v_i)$ insertions. Equation (3.7) defines a (1,1)-form. To get a number for the above equation, the energy-momentum tensors have to be folded in with vectors. It is clear from (3.7) that $\partial\bar{\partial}\ log\ \tilde{\mu}$ would vanish if the left and right movers did not mix. As shown in section III, the state generated by the string coordinates mixer left and right. Thus, the evaluation of (3.7) reduces to a similar computation for the bosonic contribution to $|W>$. A straightforward computation then proves (3.6).

In (3.4) the ghost charge has been absorbed at a single point on the surface. It is not difficult however to modify the prescription so that the vacuum charge is absorbed at more than one point. Similarly, it is possible to obtain the corresponding measure for scattering amplitudes and derive the physical state conditions as well as the behavior of the measures at the boundary of moduli space; thus making sure that the (3g-3)-form ρ is indeed the Mumford form on M_g (see [4] for details).

IV CONCLUSIONS AND OUTLOOK

The formalism explained in this lectures can also be applied to orbifold and toroidal compactifications of the bosonic string; and in principle it should also help in providing a geometrical interpretation of the discrete series of Virasoro representations [16, 23]. There is a natural extension of the operator formalism to super and heterotic strings which is currently being completed [24]. The states are constructed on super-Riemann surfaces with punctures, and follows arguments similar to those in [4], one obtains a well-defined supermeasure in supermoduli space. The ambiguities in super-string perturbation theory recently discovered by E. and H. Verlinde [25] (for detailed references and a thorough exposition of these ambiguities see [26]) seem to appear when one tries to construct the super measure at genus g starting with the super-sphere with three punctures (the "vertex") and the sewing operators ("the propagator"). The advantage however, is that one works from the beginning with conformally invariant data and the gauge fixing procedure in the space of super-frames on a super-Riemann surface can be circumvented. We believe that this approach will yield some important insights in issues concerning the finiteness and consistency of superstring perturbation theory.

Acknowledgements: I would like to thank the organizers of the school for the opportunity to present these results in which such a pleasant atmosphere. I am also grateful to C. Gomez, C. Reina, G. Moore, P. Nelson, G. Sierra and C. Vafa for countless conversations and discussions in many of the issues treated in these lectures.

References

[1] L. Alvarez-Gaumé, C. Gomez, and C. Reina: Phys. Lett. 190B (1987) 55

[2] L. Alvarez-Gaumé C. Gomez and C. Reina: "New Methods in String Theory" CERN-TH. 4775/87. Lectures Delivered at the "Trieste School on Superstrings". April 1987. World Scientific 1988.

[3] L. Alvarez-Gaummе, C. Gomez, G. Moore and C. Vafa: "Strings in the Operator Formalism" BUHEP-87/49, CERN-TH 4883/87, HUTP-87/A083, IASSNS-HEP-87/68.

[4] G. Segal and G. Wilson: Publication Math de l'I.H.E.S. 61 (1985), 1.

[5] B.A. Dubrovin, I.M. Krichever and S.P. Novikov: "Topological and Algebraic Geometric Methods in Contemporary Mathematical Physics" Sov. Sci. Reviews, 3 (1982). 1.

[6] E. Date, M. Jimbo, M. Kashiwara and T. Miwa: J. of Phys. Soc. of Japan 50, (1981), 3806.

[7] S. Saito: Tokyo Metropolitan Univ. Preprints: TMUP-HEL-8613/8615/8701/8707.

[8] N. Ishibashi, Y. Matsuo, and H. Ooguri: Univ. of Tokyo Preprint UT-499 (1986),

[9] C. Vafa: Phys. Lett. 190B (1987), 47.

[10] E. Witten: "Quantum Field Theory, Grassmannians and Algebraic Curves", PUPT-1057, May 86. To appear in C.M.P.

[11] G. Segal: Talk presented at the conference: "Links Between Geometry and Mathematical Physics", held at Schlöss Ringberg, March 1987; based on joint work with D. Quillen.

[12] C. Vafa: "Conformal Theories and Punctured Riemann Surfaces", HUTP-87/A086.

[13] A. Beilinson, Y. Manin, and Y.A. Shechtman: Localization of Virasoro and Neven-Schwartz Algebras. Preprint 1986; M. Kontsevich, Funct. Anal. and Appl. 21 (1987) 156 (=Funk. Anal. Pril. 21 (1987) 78).

[14] E. Arbarello, C. DeConcini, V. Kac and C. Procesi: "Moduli Space of Curves and Representation Theory", University of Rome and MIT preprint, 1987.

[15] D. Friedan and S. Shenker: Phys. Lett. 175B (1986) 287; Nucl. Phys. B281 (1987) 509.

[16] A. Belavin, A. M. Polyakov and A.B. Zamalodchikov: Nucl. Phys. 241 (1984), 333.

[17] D. Friedan, E. Martinec and S. Shenker: Nucl. Phys. B271, (1986), 93.

[18] A. Neven and P. West: Phys. Lett. 179B (1986) 235, ibid. 180B (1986) 34; 193B Nucl. Phys. B278 (1986) 601.
P. West: "Multiloop Ghost Vertices and the Determination of the Multiloop Measure" CERN-TH.4937/88 (1987) 187. "The Group Theoretical Approach to String Amplitudes". CERN-TH preprint. To appear in Comm. Math. Phys. and references Therein.
M.D. Freeman and P. West: "Ghost Vertices for the Bosonic String Using the Group Theoretic Approach to String Theory". CERN.TH.49 38/88 A recent review of this approach with many details and references can be found in P. West: "A REview of Duality String Vertices, Overlap Identities and the Group Theoretic Approach to String Theory" CERN-TH4819/87.

[19] A. Leclair, M.E. Peskin and C.R. Preitschopf, "String Theory on the Conformal Plane, I, II, and III", SLAC-PUB-4306, 4307, 4464.!A. LeClair: Harvard preprint HUTP-87/A033, to appear in Nucl. Phys. B., A. Leclair: "An Operator Formulation of the Superstring". Princeton preprint PUTP-1080 and references therein.

[20] P. Di Vecchia, R. Nakayama, J.L. Peterson and S. Sciuto: Nucl. Phys. B282 (1987) 103; J.L. Peterson and J.R. Sidenivs, Niels Bohr Institute preprint NBI-HE-87-35; P. Di Vecchia, M. Fran, A. Lerda and S. Scivto, Nordita-87/36 P.

[21] J. Fay "Theta Functions on Riemann Surfaces" Lecturer Notes in Mathematics Vo. 352. Springer-Verlag (1973).

[22] A. Belavin and V. Knizhnik Phys. Lett. 168B (1986) 201.

[23] D. Friedan, Z. Qiu and S. Shenker: Phys. Rev. Lett. 52 (1984) 1575.

[24] L. Alvarez-Gaumé, C. Gomez, P. Nelson, G. Sierra and C. Vafa: in preparation.

[25] E. Verlinde and H. Verlinde: Phys. Lett. 192B (1987) 45

[26] J. Atick, G. Moore and A. Sen: "Some Global Issues in String Perturbation Theory". IASSNS-HEP-87/61.
ex 0

ON STRINGS AND J.DOUGLAS VARIATIONAL PRINCIPLE

Tullio Regge
Dipartimento di Fisica Teorica
Universita di Torino
Via Pietro Giuria 1, Torino 10125, Italy

ABSTRACT. The approach of J.Douglas to Plateau's problem is here reviewed with an eye to physical applications. It is shown that the functional of Douglas is related to the wave function of the string in a manner which closely resembles the one proposed by Hawking for the wave function of the Universe.

1. The work of J.Douglas.

In the main the formalism of string theory as it has been developed in the recent years bears a strong formal resemblance to the pioneering work of Monge and Weierstrass on minimal surfaces. The final solution to Plateau's problem has been however achieved only about 60 years ago by Jesse Douglas who discovered a then totally new variational approach to the problem.

Since this approach has remained virtually unknown outside an inner circle of specialists I am going to review it briefly and discuss its relationship to physics and to the standard theory of strings. The work of J.Douglas appeared in successive stages of development starting from Dic. 1926 up to the thirties. Here I refer mainly to ref. [1,2]. In [1] the author discusses mainly the case where the surface Σ is diffeomorphic to the unit disk and is bounded by a closed Jordan curve Γ in euclidean space R^n in n dimensions. In [2] the ideas are then extended to the case where Σ has a non-trivial topology.

Douglas considers all possible ways g of representing Γ as a topological image of the unit circle S^1, i.e. $g: S^1 \longrightarrow R^n$:

$$x = g(\sigma) \quad \text{where} \quad 0 \leq \sigma \leq 2\pi\,, \; x \in R^n \quad \text{and}$$

$$g^\mu(0) = g^\mu(2\pi) \tag{1}$$

K. Bleuler and M. Werner (eds.), Differential Geometrical Methods in Theoretical Physics, 145–148.

The principal idea is then the introduction of the functional :

$$A(g) = \frac{1}{4\pi} \int_{S^1 \times S^1} \frac{|\, g(\sigma) - g(\Theta) \,|^2}{(2 \sin \frac{1}{2} (\sigma - \Theta))^2} \, d\sigma \, d\Theta \qquad (2)$$

where the integrand is simply: square of the chord of the contour Γ over square of the chord of the unit circle S^1. The double integral is improper at $\sigma = \Theta$ but can be handled easily by a limiting procedure (see [1]). In R^n with the standard euclidean metric the functional A(g) is lower semi-continous and therefore, by a theorem of Frechet (see [3]) it attains a minimum for a certain map g^*. The main point of interest is that if the representation $g^*(\sigma)$ is extended inside the unit circle according to Poisson's integral by harmonic functions of u,v :

$$x = \operatorname{Re} F(w) \quad \text{where } w = u+iv \qquad (3)$$

then:

$$\sum_{i=1}^{n} F_i'^2(w) = 0 \qquad (4)$$

Condition (4), according to Weierstrass, defines a minimal surface Σ^*. This surface is bounded by Γ and the area of Σ^* is then equal to $A(g^*)$. The remarkable fact is that the final solution to Plateau's problem, that is Σ, is not obtained by minimizing over a set of surfaces but rather on a set of maps g^*. Therefore the approach of Douglas is not directly related to any Wick rotated canonical formalism.

It is not difficult however to establish a direct connection between the functional A(g) and the quantum mechanics of the string. To this purpose we introduce the Fourier expansion of g(σ) given by:

$$a_m = \frac{1}{2\pi} \int_{S^1} g(\Theta)\, e^{-im\Theta} \, d\Theta$$

In terms of this expansion the functional A(g) can be expressed as:

$$A(g) = 2\pi \sum_{m=1}^{\infty} m \, |a_m|^2$$

Apart from a trivial numerical factor we see that $\Phi = \exp(-A(g))$ is the ground state wave function of all oscillators of the string. It has to be noted however that Φ is not a physical state and does not satisfy Virasoro's constraints except for the case $g = g^*$, that is for the minimal surface. In fact the Virasoro group Diff S^1 acts naturally on the set of all maps g and therefore on $A(g)$ and Φ and we expect that only Φ^* would have a physical meaning.

The above analysis is yet incomplete but it shows already some analogy between Φ and the wave function of the Universe in general relativity. Both of them represent quantum tunneling into nothing. In Hawking's analysis the role of S^1 (of Σ) is played by the space-like hypersurface Σ^3 (euclidean manifold M^4 with boundary Σ^3) which in most applications is restricted to be S^3 (disk D^4) for reasons of simplicity.

There is no analogue in general relativity however for Douglas functional and this suggests the remote possibility that there may exist one. Physically this would imply that we can calculate the present quantum state of the Universe by an action at a distance formalism and this implies in turn some knowledge about the general integral of Einstein's field equations. Of course we know the general integral for the motion of the string.

Finally let me remark that in [2] Douglas discusses the more general case of a non-trivial surface Σ with a boundary Γ made of a multiple set of linked closed Jordan curves. This discussion is needed if we intend to deal with the interacting string.

2. The geometry of strings.

I end the above discussion with a few remarks on the geometry of strings. I am not alone in thinking that a full understanding of the functional space of strings is needed in order to build a logically and esthetically satisfactory theory. At the moment strings are mainly represented as maps $S^1 \dashrightarrow R^n$ and their properties follow from this definition. I would prefer a new axiomatic approach working in exactly the reversed direction where the infinite dimensional functional space of strings is the primary entity and R^n appears in a convenient approximation for it when we deal with the lowest excitations of the string. In [3] the functional space is given the structure of an infinite dimensional Kähler manifold and this is certainly already a very interesting and important step. I would welcome a new step where we free ourselves entirely from the need of introducing a preliminary finite dimensional space-time and where this space-time is a relatively trivial substructure of the whole space.

An infinite dimensional world is a much more interesting object than a finite one. All the extra dimensions would show up at the big bang providede we go close enough to the T = 0 singularity.

While writing this note I received a preprint [4] with an interesting discussion on the Douglas functional A(g) and the role of the group SL(2,R).

3. References.

1.- J.Douglas, "Solution of the problem of Plateau", Trans.Am.Math.Soc., Jan. 1931.

2.- J.Douglas, "Minimal surfaces of higher topological structure", Annals of Math. Vol.40,No.1, Jan. 1939.

3.- M.J.Bowick and S.G.Rajeev, "The holomorphic geometry of closed bosonic string theory and Diff S^1 / S^1". MIT preprint, Submitted to Nuclear Physics B.

4.- V.Guillemin, B.Kostant, and S.Sternberg, "Douglas' Solution of the Plateau Problem", Preprint.

GROUPOIDS AND LIE BIGEBRAS IN GAUGE AND STRING THEORIES

MEINHARD E. MAYER
Department of Physics
University of California
Irvine, CA 92717
USA

ABSTRACT. This talk deals with two seemingly disparate mathematical concepts: Lie groupoids and algebroids, and Lie bigebras (bialgebras), also known as Lie-Hopf-algebras (LHA), which have recently found to have applications in quantum field theory, gauge theory, and string theory. Although many of the mathematical concepts have been around for some time, the interest in groupoids lies in potential application to string theories, whereas Lie bigebras appear in recent work by Drinfel'd and others on "quantum groups" and their relation to the classical Yang-Baxter equation.

Introduction

At one of the early of these conferences (Bonn 1975) [5a], I talked about attempts at formulating gauge field theories in terms of a "representation theory of the holonomy groupoid". My interest in groupoids was rekindled by a recognition that open bosonic strings, and the Witten composition law of strings [6] can be viewed, in part, as groupoid operations. I wish to emphasize the natural way in which some aspects of groupoid theory enter into the formulation of gauge theories and some models of string theory. I will start out with a review of basic definitions. While these notes were being written, I received the recent monograph by Mackenzie [18], which deals with the relation between Lie groupoids and algebroids and principal bundle theory, and contains a treatment of connections in groupoids based on the Atiyah sequence [19]. I was very pleased to see this, since I have been using this approach to the definition of connections in principal bundles in my forthcoming book [20]. Since connections in Lie groupoids turn out to be equivalent to these, this furnishes a further argument in favor of a groupoid approach to gauge theories. The existence of [18] book allowed me to reduce the amount of background material included in this lecture. The reader is advised to consult this book for topics which space does not permit me to treat here.

Several lecturers at this Conference (Fröhlich, de Vega) dealt with r-matrices and Yang-Baxter equations, but did not go into the relations these concepts have with "quantum groups" or Lie-Hopf Algebras. I have therefore decided to include a very brief overview of this new and exciting topic.

Part I of the lecture deals with some aspects of groupoid theory, as applied to gauge and string physics. It contains a brief review of the necessary mathematics.

K. Bleuler and M. Werner (eds.), Differential Geometrical Methods in Theoretical Physics, 149–164.

Part II deals with the other topic: *Lie bigebras and "quantum groups"* and their uses in quantum field theory. This topic is again related to a subject about which I talked at an earlier of these conferences (Bonn 1973[B, see [M,71, 73]), where I made an attempt to advocate the use of W*-bigebras (Hopf–von Neumann) algebras, as a way of describing the (possibly broken) symmetries in algebraic quantum field theory. Since then Doplicher and Roberts [DR] have succeeded in partially completing the Doplicher-Haag-Roberts (DHR) program [DHR 69,71- -74], so that now it can be considered as a self-contained part of general quantum field theory (still missing is the incorporation of gauge theories). Although in the intervening time my interests have shifted in a more differential-geometric direction, I have often expressed my conviction that Hopf-algebraic aspects are bound to play an important role in quantum gauge theory (private conversations with, among others, Daniel Kastler, Rudolf Haag, Detlev Buchholz, and others). Unfortunately, the mathematical machinery seemed to have been lacking. At the International Congress of Mathematicians of 1986 P. Cartier presented in Berkeley the paper "Quantum Groups" of V. G. Drinfel'd (which appeared in Russian [D 86] and reached me in May, 1986), in which Lie-Hopf algebras (LHA) or Lie bigebras (Drinfel'd uses the longer name Lie- bialgebras) are introduced and their relations to solutions of the classical Yang-Baxter equations (CYBE) are discussed, and the notion of quantum group and their relation to the quantum Yang-Baxter equations (QYBE) are explored. Part 2 consists of a review some of Drinfel'd's definitions and an outline of the way I hope LHA-s will enter quantum field theory and conformal field theory in the near future.

The references to the two parts are given in different styles: those to Part 1 (groupoids) are in the form [Number], whereas those for Part 2 are in the form [Author, Year].

1. Groupoids, Gauge Theory, and Strings

1.1. INTRODUCTION

Although the "noncommutative differential geometry" of Alain Connes [1] has had a considerable impact on mathematics (see, e. g., [2]) few physicists paid attention to this very important development. The most notable exception was Daniel Kastler, who has heroically attempted to spread the new gospel in the mathematical physics community, cf. [3] as well as, more recently, Jean Bellissard [4], who (with collaborators) has pointed out that Connes' K-theory has remarkable applications in the theory of the quantum Hall effect as well as in other areas of condensed-matter physics. Under Daniel Kastler's influence, I have made some feeble attempts to call attention to this important development [5]. However, this situation has started to change: in a recent paper [6] (and a sequel), E. Witten has attempted to discuss the implications of Connes' noncommutative geometry for superstring theory [7], in particular, the interpretation of the BRS(T) charge operator Q_{BRS} [8] as a graded derivation. He proposed a "composition law" for open strings which strongly suggests an interpretation in terms of a groupoid — a concept which also plays a fundamental role in Connes' theory, and in the theory of foliated manifolds [10–12]. Among others, I have tried for several years to espouse the point of view that the BRS-charge and the Kugo-Ojima criterion for selecting the physical subspace in quantized gauge theories should be taken seriously, and that the group of gauge transformations and its Maurer-Cartan form (the "ghost") should be given consideration already at the classical level.

Recently Leonard Gross [13] has proposed an approach to gauge theories based on "lasso forms". Although his emphasis is on finding a measure on such forms, his results are closely related to a groupoid reformulation of gauge theory, as well as to early work by H. Loos [13a] as well as to my early attempts to quantize the holonomy of gauge fields [5a], and these relations need to be investigated further.

In 1986 I translated a paper of M. B. Mensky [14] which discusses a related approach to the theory of interacting strings, in terms of the group of paths on the path group of a manifold. Although Mensky's approach and aim is different, his paper (and presumably some of his other publications) makes some valuable technical points on the relations between the representations of the path groupoid, the fundamental group and the group of free paths on a homogeneous space. I will not discuss this paper here.

I will try to explain how one could possibly use the groupoid concept for the description of the composition of open strings and how it could be extended to closed strings. Since I am not an expert on superstring theory, these remarks should be interpreted as heuristic, and they are published in the hope that the people actively engaged in research will find them useful and will be able to exploit them more fully.

Since the notions of groupoid and algebroid are not very familiar to physicists. I start in Sec. 1.2 with a brief listing of the definitions and a few examples (for details see [9, 18]). Section 1.3 is devoted to a discussion of strings in terms of groupoids, thus giving some intuitive feeling for groupoids. Sec. 1.4 returns to the definition of Lie groupoids and algebroids and gives a brief account of the theory of connections and holonomy. Section 1.5 deals with a reformulation of gauge theory in terms of Lie groupoids and algebroids and their connection with principal bundle theory.

1.2. WHAT ARE GROUPOIDS?

A *groupoid* Γ can be thought of as a collection of transformations ("*arrows*") α, β, γ, of a set B of *objects* which can be identified with the set of *units* Γ^0 of Γ. The elements of B will denoted by $x, y, z \in B$, and their images in Γ by $\tilde{x}, \tilde{y}, \tilde{z} \in \Gamma$, respectively. The "object inclusion map" which identifies elements of the *base space* B with units in Γ^0 is denoted by $\epsilon : x \mapsto \tilde{x}$. If γ is an arrow from x to y then $x := s(\gamma)$ is called its *source* and $y := t(\gamma)$ is called its *target* (or source and target *projection*, respectively. The reversed arrow is called the *inverse* of γ and is denoted by γ^{-1}. The composition of arrows $\alpha\beta$ is defined only if the target of β is the source of $\alpha : t(\beta) = s(\alpha)$ (mappings compose from right to left!). Composition is *associative* iff both terms $\alpha(\beta\gamma)$ and $(\alpha\beta)\gamma$ are defined. Obviously:

$$s(\gamma^{-1}) = t(\gamma), \ t(\gamma^{-1}) = s(\gamma), \quad \text{and} \quad \gamma^{-1}\gamma = \widetilde{s(\gamma)}, \ \gamma\gamma^{-1} = \widetilde{t(\gamma)};$$

this is the reason why the arrows $\tilde{x}$ corresponding to the objects x are called *units* or *identities*. It is easy to see that they actually behave like units, i.e., if $s(\gamma) = x$ and $\gamma\eta = \eta$, then $\gamma = \tilde{x}$, etc. The set of composable pairs of arrows is denoted by Γ^2.

For two subsets A and C of Γ one uses the notation $\Gamma_A^C = t^{-1}(C) \cap s^{-1}(A)$ and $\Gamma\mid_A = \Gamma_A^A$, the restriction of the groupoid to A. If A consists of a single element $A = \{x\}$ then $\Gamma_x^x = G(x)$ is the *isotropy group* (or *vertex group*) of Γ at $x \in B$. There are two trivial extreme cases: *a)* when a groupoid consists of units only (no arrows) it is just a space without any structure; it can be identified with its base B; *b)* when the groupoid has only one object (unit), i.e., the arrows are always composable, and the groupoid reduces to a group.

It is sometimes convenient to identify the objects with the units as identities, i.e., omit the tilde and write, when defined: $x\gamma = \gamma$ and/or $\gamma x = x$. The definition of a *subgroupoid* is the obvious one; the set of units $\Gamma^0 = \tilde{B}$ is sometimes called the *base subgroupoid*.

A groupoid is called *transitive* if the map (s, t) from Γ to $B \times B = \epsilon^{-1}(\Gamma^0 \times \Gamma^0)$ (called the *anchor* by Mackenzie [18]) is surjective (onto) and it is called *principal* if it is bijective (one-to-one). A groupoid is called *connected* if for any two units x, y there exists an element α such that $x = s(\alpha)$ and $y = t(\alpha)$. It is convenient to think of a groupoid as "paths" (more precisely, as *homotopy classes or reparametrization classes of paths*, as will be explained below): the inverse path is defined, two paths can be composed if the end of one is the beginning of the next, associativity holds only if both compositions are defined, and left and right identities are defined as the addition of a zero path at the left or right end.

A *topological groupoid* is a groupoid where composition and taking of inverses (when they exist) are continuous. The properties of topological (smooth) groupoids are easily established locally in terms of germs. It turns out that topological groupoids are equivalent to Cartan principal bundles (principal bundles which are not necessarily locally trivial [18]). Similarly one can define *smooth groupoids* and (see below, Sec. 1.4) *Lie groupoids* and their associated *Lie algebroids*, structures which are closely related to principal bundles.

A *representation* of a topological groupoid Ω over a base space B is defined as a morphism of the groupoid into an *action* of a groupoid on a fibered space (M, p, B) over the same base space B (in particular, a vector bundle over B) defined as follows. Consider the "fibered product" $\Omega \star M = \{(\gamma, m) \in \Omega \times M | s(\gamma) = p(m)\}$. An *action* of Ω on (M, p, B) is a continuous map $\rho : \Omega \star M \to M, (\gamma, m) \mapsto \gamma m \in M$ with the properties:

(*i*) $$p(\gamma m) = t(\gamma), \qquad \forall\, (\gamma, m) \in \Omega \star M,$$

(*ii*) $$\beta(\gamma m) = (\beta\gamma)m, \quad \forall\, (\beta, \gamma) \in \Omega^2,\ (\gamma, m) \in \Omega \star M,$$

(*iii*) $$\widetilde{p(m)}m = m, \qquad \forall\, m \in M.$$

The cases of interest are when (M, p, B) is a topological group bundle and $m \mapsto \gamma m$ is a topological group isomorphism, or, when (M, p, B) is a *vector bundle* and $m \mapsto \gamma m$ is a linear map between fibers. In this case we can talk about a *linear representation of the groupoid* Ω. In particular, if the fibers are finite-dimensional vector spaces, not necessarily of the same dimensions, the linear maps among them (which represent the action groupoid) are $n \times m$ matrices, and the composition is defined only if the number of rows of the matrix on the left equals the number of columns of that on the right. An infinite-dimensional representation may have Hilbert bundles as objects and unitary transformations as elements of the groupoid.

As far as physics is concerned, the C^*-algebra of a groupoid may be the most useful concept, although it requires some technical concepts related to measure theory (this part is rather sketchy, and the reader should consult the literature for details). First of all one must define local compactness on a topological groupoid. Then one defines, as usual, the space of continuous functions with compact support on Γ, which will be denoted by $C^0(\Gamma)$. In Connes' theory the notion of measurable groupoid plays an important role. Connes defines a concept of transverse measure and Haar system for locally compact groupoids, concepts which extend the integration theory on groups. In terms of this noncommutative integration $C^0(\Gamma)$ can be turned into a C*-algebra by defining a convolution $f \star g$, and an involution f^*. The representation theory of the groupoid is then defined in terms of a Hilbert bundle and its space of square-integrable sections. One of the most significant new concepts introduced by Connes is the construction of the holonomy groupoid of a foliated manifold. (The holonomy groupoid of a gauge-theory bundle should be familiar to physicists – its elements are the "Wilson parallel transporters" $\int e^A$ of lattice gauge theory along a path.)

Before continuing with the mathematical definitions of Lie groupoids, Lie algebroids, and connections, which are applicable to gauge theory I will briefly illustrate how groupoid-related ideas appear in the theory of open bosonic strings.

1.3. OPEN (BOSONIC) STRING INTERACTIONS AS GROUPOID OPERATIONS.

The elementary interactions of open and closed strings are described in the various reviews of John Schwarz [6]. Consider the most elementary interaction (of the Yang-Mills type) in which two strings break or join (Fig. 1). It is clear that since the words string and path are almost synonymous, these interactions can be interpreted directly in terms of groupoids. In other words, we have a "representation" of the path groupoid of $(D-1)$- dimensional space (the D-th dimension plays the role of time, and thus the string sweeps out a "world sheet" in D-space) by the open strings.

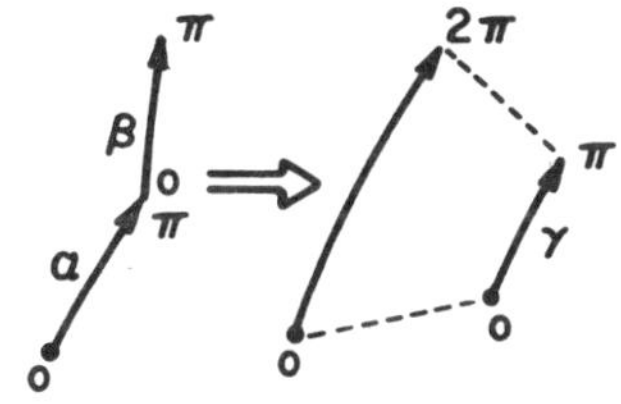

Fig. 1

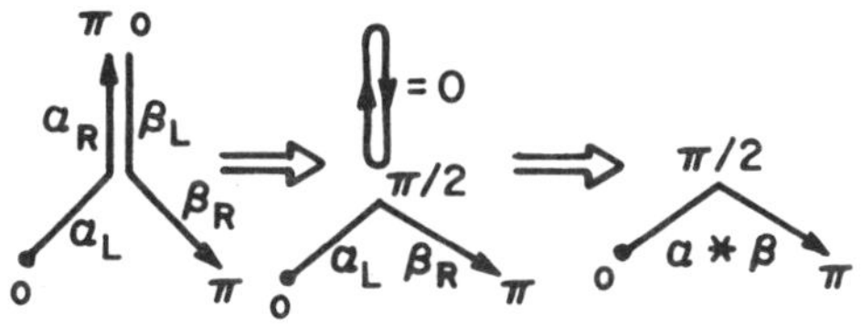

Fig. 2

On the other hand, in order to be able to accommodate the BRS-charge operator in string theory, Witten [6] has proposed a different composition law for open strings: each string is divided into a left half (parameter between 0 and $\pi/2$) and a right half (parameter running from $\pi/2$ to π). Two oriented open strings α and β can be composed if and only if the left half of the string α coincides with the right half of the second string β: these two halves are traversed in opposite directions, and are thus an "appendix" attached to the middle of the resulting string. The appendix is then "shed", resulting in a new string $\gamma = \alpha * \beta$ reparametrized with a parameter σ running again from 0 to π.

Fig. 3

Just like a chromosome, the new string resulting from the $*$-composition carries the "information of its two parents": the left half comes from string α, the right half, from string β (Fig. 2). Another type of interaction of oriented open strings is the exchange interaction: two strings cross in their midpoints and exchange right and left halves (Fig. 3). This is a new operation and can be interpreted as a groupoid operation only if we decompose the "crossing" into several steps as illustrated in Fig. 4: first the strings split at their midpoints and grow "virtual appendices" (parameters running respectively from 0 to $\pi/2$ and from $\pi/2$ to π); these appendices split, their halves are "flipped over", they recombine into new appendices ($\mu_L * \nu_R$ and $\nu_L * \mu_R$) which are then shed, resulting in the final configuration ($\alpha_L * \beta_R$ and $\beta_L * \alpha_R$). Reparametrizations are done "on the fly". This works in $(D-1)$ dimensions, but is hard to interpret in D dimensions, where the world sheets representing the strings and their appendices get hopelessly entangled.

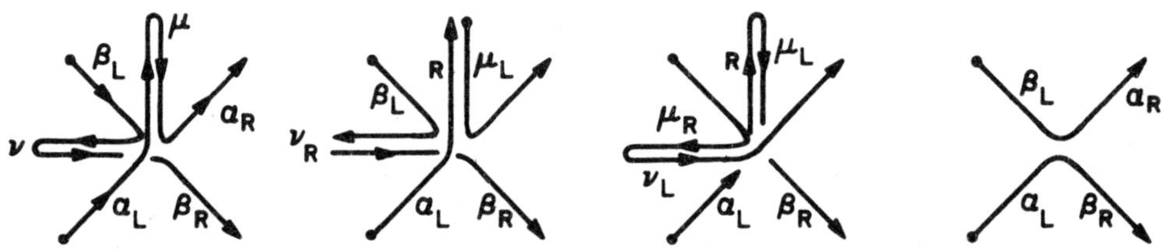

Fig. 4

It is clear that the composition $\alpha * \beta$, can be extended to three strings α, β, γ. However, these can be combined in two different ways: $\alpha * (\beta * \gamma)$ and $(\alpha * \beta) * \gamma$ (Fig. 5). As it stands this composition is not associative, since the two resulting parametrizations will be different. To get from one to the other we have to use the *homotopy* represented in Fig. 6.

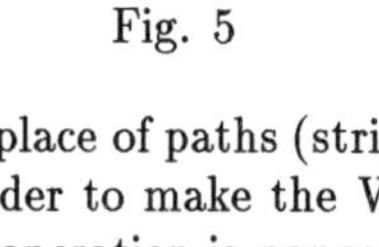

Fig. 5

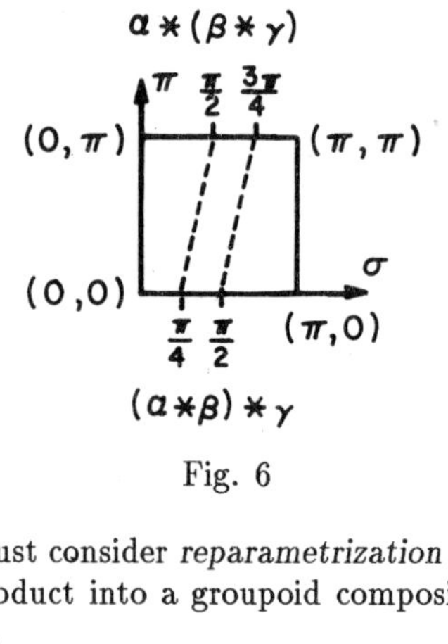

Fig. 6

This is why in place of paths (strings) one must consider *reparametrization classes* of paths (strings), in order to make the Witten $*$-product into a groupoid composition. It is also clear that this operation is noncommutative.

However, Witten's prescription cannot handle the interactions of closed strings (which is nonassociative) or the interaction of closed and open strings. In order to discuss these interactions one is forced to consider the 2-dimensional world sheets swept out by these strings in D-dimensional ambient space, as is done in the literature devoted to the field theory of strings (cf [7]). See, however, Graeme Segal's talk [15] in this volume, where the interactions of closed strings are described in semigroup language.

It should also be noted that the Q_{BRS} operator, which appears in the "kinetic" term of the Witten action $\int(A * QA + \frac{2}{3}A * A * A)$ can be given a simple interpretation in the cohomology theory of Lie algebroids. I hope to return to this question in a more detailed publication soon.

1.4. LIE GROUPOIDS AND ALGEBROIDS. CONNECTIONS AND HOLONOMY

This section contains a very brief summary of facts about Lie groupoids, Lie algebroids and their relation to principal bundles, as well as the definition of connections.

The definition of a *Lie groupoid* Γ over a base manifold B is straightforward: one first makes Γ into a manifold, and requires that the projections s and t should be smooth (surjective submersions), and the multiplication (where defined) should be smooth. One then introduces a notion of *local triviality* and defines a *Lie groupoid as a locally trivial differentiable groupoid.* Equivalently, one can define a Lie groupoid as a differentiable groupoid for which the source map $s : \Gamma \to B$ is a *fibration* (surjective submersion). The same is also true for the target map $t : \Gamma \to B$. The *fibers* $\Gamma_x = s^{-1}(x)$ and $\Gamma^y = t^{-1}(y)$ are manifolds and any element $\gamma \in \Gamma$ such that $s(\gamma) = x, t(\gamma) = y$ defines a diffeomorphism between the fibers. A vector field ξ on Γ is defined to be right-invariant if it is vertical with respect to $s : s_*\xi = 0$ and $\gamma_*\xi = \xi$. On the other hand, already at the topological level, one can prove an equivalence between locally trivial groupoids and principal bundles. Much of the theory of reduction of the structure group of principal bundles also translates into groupoid language. There is obviously a close relation between Lie groupoids and smooth principal bundles. In particular, the theory of connections extends easily to Lie groupoids, if one starts out from the Atiyah sequence, which I recall briefly.

Atiyah defines a connection in a principal bundle $P(M, G, \pi)$ [19] (cf. also the forthcoming book [20]), as a splitting of the exact sequence of vector bundles $\mathcal{A}(P)$ (*Atiyah sequence*) defined as follows. Let $AdP = P \times_{Ad} \mathcal{G}$ be the bundle associated to P with fiber the Lie algebra $\mathcal{G}$ of the structure group G and the adjoint action of G on $\mathcal{G}$; let TP/G denote the quotient of the tangent bundle of P with respect to the right action of G, considered as a vector bundle over TM with projection the tangent map π_* of π, and let ι denote the inclusion map of $Ad(P)$ in TP/G (the identification of elements of $\mathcal{G}$ with fundamental vector fields. Then the Atiyah sequence $\mathcal{A}(P)$:

$$0 \longrightarrow Ad(P) \xrightarrow{\iota} TP/G \xrightarrow{\pi_*} TM \longrightarrow 0 \tag{1.1}$$

is exact, and a *connection is defined as a splitting of* (1.1), i.e, a vector bundle morphism $\chi : TM \longrightarrow TP/G$ such that $\pi_* \circ \chi = \mathrm{Id}_{TM}$. This induces a morphism $\omega : TP/G \longrightarrow Ad(P)$ such that $\iota \circ \omega = \mathrm{Id}_{TP/G}$ (to distinguish ω from the connection one-form sometimes denoted by the same letter, Mackenzie calls χ the connection and ω a *back-connection.* This splitting leads to the well-known direct-sum decomposition of the tangent spaces to P into horizontal and vertical parts, to the identification of ω with the $\mathcal{G}$-valued connection form defined as a section of $Ad(P)$, or to the definition of a connection in terms of a family of $\mathcal{G}$-valued one-forms over the open sets U_i of a trivializing atlas of M, with the known properties. Smoothness and equivariance under the right action of G completes the picture. The *curvature* is then defined in terms of a bracket for χ, or as a measure of failure of the connection form to be Maurer-Cartan, and the usual structure equations follow.

1.4.1.· *Lie Algebroids and Lie-Algebra Bundles.* In the same way that the Lie algebra of a Lie group determines its local properties and, in most interesting cases, is sufficient for physical applications, the Lie algebroid of a Lie groupoid contains much of the local information about the Lie groupoid (and because of the existence of an exponential map, exhibits some similarities to the properties of the Lia algebra of a Lie group). Here is the definition:

A *Lie algebroid* on a manifold B is a vector bundle (A, p, B) together with a vector bundle map $q : A \longrightarrow TB$ (called the *anchor* of A by Mackenzie), and a bracket operation on sections, $[\ ,\ ] : \Gamma(A) \times \Gamma(A) \to \Gamma(A)$ ($\Gamma(A)$ is the vector space of smooth sections of A, not to be confused with the groupoid Γ) which is bilinear and skew-symmetric, satisfies the Jacobi identity and the two properties:

$$q([X,Y]) = [q(X), q(Y)], \qquad X, Y \in \Gamma(A); \tag{1.2a}$$

$$[X, uY] = u[X,Y] + q(X)(u)Y, \quad X, Y \in \Gamma(A), u \in C^\infty(B). \tag{1.2b}$$

(Mackenzie calls a Lie algebroid A *transitive* if q is a submersion, *regular* if q is of constant rank, and *totally intransitive* if $q = 0$.

A *Lie algebra bundle* (LAB) is a vector bundle (E, p, B) with a field of Lie brackets such that E admits an atlas $\{\psi_i : U_i \times \mathcal{G} \to L|_{U_i}\}$, ψ_i being Lie-algebra isomorphisms. Any LAB is a totally intransitive Lie algebroid, but the converse is not necessarily true. The most interesting example is the *Lie algebroid of covariant differential operators on a vector bundle* (see [18], Sec. 3.2).

It turns out that transitive Lie algebroids are the middle term of an Atiyah sequence of the type (1.1), and one is thus naturally led to the definition of a splitting, which defines an *infinitesimal connection*. Since the representation of the fundamental groupoid of the base manifold in a Lie groupoid defines a notion of *parallel transport*, and hence a *path connection*, it is necessary to distinguish these two notions.

1.4.2. *Infinitesimal Connections and Holonomy.* It is surprising that the relation between connections in a groupoid and connections in principal bundles, which was known for a number of years, has not caught the attention of many differential geometers and even less that of us physicists interested in gauge theories.

An *infinitesimal connection* in a Lie groupoid is defined as a splitting of the sequence $\mathcal{A} : L \longrightarrow A \longrightarrow TB$ associated to its Lie algebroid A (see [18], Ch. III), where L is the *adjoint Lie-algebra bundle* which makes the sequence exact. Thus a connection $\chi : TB \to A$ is a linear map such that $q \circ \chi = \mathrm{Id}_{TB}$. The theory of infinitesimal connections in Lie groupoids is practically identical to connection theory as outlined above. The curvature is defined either in terms of the Schouten-Nijenhuis bracket [20, L 83] for vector-bundle valued forms, or, simpler still, in terms of the brackets defined above, as a skew-symmetric vector bundle morphism $R_\chi : TB \oplus TB \to L$ defined by

$$\iota(R_\chi(X,Y)) = \chi[X,Y] - [\chi X, \chi Y], \quad X, Y \in \Gamma(TB). \tag{1.3}$$

The most promising aspect of the groupoid approach lies in the relation between infinitesimal connections and the action of its *holonomy groupoid.* The latter is defined in terms of a *path connection*, i.e., a mapping from the continuous paths in B (or reparametrization classes of paths) into piecewise smooth paths in Γ. Space does not permit me to go into all details, which can be found in [18]. The important result is that the holonomy groupoid of a path-connection in a groupoid is a reduction of the groupoid (this is related to the Ambrose-Singer theorem for connections in principal bundles). In the Lie-groupoid-algebroid context the proof of the Ambrose-Singer theorem on the relation of curvature and holonomy is seen in a new light and becomes very simple. Since the monograph [18] treats this in detail, I will not dwell on this here. I have changed my terminology slightly from that used in the oral presentation of the talk, to bring it into conformity with the one used by Mackenzie. A brief outline of how gauge theory can be translated into this language (and what, if anything, can be learned from this) follows.

1. 5. A GROUPOID FORMULATION OF GAUGE THEORY.

A reformulation of gauge theories in terms of the holonomy groupoid is only a slight generalization of the formulation in terms of the holonomy group (Wilson loops – in lattice gauge theory). Thus, we replace the principal bundle $P(G, M, \pi)$ with structure group G and base space M by a Lie groupoid Γ and its Lie algebroid $A(\Gamma)$, and in place of the connection form ω and its curvature $\Omega = d\omega + \frac{1}{2}[\omega, \omega]$ we use the connection χ in the algebroid and its curvature defined as in Eq. (1.3) (or in terms of the Schouten-Nijenhuis bracket).

The more global point of view, in terms of representations of the *holonomy groupoid* seems more promising, and the rest of this section is devoted to a quick survey of this. First, it is necessary to distinguish between the notions of a *curve* c_x^y between two points $x, y \in B$ considered as a one-dimensional submanifold of B with a *fixed parametrization*: $c_x^y : [0,1] \ni \sigma \mapsto c_x^y(\sigma) \in B, c_x^y(0) = x, c_x^y(1) = y$ and a *path* (or *path-class*) $\hat{c}_x^y$ is an equivalence class of curves differing from each other by a reparametrization $\sigma \mapsto \sigma(\tau)$ or the attaching of an *appendix*, i.e., another curve attached to any point and traversed back and forth. An appendix may also be attached to the endpoint of a curve. However, attaching a *loop* at the end of a curve is to be considered a distinct operation since the "spoon"[13a] or "lasso"[13] formed in this way can pick up an element of curvature, and the composition of a path with a loop may produce a different kind of *parallel transport* from x to y than the path $\hat{c}_x^y$. It is clear that the juxtaposition of paths (followed by a reparametrization of the resulting path) leads to a groupoid, where the inverse is the same path traversed in the opposite direction, and the source and target maps are the points x, y or appendices attached to them. I will denote this *path groupoid* by Π. The isotropy subgroup $G(x) = \Pi_x^x$ is identical to the loop group at the point x (reparametrization classes of loops based at this point), and is not to be confused with the *fundamental group*, which consists of homotopy classes of loops.

In order to describe a gauge theory, we consider a Lie groupoid Γ together with its Lie algebroid over B and associated families of vector bundles, and a representation of the path groupoid Π in Γ, as well as the induced action (*parallel transport*) in the associated vector bundles. This is implemented by means of the exponential map, and leads to a *product-integral (path-ordered exponential of the integral) of covariant differential operators* in these bundles. In other words, in the groupoid approach, we replace the connection one-form $A_\mu(x)dx^\mu$ by the action groupoid of parallel transport induced in associated vector bundles. The role of path-connection and holonomy becomes more prominent, and the theory is open to various generalizations, although at this level it is completely equivalent to conventional gauge theory in terms of connections and curvatures in principal bundles.

The group of gauge transformations, identified with the group of sections of the associated bundle $AdP = P \times_{AdG} G$ is seen in a new light in the groupoid interpretation, namely as an intertwining of the groupoid action with the adjoint action. One is naturally led to the BRS cohomology as part of the cohomology theory of the corresponding Lie algebroid. Details of this will be found in my forthcoming book [20].

The new insight one gains on gauge theory by switching from principal bundles to Lie groupoids and algebroids is the renewed emphasis on *paths and parallel transport* of objects along paths in the base space. This becomes even more important in Kaluza-Klein type theories, where the dimensional reduction and compactification of parts of the fiber leads to a new fibered structure of the base space which becomes a locally a product of a space-time neighborhood and a compact homogeneous space. Thus a path on the base space might be thought of as a path in this fibered space, where the "appendix" which lies entirely in the homogeneous space is "macroscopically null-homotopic", and thus does not contribute to the holonomy, except at the scale of the compactified homogeneous manifold.

Another interesting aspect needing further study is the problem of higher-dimensional extended objects. We have already seen that open strings have a natural interpretation in terms of groupoids. The problem of defining a Stokes theorem [13, 14] for noncommutative differential forms leads to interesting extensions of groupoid theory to "ordered membranes" or higher-dimensional analogues.

2. Lie Bigebras and Quantum Groups

2.1. WHAT ARE BIGEBRAS ?

This section contains a brief review of the basic facts about algebras, cogebras, and bigebras (a.k.a. bialgebras or Hopf algebras). By *algebra* will be mean an associative algebra A over the field K ($K = \mathbf{R}$ or $\mathbf{C}$) with unit $\mathbf{1}$, and multiplication denoted $m : (x, y) \mapsto xy,\ x, y \in A$. The natural embedding of K in A (given by the unit) will be denoted by $u : K \hookrightarrow A$. The tensor product $\otimes_K$ will be denoted simply by $\otimes$.

Then multiplication can be viewed as a morphism $m : A \longrightarrow A \otimes A$ which makes the following diagram commutative (this simply expresses the associativity of the algebra A):

$$
\begin{array}{ccc}
A \otimes A \otimes A & \stackrel{\mathrm{m}\otimes\mathrm{Id}}{\longrightarrow} & A \otimes A \\
\downarrow{\scriptstyle \mathrm{Id}\otimes\mathrm{m}} & & \downarrow{\scriptstyle \mathrm{m}} \\
A \otimes A & \stackrel{\mathrm{m}}{\longrightarrow} & A
\end{array}
\tag{2.1}
$$

The embedding of K in A can be described by the following commutative diagram:

$$
\begin{array}{ccccc}
K \otimes A & \stackrel{\mathrm{u}\otimes\mathrm{Id}}{\longrightarrow} & A \otimes A & \stackrel{\mathrm{Id}\otimes\mathrm{u}}{\longleftarrow} & A \otimes K \\
\| & & \downarrow{\scriptstyle \mathrm{m}} & & \| \\
A & \stackrel{\mathrm{Id}}{\longrightarrow} & A & \stackrel{\mathrm{Id}}{\longleftarrow} & A
\end{array}
\tag{2.2}
$$

A *cogebra* (or coalgebra) is defined by simply turning all the arrows around in Eqs. (2.1) and (2.2). In other words, a vector space C over K is a *coassociative cogebra* with *comultiplication* $\Delta : C \longrightarrow C \otimes C$ (Δ stands for *diagonal map*) and *counit* ϵ (also called augmentation), so that the following "dual" diagrams are commutative :

(2.3)
$$\begin{array}{ccc} C\otimes C\otimes C & \xleftarrow{\Delta\otimes \mathrm{Id}} & C\otimes C \\ \uparrow{\scriptstyle \mathrm{Id}\otimes\Delta} & & \uparrow{\scriptstyle \Delta} \\ C\otimes C & \xleftarrow{\Delta} & C \end{array}$$

and

(2.4)
$$\begin{array}{ccccc} K\otimes C & \xleftarrow{\epsilon\otimes \mathrm{Id}} & C\otimes C & \xrightarrow{\mathrm{Id}\otimes\epsilon} & C\otimes K \\ \Vert & & \uparrow{\scriptstyle \Delta} & & \Vert \\ C & \xleftarrow{\mathrm{Id}} & C & \xrightarrow{\mathrm{Id}} & C \end{array}$$

A vector space which is both an algebra and a cogebra is called a *bigebra or bialgebra* (I prefer the shorter term "bigebra", used by Bourbaki, Dieudonné, and Cartier in the early papers on the subject). Some authors call such an object a *Hopf algebra*, but following Sweedler and others this term will be reserved for bigebras with an additional structure element, related to the group inverse, called an *antipode*.

The notion of antipode is motivated by considering the algebra $A = Fun(G)$ of functions on a group G (continuous functions for a locally compact group, smooth functions for a Lie group, etc.) with pointwise addition and multiplication.

We consider, for example, the algebra $F = L^\infty(G)$ with multiplication pointwise multiplication of functions on G, and the unit defined as the function $f = 1$. The algebra F is also a cogebra under the *comultiplication*

(2.5) $$\Delta : F \longrightarrow F\otimes F : \Delta(f)(s,t) = f(st), \qquad f(st) \in L^\infty(G\times G) = F\otimes F, \quad f \in F,$$

with the counit naturally defined as evaluation at the group identity $\epsilon : f \mapsto f(e) \in \mathrm{R}$. This makes F into a *bigebra* or *bialgebra*, i.e., a vector space which is both an algebra and a cogebra. In our example the algebra is commutative, but the cogebra is not *cocommutative*, since $st \neq ts$.

The existence of the inverse g^{-1} in the group leads to the definition of the *antipode* (or *involution* j, whose properties can be derived from the example under consideration. Define the *involution* in F by $j(f)(g) = f(g^{-1})$. It is *involutive*: $j^2 = j\circ j = \mathrm{Id}$, and since $(gh)^{-1} = h^{-1}g^{-1}$, its interaction with Δ is given by $\sigma\circ\Delta\circ j = (j\otimes j)\circ\Delta$, where $\sigma : F\otimes F \to F\otimes F : f_1\otimes f_2 \mapsto f_2\otimes f_1$ permutes the two algebras.

The interaction of Δ, j, u, and m can be summarized in the following commutative diagram:

(2.6)
$$\begin{array}{ccccccc} A & \xrightarrow{\Delta} & A\otimes A & \underset{\mathrm{Id}\otimes j}{\xrightarrow{j\otimes \mathrm{Id}}} & A\otimes A & \xrightarrow{m} & A \\ {\scriptstyle \epsilon}\downarrow & & & & & & \uparrow{\scriptstyle u} \\ K & & & & & & K \end{array}$$

where the mappings above and below the horizontal arrow correspond to the two orderings of factors in the tensor product, and the two copies of the field $K = \mathbf{R}$ or $\mathbf{C}$ at the bottom should be identified.

The set $(F, m, u, \Delta, \epsilon, j)$ is an example of a *commutative Hopf algebra* or *commutative bigebra*. The general definition follows by abstracting from this example, i.e., an *involutive bigebra* or *Hopf algebra* (not necessarily commutative or cocommutative) will be a vector space F endowed with a multiplication, unit, comultiplication, counit, and involution satisfying the diagrams (2.1)–(2.6). In the noncommutative case the definition of involution or antipode is not unique — one can introduce a second antipode, j', corresponding to the other order of multiplication but the same comultiplication. The two antipodes are "inverses" in the sense that $j'j = jj' = \mathrm{Id}$, but we no longer have $j^2 = \mathrm{Id}$ so that the term involution is no longer appropriate.

2.2. LIE BIGEBRAS AND THEIR APPLICATIONS

A *Lie bigebra* (or *Lie Hopf algebra* = LHA) is defined as a vector space $\mathcal{G}$ which has the structure of a Lie algebra and of a (Lie) cogebra, such that the map Δ —which in this context may be called a *cocommutator or cobracket* – - is a 1-cocycle (since $\mathcal{G}$ acts on $\mathcal{G} \otimes \mathcal{G}$ via the adjoint representation). Drinfel'd proved that the category of finite-dimensional Lie bigebras is equivalent to the category of connected simply connected Poisson-Lie groups.

A *Poisson-Lie group* (PLG) is a Lie group G with a Poisson manifold structure (cf., e.g., [L 83], [W 83]), on $Fun(G)$, so that the comultiplication $\Delta : Fun(G) \to Fun(G \times G) = Fun(G) \otimes Fun(G)$ is a Lie algebra homomorphism. A *quantization* of a Poisson algebra is defined as a *deformation* depending on a parameter $\hbar$ (cf. [L83], where additional references can be found).

Drinfel'd proves a number of other beautiful results on Poisson and co-Poisson algebras. He also establishes a remarkable relation with the r-matrices which are solutions of the classical Yang-Baxter equation, which was discussed extensively in the lectures by Fröhlich, de Vega, and others at this Conference. The motivation for this work came from the quantum inverse scattering approach to integrable dynamical systems.

Another, and even more exciting, suggestion made by Drinfel'd is the *quantization of universal enveloping algebras* — QUE-algebras — and the so-called QFSH (quantum formal series Hopf) algebras.

As far as I know at present no serious attempts have been made to relate Drinfel'd's work to the Doplicher-Roberts [DR 72, 84, 87] work, where Hopf algebras also play a prominent role. I hope to be able to return to this problem in the near future, and also explore how this machinery could be used in a quantum theory of gauge fields.

3. Outlook

It is safe to predict that the uses of Lie groupoids, their algebroids and connections, and their representations in C*- and W*-algebras will make some inroads into mathematical physics. The Hopf-algebra techniques which have appeared in two disjoint physics contexts — the Doplicher-Haag-Roberts approach to superselection sectors and symmetries in algebraic field theory, and the Drinfel'd-Belavin-Cherednik-Sklyanin papers, dealing with applications to Yang-Baxter equations, r-matrices, and inverse scattering —will also spread.

One can hope that these two directions will converge. One possible area of possible overlap is the circle of ideas discussed by Fröhlich in his talk, coming from two-dimensional field theories and leading to knot theory and braid groups.

Although it is not clear at present whether there is any direct link between the two seemingly disparate topics discussed in this talk, allow me to get somewhat speculative and express the hope that both Lie groupoids and Lie bigebras will find uses in a future theory which subsumes all the interesting aspects of gauge theories and string field theory. I will try to outline a vague "scenario" of how this could work. I also want to point out that these approaches should be tried out in another area where progress has been slow: quantum theory in gravitational fields and quantum theory of the gravitational field itself.

A future quantum theory of extended objects (be it strings, membranes, or theories which involve an underlying manifold which itself has a bundle structure with space-time as base and some "compactified Kaluza-Klein space" as fiber) has to incorporate two major ingredients: the differential geometry/topology of the underlying structures and the operator-algebraic aspects of fields, observables, states, and generalized superselection structure. Clearly the mathematical objects discussed in this talk will, in their present form or a future refined form, be necessary to describe these objects. Let me try to sketch a model of such a structure, in a very vague and qualitative way. Suppose that the quantum theory of such "extensons" (a temporary name for such a generic *extended but elementary* objects, such as strings, membranes Kaluza-Klein spaces) is described by some big algebra $\mathcal{F}$ — the *field algebra* and that there is a distinguished subalgebra $\mathcal{A}$ — the *observables* and an associated set of *states* $\mathcal{S}$, a subset of which can be given a manifold structure and identified with the space-time manifold. One of the most elementary notions which arises in this context is that of *holonomy*, i.e., the representation of reparametrization classes of paths in either the field algebra $\mathcal{F}$, or the space of states $\mathcal{S}$. In other words we are dealing with a *representation of the holonomy groupoid* of the "classically underlying structure" by the *groupoid of intertwiners* [DR 84, 86] of the appropriate algebras. On the other hand, these objects have a Hopf-algebra structure, since they also embody the duality of the internal symmetry (group) of the theory. A further study of these aspects might thus establish a link between the two seemingly disjoint structures discussed in the body of this talk.

Another aspect which has not been mentioned here at all (except that it is somehow implicit in superstring theories) is gravity, either as a quantum theory of gravity itself, or, more modestly, as a consistent theory of quantum fields in a gravitational background. This difficult subject has formed the object of much research, ranging from treatments which make little sense, dealing with such incomprehensible notions as "**the** *wave function of the Universe*" to some attempts at incorporating curved space into the algebraic approach to quantum field theory (see, e.g., [HNS 84], [FH 86], and remarks on using the tangent bundle I made at a previous Como gathering two years ago (last reference in [5])). Here too, we might want to see if progress could not be made by describing the effects of the curved background by its holonomy groupoid represented as intertwiners of the field algebras defined over different tangent spaces. This is a topic which deserves special attention, and I intend to return to it in the near future.

I hope these remarks, some of them perhaps too vague and cryptic, will have persuaded some of you that the subject deserves to be pursued further, and that both groupoids and algebroids and bigebras might play a role in future developments of unified theories of particle interactions.

REFERENCES

Part I. This bibliography contains selected references to facilitate further study; it is by no means complete. For additional mathematical references the reader should consult the extensive bibliographies in Refs. 2, 10, 11, 18.

1. A. Connes, 'Sur la théorie noncommutative de l'intégration,' in LNM 725, Berlin-Heidelberg-New York: Springer Verlag, 1979, pp. 19–143. 'Feuilletages et algèbres d'opérateurs,' in *Séminaire Bourbaki* 79/80, LNM 842, Berlin-Heidelberg-New York: Springer Verlag, 1980, exp. 551. 'A Survey of Foliations and Operator Algebras,' in Ref. 2, pp. 521-632.

2. R. Kadison, Ed., *Operator Algebras and Applications*, 2 vols. Proc. Symp. Pure Math. vol 38, Providence: AMS, 1982.

3. D. Kastler, 'Lecture Notes on Connes' Cyclic Cohomology,' U. C. Irvine, 1984, and private communications. 'Cyclic Cohomology within the Differential Envelope,' Preprint CPT-84/P.1697 (To be published as a book by Hermann, Paris). 'An Invitation to Alain Connes' Cyclic Cohomology,' ZiF Preprint 132b, Bielefeld, 1985. Private notes, received January 1988.

4. J. Bellissard, 'K-Theory of C*-Algebras in Solid-State Physics,' Lecture at the Conference on *Statistical Mechanics and Field Theory: Mathematical Aspects*, Groningen, 26–30 August, 1985, Preprint CPT-85/P.1816, December 1985 (Contains an extensive Bibliography).

5. M. E. Mayer, 'Quantization of Gauge Theories,' vv *Acta Phys. Austriaca*, Suppl. **XXIII**, 481 (1981); 'A Comment on "Ghosts and Geometry" ,' *Phys. Lett.* **B59**, 355 (1983). 'Quantum Theory in Vector Bundles,' in: *Fundamental Aspects of Quantum Theory*, A. Frigerio and V. Gorini, eds., NATO ASI Series B: Physics Vol. 144, New York: Plenum Press, 1986, pp. 403–410.

5a. M. E. Mayer, 'Gauge Fields as Quantized Connection Forms,' in: *Differential Geometrical Methods in Mathematical Physics (Bonn 1975)*, K. Bleuler and A. Reetz, eds., LNM vol 570, Berlin-Heidelberg-New York: Springer Verlag, 1977 pp. 307–349.

6. E. Witten, 'Noncommutative Geometry and String Field Theory,' *Nucl. Phys.* **B268**, 253–294 (1986). 'Interacting Field Theory of Open Superstrings,' *Nucl. Phys.* **B276**, 291–324 (1986).

7. See, e. g., J. H. Schwarz, 'Superstrings,' *Phys. Reports* **89** (1982); Lectures at: Jerusalem Winter School, December 1984; Scottish Universities Summer School, June 1985; Ecole Normale Supérieure, August 1985; UC Irvine, 1986. M. B. Green, J. H. Schwarz, and E. Witten, *Superstring Theory*, 2 vols., Cambridge Univ. Press, 1987. See also J. Schwarz (Ed.), *Superstrings: The First Fifteen Years*, 2 vols., World Scientific, Singapore, 1985.

8. R. Kugo and I. Ojima, 'Local Covariant Operator Formalism of Non-Abelian Gauge Theories and Quark Confinement Problem,' *Prog. Theor. Phys.* (Kyoto) Suppl. **66**(1979). Cf. also Ref. 5 (the second reference contains an extensive bibliography).

9. J. Westman, 'Harmonic Analysis on Groupoids,' *Pac. J. Math.* **27**, 621 (1968); 'Cohomology For Ergodic Groupoids,' *Trans. AMS*, **146**,465 (1969); Several Preprints, Univ. of Calif., Irvine, 1969–1975. A. Ramsay, 'Virtual Groups and Group Actions,' *Advances in Math.* **6**, 253–322 (1971), and earlier work by G. W. Mackey quoted there.

10. B. L. Reinhart, *The Differential Geometry of Foliations*, Ergebnisse der Mathematik, Vol. 99 , Berlin-Heidelberg-New York: Springer Verlag, 1983. [Contains a detailed bibliography.] A. Kumpera and D. Spencer, *Lie Equations, Vol. I*, Annals of Math. Studies, Nr. 73, Princeton University Press, Princeton, 1972. Appendix.

11. J. N. Renault, 'C*-Algebras of Groupoids and Foliations,' In Ref. 2, Part 1, P 339; *A Groupoid Approach to C*-Algebras*, Lecture Notes In Mathematics No. 79., Springer Verlag, Berlin-Heidelberg-New York, 1980.

12. B. Lawson, *The Quantitative Theory of Foliations*, AMS, Providence 1977; R. Bott, *Lectures on Characteristic Classes and Foliations*, LNM 279, Berlin-Heidelberg-New York: Springer Verlag, 1972, Pp. 1 – 94. For a discussion of the fundamental groupoid, see E. H. Spanier, *Algebraic Topology*, Sec. 7, New York: McGraw Hill, 1966; also Ref. 18.

13. L. Gross, 'A Poincaré Lemma for Connection Forms,' *J. Funct. Anal.*, **63**, 1–46 (1985); Private Communication, Nov. 1986.

13a. H. G. Loos, 'Internal Holonomy Groups of Yang-Mills Fields,' *J. Math. Phys.* **8**, 2114–2124 (1967).

14. M. B. Menskiĭ, 'The Path Group and the Interaction of Quantized Strings,' *Zh. Eksp. Teor. Fiz.* **90**, 416 (1986); [*Sov. Phys. JETP* (1987)] and additional papers of the same author quoted there (in particular, in English: *Lett. Math. Phys.* **3**, 513 (1979).

15. G. Segal, *this volume page* .

16. F. Kamber and P. Tondeur, *Foliated Bundles and Characteristic Classes*, LNM 493, Berlin-Heidelberg-New York: Springer Verlag, 1975.

17. Yu. I. Manin, *Gauge Theory and Complex Geometry*, Nauka, Moscow, 1984 .

18. K. Mackenzie, *Lie Groupoids and Lie Algebroids in Differential Geometry*, LMS Lecture Notes 124, Cambridge University Press, 1987.

19. M. F. Atiyah, 'Complex-analytic Connections in Fibre Bundles,' *Trans. AMS* **85**, 181–207 (1957); H. K. Nickerson, 'On Differential Operators and Connections,' *Trans. AMS* **99**, 509–539 (1961).

20. M. E. Mayer, *Vector Bundles and Gauge Theory* (to be completed in 1988, Springer 1989?), in particular Chapter 4

Part II. This is a very incomplete bibliography on a vast subject, compiled in a hurry; it comes with my sincere apologies to authors who have not been cited.

[D 86] V. G. Drinfel'd, 'Kvantovye Gruppy (Quantum Groups),' *Zapiski Nauchnykh Seminarov LOMI* T. 155, Leningrad, 1955, pp. 18–49. [English version: *Proceedings of the International Congress of Mathematicians, Berkeley, 1986.* A. M. Gleason, Ed., AMS, Providence, 1987, pp. 798–820.]

[DHR 69] S. Doplicher, R. Haag, and J. Roberts, 'Fields, Observables, and Gauge Transformations,' *Commun. Math. Phys.* **13**, 1 ; **15**, 173 (1969).

[DHR 71-74] S. Doplicher, R. Haag, and J. Roberts, 'Local Observables and Particle Statistics,' *Commun. Math. Phys.* **23**, 199 (1971); **35**, 49 (1974).

[DR 72] S. Doplicher and J.E. Roberts, 'Fields, Statistics and Nonabelian Gauge Groups,' *Commun. Math. Phys.* **28**, 331 (1972).

[DR 84] S. Doplicher and J.E. Roberts, 'Compact Lie Groups Associated with Endomorphisms of C*-Algebras,' *Bull. AMS* **11**, 333 (1984).

[DR 86] S. Doplicher and J. Roberts, 'C*-Algebras and Duality for Compact Groups: Why there is a Compact Group of Symmetries in Particle Physics,' Preprint, to appear in *Proc. of Intern. Conf. on Mathematical Physics*, Marseille, July 16–25, 1986.

[FH 86] K. Fredenhagen and R. Haag, 'Generally Covariant Quantum Field Theory and Scaling Limits,' *DESY Preprint* 86-066, June 1986.

[HNS 84] R. Haag, H. Narnhofer, and U. Stein, 'On Quantum Field Theory in Gravitational Background,' *Commun. Math. Phys.* **94**, 219–238 (1984).

[L 83] A. Lichnerowicz, 'Quantum Mechanics and Deformations of Geometrical Dynamics,' in *Quantum Theory, Groups, Fields, and Particles*, A. O. Barut, ed., pp. 3–82, Dordrecht: Reidel, 1983.

[M 71] M. E. Mayer, 'Automorphisms of C*-Algebras, Fell Bundles, W*-Bigebras, and the Description of Internal Symmetries in Algebraic Quantum Theory,' *Acta Phys. Austriaca* Suppl. **VIII**, 177–226 (1971).

[M 73] M. E. Mayer, 'The Uses of Group–Theoretical Duality Theorems in Quantum Theory,' in *Proceedings of the Conference on Differential-Geometric Methods in Physics*, K. Bleuler and A. Reetz, Eds., Bonn 1973.

[R 76] J. E. Roberts, 'Local Cohomology and Superselection Structure,' *Commun. Math. Phys.* **51**, 107 (1976); Lecture at the RCP No. 25, Strasbourg., December 1976.

[W 83] A. Weinstein, 'The Local Structure of Poisson Manifolds,' *J. Diff. Geom.* **18**, 523–557 (1983).

THE DEFINITION OF CONFORMAL FIELD THEORY

G. B. Segal
Mathematical Institute
24-29 St. Giles
Oxford OX1 3LB
England

I shall propose a definition of 2-dimensional conformal field theory which I believe is equivalent to that used by physicists.

1. THE CATEGORY $\mathcal{C}$

The category $\mathcal{C}$ is defined as follows. There is a sequence of objects $\{C_n\}_{n\geq 0}$, where C_n is the disjoint union of a set of n parametrized circles.

A morphism $C_n \to C_m$ is a Riemann surface X with boundary ∂X, together with an identification $i : C_m - C_n \to \partial X$. (We identify morphisms $(X,i),(X',i')$ if there is an isomorphism $f : X \to X'$ such that $f \circ i = i'$. Notice that the boundary of a Riemann surface is canonically oriented. The identifications i are supposed to be orientation-preserving, and $C_m - C_n$ means the union $C_m \amalg C_n$ with the orientation of C_n reversed.)

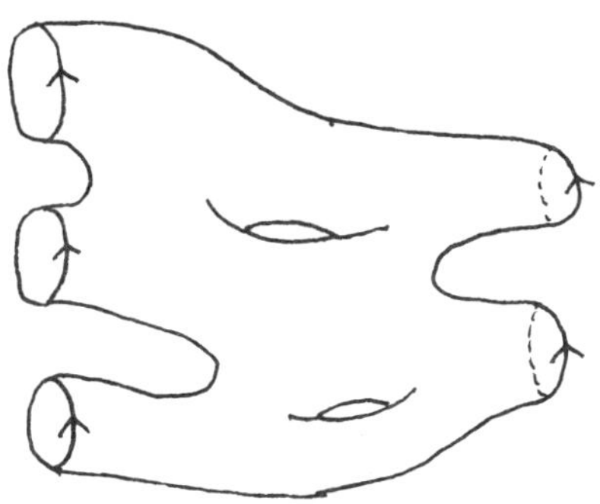

A morphism $C_3 \to C_2$.

K. Bleuler and M. Werner (eds.), Differential Geometrical Methods in Theoretical Physics, 165–171.

2. THE SEMIGROUP $\mathcal{E}$

The semigroup of morphisms $C_1 \to C_1$ falls into connected components corresponding to the genus and number of connected components of X. The "identity" component $\mathcal{E}$ is the space of surfaces which are cylinders, i.e. homeomorphic to $[0,1] \times S^1$. The central idea of this note is that the semigroup $\mathcal{E}$ should be regarded as the complexification of the group $\mathrm{Diff}(S^1)$. (It is well known that there is no group which is a complexification of $\mathrm{Diff}(S^1)$.) We shall see in a moment why $\mathrm{Diff}(S^1)$ is the Shilov boundary of $\mathcal{E}$. A simple example of the relationship which exists between $\mathcal{E}$ and $\mathrm{Diff}(S^1)$ is afforded by the semigroup

$$\{g \in GL_n(\mathbb{C}) : \|g\| < 1\} :$$

this is a bounded domain in $\mathbb{C}^{n^2}$ whose Shilov boundary is the group U_n. Another, more relevant, example is provided by the sub-semigroup of $PSL_2(\mathbb{C})$ consisting of transformations of the Riemann sphere which map the disc $\{z \in S^2 : |z| \le 1\}$ into itself. This has the subgroup $PSU_{1,1}$ of $\mathrm{Diff}(S^1)$ as its Shilov boundary, and is itself a subsemigroup of $\mathcal{E}$.

Let us now give a more precise description of $\mathcal{E}$. Let

$$D_0 = \{z \in \mathbb{C} : |z| \le 1\}$$

$$D_\infty = \{z \in \mathbb{C}\cup\{\infty\} : |z| \ge 1\} .$$

Then $\mathcal{E}$ is the space of pairs (f_0,f_∞), where $f_0 : D_0 \to S^2$ are holomorphic embeddings such that

$$f_0(0) = 0, \quad f_\infty(\infty) = \infty, \quad f'(\infty) = 1, \tag{2.1}$$

and $f_0(D_0)$ and $f_\infty(D_\infty)$ are disjoint. (To say that $f'_\infty(\infty) = 1$ means that f_∞ has residue 1 at $z = \infty$.)

To see this, observe that for given (f_0,f_∞) the surface $S^2 - f_0(\mathring{D}_0) - f_\infty(\mathring{D}_\infty)$ is a cylinder with parametrized ends; on the other hand if X is a cylinder with parametrized ends then one can form a closed surface $\hat{X} = D_0 \cup X \cup D_\infty$, and the Riemann mapping theorem identifies this with the standard S^2 uniquely up to an element of $PSL_2(\mathbb{C})$ which is fixed by the normalization (2.1).

$\mathcal{E}$ can be identified with an open set in an affine hyperplane in $F \oplus F$, where F is the vector space of holomorphic maps $f : D_0 \to \mathbb{C}$ such that $f(0) = 0$. This open set has a Shilov boundary which consists of the pairs (f_0,f_∞) such that $f_0(S^1) = f_\infty(S^1)$, where $S^1 = \partial D_0 = \partial D_\infty$. The boundary is identified with $\mathrm{Diff}(S^1)$ by

$$(f_0, f_\infty) \mapsto (f_\infty^{-1} \circ f_0 | S^1) \ .$$

There is a holomorphic map $q : \mathcal{E} \to \{z \in \mathbb{C} : |z| < 1\}$ whose boundary value is the rotation number

$$q : \mathrm{Diff}(S^1) \to S^1 \ .$$

In fact

$$q(X) = e^{2\pi i \tau} \ ,$$

where τ is the modulus of the torus obtained from $X \in \mathcal{E}$ by identifying its two boundary circles. (The modulus is well-defined up to an additive integer, and not just up to the action of $PSL_2(\mathbb{Z})$ on the upper-half-plane, because the torus has a preferred generator of π_1.)

3. SOME EXAMPLES

(i) One connected component of the space of morphisms $C_0 \to C_1$ consists of surfaces which are topologically discs. This component can be identified with the complex manifold $\mathrm{Diff}(S^1)/PSL_2(\mathbb{R})$.

(ii) The semigroup of morphisms $C_0 \to C_0$ is the space of isomorphism classes of (not necessarily connected) Riemann surfaces, the operation being disjoint union.

(iii) For any m,n the space of morphisms $C_m \to C_n$ is a complex manifold with one connected component for each topological type of surface X.

(iv) One component of the morphisms $C_1 \to C_2$ corresponds to pairs of simple closed curves in D:

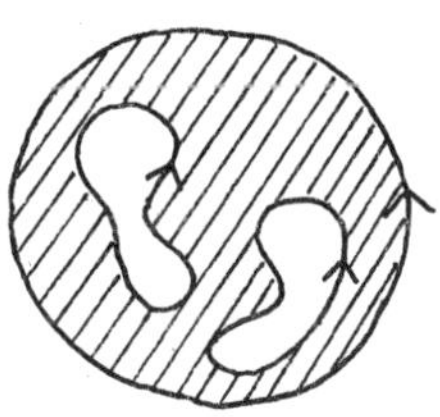

The Shilov boundary of this component consists of graphs Γ together with piecewise-smooth maps $S^1 \to \Gamma$, $(S^1 \quad S^1) \to \Gamma$:

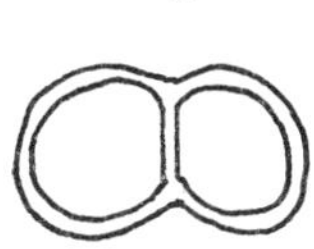

or

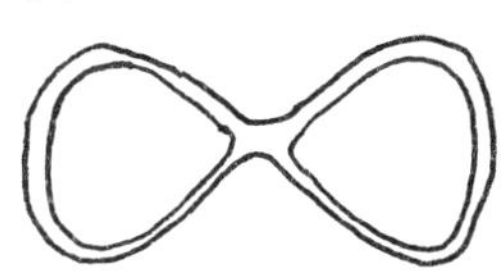

corresponding to "splitting" of strings.

4. THE DEFINITION

Definition (4.1).

A conformal field theory is a representation of $\mathcal{C}$, i.e. a continuous functor T from $\mathcal{C}$ to complex Hilbert spaces such that

(i) $T(C_n) = \underbrace{H \otimes \ldots \otimes H}_{n} = H^{\otimes n}$ for some Hilbert space H;

(ii) for each morphism $\xi : C_m \to C_n$ the operator $T(\xi) : H^{\otimes m} \to H^{\otimes n}$ is of the trace class;

(iii) T is a *-functor, i.e. $T(\xi)^* = T(\bar{\xi})$ for each $\xi : C_m \to C_n$, where $\bar{\xi} : C_n \to C_m$, is defined by the Riemann surface complex-conjugate to ξ;

(iv) T has the crossing property that if a morphism $\xi : C_m \to C_n$ is made into a morphism $\tilde{\xi} : C_{m-1} \to C_{n+1}$ by reversing the orientation of an incoming boundary circle then $T(\xi)$ and $T(\tilde{\xi})$ are adjoint (i.e. correspond to the same element of $H^{\otimes(m+n)}$);

(v) T has the collapsing property that if a morphism $\xi : C_{m+r} \to C_{n+r}$ is made into a morphism $\check{\xi} : C_m \to C_n$ by attaching the first r outgoing circles to the first r incoming circles, then $T(\check{\xi}) = \text{trace } T(\xi)$, where the trace is taken over $H^{\otimes r}$.

5. THE MORPHISM ρ_q AND THE PARTITION FUNCTION

If $q \in \mathbb{C}$ with $0 < |q| < 1$ let ρ_q denote the morphism $C_1 \to C_1$ represented by the annulus $\{z \in \mathbb{C} : |q| \le |z| \le 1\}$ with its boundary circles parametrized by $\theta \mapsto e^{i\theta}$ and $\theta \mapsto qe^{i\theta}$. The ρ_q form a subsemigroup of $\mathcal{E}$: $\rho_{q_1}\rho_{q_2} = \rho_{q_1 q_2}$.

Definition (5.1).

The partition function Z_T of the theory T is defined by $Z_T(q) = \text{trace } T(\rho_q)$.

Condition (v) above implies trivially

Proposition (5.2).

For any theory T the partition function Z_T is modular, i.e. if $q = e^{2\pi i z}$ with $\text{Im}(z) > 0$ then $Z_T(q)$ is invariant when z is transformed by $PSL_2(\mathbb{Z})$,

6. WICK ROTATION

To motivate our definition, consider a 2-dimensional Minkowskian space-time Σ in which space is a circle. (Thus topologically $\Sigma \cong S^1 \times \mathbb{R}$.) The group $\mathcal{D}$ of conformal diffeomorphisms of X is a $\mathbb{Z}$-fold covering group of $\mathrm{Diff}(S^1) \times \mathrm{Diff}(S^1)$, for each element of $\mathcal{D}$ permutes the circle of left-moving light-curves and also the circle of right-moving light-curves. We expect a conformal field theory to be at least a projective unitary representation H of $\mathcal{D}$. Let us restrict attention for the moment to the translation subgroup $P = (\mathbb{R} \times \mathbb{R})/2\pi\mathbb{Z}$ inside $\mathcal{D}$, where $(\xi,\eta) \in P$ acts on $\mathbb{R} \times S^1$ by $(t,\theta) \mapsto (t+\xi+\eta, \theta+\xi-\eta)$. The positivity of the energy is expressed by the fact that the unitary action of P on H is the boundary value of a holomorphic representation by contraction operators of the semigroup $P^+_{\mathbb{C}} = \{(\xi,\eta) \in (\mathbb{C} \times \mathbb{C})/2\pi\mathbb{Z} : \mathrm{Im}(\xi) > 0, \mathrm{Im}(\eta) > 0\}$ contained in the complexification $P_{\mathbb{C}}$ of P. This holomorphic contraction representation restricts to a representation T of the semigroup $Q = \{q \in \mathbb{C} : 0 < |q| < 1\}$, where $e^{i\xi} \in Q$ is identified with $(\xi,-\xi) \in (\mathbb{C} \times \mathbb{C})/2\pi\mathbb{Z}$, with the "reflection-positivity" property that $T(\bar{q}) = T(q)^*$. Conversely a reflection-positive contraction representation of Q on H "Wick-rotates" to a unitary action of P on H.

In our situation P is replaced by $\mathcal{D}$ and Q by the semigroup $\mathcal{E}$. We have

<u>Proposition (6.1)</u>.

For any theory T the action of $\mathcal{E}$ on H can be continued analytically to a projective unitary representation of $\mathcal{D}$ (i.e. of $\mathrm{Diff}(S^1) \times \mathrm{Diff}(S^1)$) on H.

7. PROJECTIVE FIELD THEORIES

Examples of conformal field theories are not easy to produce, and in practice one constructs them out of simpler objects which - in physicists' language - possess a <u>conformal anomaly</u>. I shall call these <u>projective</u> theories, following the terminology of group representation theory.

A projective representation of a group is a representation of a central extension of the group. For a category we replace this by:

<u>Definition (7.1)</u>.

A central extension of a category $\mathcal{C}$ is the assignment of a complex line L_ξ to each morphism ξ of $\mathcal{C}$, together with coherent isomorphisms $L_\xi \otimes L_\eta \cong L_{\xi\circ\eta}$ whenever $\xi\circ\eta$ is defined.

When $\xi \mapsto L_\xi$ is given we shall write $\tilde{\mathcal{C}}$ for the category with the same objects as $\mathcal{C}$, in which the morphisms are pairs (ξ,λ), with ξ a morphisms in $\mathcal{C}$ and $\lambda \in L_\xi$.

For the category $\mathcal{C}$ of interest to us there is essentially just one central extension. It assigns to a Riemann surface X with boundary the determinant line Det_X (in the sense of Quillen) of the $\bar{\partial}$-operator acting on functions on X, with its natural boundary conditions. In fact Det_X can be identified canonically with the top exterior power of the space of holomorphic differentials on the closed surface $\hat{X}$ formed by adding a standard disc to each boundary circle of X; but that does not illuminate the crucial isomorphism $Det_{X\cup Y} = Det_X \otimes Det_Y$.

The determinant line is essentially the only central extension of because of the following theorem of Mumford.

Proposition (7.2).

Any central extension of $\mathcal{C}$ is of the form $X \mapsto \overline{Det}_X^{\otimes p} \otimes Det_X^{\otimes q}$ for some $p,q \in \mathbb{C}$ such that $p-q \in \mathbb{Z}$.

We can now define a projective conformal field theory as a representation of a central extension of $\mathcal{C}$. We may as well assume that the central extension is the one given by $X \mapsto Det_X$, and that the operator $T(X,\lambda)$ for $\lambda \in Det_X$ satisfies

$$T(X,a\lambda) = \bar{a}^p a^q T(X,\lambda)$$

for $a \in \mathbb{C}^\times$. The pair (p,q) is called the central charge of the theory. (In the natural examples one has $p = q$.) The properties demanded of a functor in Definition (4.1) all make sense for projective functors, for there is a canonical isomorphism $Det_X \cong Det_{\check{X}}$ when $\check{X}$ is formed from X by attaching an outgoing to an incoming boundary circle.

For a cylinder X, in particular for the standard ρ_q, there is a canonical element $\varepsilon \in Det_X$. We can therefore define the partition function of a projective theory as the trace of the operator associated to (ρ_q,ε). We must be careful, however, because the element of $Det_{\check{X}}$ obtained from ε does not depend only on $\check{X}$. When this is taken into account we have

Proposition (7.3).

The partition function Z of a projective theory of central charge (p,q) has the modularity property

$$Z(-\tau^{-1}) = (-\bar{\tau})^p(-\tau)^q Z(\tau) .$$

8. FIELD OPERATORS

To conclude this talk I shall explain very briefly how one obtains the usual formalism of field operators from my definition.

The semigroup $\mathcal{E}$ has a subsemigroup $\mathcal{E}^+$ which consists of the holomorphic maps $f : D_0 \to D_0$ such that $f(D_0) \subset \mathring{D}_0$ and $f(0) = 0$. For such a map determines a cylinder $X_f = D_0 - f(\mathring{D}_0)$. The semigroup $\mathcal{E}^+$ acts on the Hilbert space H of any theory T. Indeed this is the case even for a projective theory, for if ε is the canonical element of the determinant line of X_f then

$$(X_g,\varepsilon).(X_f,\varepsilon) = (X_{gf},\varepsilon) .$$

<u>Definition (8.1)</u>.

A vector $\psi \in H$ is a <u>primary field</u> of the theory T if it is an eigenvector of the semigroup $\mathcal{E}^+$.

We then have $T(\rho_q).\psi = \bar{q}^{-a}q^{b}\psi$ for some positive real numbers a,b such that $a-b \in \mathbb{Z}$. We call (a,b) the <u>bidegree</u> of ψ.

A primary field ψ gives rise to an operator-valued differential form on any Riemann surface with boundary. Suppose, for example, that X is a surface whose boundary consists of two circles, i.e. X is a morphism $C_1 \to C_1$. For each triple (ψ,z,dz), where $z \in X - \partial X$, dz is a holomorphic differential at z, and $\psi \in H$ is a primary field of bidegree (a,b), we can define a trace-class <u>creation</u> operator $\psi(z) : H \to H$ such that $\psi(z)dz^{a}d\bar{z}^{-b}$ is an operator-valued differential form on X.

To construct $\psi(z)$, choose a local parameter ζ at z such that $d\zeta(z) = dz$. Let $D_\varepsilon = \{x \in X : |\zeta(x)| < \varepsilon\}$. Then $X-D_\varepsilon$ is a morphism $Y : C_2 \to C_1$, so it induces $T(Y) : H \hat{\otimes} H \to H$. Define $\psi(z) : H \to H$ by $\psi(z).v = T(Y).(\psi \otimes v)$. This is independent of the local parameter ζ because ψ is primary.

STATISTICS AND MONODROMY IN TWO- AND THREE-DIMENSIONAL QUANTUM FIELD THEORY

JUERG FROEHLICH
Theoretical Physics
ETH-Hönggerberg
CH-8093 Zürich

ABSTRACT. In Section 1, we review results on the statistics of local fields in two-dimensional relativistic quantum field theory and on the monodromy properties of Euclidean Green functions in conformal field theory. In Section 2, we describe three-dimensional gauge theories with particles of arbitrary real spin and intermediate statistics of interest in two-dimensional condensed matter physics.

1. STATISTICS IN TWO-DIMENSIONAL QUANTUM FIELD THEORY

In [1], the author has outlined a general analysis of statistics of local fields in two-dimensional relativistic quantum field theory. It was shown there that there is a close connection between statistics of fields and representations of the braid (or pure braid) groups on n strings, determined by the n-point Euclidean Green functions. These representations are given in terms of special solutions of the Yang-Baxter equations which are, in fact, the statistics matrices of the local fields of the theory.

The findings of [1] can be summarized as follows: Let $\{\Phi_\alpha(x)\}_{\alpha=1}^{p}$ be a minimal set of local, unobservable fields of some quantum field theory in two space-time dimensions sufficient to describe the structure of superselection sectors of the theory$_2$. Let x and y be space-like separated points in Minkowski space $\mathbb{M}^2$, and let $\sigma = \text{sign}(\vec{x}-\vec{y})$, where $\vec{x}$ is the space component of x. In two dimensions, σ is

K. Bleuler and M. Werner (eds.), Differential Geometrical Methods in Theoretical Physics, 173–186.

Lorentz-invariant. Then there is a $p^2 \times p^2$ matrix $R(\sigma) = (R^{\gamma\delta}_{\alpha\beta}(\sigma))$, $\alpha,\beta,\gamma,\delta = 1,\ldots,p$, called statistics matrix, such that

$$\Phi_\alpha(y)\ \Phi_\beta(x) = R^{\gamma\delta}_{\alpha\beta}(\sigma)\ \Phi_\gamma(x)\ \Phi_\delta(y)\ . \tag{1.1}$$

The properties of R are as follows:

(1) $$R(\sigma)\ R(-\sigma) = \mathbb{1} \tag{1.2}$$

(2) Let $V_i \cong \mathbb{C}^p$, $i = 1,2,3$. We set

$$R_{12}(\sigma) = R(\sigma)\Big|_{V_1 \otimes V_2} \otimes \mathbb{1}\Big|_{V_3}\ , \quad \text{and} \tag{1.3}$$

$$R_{23}(\sigma) = \mathbb{1}\Big|_{V_1} \otimes R(\sigma)\Big|_{V_2 \otimes V_3}\ . \tag{1.4}$$

Let x,y,z be three pairwise space-like separated points in $\mathbb{M}^2$, and $\sigma_1 = \mathrm{sign}(\vec{y} - \vec{x})$, $\sigma_2 = \mathrm{sign}(\vec{z} - \vec{x})$, $\sigma_3 = \mathrm{sign}(\vec{z} - \vec{y})$. Then, as shown in [1], the associativity of the field algebra generated by $\{\Phi_\alpha\}^p_{\alpha=1}$ and (1.1) imply

$$R_{12}(\sigma_1)\ R_{23}(\sigma_2)\ R_{12}(\sigma_3) = R_{23}(\sigma_3)\ R_{12}(\sigma_2)\ R_{23}(\sigma_1)\,. \tag{1.5}$$

This is a special case of the Yang-Baxter equation. A related observation was made earlier by Zamolodchikov [2]. Further properties of R-matrices, in particular a spin-statistics connection, are established in [1].

We now show that R-matrices define representations of the braid groups, B_n, on n strings. Geometrically, these groups are defined as follows: One chooses n distinct points $x_1,\ldots,x_n$ in $\mathbb{R}^2$ and considers n maps, $\gamma_1,\ldots,\gamma_n$, from [0,1] into the slab $\mathbb{R}^2 \times [0,1]$ in $\mathbb{R}^3$ such that

(a) $\gamma_i(0) = x_i$, $\gamma_i(1) = x_{\pi(i)}$,

where π is a permutation of $\{1,\ldots,n\}$;

(b) $(\frac{d}{dt}\gamma_i^3)(t) > 0$, for all $t \in [0,1]$, where x^3 is the 3-component of $x \in \mathbb{R}^3$;

(c) for $i \neq j$, $\gamma_i(t) \neq \gamma_j(s)$, for all $t \in [0,1]$ and all $s \in [0,1]$.

B_n is defined to be the group of all equivalence classes, under ambient isotopies, of maps $\gamma_1,\ldots,\gamma_n$ with properties (a), (b), (c) . The curves $\gamma_i(t)$, $t \in [0,1]$, $i = 1,\ldots,n$, are called strings.

Let $\tau_{i,i+1} \in B_n$ be the operation of once braiding the ith with the $i+1^{st}$ string, as shown in Fig. 1.

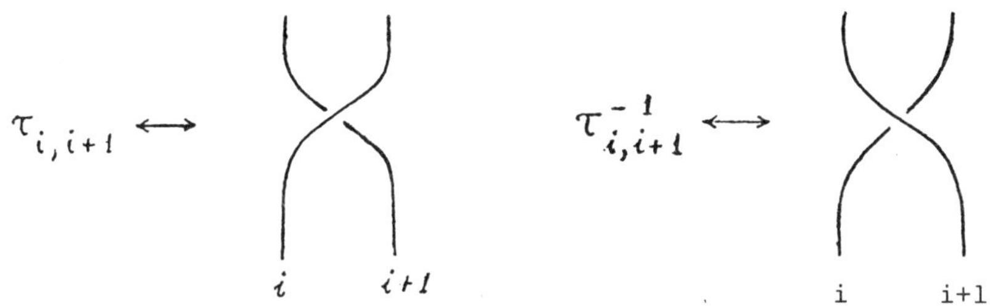

The operations $\tau_{i,i+1}$, $i = 1,\ldots,n-1$ are generators of B_n . They obey the following relations:

(i) $\tau_{i,i+1}\,\tau_{i+1,i+2}\,\tau_{i,i+1} = \tau_{i+1,i+2}\,\tau_{i,i+1}\,\tau_{i+1,i+2}$,

(ii) $\tau_{i,i+1}\,\tau_{j,j+1} = \tau_{j,j+1}\,\tau_{i,i+1}$, for $|i - j| \geq 2$.

A general element $b \in B_n$ can be written as a word in the generators $\tau_{i,i+1}$, $i = 1,\ldots,n-1$:

$$b = \prod_{k=1}^{m} \tau_{i(k),i(k)+1}^{\sigma(k)} , \quad \sigma(k) = \pm 1 . \tag{1.6}$$

Let $V_i \cong \mathbb{C}^p$, $i = 1,\ldots,n$. We define

$$R(\tau^{\sigma}_{i,i+1}) = \mathbb{1}\Big|_{V_1} \otimes \ldots \otimes R(\sigma)\Big|_{V_i \otimes V_{i+1}} \otimes \ldots \otimes \mathbb{1}\Big|_{V_n} \tag{1.7}$$

$\sigma = \pm 1$, and

$$R(b) = \prod_{k=1}^{m} R\left(\tau^{\sigma(k)}_{i(k),i(k)+1}\right) , \tag{1.8}$$

for $b \in B_n$ as in (1.6) .

It is easy to verify, using (1.2) and (1.5), that $B_n \ni b \longmapsto R(b) \in \mathrm{End}(V_1 \otimes \ldots \otimes V_n)$ defines a representation of B_n on $V_1 \otimes \ldots \otimes V_n$.

Let $\xi_1, \ldots, \xi_n$ be n points in two-dimensional Euclidean space, $\mathbb{E}^2$, with $\xi_1^0 < \xi_2^0 < \ldots < \xi_n^0$, where ξ^0 is the 0-component of $\xi \equiv (\xi^0, \vec{\xi}) \in \mathbb{E}^2$. The Euclidean Green functions of the fields $\Phi_1(x), \ldots, \Phi_p(x)$ are given by

$$G(\alpha_1, \xi_1, \ldots, \alpha_n, \xi_n) = \langle \Omega, \Phi_{\alpha_1}(0, \vec{\xi}_1) e^{-(\xi_2^0 - \xi_1^0)H} \Phi_{\alpha_2}(0, \vec{\xi}_2) \ldots \Phi_{\alpha_{n-1}}(0, \vec{\xi}_{n-1}) e^{-(\xi_n^0 - \xi_{n-1}^0)H} \Phi_{\alpha_n}(0, \vec{\xi}_n)\Omega \rangle, \tag{1.9}$$

where H is the Hamilton operator of the quantum field theory. By the spectrum condition, H is a positive operator, so that, heuristically, the r.h.s. of (1.9) is well defined if $\xi_1^0 < \xi_2^0 < \ldots < \xi_n^0$, as assumed. [For a more precise definition of $G(\alpha_1, \xi_1, \ldots, \alpha_n, \xi_n)$, see [3].]

Let D be the set of n-tuples of points for which $\xi_i = \xi_j$, for some $i \neq j$. We define

$$M_n^s = (\mathbb{E}^2)^{\times n} \setminus D / \mathcal{S}_n, \tag{1.10}$$

where $\mathcal{S}_n$ is the group of permutations of $\{1, \ldots, n\}$. The space M_n^s is not simply connected. It is easy to see, geometrically, that

$$\pi_1(M_n^s) = B_n. \tag{1.11}$$

Let $\tilde{M}_n^s$ denote the universal covering space of M_n^s. Points in $\tilde{M}_n^s$ can be written as $[b; \xi_1, \ldots, \xi_n]$, with $(\xi_1, \ldots, \xi_n) \in M_n^s$, $b \in B_n$.

In [1], a proof of the claim has been sketched that $G(\alpha_1, \xi_1, \ldots, \alpha_n, \xi_n)$, with $\xi_1^0 < \xi_2^0 < \ldots < \xi_n^0$, extends to a single-valued function $G(b; \alpha_1, \xi_1, \ldots, \alpha_n, \xi_n)$ on $\tilde{M}_n^s$ with the property that

$$G(b; \alpha_1, \xi_1, \ldots, \alpha_n, \xi_n) = R_{\alpha_1 \ldots \alpha_n}^{\alpha_1' \ldots \alpha_n'}(b^{-1} b') \, G(b'; \alpha_1', \xi_1, \ldots, \alpha_n', \xi_n), \tag{1.12}$$

$b,b' \in B_n$, where $R(b)$ is given by (1.8), for every $b \in B_n$; (summation over repeated indices is understood). In general, G does not project down to a single-valued function on M_n^s; (G is single-valued on $M_n = (\mathbb{E}^2)^{\times n} \setminus D$ if $R(\sigma)^2 = \mathbb{1}$, i.e. if all fields have ordinary Bose- or Fermi statistics).

We now assume that we are dealing with a two-dimensional <u>conformal</u> field theory.

We set

$$z = i\xi^0 + \vec{\xi}\ ,\ \bar{z} = -i\xi^0 + \vec{\xi}\ , \tag{1.13}$$

and

$$G(\cdot;\ \alpha_1,\xi_1,\ldots,\alpha_n,\xi_n) \equiv H(\cdot;\ \alpha_1,z_1,\bar{z}_1,\ \ldots\ ,\alpha_n,z_n,\bar{z}_n).$$

In conformal field theory, H is the restriction of a function $H(\cdot;\ \alpha_1,z_1,\bar{z}'_1,\ldots,\alpha_n,z_n,\bar{z}'_n)$, holomorphic in $(z_1,\ldots,z_n)$ and anti-holomorphic in $(z'_1,\ldots,z'_n)$, to the subspace $\{z_i = z'_i : i=1,\ldots,n\}$. Keeping $z'_1,\ldots,z'_n$ fixed, one would like to extend the definition of

$$h(\cdot;\ z_1,\ldots,z_n) \equiv H(\cdot;\alpha_1,z_1,\bar{z}'_1,\ldots,\alpha_n,z_n,\bar{z}'_n) \tag{1.14}$$

to $M_n^{\mathbb{C}} = \mathbb{C}^n \setminus D$, with D given by $\{z_1,\ldots,z_n : z_i = z_j$, for some $i \neq j\}$; (we have dropped the indices $\alpha_1,\ldots,\alpha_n$ from our notation which, in conformal field theory, label primary fields). However, the space $M_n^{\mathbb{C}}$ is not simply connected:

$$\pi_1\ (M_n^{\mathbb{C}}\ )\ =\ P_n\ , \tag{1.15}$$

where P_n is the pure braid group on n strings, i.e. the subgroup of B_n for which the strings γ_i satisfy $\gamma_i(0) = \gamma_i(1)$, for all i [But note that $\pi_1(M_n^{\mathbb{C}}/\mathcal{S}_n) = B_n$.] Hence, $h(\cdot;\ z_1,\ldots,z_n)$ will, in general, be a multivalued, holomorphic function on $M_n^{\mathbb{C}}$, but a single-valued function on $\tilde{M}_n^{\mathbb{C}}$, i.e. $h(b;\ z_1,\ldots,z_n)$, $b \in B_n$, is a single-valued function on $\tilde{M}_n^{\mathbb{C}}$. For $(z_1,\ldots,z_n)$ near a point $(\zeta_1,\ldots,\zeta_n) \in M_n^{\mathbb{C}}$, $h(b;\ z_1,\ldots,z_n)$ is given by a convergent Taylor series

$$h(b;\ z_1,\ldots,z_n) = \sum_{\underline{k}} a_{\underline{k}}(b) \prod_{j=1}^{n} (z_j - \zeta_j)^{k_j}\ , \tag{1.16}$$

with $\underline{k} = (k_1,\dots,k_n)$, $a_{\underline{0}}(b) = h(b;\ \zeta_1,\dots,\zeta_n)$. The connection between $h(b;\ z_1,\dots,z_n)$ and $h(b';\ z_1,\dots,z_n)$, with $b,b' \in B_n$, is given by an operator $R_z(b^{-1}b')$ which maps the Taylor coefficients $\{a_{\underline{k}}(b')\}$ to the Taylor coefficients $\{a_{\underline{k}}(b)\}$. The operators $R_z(b)$, $b \in B_n$, define a representation of B_n. They describe the monodromy of the functions $h(\cdot;\ z_1,\dots,z_n)$.

The same arguments, keeping $z_1,\dots,z_n$ fixed and varying $z'_1,\dots,z'_n$, lead one to find a representation $R_{z'}(b)$, $b \in B_n$, describing the monodromy of

$$h'(b;\ \bar{z}'_1,\dots,\bar{z}'_n) = H(b;\alpha_1,z_1,\bar{z}'_1,\dots,\alpha_n,z_n,\bar{z}'_n)\Big|_{\underline{z}\text{ fixed}}$$

If we vary $z_i,\bar{z}'_i$ simultaneously, setting $z'_i = z_i$, $i=1,\dots,n$, and then use (1.12) we arrive at the constraint

$$R_z(b)\ R_{z'}(b) = R(b)\ , \tag{1.17}$$

where $R(b)$ is given in terms of the statistics matrix, R, of the theory as in (1.8). If the statistics matrix, R, is trivial ($R(\sigma) = \mathbb{1}$) and the fields $\Phi_1,\dots,\Phi_p$ are hermitian then

$$R_{z'}(b) = R_z(b)^*\ ,$$

and, by (1.17),

$$R_z(b)\ R_z(b)^* = \mathbb{1}\ ,$$

i.e. R_z is a unitary representation of B_n.

An important problem in conformal field theory is to determine the representation R_z of B_n, $n = 2,3,4,\dots,$ as explicitly as possible, as this will help in constructing explicit formulas for the Euclidean Green functions. There are different strategies to attack this problem.

(1) Exploiting conformal Ward identities and operator product expansions, one can derive differential equations [4] for the functions $h(\cdot;\ z_1,\dots,z_n)$. These equations should then determine the monodromy properties of the h-functions. See [4,5]. This approach is successful in the example of the Wess-Zumino-Witten models [6] and some of the minimal models, [4,5].

The Euclidean Green functions of models for which the approach of [4] is successful can be expressed in terms of "conformal blocks" :

$$G(\cdot;\ \xi_1,\dots,\xi_n) = \sum_{\alpha,\beta=1}^{r} F^{\alpha}(z_1,\dots,z_n)\ k_{\alpha\beta}\ \bar{F}^{\beta}(\bar{z}_1,\dots,\bar{z}_n), \tag{1.18}$$

where r is the number of independent solutions of the differential equations satisfied by h [4]. One now attempts to determine the monodromy properties of the conformal blocks F^{α}. These properties and equation (1.17) then yield important constraints on $k_{\alpha\beta}$. In addition, knowing the singularity structure of the functions F^{α} and the matrix $k_{\alpha\beta}$ would enable one to derive the so-called fusion rules, [4,6].

(2) In [5] a "Coulomb" gas representation" for the conformal blocks, F^{α}, is derived from the representation theory of the Virasoro algebra. It permits one to determine the monodromy properties of the conformal blocks, $F^{\alpha}(z_1,\dots,z_4)$, for a variety of minimal models. This is an important step in the calculation of the exact four-point Euclidean Green functions. It turns out that it is enough to know the representations, R_z and $R_{z'}$, of B_4 determined by all four-point functions, in order to calculate the monodromy of general n-point functions.

(3) One may decompose the functions $h(b;z_1,\dots,z_n)$, see (1.14), over the spectrum of L_0 (mod. $\mathbb{Z}$), with L_n, $n \in \mathbb{Z}$, the generators of the Virasoro algebra:

$$h(b;\, z_1,\dots,z_n) = \sum_{d_1,\dots,d_{n-1}} h(b;z_1,d_1,\dots,z_{n-1},d_{n-1},z_n) \tag{1.19}$$

One then attempts to construct matrices, $A_\tau(d,d')$, such that

$$\sum_{d_i} h(b;\dots d_{i-1},z_{i+1},d_i,z_i,d_{i+1}\dots)$$
$$= A_\tau(d_{i-1},d_{i+1}) \sum_{d_i} h(b\cdot\tau_{i,i+1};\dots d_{i-1},z_i,d_i,z_{i+1},d_{i+1}\dots) \tag{1.20}$$

when z_i and z_{i+1} are exchanged along curves which form a positively oriented loop. [If the loop were negatively oriented, A_τ would be replaced by $A_{\tau^{-1}} = A_\tau^{-1}$.] If one makes the ansatz that the only non-vanishing terms contributing to the r.h.s. of (1.19) have the property that, given d_j, d_{j+1} can only have two values, for all j, ($d_j = h_{1,\ell}(c)$, $d_{j+1} = h_{1,\ell\pm 1}(c)$, with $h_{n,m}(c)$ given by the Kac formula) then one can determine candidate matrices, $A_\tau(d,d')$, for (1.20) with the help of the Yang-Baxter equations, (1.5), which, of course, just express the condition that the matrices $A_\tau(d,d')$ generate representations of the braid groups. These A-matrices are well known from the exact solution of the (restricted) solid-on-solid models. [This is not entirely surprising in view of the constraint on the values of d_j contributing to the r.h.s. of (1.19).] The solutions, $A_\tau(d,d')$, can be related to known classifications of minimal models with $c < 1$. This approach has been developed in [7]. A related approach has been analyzed, independently, by the author. One can show, using the representation theory of the Virasoro algebra, as in (1.19), that the monodromy of n-point functions can be determined in terms of the monodromy of general four-point functions. The four-point functions of minimal models with $c < 1$ are analyzed in [5]. The goal of the analysis is to classify minimal conformal field theories in terms of the representations of the braid groups determined by their Green functions and to make use of those representations in explicit calculations of Green functions. [For a different classification of minimal models with central charge $c < 1$, the so-called A-D-E classification, see [8].]

Explicit calculations of n-point Green functions, with $n > 4$ have only been accomplished in some very special models such as the Ising model. It is hoped that the program outlined above will lead to further progress in that direction.

2. SPIN AND STATISTICS IN THREE-DIMENSIONAL RELATIVISTIC QUANTUM PHYSICS

Following the algebraic approach to relativistic quantum physics developed in [9,10], it has been shown in [11,12], in complete generality, that in four space-time dimensions, particles and fields exhibit either Bose- or Fermi-statistics and carry either integral or half-integral spin. Moreover, the usual connection between spin and statistics holds: Particles and fields with integral spin obey Bose statistics, those with half-integral spin obey Fermi statistics, [12].

The notion of "field" employed in these investigations is a very general one: A "field" is, in general, not localizable in a point or a compact region in space-time, but in a space-like cone of arbitrarily small opening angle [11]. [Of course fields corresponding to local observables are local in the usual sense of the word [13,14].] Those localization properties of general "fields" follow from very general properties (locality, relativistic spectrum condition, existence of one-particle states) of relativistic quantum physics [11]. It is expected that there are gauge theories possessing physical (gauge-invariant) fields localizable in space-like cones which cannot be replaced by an equivalent set of strictly local fields. [A variety of lattice gauge theory models with these properties have been exhibited.] In spite of the rather weak locality properties of general fields (in the sense of [11]) no pathologies with spin and statistics can arise in space-times of dimension four or more.

The situation is different in three space-time dimensions. The Poincaré group in three dimensions is given by

$$\mathbb{P}_+^\uparrow = \mathbb{R}^3 \circledS SO(2,1) .$$

Relativistic one-particle states are characterized by unitary irreducible representations of the quantum mechanical Poincaré group, $\tilde{\mathbb{P}}_+^\uparrow$, of mass $m \geq 0$. Here $\tilde{\mathbb{P}}_+^\uparrow = \mathbb{R}^3 \circledS \widetilde{SO(2,1)}$ is the universal covering group of $\mathbb{P}_+^\uparrow$. Topologically,

$$SO(2,1) \simeq S^1 \times \mathbb{R}^2 ,$$

where S^1 corresponds to the group of rotations of two-dimensional space. It follows that

$$\pi_1(SO(2,1)) = \mathbb{Z} ,$$

and

$$\widetilde{SO(2,1)} \simeq \mathbb{R} \times \mathbb{R}^2 .$$

By the Wigner-Bargmann-Mackey theory, all unitary irreducible representations of $\tilde{\mathbb{P}}_+^\uparrow$ are characterized by two numbers, m (m = mass) and s (spin), where s labels the unitary irreducible representations of the subgroup of SO(2,1) which leaves invariant a vector $p \in \mathbb{M}^3$, with $p^2 = m^2$. For $m \geq 0$, that subgroup, the so-called little group, is isomorphic to the additive group of the real line, $\mathbb{R}$. Hence s labels characters of $\mathbb{R}$, i.e. is an arbitrary real number.

We thus expect that, in three space-time dimensions, there are relativistic particles of arbitrary, real spin s .

Let us first imagine, that a relativistic one-particle state can be created by applying a (possibly not observable but) strictly local field to the vacuum, Ω. Then it has necessarily integer or half-integer spin. This follows from a more general result: In three space-time dimensions, local fields in the sense of [13,14] carry integer or half-integer spin and obey Bose- or Fermi statistics. The usual spin-statistics connection is valid. See [9,10,14,1].

The results of Buchholz and Fredenhagen [11], however, still imply that any relativistic one-particle state of mass $m > 0$ and any spin s in a three-dimensional relativistic theory can be created by applying a field localized in a space-like cone of arbitrarily small opening angle to the vacuum, Ω. Thus if $s \notin \frac{1}{2}\mathbb{Z}$, that field cannot be strictly local. The existence of such fields is typical of gauge theories. We therefore expect to find particles with spin $s \notin \frac{1}{2}\mathbb{Z}$ in three-dimensional gauge theories. Gauge theory models describing such particles have, in fact be found [15-17] .

Returning to the general analysis presented in [11,12] , one observes that, in three space-time dimensions, for fields only localized in space-like cones which are not local in the usual sense, the statistics analysis of [10,11] does not lead to any physically relevant conclusions, and the spin-statistics connection, proven in [12] for four-dimensional theories, becomes void.

Operationally, a physically meaningful concept of statistics in three-dimensional theories manifests itself in properties of bound-state spectra (e.g. [15,18]), asymptotic states and scattering amplitudes [17], or in properties of the free energy of the system at positive temperature [18,19]. In the following, an attempt is made to elucidate a notion of statistics which is relevant for some three-dimensional gauge theories and, perhaps, for three-dimensional quantum gravity. We start by briefly reviewing the main features of a quantum field theory of anyons [17]. We consider a three-dimensional U(1)-Higgs model with a Chern-Simons term

$$\frac{i\mu}{2\pi}\int A \wedge dA \tag{2.1}$$

in the Euclidean action which breaks parity invariance. For suitable choices of the gauge coupling constant, e, and of the parameters in the Higgs potential, this theory has superselection sectors labelled by a magnetic flux number Φ, with

$$e\Phi = 2\pi n\,, \quad n \in \mathbb{Z}\,. \tag{2.2}$$

The sectors with $e\Phi = \pm 2\pi$ contain massive one-particle states. [This can be proven rigorously in the lattice approximation [17].] The corresponding particles are called vortices. The Buchholz-Fredenhagen theory [11] can be applied in this situation and tells us that vortices can be created by applying fields localized in space-like cones to the vacuum. This is born out explicitly in the analysis of [17]. A peculiarity of the vortices in a U(1)-Higgs model with Chern-Simons term (2.1) is that they also carry an electric charge, q, given by

$$q = \pm\ 2\mu\,. \tag{2.3}$$

Such particles have been called "anyons" [15]. It is quite simple to calculate the spin, s, of a one-anyon state in the U(1)-Higgs model with Chern-Simons term by explicitly performing a 2π-rotation. The result is [17]

$$s = \frac{1}{4\mu}\, q^2 = \mu\,, \tag{2.4}$$

where we have set $e = 1$. Performing a 2π rotation of a state with magnetic flux $\Phi = 2\pi n$ and electric charge nq one finds [17] that its angular momentum, L_n, is given by

$$L_n = n^2\mu \ (\text{mod.}\ \mathbb{Z}) = ns + \frac{n(n-1)}{2}\, 2\mu \ (\text{mod.}\ \mathbb{Z}) \tag{2.5}$$

On the basis of results in [10,11] one can develop a Haag-Ruelle collision theory for anyons. It permits one to define a Hilbert space, $\mathcal{H}_{as}$, of asymptotic anyon states whose dynamics is the free dynamics of non-interacting, relativistic particles. One can then ask what properties an asymptotic n-anyon state has. Thus, let

$$f_n \equiv f(p_1,q_1,\ldots,p_n,q_n) \tag{2.6}$$

be the momentum space wave function describing n asymptotic anyons with 3-momenta $p_1,\ldots,p_n$, $p_i \in V_m \equiv \{p:p^2=m^2, p^o > 0\}$, for all i, and charges $q_i = \pm 2\mu$, $i = 1,\ldots,n$. Here m is the mass of an asymptotic anyon, and $s = \mu$ is its spin.

Rotating f_n through an angle 2π we find, using (2.5), that

$$U(2\pi)\, f_n = \exp\left[i\, 2\pi\left(\frac{1}{4\mu}\left(\sum_{j=1}^{n} q_i\right)^2 - ns\right)\right] f_n \tag{2.7}$$

Thus, for $\mu \notin \frac{1}{2}\mathbb{Z}$, f_n cannot be a single-valued function on momentum space $(V_m)^{xn}$. Let

$$\mathcal{M}_n \equiv (V_m)^{xn} \setminus D ,$$

where

$$D \equiv \{p_1,\ldots,p_n : p_i = p_j, \text{ for some } i \neq j\} .$$

The fundamental group of $\mathcal{M}_n$ is the pure braid group, P_n. Let $\tilde{\mathcal{M}}_n$ denote the universal cover of $\mathcal{M}_n$. One can show that f_n is a single-valued function on $\tilde{\mathcal{M}}_n$. The Hilbert space of n-anyon wave functions with charges $q_1,\ldots,q_n$ carries a representation, R, of the braid group, B_n, given by a character:

$$R(\tau_{i\,i+1}) = e^{i2\pi\mu\varepsilon_i\varepsilon_{i+1}} , \tag{2.8}$$

where $\varepsilon_i = q_i/2\mu$ $(= \pm 1)$. Eqn. (2.8) is essentially equivalent to eqn. (2.7); (2π rotations correspond to special elements of B_n for which (2.8) must reduce to (2.7). The converse is slightly more subtle).

Our results show that if μ is an integer f_n obeys ordinary Bose statistics, while if μ is half-integer f_n obeys ordinary Fermi statistics. For $\mu \notin \frac{1}{2}\mathbb{Z}$, f_n has what has been called "intermediate (or θ-) statistics", [15,16].

The properties of asymptotic anyon wave functions are vaguely related to some "generalized monodromy properties" of anyon Euclidean Green functions derived in [17]. The discussion of those properties requires somewhat detailed definitions and calculations which cannot be reviewed here; but see [17]. The bottom line of that discussion is that branches of Euclidean Green functions corresponding to n anyons carrying electric charge 2μ and n anti-anyons of charge -2μ can be labelled by arbitrary links with $\leq n$ components; (their phases depend on the writhe of those links).

From the results in [17] a concept of "field bundle" can be extracted: In theories like the three-dimensional U(1)-Higgs model with a Chern-Simons term in the action or the related O(3) non-linear σ-model with a Hopf term in the action [15], one can construct operator-valued sections, $\Phi(x_1,E_1,\ldots,x_n,E_n)$, of some field bundle over the manifold of pairwise space-like separated n tuples of points, $x_1,\ldots,x_n$, in three-dimensional Minkowski space such that $\Phi(x_1,E_1,\ldots,x_n,E_n)$ has non-vanishing matrix elements between the vacuum sector and a sector of electric charge $2\mu\left(\sum_{i=1}^{n}\varepsilon_i\right)$. The symbols "$E_i$" stand for distributions of electric flux with source at x_i and total electric charge $2\mu\,\varepsilon_i$, $i = 1,\ldots,n$. The field bundle carries a flat connection whose holonomy is given by (2.8).

This notion of field bundle bears some degree of arbitrariness and should not be taken for a "kinematical" explanation of intermediate statistics. It may nevertheless be a useful concept; for example in the analysis of models of three-dimensional quantum gravity. A special feature of gravity in three dimensions is that space-time curvature is concentrated in regions where matter is present and that there are no dynamical gravitons [20]. In other words, curvature is concentrated along the world lines of "virtual particles". It is conceivable that one might be able to define a model of three-dimensional quantum gravity as a gas of interacting disclination loops and -lines tied to matter loops and -lines, just like the three-dimensional U(1)-Higgs model with Chern-Simons term can be formulated as a gas of interacting magnetic flux loops and -lines tied to electric flux loops and -lines. [This makes good mathematical sense in the lattice approximation [17].] At present, this is of course rather speculative.

An even wilder speculation would be that the curvature of four-dimensional space-time is concentrated on world sheets of "strings". [Since "curvature bones" of piecewise flat four-dimensional space-time manifolds are two-dimensional, the speculation is, a priori, not entirely non-sensical.]

Three-dimensional field theories of anyons, i.e. of excitations with fractional spin and intermediate statistics, although uninteresting for particle physics, might have interesting applications in condensed matter physics, e.g. to the fractional quantum Hall effect [21] or to models of two-dimensional anti-ferromagnets [22].

Acknowledgments. Useful discussions with G. Felder, B. Schroer and, expecially, P.-A. Marchetti are gratefully acknowledged. I also thank A. Connes, V. Jones, T. Kohno and G. Mack for useful conversations.

References

1. Jürg Fröhlich, "Statistics of Fields, the Yang-Baxter Equation, and the Theory of Knots and Links", to appear in the proceedings of the Cargèse summer school on string theory, conformal field theory and renormalization theory, 1987

2. A.B. Zamolodchikov, as quoted in: "Spectrum and Scattering of Excitations in the One-Dimensional Isotropic Heisenberg Model", by L.D. Faddeev and L.A. Takhtadzhyan

3. K. Osterwalder and R. Schrader, Commun.Math.Phys. 31, 83 (1973), 42, 281 (1975); V. Glaser, Commun.Math.Phys. 37, 257 (1974)

4. A.A. Belavin, A.M. Polyakov and A.B. Zamolodchikov, Nucl.Phys. B 241, 333 (1984)

5. V.S. Dotsenko and V.A. Fateev, Nucl. Phys. B 240 [FS 12], 312 (1984), B 251 [FS 13], 691 (1985); P. Christe and R. Flume, Nucl. Phys. B282, 466 (1987); P. Christe, Ph.D. thesis, Bonn 1986

6. V.G. Knizhnik and A.B. Zamolodchikov, Nucl. Phys. B 247, 83 (1984); A. Tsuchiya and Y. Kanie, Lett. Math.Phys. 13, 303 (1987); T. Kohno, "Linear Representations of Braid Groups and Classical Yang-Baxter Equations" in: "Contemporary Mathematics: Artin's Braid Group", Santa Cruz, 1987

7. K.H. Rehren and B. Schroer, "Quasiprimary Fields: An Approach to Positivity of 2D Conformal Quantum Field Theory", Preprint 1987; K.H. Rehren, "Locality of Comformal Fields in Two Dimensions: Exchange Algebra on the Light-Cone", Preprint 1987, to appear in Commun. Math. Phys.

8. A. Capelli, C. Ithykson and J.B. Zuber, Nucl. Phys. B 280, 445 (1987); Commun. Math. Phys. 113, 1 (1987)

9. R. Haag and D. Kastler, J. Math. Phys. 5, 848 (1964)

10. S. Doplicher, R. Haag and J.E. Roberts, Commun. Math. Phys. 13, 1 (1969); 15, 173 (1969); 23, 199 (1971); 35, 49 (1974)

11. D. Buchholz and K. Fredenhagen, Nucl. Phys. B 154, 226 (1979); Commun. Math. Phys. 84, 1 (1982)

12. D. Buchholz and H. Epstein, Fisika, 17, 329-43 (1986)

13. R.F. Streater and A.S. Wightman, "PCT, Spin and Statistics and All That", Reading, Mass.: Benjamin 1978

14. R. Jost, "The General Theory of Quantized Fields", Providence, R.I.: Publ. Amer. Math. Soc. 1965

15. F. Wilczek, Phys. Rev. Lett. 49, 957 (1982); F. Wilczek and A. Zee, Phys. Rev. Lett. 51, 2250 (1983)

16. Y.-S. Wu, "Fractional Quantum Statistics in Two-Dimensional Systems", in: Proc. 2nd Int. Symp. Foundations of Quantum Mechanics, Tokyo 1986, pp. 171-180, and refs. given there

17. J. Fröhlich and P.-A. Marchetti, "A Gauge Theory of Anyons", Preprint 1988

18. S. Kind, diploma thesis, ETH 1988

19. D.A. Arovas, R. Schrieffer, F. Wilczek and A. Zee, Nucl. Phys. B 251 [FS 13] , 117 (1985)

20. S. Deser, R. Jackiw and G. 't Hooft, Ann. Phys. 152, 220 (1984); S. Giddings, J. Abbott and K. Kuchar, Gen. Rel. and Grav. 16, 751 (1984)

21. R.B. Laughlin, Phys. Rev. Lett. 50, 1395 (1983); B.I. Halperin, Phys. Rev. Lett. 52, 1583 (1984); D.A. Arovas, R. Schrieffer and F. Wilczek, Phys. Rev. Lett. 53, 722 (1984); R. Tao and Y.-S. Wu, Phys. Rev. B 31, 6859 (1985)

22. P.B. Wiegmann, "Super conductivity in Strongly Correlated Electronic Systems and Confinement vs. Deconfinement Phenomena", Preprint 1987

QUANTUM GROUPS (YBZF ALGEBRAS) AND INTEGRABLE THEORIES : AN OVERVIEW

H.J. de Vega

Laboratoire de Physique Théorique et Hautes Energies* - Paris

Abstract. Integrable QFT and statistical models are briefly reviewed paying special attention to the algebric structures responsable of their solvability : YBZF algebras (also called Quantum groups)

The light-cone lattice approach here exposed naturally associates an integrable massive QFT to a gapless lattice model in the scaling limit. Alternative scaling limits yielding conformal invariant theories are also reviewed

The construction of exact solution of two dimensional integrable theories made impressive progress in the last years. By integrable theories we mean models possessing as many commuting and conserved physical

---

* Postal adress

Universite Pierre et Marie Curie, Tour 16, 1er etage

4, place Jussieu

75252 Paris Cedex 05 - France.

K. Bleuler and M. Werner (eds.), Differential Geometrical Methods in Theoretical Physics, 187–210.

magnitudes as degrees of freedom. That is an infinite number for field theories or statistical models.

Integrable theories are interesting since they are usually exactly solvable in a large extent. One obtains detailed information about the physics of the models without relying in any approximation. In addition, integrable models happen to be realistic for different phenomena in condensed matter physics. In particle physics, two-dimensional QFT are interesting laboratories to understand four-dimensional physics and they can probably be used to build string models.

Eigenvalues and eigenvectors of the hamiltonian, the momentum and higher conserved magnitudes are usually explicitly calculable in an integrable model. In this way exact mass spectra and S-matrices for QFT and the free energy in statistical models are derived. Moreover, form factors and one-point functions (order parameters) can be also computed explicitly.

These results are obtained through the use of the Bethe Ansatz (BA) and its different generalizations [1-3]. Actually the BA is not merely an Ansatz but it provides the exact solution of the models. To be more precise, the investigations of these last years show that the structure underlying all integrable theories (both QFT and statistical models) is the so-called Yang-Baxter-Zamolodchikov-Faddeev algebra (YBZF). (sometimes also called Quantum Groups) These structures are new in mathematics and their last mathematical meaning and scope is still under investigation.

Integrable theories in more than two-dimensions are known. As classical field theories one can mention self-dual Yang-Mills in 4 + 0 [4] or (3 + 1) dimensions [5], supersymmetric Yang-Mills in 10 dimensions [6] and $\mathcal{N} = 3$ and 4 extended susy Yang-Mills in four dimensions [7]. The integrability structure of these models seems to be deeply connected with twistors and supertwistors [8]. In three-dimensional statistical mechanics, the tetrahedron equations and Zamolodchikov's solution [9] are three dimensional extensions of the 2-D YBZF equations.

A YBZF algebra can be defined as follows. Let $T_{ab}(\Theta)$ be a set of quantum operators for $1 \leqslant a, b \leqslant q$ and $\Theta \in \mathbb{C}$ acting on a quantum space $\mathcal{V}$. $q \geqslant 2$ is a given number that defines the dimensionality of the auxiliar space (A) where T_{ab} acts as a q x q matrix. Θ is called the spectral parameter due to the connection with spectral problem in the inverse scattering method. Now the YBZF algebra is defined by the set of relations

$$R(\Theta - \Theta')\left[T(\Theta) \otimes T(\Theta')\right] = \left[T(\Theta') \otimes T(\Theta)\right] R(\Theta - \Theta') \qquad (1)$$

where the tensor product notation

$$(A \otimes B)_{ab,\, cd} = A_{ac}\, B_{bd}$$

in the auxiliary space is used. More explicitly eq. (1.1) reads

$$R(\Theta - \Theta')_{ab,ef}\, T_{ec}(\Theta)\, T_{fd}(\Theta') = T_{ae}(\Theta')\, T_{bf}(\Theta)\, R_{ef;cd}(\Theta - \Theta') \quad (2)$$

where we sum over repeated indices. Here $R_{ab,cd}(\Theta - \Theta')$ are c-numbers in the quantum space $\mathcal{V}$. That is, R is a matrix in $A \otimes A$. In eq. (1) and (2), it is understood an operational product for the $T_{ab}(\Theta)$ acting on $\mathcal{V}$. The $R_{ab,cd}(\Theta - \Theta')$ form the so-called R-matrix that defines the YBZF algebra. They can be thought as "structure constants" and the $T_{ab}(\otimes)$ play the rôle of "generators".

The link between a YBZF algebra (1) and integrability (in the above defined sense) is immediate. Left multiplying (1.1) by $R(\Theta - \Theta')^{-1}$ yields

$$T(\Theta) \otimes T(\Theta') = R(\Theta - \Theta')^{-1} [T(\Theta') \otimes T(\Theta)] R(\Theta - \Theta')$$

Taking now trace on $A \otimes A$ gives

$$[\tau(\Theta), \tau(\Theta')] = 0 \tag{3}$$

where

$$\tau(\Theta) \equiv \sum_{a=1}^{q} T_{aa}(\Theta) \tag{4}$$

and we used the property

$$tr(A \otimes B) = tr A \; tr B$$

Therefore we have a family of commuting transfer matrix $\tau(\Theta)$ and an infinite number of operators C_n defined by

$$\log \tau(\Theta) = \sum_{n} C_n \Theta^n, \quad [C_n, C_m] = 0 \tag{5}$$

These C_n are usually conserved charges since the hamiltonian and the momentum are connected with the lower C_n [1-3] or with $\log \tau(\Theta)$ at some special Θ. In this way the connection of YBZF with integrable theories is straightforward.

The associativity requirement for the product of operators $T_{ab}(\Theta)$ as defined by eq.(1) constraints the "structure constants" $R_{ab,cd}(\Theta - \Theta')$. Taking a product of three operators $T(\Theta_1) \otimes T(\Theta_2) \otimes T(\Theta_3)$ and reordering with the help of eq.(1) in two inequivalent ways leads to the triangle lations or Yang-Baxter equations

$$S^{(12)}(\Theta_1-\Theta_2)\ S^{(13)}(\Theta_1-\Theta_3)\ S^{(23)}(\Theta_2-\Theta_3) =$$
$$= S^{(23)}(\Theta_2-\Theta_3)\ S^{(13)}(\Theta_1-\Theta_3)\ S^{(12)}(\Theta_1-\Theta_2) \qquad (6)$$

here the matrix $S^{(ij)}$ $(1 \leqslant i < j \leqslant 3)$ acts in the tensor product of spaces $\otimes A_j$ as $PR(\Theta_i - \Theta_j)$ and it is the unit matrix in the remaining space.

ere

$$P_{ab,cd} = \delta_{ad}\,\delta_{bc} \qquad (7)$$

he algebraic equations (6) are a sufficient condition for the associativity f the YBZF algebra (see [1-3] for more details). It must be stressed that he Yang-Baxter equations (6) (or factorization equations) are a heavily verdetermined set since they contain a priori q^6 equations and only q^4 nknowns. Despite of this fact a rich set of solutions is known. All of them ossess at least a discrete symmetry $Z_q \otimes Z_q$ that reduces the number of ndependent equations and probably permit the very existence of non-trivial solutions.

The main property of the YBZF algebras is the following reproduction roperty. If the operators $T_{ab}(\Theta)$ acting on $\mathcal{V}$ obey eq. (1) - (2), so does

$$T^{(K)}_{ab}(\theta,\{\mu,g\}) = \sum_{a_1\cdots a_{K-1}} [g_1 T(\theta-\mu_1)]_{aa_1} \otimes [g_2 T(\theta-\mu_2)]_{a_1 a_2} \otimes \cdots \otimes [g_K T(\theta-\mu_K)]_{a_{K-1} b} \qquad (8)$$

on $\mathcal{V} \otimes \mathcal{V} \otimes \ldots \otimes \mathcal{V}$ (K times) with the same R-matrix. Here $g_i \in \mathcal{G}$ $(1 \leqslant i \leqslant K)$ and the μ_i are arbitrary parameters. We have

$$R(\theta-\theta')\left[T^{(K)}(\theta,\{\mu,g\}) \otimes T^{(K)}(\theta',\{\mu,g\})\right]=\left[T^{(K)}(\theta',\{\mu,g\}) \otimes T^{(K)}(\theta,\{\mu,g\})\right]R(\theta-\theta') \tag{9}$$

Eq.(7) follows inserting eq. (6) and using repeatedly eq. (1). So, starting from a given YBZF algebra on $\mathcal{V}$ it is always possible to construct other representations on $(\mathcal{V})^K$. One can also construct still other representations $\tilde{T}_{ab}^{(K)}(\Theta, \{\mu, g\})$ multiplying the generators $g_i\, T(\Theta - \mu_i)$ from right to left as

$$\tilde{T}^{(K)}_{ab}(\theta,\{\mu,g\})=\sum_{a_1\cdots a_{K-1}}\left[g_1 T(\theta-\mu_1)\right]_{a_1 b}\otimes\left[g_2 T(\theta-\mu_2)\right]_{a_2 a_1}\otimes\cdots\otimes\left[g_K T(\theta-\mu_K)\right]_{a\, a_{K-1}} \tag{10}$$

It must be stressed that any solution $R(\Theta)$ of the YB equations (1.5) yields a YBZF algebra. Setting

$$\left[t_{ab}(\theta)\right]_{cd} \equiv R^{bd}_{ca}(\theta)$$

one finds from eq. (6)

$$R(\Theta - \Theta')\ [t(\Theta) \otimes t(\Theta')] = [t(\Theta') \otimes t(\Theta)]\ R(\Theta - \Theta') \tag{11}$$

As we see, the representation of YBZF algebras sounds as a natural generalization of Lie Algebras. This assertion is actually correct in the sense that YBZF algebras are deformations of Lie Algebras as quantum mechanics is a generalization of classical mechanics [10,11]. It can be noticed that the reproduction property (8) in the particular case K = 2 defines a comultiplication of the generators leading to the structure of a Hopf algebra.

Let us now consider physical applications of the YBZF algebra concepts first in two dimensional statistical mechanics and then in field theory.

The matrix elements of the generators $t_{ab}(\Theta)$ can be defined as statistical weights of a vertex configuration [fig. 1]. The indices a,b ($1 \leqslant a,b \leqslant q$) label the states of the horizontal bonds and the indices α, β those of the vertical bonds ($1 \leqslant \alpha, \beta \leqslant \dim \mathcal{V}$). For meaningful statistical models one needs $[t_{ab}]_{\alpha}^{\gamma}(\Theta) \geqslant 0$. However, models with some negative and complex weights have interesting mathematical structure too [3,12,13] For a rectangular N x L lattice, the partition function follows by the usual definition :

$$Z = \sum_{\substack{\text{all} \\ \text{configurations}}} \prod_{\substack{\text{whole} \\ \text{lattice} \\ 1 \leqslant i \leqslant N,\, 1 \leqslant \ell \leqslant L}} t_{a_i b_{i+1}}(\Theta)_{\alpha_\ell}^{\beta_{\ell+1}} \tag{12}$$

This Z can be written in terms of the transfer matrix $\tau(\Theta)$ associated by eq. (4) to the YBZF generator

$$T_{ab}(\Theta) = \sum_{a_1 \cdots a_{L-1}} t_{aa_1}(\Theta) \otimes t_{a_1 a_2}(\Theta) \otimes \cdots \otimes t_{a_{L-1} b}(\Theta) \tag{13}$$

One finds

$$Z = \mathrm{Tr}\,[\tau(\Theta)^N] \tag{14}$$

where $\tau(\Theta) = \sum_{a=1}^{q} T_{aa}(\Theta)$,

for periodic boundary conditions in both directions. Therefore, the free energy in the thermodynamic limit is given by the maximum eigenvalue of $\tau(\Theta)$, $\lambda_{max}(\Theta)$

$$f = -\lim_{\substack{N \to \infty \\ L \to \infty}} \frac{1}{NL} \log Z = -\lim_{L \to \infty} \frac{1}{L} \log \lambda_{max}(\Theta) \tag{15}$$

This relation shows that the physical properties of the system are directly related to the YBZF algebra. The fact that $\tau(\Theta)$ is a commuting family [eq. (3)] allows to diagonalize it. This can be done just using the YBZF algebra as a tool to build the eigenvectors and eigenvalues [1-3].

Besides 2-D vertex models, two dimensional classical spin models (IRF models) and solid-on-solid models possess YBZF algebras. In one-dimension, quantum magnetic hamiltonians as the Heisenberg model and generalizations can be built from these algebras [1-3,13,14].

Implicit and explicit connections between YBZF algebras and 2-D QFT are known since some time [1-3]. Let us expose here the light-cone approach [15,16]. This approach is the more general and precise way to construct integrable QFT and conformal invariant theories. One starts from integrable lattice models like vertex models [fig. 1] on a diagonal lattice

[16]. This diagonal lattice is a discretization of Minkowski space-time in light-cone coordinate $X_{\pm} = X \pm t$. The matrix elements $[t_{ab}(\Theta)]_{\alpha}{}^{\beta}$ are now interpreted as quantum mechanical transition amplitudes of bare particles. propagating to the right or to the left by the bonds at the speed of light. In the simplest case dim $\mathcal{V} = q = 2$ and we interpret these particles as bare fermion without internal degrees of freedom. The allowed microscopical amplitudes assuming a U(1) charge conservation are depicted in fig. 2. This correspond to the six-vertex model in the statistical mechanical language of Fig. 1. We have only three independent amplitudes because we assumed parity invariance. We can organize the microscopic amplitudes at a site into a unitary scattering matrix

$$R^{\alpha\beta}_{\alpha'\beta'} =$$

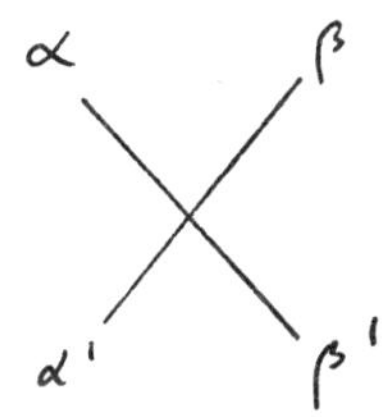

$$= \begin{pmatrix} 1 & 0 & 0 & 0 \\ 0 & c & b & 0 \\ 0 & b & c & 0 \\ 0 & 0 & 0 & \omega \end{pmatrix} \qquad (16)$$

We can build now the operators describing the evolution by one lattice step in the diagonal directions

$$U^R =$$

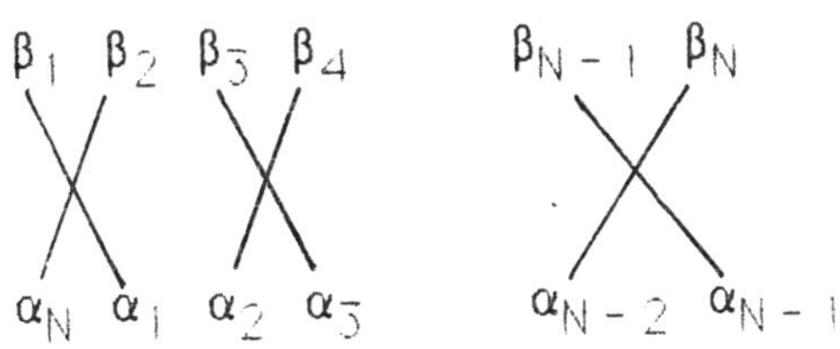

(17)

$U_L =$ 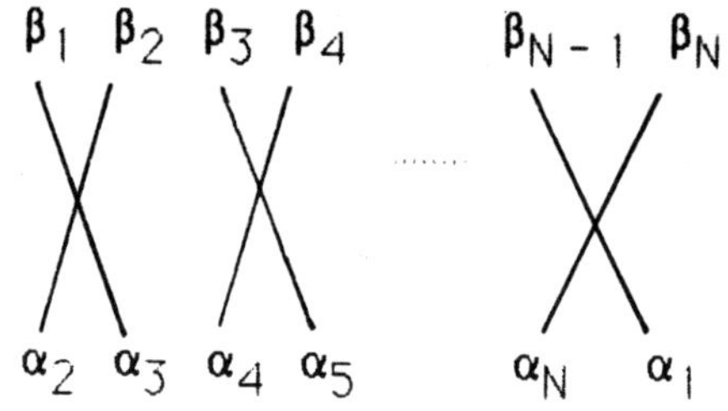

where the numbers 1, ..., 2N label the sites.

A second quantized formalism can be introduced defining fermion operators

$\psi_{R,n}$ and $\psi_{L,n}$

They are associates to the links stemming upwards of each site to the right and left respectively [fig. 3]. They fulfil canonical anticommutations rules.

$$\{\psi_{A,n}, \psi_{B,m}\} = 0, \{\psi_{A,B}, \psi^{\dagger}_{B,m}\} = \delta_{AB}\,\delta_{nm} \qquad (18)$$

$$1 \leqslant n,m \leqslant N$$

$$A,B = R,L.$$

$\psi_{R,n}$ and $\psi_{L,n}$ can be assembled into a two-component spinor. This provides a diagonal representation for γ_5, it is the Chiral representation. We avoid species doubling since our lattice hamiltonian and momentum defined here from U_R and U_L are non local operators. It is convenient to write

$$\psi_{R,m} = \psi_{2m}\ ,\ \psi_{L,m} = \psi_{2m-1}\ ;\ 1 \leq m \leq N$$

Now the R matrix (16) can be expressed in second-quantized language as [16]

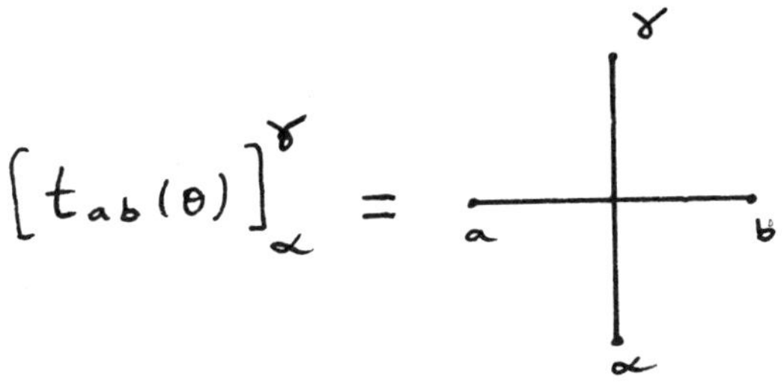

Fig. 1

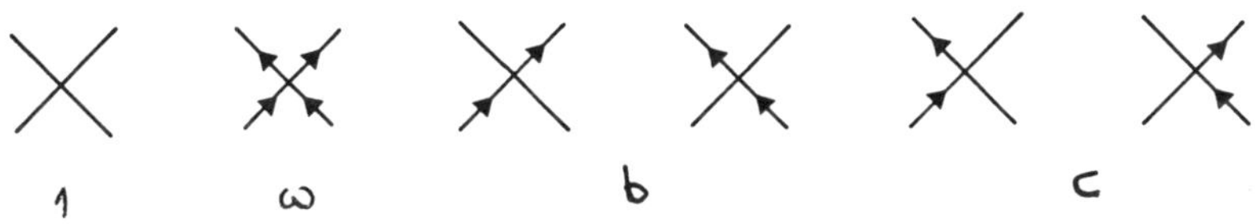

Fig. 2

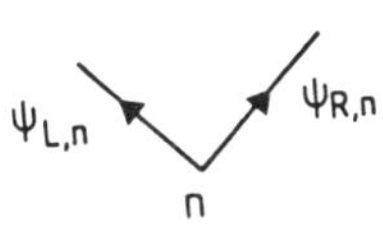

Fig. 3

$$R_{m,m+1} = 1 + b\, K_{m,m+1} + (c-1)\, K^2_{m,m+1} + \qquad (19)$$

where
$$+ (w-1)\, \psi^+_m \psi_m \psi^+_{m+1} \psi_{m+1} \;,$$

$$K_{n,m} = \psi_n^+ \psi_m + \psi_m^+ \psi_n \qquad (20)$$

It is now possible to derive the lattice equations of motion for the fermion operators ψ_n. We find using eqs. (17)-(20) [16]

$$U_R\, \psi_{2m-2}\, U_R^+ = U_L\, \psi_{2m}\, U_L^+ = b^*\, \psi_{2m} + c^*\, \psi_{2m-1} + \qquad (21)$$
$$+ (c/w - c^*)\, \psi^+_{2m} \psi_{2m} \psi_{2m-1} - (b/w + b^*)\, \psi^+_{2m-1} \psi_{2m-1} \psi_{2m}$$

$$U_L\, \psi_{2m+1}\, U_L^+ = U_R\, \psi_{2m-1}\, U_R^+ = b^*\, \psi_{2m-1} + c^*\, \psi_{2m} + \qquad (22)$$
$$+ (c/w - c^*)\, \psi^+_{2m-1} \psi_{2m-1} \psi_{2m} - (b/w + b^*)\, \psi^+_{2m} \psi_{2m} \psi_{2m-1} \,.$$

We can interpret in eqs. (21)-(22) the b-terms as kinetic energy, the c-terms as mass terms and the trilinear pieces as interactions. It must be remarked that the equations of motion are <u>local</u> on the lattice although the lattice hamiltonian is <u>totally</u> non-local.

The continuum limit eqs. (21)-(22) is carefully analyzed in ref. (16) where the continuum field hamiltonian H and momentum P are derived. We find that as the lattice spacing tends to zero

$$U_{R,L} = e^{2i\mu Q} \left[1 - \frac{ia}{2} (H \pm P) + O(a^2) \right] \qquad (23)$$

provided the weights b and c behave for $a \to 0$ as [16]

$$b \underset{a\to 0}{=} e^{i\mu} \left[1 + O(a^2) \right] \qquad (24)$$

$$c \underset{a\to 0}{=} - \frac{i}{2} e^{i\mu}\, m_0 a \left[1 + O(a^2) \right]$$

where μ and m_0 are fixed parameters. We find for H and P those of the massive Thirring model (MTM)

$$P = -i\int dx\, \psi^{+}\partial_x \psi$$

$$H = \int dx\left[-i\,\psi^{+}\gamma_5\partial_x\psi + m_0\bar{\psi}\psi + \frac{g}{2}(\bar{\psi}\gamma_\mu\psi)^2\right] \tag{25}$$

with $g = -2\,\mathrm{cotg}(\mu-\mu_0)$, $\omega = e^{2i\mu_0}$, $\psi = \begin{pmatrix}\psi_R\\ \psi_L\end{pmatrix}$.

Eq. (25) gives the bare operators and eq. (24) defines the bare scaling limit. This is different to the renormalized scaling limit giving the physical sector of the Fock space [15,16]. Both the particle spectrum and physical S-matrices follow rigorously in the renormalized scaling limit computed in this light-cone approach. The bare limit (24)-(25) tells us which model one is actually solving. We use the word "rigorous" since we solved in this approach a lattice model exactly, then we take the infinite volume limit and finally the a → 0 (scaling) limit. In other words here one solves (exactly) a model with both UV and volume cutoffs and then lets the cutoffs to infinity. This is clearly better than coordinate Bethe-Ansatz (CBA) where the UV cutoff is introduced <u>after</u> the obtention of the solution. For the MTM and the Chiral Gross-Neven model the results of the CBA coincide with the light-cone approach for on-shell magnitudes. Hence the CBA works well in these cases. This is not the case of the multiflavor Chiral model treated in ref. [17] by CBA. As it is shown in ref [18] the results of ref. [17] are not correct.

The lattice light-cone approach was extended in ref [18] to Chiral fermionic models with any simple Lie group of symmetry and lagrangian

$$\mathcal{L} = i\bar{\Psi}\partial\Psi - g\,(\bar{\Psi}\gamma_\mu T^\alpha\Psi)(\bar{\Psi}\gamma^\mu T_\alpha\Psi) \tag{26}$$

Here Ψ transforms under an irreducible representation ρ of G and T_α are the G-generators in that representation. The light-cone lattice approach works here as follows. The lattice H and P are defined as

$$H \pm P = \frac{2i}{a}\,\log U_{R,L}(\Theta) \tag{27}$$

where a is the lattice spacing and $U_{R,L}(\Theta)$ are given by eq. (13) from the rational R-matrix invariant under G taken in the ρ representation. For large Θ we have

$$R(\Theta) = P\left[1 + \frac{\Pi + \lambda}{i\Theta} + O(1/\Theta^2)\right] \tag{28}$$

where λ is a numerical constant, the exchange operator P was defined in eq. (7) and

$$\Pi = \sum_{\alpha=1}^{\dim G} T_\alpha \otimes T^\alpha \tag{29}$$

We then introduce the lattice operator

$$T_n^\alpha \equiv 1 \otimes \cdots \otimes \overbrace{T^\alpha}^{n\text{th site}} \otimes \cdots \otimes 1 \tag{30}$$

Using eqs. (13), (28) and 29) and the Lie algebra commutators

$$[T_\alpha, T_\beta] = i f_{\alpha\beta}{}^{\gamma} T_\gamma$$

we can show that T_n^α obey local equation of motion on the lattice [18]

$$U_R T_{2m-2}^\alpha U_R^+ = U_L T_{2m}^\alpha U_L^+ = T_{2m}^\alpha + \frac{2i}{\Theta} f^\alpha{}_{\beta\gamma}\, T_{2m-1}^\beta T_{2m}^\gamma + O\left(\frac{1}{\Theta^2}\right) \tag{31}$$

$$U_R T_{2m-1}^{\alpha} U_R^{+} = U_L T_{2m+1}^{\alpha} U_L^{+} = T_{2m-1}^{\alpha} - \frac{2i}{\Theta} f^{\alpha}{}_{\beta\gamma} T_{2m-1}^{\beta} T_{2m}^{\gamma} + O\left(\frac{1}{\Theta^2}\right)$$

The bare scaling limit is now defined as $a \to 0$, $\Theta \to \infty$, $x = n\,a$ fixed. We get

$$\partial_\mu J^{\mu\alpha} = 0$$
$$\partial_0 J_1^{\alpha} - \partial_1 J_0^{\alpha} + i g f^{\alpha}{}_{\beta\gamma} [J_0^{\beta}, J_1^{\gamma}] = 0 \tag{32}$$

where

$$J_R^{\alpha}(x) = \frac{1}{g a \Theta} T_{2m}^{\alpha} \quad , \quad J_L^{\alpha}(x) = \frac{1}{g a \Theta} T_{2m-1}^{\alpha} .$$

Therefore we have a lattice version of the G-algebra currents $J_\mu^{\alpha}(x)$ associated to an exactly solvable discretization of the field models.

Let us now discuss the renormalized scaling limit. The operators U_R and U_L, as explained before, are the light-cone evolution operators in a discretized Minkowski space-time. The hamiltonian and momentum in the continuum theory are defined by the $a = 0$ limit of eq. (27). This limit is to be taken such that the physical masses are finite. We derive the form of $\Theta = \Theta(a)$ from the spectrum of $U_R(\Theta)$ and $U_L(\Theta)$ on the lattice in order to obtain such finite mass spectrum for $a \to 0$.

The light-cone transfer matrices U_R and U_L express in terms of the row-to-row transfer matrix $\tau(\Theta, \{\mu\}) = \sum_{a=1}^{q} T_{aa}^{(N)}(\Theta, \{\mu, 1\})$ that follows from eq. (8) when $g_1 = \ldots = g_K = 1$ and $\mu_k = (-1)^{k+1}\,\Theta$. Then we showed [18] that

$$U_L(\Theta) = \tau(\Theta, \{\mu_k = (-1)^{k+1}\Theta\}) \qquad (33)$$

$$U_R(\Theta) = \tau(-\Theta, \{\mu_k = (-1)^{k+1}\Theta\})^{+}$$

Then, the spectrum of $\tau(\Theta_0 ; \{\Theta_k\})$ provides that of $U_R(\Theta_0)$ and $U_L(\Theta_0)$. The eigenvectors and eigenvalues can be exactly computed by the Bethe-Ansatz and its nested generalizations [3,14,19]. The antiferromagnetic ground state eigenvector provides in the scaling limit the vacuum state of the QFT. The particle states follow from the lowest lying excitations of the lattice vertex model. Since there is a factor a^{-1} in eq. (27), only gapless lattice models may yield finite energy states in this scaling limit.

Let us take the G-invariant fundamental vertex models defined by eqs. (28)-(32). The low-lying excitation over the antiferromagnetic ground state are holes with large rapidity ϕ. The large ϕ behavior of the $\tau(\Theta_0, \{\Theta_k\})$ eigenvalues is independent of Θ_j and given by [20]

$$f_\ell(\varphi) = -\lim_{N\to\infty} \frac{\lambda_\ell(\varphi, \Theta_0, \{\Theta_k\})}{\lambda_{max}(\Theta_0, \{\Theta_k\})} \underset{\varphi+\Theta_0 \to \pm\infty}{=} \frac{i m_\ell}{\pi} e^{\mp\kappa(\varphi+\Theta_0)} + o(e^{\mp 2\kappa(\varphi+\Theta_0)}) \qquad (34)$$

Here $\lambda_\ell(\phi, \Theta_0, \{\Theta_k\})$ is the contribution of a hole in the ℓth branch $(1 \leq \ell \leq \text{rank } G)$ to the eigenvalue of $\tau(\Theta_0, \{\Theta_k\})$. The dimensionless parameters κ and m_ℓ are given in Table 1.

Combining eq. (34), with eq. (33) and (27) yields the dispersion relation for $\Theta \to +\infty$

Lie Algebra	Dynkin's Diagram	κ	m_k
A_n	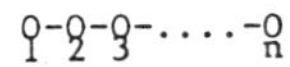	$2\pi/n+1$	$\sin(\pi k/n+1)$, $1 \leqslant k \leqslant n$
B_n		$\pi/2n-1$	$\sin(\pi k/2n-1)$, $1 \leqslant k \leqslant n-1$; $m_n = \frac{1}{2}$
C_n		$\pi/n+1$	$\sin\big(\pi k/2(n+1)\big)$, $1 \leqslant k \leqslant n$
D_n	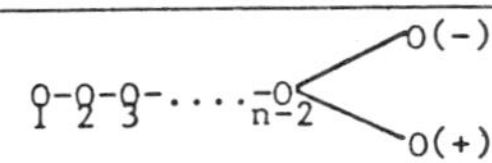	$\pi/n-1$	$\sin\big(\pi k/2(n-1)\big)$, $1 \leqslant k \leqslant n-2$; $m_{\pm} = \frac{1}{2}$
E_6	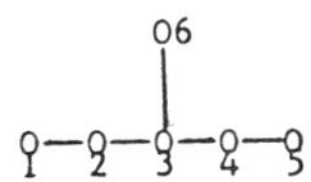	$\pi/6$	$m_1 = m_5 = \frac{m_6}{2} = \frac{\sqrt{3}}{2}$; $m_2 = m_4 = \frac{3+\sqrt{3}}{2}$; $m_3 = \frac{3+\sqrt{3}}{\sqrt{2}}$
E_7	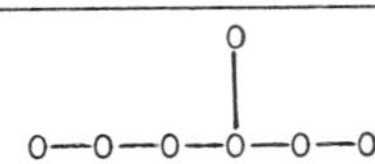	$\pi/9$	(*)
E_8	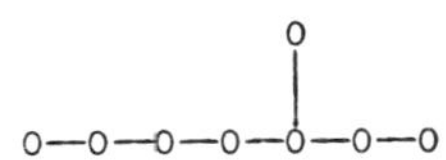	$\pi/15$	(*)
F_4	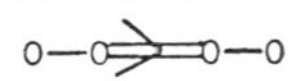	$\pi/9$	(*)
G_2		$\pi/6$	(*)

(*) These values can be (very laboriously) extracted from Refs. [27]

TABLE I

$$E_\ell(\varphi) = \frac{e^{-\varkappa\Theta}}{\pi a} m_\ell \cosh(\varkappa\varphi) + O(e^{-2\varkappa\Theta}) \quad (35)$$

$$p_\ell(\varphi) = \frac{e^{-\varkappa\Theta}}{\pi a} m_\ell \sinh(\varkappa\varphi) + O(e^{-2\varkappa\Theta})$$

It is then natural to define the scaling limit according to

$$a \to 0, \Theta \to \infty, \mu = \frac{e^{-\varkappa\Theta}}{\pi a} = \text{fixed} \quad (36)$$

μ is the renormalized or physical mass and the particle mass spectrum of these integrable QFT'S is given by

$$M_\ell = \mu m_\ell \quad (37)$$

We recognize in eq. (35) $\varkappa\phi$ as the physical particle rapidity

This is a very general way of constructing integrable QFT's. The operators H and P given by Eq. (27) are well defined on the lattice as well as all the higher conserved charges. In the continuum limit $a \to 0$, they provide the energy and momentum of a relativistic invariant QFT, as long as the spectrum of the original vertex model is gapless. This is usually the case for statistical weights $\left[t_{ab}(\Theta)\right]_c^d$ which are rational or trigonometric functions of the spectral parameter Θ. In addition to the particle spectrum, the S-matrix is exactly calculable from the BA equation by standard methods [1]-[3].

The field theoretic models discussed up to here correspond to finite values of q, namely, a finite dimensional vector space for each link in the light-cone lattice. This is clearly appropriate for fermionic fields. Since there exist representations of the YB algebra for $q = \infty$, also bosonic QFT's may be described in this framework.

The $S = \infty$ representation of the XXX magnet relates to the SU(2) Principal Chiral σ-Model (PCM), as was developped in ref. [21]. Let us recall that the physical particle states of this model transform under the $SU(2)_L \times SU(2)_R$ group. The counting of states in the BA equations [21] and our derivation [22] show that only the $SU(2)_L$ singlet sector of the model is described by the H and P associated through eq. (27) to the infinite S limit of the R-matrix[23]

$$R_{12}(\theta) = \frac{\Gamma(2s+1+i\theta)\,\Gamma(\mathbb{J}+1-i\theta)}{\Gamma(2s+1-i\theta)\,\Gamma(\mathbb{J}+1+i\theta)} \tag{38}$$

Here the operator $\mathbb{J}$ is defined by

$$\mathbb{J}(\mathbb{J}+1) = 2s(s+1) + 2\,\vec{S_1}\cdot\vec{S_2} \tag{39}$$

$\vec{S}_1$ and $\vec{S}_2$ are spin-S operators $[\vec{S}_1^{\,2} = \vec{S}_2^{\,2} = S(S+1)]$ acting on the horizontal and vertical spaces respectively.

In other words, the Hamiltonian of the quantum PCM <u>does not follow</u> from the vertex constructions (27) and (38) even in the scaling limit. Only at the classical level can an equivalence be established between the respective classical analogues [21]. Although (27) and (38) do not provide at $S = \infty$ the full PCM Hamiltonian, they correctly reproduce its restriction to the $SU(2)_L$ -singlets, and this is enough to calculate all particle masses as well as the invariant S-matrix amplitudes.

The lattice current construction, eqs. (30)-(32), also applies to the PCM. For large Θ the R-matrix (38) admits a semiclassical expansion of the type (28). Therefore the whole construction holds. It must be noted that we have once again only one conserved and curvature-less matrix current either the one associated to $SU(2)_R$ or that associated to $SU(2)_L$

In conclusion, the light-cone transfer matrices U_R and U_L associated to each integrable gapless vertex model, yield integrable and massive QFT's in the continuum limit. Since the scaling limit can sometimes be performed in several inequivalent ways, one can construct different QFT's from a single vertex model.

Besides the scaling limits leading to massive QFT as eq. (36), there exists the scaling limit

$$a \to \infty \qquad \Theta = \text{fixed} \tag{40}$$

yielding a conformal invariant theory

The conformal invariant theory describes only the long range properties of the integrable lattice model. In this sense integrable models have much more structure than conformal field theories.

The finite size resolution of Bethe-Ansatz equations of the lattice model following the methods of ref (24), gives the values of the central charge C and of the conformal dimensions $(h, \bar{h})$ of operators for a large class of models [20,25]. Branching coefficients can be related to one-point functions [26]

The large size behavior of the logarithm of the transfer matrix eigenvalues close to the ground state λ_n takes the form(for periodic boundary conditions)

$$\frac{f_{m,N}(\Theta) - f_{m,\infty}(\Theta)}{v} = -\frac{\pi}{6N^2}c + \frac{2\pi}{N^2}x_m + o\left(\frac{1}{N^2}\right) \tag{41}$$

$f_{m,N} \equiv - \frac{\mathrm{Re} \log \Lambda_m}{N}$, v is the speed of sound. Namely the dispersion tion for the excitations behave as

$$E = v p + O(p^3) \qquad (42)$$
$$p \to 0$$

re p stands for the momentum.

The values of c and x_n follow from the BA equations and comparison ı conformal field theory tells us that c = center of Virasoro algebra and $h_n + \bar{h}_n$ (for the groundstate $x_n = 0$).

The value of c is known for all fundamental vertex models with YB bras associated to simply-laced Lie algebras [5,6] and for the spin -S ?) model [28]. These results indicate the general formula for c [18]:

$$c = \frac{x \dim G}{x + \tilde{h}} \qquad (43)$$

dim G in the dimension of the Lie algebra and $\tilde{h}$ the dual Coxeter ber. The Sugawara construction of ref.(29) also fulfils eq. (43). We eve that the gapless integrable theories associated to a Lie algebra G ide through their long-range behavior, an alternative realization of the ormal algebra.

The conformal dimensions for the six-vertex model result [25]

$$x_{p,q} = \Delta + \bar{\Delta} = \frac{1}{2}\left(1 - \frac{\gamma}{\pi}\right) p^2 + \frac{q^2}{2\left(1 - \frac{\gamma}{\pi}\right)} \;, \quad p, q \in \mathbb{Z}$$

$$s_{p,q} = \Delta - \bar{\Delta} = p q$$

Here s_{pq} is the spin of the operator A associated to the excitated state (that is $\langle \varphi_n | A | 0 \rangle \neq 0$). The value of S can be extracted from the finite size correction to the momentum

$$P_N = P_\infty + \frac{2\pi S}{N^2} + o\left(\frac{1}{N^2}\right)$$

and

$$P_N = - \mathrm{Im} \, \mathrm{Log} \, \tau \big|_{\theta = 0}$$

For the critical q-states Potts model we find [25]

$$c = 1 - \frac{6}{m(m+1)}$$

where $q = 4 \cos^2[\pi/(m+1)]$; $m = 2, 3, 4, \ldots$ The conformal dimensions fulfil Kac's formula [25]. The conformal dimensions for models with many states per bond like those considered in ref. [3,13] indicates the presence of an extended conformal algebra (in the sense of [30]) for long distances [31].

References

[1] L.D. Faddeev and L.A. Takhtadzhyan, Russ. Math. Surveys, 34, 11 (1979).

L.D. Faddeev, les Houches lectures, (1982), North Holland.

[2] P.P. Kulish and E.K. Sklyanin, in : Springer Lectures in Physics, vol. 151 (1981).

[3] H.J. de Vega. Lectures in the Sâo Paulo School, september 1987,

LPTHE preprint 87-54. To appear in the Proceedings (World Scientific)

[4] R. Ward, Phys. Lett. 61A, 81 (1977)

A.A. Belavin and V.E. Zakharov, Phys. Lett. 73B, 53 (1978)

M.F. Atiyah, "Geometry of Yang-Mills Fields", Lezione Fermiane, Pisa (1979).

[5] H. J. de Vega, CERN preprint TH 4817 to appear in Comm. Math. Phys.

[6] E. Witten, Nucl. Phys. B 266, 245 (1985)

[7] E. Witten, Phys. Lett. 77 B, 394 (1978)

[8] J. Avan and H.J. de Vega, LPTHE preprint 86-23 to appear in Int. J. of Mod. Phys.

[9] A.B. Zamolodchikov, JETP 52, 325 (1980) and Comm. Math. Phys. 79, 489 (1981)

R.J. Baxter, Comm. Math. Phys. 88, 185 (1983)

[10] V.G. Drinfeld, Dokl. Akad. Nauk. 32, 254 (1985)

[11] H.J. de Vega, Phys. Lett. B 177, 171 (1986)

[12] A.A. Belavin, Nucl. Phys. B 180, 189 (1981)

[13] O. Babelon, H.J. de Vega and C.M. Viallet, Nucl. Phys. B 190, 542 (1981).

[14] O. Babelon, H.J. de Vega and C.M. Viallet, Nucl. Phys. B 200, 266 (1982).

[15] T.T. Truong and K.D. Schotte, Nucl. Phys. B 220, 77 (1983) and B 230, 1 (1984).

M. F. Weiss and K.D. Schotte, Nucl. Phys. B 225, 247 (1983).

[16] C. Destri and H.J. de Vega, Nucl. Phys. B 290, 363 (1987).

[17] L.V. Avdeev and M.V. Chizov, Phys. Lett. B 184, 363 (1987).

[18] C. Destri and H.J. de Vega, CERN-TH preprint 4895/87.

[19] L.A. Takhtadzhyan, Journal of Sov. Math. 2470 (1983).
N.Y. Reshetikhin, Theor. Math. Phys. 63, 555 (1986).
H.J. de Vega and M. Karowski, Nucl. Phys. B 280, 225 (1987).

[20] H.J. de Vega, J. Phys. A 20, 6023 (1987).

[21] L. D. Faddeev and N.Y. Reshetikhin, Ann. Phys. 167, 227 (1986)

[22] C. Destri and H.J. de Vega, CERN preprint TH 4893/87 to be published in Phys. Lett. B.

[23] P.P. Kulish, N.Y. Reshetikhin and E.K. Sklyanin, LMP 5, 393 (1981).

[24] H.J. de Vega and F. Woynarovich, Nucl. Phys. B 251, 439 (1985).
H.P. Eckle and F. Woynarovich, J. Phys. A 20, L97 (1987).

[25] H.J. de Vega and M. Karowski, Nucl. Phys. B 285, 619, (1987).
F. Woynarovich, Phys. Rev. Lett. 59, 259 (1987).
M. Karowski, Brasov Lectures (1987).
C.J. Hamer, G.R.W. Quispel and M. J. Batchelor, Canberra preprint 9/87.
See also ref [31].

[26] See for a review E. Date, M. Jimbo, T. Miwa and M. Okado, RIMS-590, Kyoto preprint, 1987.

[27] E. Ogievetsky, and P.B. Wiegmann, Phys. Lett. B 168, 360 (1986)

[28] H. M. Babujian, Nucl. Phys. B 215, 317 (1983).
I. Affleck, PRL 56, 746 (1986).

[29] P. Goddard and D. Olive, Int. J. of Mod. Phys. A1, 303 (1987).

[30] V.A. Fateev and A. B. Zamolodchikov, Nucl. Phys. B 280, 644 (1987).

[31] H.J. de Vega, LPTHE preprint 88/03.

INTEGRABLE SYSTEMS AND CONFORMAL INVARIANCE

M. Karowski
Theoretical Physics, ETH-Hönggerberg
CH-8093 Zürich, Switzerland

ABSTRACT. Statistical systems at second order phase transition points should exhibit conformal invariance for long distances. Therefore they will yield conformal quantum field theories in the scaling limit. Their conformal properties can be analysed using finite size scaling behaviour. For integrable lattice models in two dimensions methods are proposed to calculate from the Bethe ansatz solution the "conformal anomaly" and scaling dimensions. As an application results for the q-state Potts model and modified six-vertex models are presented. It is conjectured that the latter exhaust the class of all unitary conformal quantum field theories with "central charge" $c < 1$.

1. Conformal Invariance.

Conformal invariance of two dimensional quantum field theories has become a subject of main interest because of its connection to string theory. I would like to report on results obtained for two-dimensional integrable models. Some of the investigations rely on a collaboration with H. de Vega [1]. Here I will explain only the main results which are obtained by the "Bethe ansatz" method. The detailed calculations can be found in ref. [2]. Statistical mechanical systems at second order phase transition points (i.e. critical systems) should possess conformal invariance [3]. We shall use this fact to obtain information on conformal quantum field theories.

In two dimensions conformal invariance is a rather strong restriction since the transformation group is infinite dimensional, in contrast to higher dimensions. It is related to the Virasoro algebra [4]

$$[L_n, L_m] = (n-m)L_{n+m} + \tfrac{1}{12}c(n^3-n)\delta_{n,-m} \tag{1}$$

$$(c = \text{"central charge"})$$

for which a well developed theory exists [5]. Belavin, Polyakov and Zamolodchikov [6] classified conformal quantum field theories in terms of the central charge

$$c = 1 - \frac{6}{\nu(\nu-1)} \tag{2}$$

K. Bleuler and M. Werner (eds.), Differential Geometrical Methods in Theoretical Physics, 211–218.

where ν is a rational number if there exists only a finite number of "primary fields", i.e. fields transforming under conformal transformations

$$z = x + iy \rightarrow w(z)$$

$$\bar{z} = x - iy \rightarrow \bar{w}(\bar{z}) \tag{3}$$

like

$$A(z,\bar{z}) \rightarrow \left(\frac{dw}{dz}\right)^{\Delta} \left(\frac{d\bar{w}}{d\bar{z}}\right)^{\bar{\Delta}} A(w,\bar{w}). \tag{4}$$

The conformal dimensions can assume values given by the Kac-formula [5]

$$\Delta, \bar{\Delta} = h_{pq} = \frac{(\nu p - (\nu - 1)q)^2 - 1}{4\nu(\nu - 1)} \tag{5}$$

$$1 \leq q \leq p < \nu - 1 \qquad p, q = \text{integer}.$$

Friedan, Qiu and Shenker [7] have shown that unitarity restricts the number ν to

$$\nu = integer \geq 2 \quad \text{or} \quad c \geq 1. \tag{6}$$

It would be interesting to find integrable lattice models which yield these conformal quantum field theories in the scaling limit. Therefore it is useful to calculate the central charge and the conformal dimensions for integrable critical statistical models. This can be done by analysing finite size effects.

2. Finite Size Behaviour.

For a statistical system on a periodic $M \times N$ square lattice one introduces a transfer matrix τ (defined on an N-chain). The partition function can be written as

$$Z = \operatorname{tr} \tau^M = \sum_i \lambda_i^M \tag{7}$$

where the sum extends over all eigenvalues of the transfer matrix

$$\tau\psi = \lambda\psi. \tag{8}$$

Cardy [8] has shown that conformal invariance implies for the maximal eigenvalue

$$\lambda_{max} \approx \exp(-Nf + \frac{1}{N}\frac{\pi}{6}c) \qquad \text{for} \quad N \rightarrow \infty \tag{9}$$

where f is the free energy per site (in the thermodynamic limit) and c is the central charge of the Virasoro algebra (cf. eq. (1)).

Introducing in addition the "crossed" transfer matrix $\hat{\tau}$ (defined on an M-chain) we obtain for $1 \ll N \ll M$ two expressions for the partition function

$$Z \approx \exp(-NMf) \sum_i (\hat{\lambda}_i / \hat{\lambda}_{max})^M \tag{10a}$$

$$Z \approx \exp(-NMf + \frac{M}{N}\frac{\pi}{6}c) \tag{10b}$$

where only the "low energy excitations" with $\hat{\lambda}_i \approx \hat{\lambda}_{max}$ contribute to the sum in eq. (10a). Hence two methods to investigate the partition function on a strip $1 \ll N \ll M \to \infty$ are availble:

(a) one considers the transfer matrix $\hat{\tau}$ for an infinite chain ($M \to \infty$) and calculates the large N corrections to the sum in eq. (10a) or

(b) one considers the transfer matrix τ on an N-sites chain and calculates the large N corrections to the maximal eigenvalue λ_{max} according to eq. (10b).

By these procedures one obtains the central charge of the model.

In addition the conformal dimensions of operators of the model can be determined if one looks for excited states with energy $E = -\Re \log \lambda$ and momentum $P = -\Im \log \lambda$. As is argued in ref. [6].one has for $N \to \infty$

$$E_i - E_0 \approx \frac{2\pi}{N} d_i \tag{11}$$

$$P_i - P_0 \approx \frac{2\pi}{N} s_i$$

where $d_i = \Delta + \bar{\Delta}$ $(s_i = \Delta - \bar{\Delta})$ is the "scaling dimension" ("spin") of the operator A associated to the excitation ψ_i, i.e. $\langle 0|A|\psi_i\rangle \neq 0$.

Since a direct calculation of the central charge and the conformal dimensions for specific models is usually not simple, it seems worthwhile to determine them for integrable models from the Bethe ansatz solution. This will be done in the following for the six-vertex model (the critical 8-vertex model) and for some modified six vertex models related to Potts models and other models of interest. Method (a) has been developed in ref. [1], method (b) is applied in sect. 4 and ref. [2].

3. Integrable Models.

By means of the technique known as Yang-Baxter algebra method, algebraic Bethe ansatz or quantum invers method [9] the eigenvalue problem eq. (8) of the transfer matrix is solved. The eigenstates appear as "n-pseudoparticle" states, where the momenta p_j of the pseudoparticles have to fulfil the "Bethe ansatz equations"

$$\exp(iNp_j) \prod_{\substack{i=1 \\ i \neq j}}^{n} \exp(\Phi_{ji}) = 1 \tag{12}$$

$$j = 1, \ldots, n \qquad 0 \leq n \leq N/2.$$

These eqations express the condition that the total phase factor around one period N has to be equal to one. This factor consists of the free part $\exp(iNp_j)$ and the scattering parts with respect to all other pseudoparticles $\exp(i\Phi_{ji})$.

For the six-vertex model (cf. [9,10]) parametrized by

$$\omega_1 = \omega_2 : \omega_3 = \omega_4 : \omega_5 = \omega_6 = \sin(\gamma - \theta) : \sin\theta : \sin\gamma \tag{13}$$

the "phase shifts" $\Phi_{ji} = \Phi(q_j - q_i)$ are given by

$$e^{ip(q)} = \frac{\sinh\frac{1}{2}(q - i\gamma)}{\sinh\frac{1}{2}(q + i\gamma)} \qquad e^{i\Phi(q)} = \frac{\sinh\frac{1}{2}(q + 2i\gamma)}{\sinh\frac{1}{2}(q - 2i\gamma)}. \tag{14}$$

There exist detailed investigations of the Bethe ansatz equations (12) (see e.g. ref. [11]). The analysis simplifies for the discrete values of the "coupling"

$$\gamma = \pi/\nu \qquad \nu = 2, 3, 4, \ldots. \tag{15}$$

We are interested in these values. So we assume eq. (15) in the following.

Any set of roots of eq. (12) $\{q_1, \ldots, q_n\}$ defines an eigenstate and an eigenvalue $\lambda(\theta,\gamma) = \lambda_A + \lambda_D$ of the transfer matrix $\tau(\theta,\gamma)$

$$\log\lambda_A(\theta,\gamma) = N\omega_1 - i\sum_{i=1}^{n}\left(p\left(q_i + 2i\theta\right) - \pi\right) \tag{16}$$
$$\lambda_D = (\omega_1/\omega_3)^N \lambda_A^*(\gamma - \theta^*, \gamma).$$

Taking the logarithm of eq. (12)

$$2\pi I_j = Np(q_j) + \sum_{i=1}^{n}\Phi(q_j - q_i) \tag{17}$$

we observe that sets of roots $\{q_1, \ldots, q_n\}$ are given by sets of numbers $\{I_1, \ldots, I_n\} \subset \mathbf{Z} + \frac{1}{2}$ [1]. For the "ground state" (corresponding to the maximal eigenvalue λ_{max}) the number of real roots is maximal. This means $n = \frac{1}{2}N$ (N even) and the numbers I_j accept all possible values in an interval $I_- \leq I_j \leq I_+$. For "excited states" there may be "holes" in the sequence of the I_j. In addition there may appear complex roots q: the "k-strings" (i.e. "bound states" of k pseudoparticles)

$$q = \text{real} + i\gamma(-k+1, -k+3, \ldots, k-1) \qquad 1 < k \leq \nu - 1 \tag{18a}$$

and the "anti-pseudoparticle"

$$q = \text{real} + i\pi. \tag{18b}$$

---

[1] The $\frac{1}{2}$ is due to $\Phi(0) = \pi$.

If there are only finitely many holes, in the limit $N \to \infty$ the roots become dense on the whole real line and the "last roots" near to $\pm\infty$ behave like

$$q_\pm \approx \pm\tfrac{1}{2}\nu \log N\,. \tag{19}$$

4. Results for Six-Vertex and Potts Models.

The Potts models are a generalizations of the Ising model where the spin at each site can accept q values instead of two. For $q = 2,3,4$ the models posses a critical point which we consider in the following. The partition function on a cylinder [10] can be written in terms of a six-vertex model with a "seam" from the bottom to the top of the cylinder. This seam carries extra Boltzmann weights which produce extra terms $-i\gamma m$ and $2\gamma m$ with $m = 1$ on the r.h.s. of eqs. (16) and (17), respectively. The number q $(q = 2,3,4)$ is related to the six-vertex coupling $\gamma = \pi/\nu$ $(\nu = 4,6,\infty)$ by

$$q = 4\cos^2\gamma \tag{20}$$

In the following the partition function Z_m and other formulas apply for $m = 0$ to the ordinary six-vertex model and for $m = 1$ to the Potts model

$$Z_0 = Z^{6-vertex} \qquad Z_1 = Z^{Potts}. \tag{21}$$

In refs. [12,13,1,2] techniques were developed to calculate finite size corrections for a large chain of length N. Using these methods we find [1,2] for the "ground state"

$$\log\lambda_{max} \approx -Nf + \frac{1}{N}\frac{\pi}{6}\left(1 - \frac{6m^2}{\nu(\nu-1)}\right)\sin\nu\theta. \tag{22}$$

The vertex weights (13) fulfil $\omega_1 = \omega_2 = \omega_3 = \omega_4$ for $\nu\theta = \pi/2$ where we expect isotropy and conformal invariance in the scaling limit. Comparing eqs. (22) and (9) we conclude that the central charges of the Virasoro algebra of the six-vertex and the q-state Potts models are

$$c^{6-vertex} = 1 \tag{23a}$$

$$c^{Potts} = 1 - \frac{6}{\nu(\nu-1)}. \tag{23b}$$

As already mentioned in sect. 3 the ground state correspond to the maximal set of real roots $q_1,\ldots,q_{N/2}$. Excitations possess "holes" and complex roots. We obtain [2] for these states taking eqs. (11) into account[2]

$$\log\lambda_i - \log\lambda_{max} \approx -\frac{2\pi}{N}\{(\Delta+\bar{\Delta})\sin\nu\theta + i(\Delta-\bar{\Delta})\cos\nu\theta\}. \tag{24}$$

[2]Note that the imaginary term correspond to the momentum since $\tau(\theta = 0) = \exp(-iP)$.

The general expressions for the conformal dimensions Δ and $\bar{\Delta}$ are rather complicated. But the primary fields correspond to low energy excitations with a small number of holes $h_{\pm}$ at the ends of the real roots "near to $\pm\infty$". In addition there may be a number $n_{\pm}^{(k)}$ of k-strings "near to $\pm\infty$".[3] For these configurations of roots we obtain [2]

$$\Delta = \frac{\left(\nu\left(h_{+} - 2\sum_k n_{+}^{(k)}(k-1)\right) - \Sigma + m\right)^2 - m^2}{4\nu(\nu-1)} + \text{integer} \tag{25a}$$

$$\bar{\Delta} = \frac{\left(\nu\left(h_{-} - 2\sum_k n_{-}^{(k)}(k-1)\right) - \Sigma - m\right)^2 - m^2}{4\nu(\nu-1)} + \text{integer} \tag{25b}$$

where $\Sigma = \frac{1}{2}h - \sum_k n^{(k)}(k-1)$ is the "spin"[4] of the Bethe state and h ($n^{(k)}$) is the total number of holes (k-strings). We have $\Sigma = 0$ since we consider fixed boundary conditions at the top and the bottom of the cylinder. It turns out [2] that we get minimal values for Δ and $\bar{\Delta}$ (i.e. the integers on the r.h.s. of eqs. (25) vanish) if there are no holes inside $[q_-, q_+]$, $n_{+}^{(k)} = 0$ for all k, $n_{-}^{(k)} = 1$ for only one k and the number of holes have to fulfil $h_+ = h_- = k - 1$. For the six-vertex model we obtain the conformal dimensions from eqs. (25) with $m = 0$. For the q-state Potts model the extra weights on the seam imply [2] that for the minimal Δ and $\bar{\Delta}$ only the k-strings with

$$k = 2, 3, \ldots, \nu - m - 1 \tag{26}$$

contribute (i.e. for $m = 1$ the $\nu - 1$-string and the "antipseudoparticle" do not appear).

Comparing eqs. (25) for $m = 1$ and $\Sigma = 0$ with the Kac-formula (5) we find for the Potts model the conformal dimensions

$$\Delta = \bar{\Delta} = h_{k,1} \qquad k = 1, 2, \ldots, \nu - 2. \tag{27}$$

These correspond to the "energy operators" of the Potts model (cf. [14,8,15]). The equivalence of the Potts model and a six-vertex model with "seam" for $m = 1$ holds for the partition functions (cf. eq. (21)) but not for the operator content. The order parameter of the Potts model, for example, appears in a six-vertex model with a "seam" for $m = \nu/2$ (cf. [15]).

If we consider a partition function of a generalized model

$$Z = Z_1 + \sum Z_m \tag{28}$$

---

[3]The k-strings which do not approach $\pm\infty$ like the last real roots $q_{\pm}$ as $N \to \infty$ will not affect Δ and $\bar{\Delta}$.

[4]Σ must not be confused with the "spin" s of the field operator (cf. eq. (11)).

where the sum extends over some values of $m \in [2, \nu - 1]$ we obtain the dimensions corresponding to Bethe states with a k-string

$$\Delta = \bar{\Delta} = h_{k+m-1,m} \qquad k = 1, \ldots, \nu - m - 1 \quad 1 \leq m \leq \nu - 1. \tag{29}$$

Note that the ground state always belongs to Z_1. The "magnetization" operators of the q = 3-state Potts model ($\nu = 6$) correspond to $m = \nu/2 = 3$ (cf. [15]). Their dimensions are therefore (in agreement with numerical estimates [15] and conjectures [14,8])

$$\Delta = \bar{\Delta} = h_{k+2,3} \qquad k = 1, 2. \tag{30}$$

The $1/q$-para fermion operators [16] correspond to "boundary conditions" $m = \nu/q = 2$ for $q = 2, 3$ [16,15]. The Bethe states have spin $\Sigma = 1$ and a symmetric hole configuration $h_+ = h_- = 1$. For $\Sigma \neq 0$ the integers on the r.h.s. of eqs. (25) only vanish, if there are no higher strings. Hence we obtain from eq. (25) the scaling dimensions and spin

$$d = \Delta + \bar{\Delta} = h_{21} + h_{23} = \frac{\nu - 1}{2\nu} + \frac{\nu^2 - q^2}{2q^2\nu(\nu - 1)} \tag{31}$$

$$s = \Delta - \bar{\Delta} = h_{21} - h_{23} = 1/q$$

in agreement with ref. [16,15]. Finaly let us look for the "chiral" operators of the 3-state Potts model (cf. [14,8]). Their conformal dimensions are given by the "boundary conditions" $m = \nu/2 = 3$ and Bethe states with $\Sigma = 2$ or four holes. For $h_+ = 0, 1, 2, 3$ and $h_- = 4 - h_+$ one finds from eq.(25) respctively

$$\Delta = h_{11}, h_{21}, h_{31}, h_{41} = 0, \tfrac{2}{5}, \tfrac{7}{5}, 3$$

$$\bar{\Delta} = h_{15}, h_{25}, h_{35}, h_{45} = 3, \tfrac{7}{5}, \tfrac{2}{5}, 0.$$

At the end let us consider a partition function given by eq (28) where the sum extends over all $m = 2, \ldots, \nu - 1$. The dimension formula (29) obviously describes the dimensional spectrum of the RSOS models [17], in the language of ref. [18] the $(A_\nu, A_{\nu-1})$ series of conformal quantum field theories. For this interpretation of eq. (28) the number m is not connected to extra weights on a "seam" or to "boundary conditions", but it is related [2] to a "Θ-vaccum" like angle appearing in the construction of Bethe states [9]. The RSOS models of ref. [17] can be generalized [19] to models corresponding to all modular invariant conformal quantum field theories with $c < 1$ [18]. All these models are related to modified six-vertex models. Results for these models will be published elsewhere [20].

References.

[1] H.J. de Vega and M. Karowski, Nucl. Phys. B285 [FS19] (1987) 619;

M. Karowski, in 'String Theory Quantum Cosmology and Quantum Gravity Integrable and Conformal Invariant Theories', Proceedings of the Paris-Meudon Colloquium 22-26 sept. 1986, ed. H.J. de Vega et al., World Scientific Pub. Co (1987);
M. Karowski, to be published in the Proceedings of the 'XXVI Internationale Universitätswochen für Kernphysik', Schladming, Feb. 1987.
[2] M. Karowski, to be published.
[3] A.M. Polyakov, ZhETF 55 (1969) 1026;
A.A. Migdal, ZhETF 55 (1969) 1964;
K. Wilson, Phys. Rev. 179 (1969) 1499.
[4] M.A. Virasoro, Phys. Rev. D10 (1970) 2933.
[5] V.G. Kac, Lecture Notes in Physics 94 (1979) 441;
B.L. Feigin and D.B. Fuchs, Fuct. Anal. Appl. 16 (1982) 114;
A. Rocha-Caridi, in 'Vertex operators in mathematics and physics', ed. J. Lepowsky et al. (Springer, NY, 1984) p. 451.
[6] A.A. Belavin, A.M. Polyakov and A. B. Zamolodchikov, Nucl. Phys. B241 (1984) 333.
[7] D. Friedan, Z. Qiu and S. Shenker, Phys. Rev. Lett. 52 (1984) 1575.
[8] J. L. Cardy, Nucl. Phys. B270 [FS16] (1986) 186.
[9] L. A. Takhtadzhan and L. D. Faddeev, Russian Math. Surveys 34 (1979) 186.
[10] R. J. Baxter, Exactly solvable models (Academic Press, 1982).
[11] M. Takahashi and M. Suzuki, Prog. Theor. Phys. 48 (1972) 2187.
[12] H. J. de Vega and F. Woynarovich, Nucl. Phys. B251 (1985) 439;
F. Woynarovich and H. P. Eckle, J. Phys. A20 (1987) L97.
[13] C. J. Hamer, J. Phys. A19 (1986) L3335;
C. J. Hamer, J. Stat. Phys. 47 (1987) 331;
L. V. Ardeev and B. D. Dörfel, J. Phys. A19 (1986) L13;
H. J. Hamer, C. R. W. Quispel and M. T. Batchelor, Canberra, preprint 9-1987.
[14] V. S. Dotsenko, Nucl. Phys. B235 [FS11] (1984) 54.
[15] F. C. Alcaraz, M. N. Barber and M. T. Batchelor, Phys. Rev. Lett. 58 (1987) 771.
[16] B. Nienhuis and H. J. F. Knops, Phys. Rev. B32 (1985) 1872.
[17] G. E. Andrews, R. J. Baxter and P. J. Forester, J. Stat. Phys. 35 (1984) 193.
[18] A. Cappelli, C. Itzykson and J. B. Zuber, Nucl. Phys. B280 [FS18] (1987) 445.
[19] V. Pasquier, Nucl. Phys. B285 [FS19] (1987) 162.
[20] M. Karowski, to be published.

Algebraic Aspects of Non-Perturbative Quantum Field Theories*

Bert Schroer

FU Berlin, Institut für Theorie der Elementarteilchen,
Arnimallee 14, D-1000 Berlin (West) 33

Based on a lecture delivered at the "Centro Volta",

Como, Sept. 1987

A b s t r a c t

We review recent progress in quantum field theory which originates from a careful analysis of causality. In conformal invariant theories the "exchange algebra" is directly related to the dimensional spectrum. Arguments are given that this new algebraic structure transcends conformal invariance and constitutes a genuine non-perturbative non-commutative algebraic aspect of quantized fields.

February 1988

K. Bleuler and M. Werner (eds.), Differential Geometrical Methods in Theoretical Physics, 219–259.

1. Introduction

> After almost half a century of existence the main question about quantum field theory seems still to be: what does it really describe? and not yet: does it provide a good description of nature?
>
> J. A. Swieca

Ever since quantum field theory has been applied to strong interactions, physicists have tried to obtain a nonperturbative understanding. Dispersion theoretic sum rules, the S-matrix bootstrap, the dual models (and their reformulation in string language) and the conformal bootstrap of the 70^s are prominent cornerstones on this thorny path. Furthermore instantons and topological solitons have shed some light on the nonperturbative vacuum structure respectively on the existence of nonperturbative "charge" sectors.

To these attempts an additional one was recently added[1], which is yet not easily describable in terms of one "catch phrase". Different from previous attempts, it is almost entirely based on new noncommutative algebraic structures: "exchange algebras" whose "structure constants" are braid matrices which generate a homomorphism of the infinite (inductive limit) Artin braid group B_∞ into a von Neumann algebra. Mathematically there is a close relation to recent work of Jones[2]. Its physical origin is the result of a subtle analysis of Einstein causality expressed in terms of local commutativity of space-like separated fields. It is most clearly recognizable in conformal invariant quantum field theories. It has been known for a long time[3] that the resolution of the "Einstein causality paradox" in globally conformal invariant interacting theories[4] yields a very interesting decomposition of a local field into irreducible conformal pieces, which are best viewed as a kind of generalization of creation and annihilation operators. The space-like commutativity forces these irreducible components to obey an "exchange algebra". The form of the braid matrices (in physics known as: ≡ constant Yang - Baxter matrices) which feature in this algebra is restricted by the eigenvalues of the unitary operators $Z_{L,R}$ which generate the nontrivial center of the global conformal group. These eigenvalues also determine the physical interpretation of the eigenvalues of the braid matrices. The analysis for two-dimensional field theories has the additional simplicity that this central sectors relate in a simple manner with the representation sectors of the observables, i.e. the sectors of the Virasoro algebra with a fixed Virasoro-charge c, respectively those of the current algebra with a fixed level k. The irreducible pieces of local fields turn out to be intertwining fields of the representation sectors. In this way contact is made with the conformal block formalism of Belavin, Polyakov and Zamolodchikov[5]. The global decomposition formalism is very powerful. Among other things it allows for a calculation of the Rocha-Caridi[6] and Peterson-Kac[7] characters with the help of the field-theoretical Kubo-Martin-Schwinger boundary condition[8]. As a result it offers hope for a deeper physical understanding of the mysterious modular properties. One should like to derive these properties from the exchange algebra structure and the "soliton completeness" of the local field. In the discussed examples it is possible to construct just one local field which creates all sectors and whose

KMS-state fulfills Nelson's symmetry property[9]. The various ways of achieving this with a given exchange algebra are classified by the ADE classification of Capelli, Itzykson and Zuber[10].

From the point of view of the global decomposition theory and its accompanying exchange algebra, the success of the two dimensional conformal bootstrap program by the block formalism of Belavin, Polyakov and Zamolodchikov[5] turns out to be the tip of an ice-berg. The global decomposition theory with its exchange algebra structure also works in higher dimensions. However, here the central sectors constitute a refinement of the observable sectors of Doplicher, Haag, and Roberts[11] The physical origin of this is quite clear: <u>in higher dimensions observables of conformal field theories as canonical currents and the energy momentum tensor do not fulfill Huygens' principle</u> (apart from free field theories). They already by themselves contain an interaction whose strength is measured by their nonvanishing time-like commutators. Here exchange algebras restrict the structure for time-like separations. The situation would be extremely complicated, if there would be no possibility of relating higher-dimensional conformal field theories to two-dimensional ones. A tempting idea is to do this via the "longitudinal scaling process" similar to the field theoretic scaling limits in the discussion of the Hawking effect[12].This is vaguely related to the "$p \to \infty$ frame" and the old light cone quantization.

The disparity of the observable sectors and the central Z sectors of the conformal decomposition theory in higher dimensions also makes it plausible that the introduction of finite correlation length ("mass-like" perturbation) is only reasonably simple in two-dimensional quantum field theories. For d > 2 one needs a radical new idea which is expected to be akin to a string idea. The present day string theory which has developed apart from quantum field theory, does not serve this purpose and we will take a critical look at this situation in the last section.

A good understanding of the present framework is obtained if one recalls ordinary renormalized perturbation theory in the spirit of Bogoljubov and Shirkov[13] without functional integrals. The Einstein causality postulate and unitarity (positivity) can be expressed in terms of nonlinear equations similar to the well known Schwinger-Dyson equations. Starting with a set of free fields and iterating, one finds that the various renormalizable possibilities can be parametrized in terms of local couplings of these fields as expressed by an interaction Lagrangian L_{int}. So a perturbation theoretical treatment of Einstein causality makes contact with "classical" Lagrangians. The exchange algebras on the other hand give a nonperturbative classification and here the contact with classical notions as Lagrangians becomes very remote. Even in d=2 a Lagrangian presentation is only known for a very few models and even in these few cases this Lagrangian plays no significant rôle in the solution of the model. The important feature is the exchange algebra and its vacuum representation. In some sense this algebra linearizes the untreatable aspects of Schwinger-Dyson type bootstrap equations so that the nonlinear equations become identities.

Noncommutative algebra and their representation theory is presently much more powerful than euclidean functional integrals. Even in quantum mechanics, for example in the problem of understanding why the quasiclassical approximation for a particle moving on a group manifold turns out to be exact, the Feynman path analysis requires the study of difficult subtleties of the topology and differential geometry of infinite-dimensional loop spaces and an understanding of the "Duistermaat-Heckman mechanism" for infinite dimensional Kähler manifolds[14]. The Feynman mash, necessary for the analytic understanding, is not the one of Monte-Carlo lattice techniques but rather a mash resulting from the intrinsic geometric structure of the infinite dimensional loop space.

Algebraic properties played a significant rôle in the formulation of the principles of quantum field theory due to Wightman, Haag, Jost and Lehmann, Symanzik and Zimmermann. However, these principles were too rough in order to permit a classification of concrete models. The euclidean functional approach, which formally allows to use more detailed information with the help of statistical mechanics methods, was thought to be helpful. It did permit a mathematical control in certain cases (constructive quantum field theories) as well as the incorporation of the Θ-vacuum structure and (notably two-dimensional) topological solitons[15], however, a convincing non-perturbative success was not achieved. The exchange algebra approach, which is the main topic of these lectures, has turned the situation. The first success was the recognition that the systematics of the subtle rational numbers appearing as critical indices of two-dimensional quantum field theories is most easily understood in terms of the braid structure of exchange algebras which result from an analysis of the deepest principle of relativistic quantum field theory: Einstein causality.

It is informative to compare the present development with new trends in Geometry. Here the power of noncommutative algebra for the understanding of geometrical problems has been amply demonstrated notably in the work of Connes[16]. Quantum physics was noncommutative from its outset. Its "classical links" have been described later in form of Feynman path integrals, which are an ideal tool to obtain a covariant perturbative understanding. Introducing exchange algebras amounts to a return to the roots of relativistic quantum physics.

In ordinary quantum mechanics where the Einstein causality principle is absent, the algebraic structure is not enough, one also needs hamiltonians. In relativistic quantum field theory the Einstein causality postulate, however, transcends kinematics and the representation theory of exchange algebras constitutes a novel non-Hamiltonian and non-Lagrangian approach to relativistic field theory.

Although we will not use the language of C* algebras, the exchange algebra is most profitably viewed as an extension of the Haag-Kastler[17] framework of quantum field theory. Among the pure algebraic postulates of Haag and Kastler there is no property which distinguishes quantum physics from classical physics. The exchange algebras on the other hand are by their very nature noncommutative algebras. In this sense they are analogous to the Heisenberg commutative relations and the canonical structure in nonrelativistic quantum physics.

There are many exchange algebras and their classification is one principle task of the new algebraic approach. One set of braid matrices may belong to different exchange algebras depending on the nature (transformation properties, symmetries) of the "exchange fields". The dimensional spectrum of the local fields is determined by the structure of the exchange algebra. The second principle task is the construction of the representation theory of the exchange algebras in particular of the vacuum representation.

To the extent that every consistent field theory has a scale invariant Gell-Mann Low fixed point (we expect this picture since it has a very high degree of plausibility), a subgroup of the abelian centre $Z_{L,R}$ as well as a subalgebra of the exchange algebra (depending on the direction in which one leaves the fixed point) will survive. In terms of correlation functions this means that the analytic continuation into a larger domain than the BWH domain[18] is possible in which the analytic functions of several variables become multivalued. However, it is easier and conceptually clearer to express this property in terms of the exchange algebras.

Artin braids appear at several places in recent physics. The generalized statistics algebra of Fröhlich[19] is another application which is rather closely related to exchange algebras. Another application is obtained in 3-dimensional field theories with long ranged fields (nonabelian gauge fields, gravitational fields). The general framework is that of Buchholz and Fredenhagen in which the 3-dimensional fields localized in cones appear with braid structure[20]. These braids presumably appear for Lagrangians with Chern-Simons terms[21]. They are different from the braids of this paper which appear in conventional field theories without long range (gauge) interactions.

2. The Exchange Algebra in Conformal QFT_2

The simplest illustration of new nonperturbative algebraic structures in quantum field theory is provided by two-dimensional conformally invariant field theories. Here the 6-dimensional group generated by Poincarè transformations, dilations and proper conformal transformations factorizes into $SL_L(2,R) \times SL_R(2,R)$ acting as linear fractional transformations on the two light-cone variables $u=x+t$, $v=x-t$. The fact that the whole diffeomorphism group of the two light-cones is unitarily represented is also important[23]. We will, however, try to de-emphasize this aspect since it has no generalization to higher dimensions. It has been known for some time and will not be repeated here[23,24] that there is an interesting clash between Einstein causality for quantum fields and their formal global conformal transformation law of such objects if they would transform irreducibly. This "paradox of Einstein causality" with the global transformation finds its resolution in a very subtle global decomposition theory. An alternative but equivalent statement is that the use of SL(2,R) in quantum field theory requires to work with its universal covering group which has a centre generated by a unitary operator Z for each light-cone. Interacting causal fields in conformal quantum field theories are such fields which violate Huygens' principle: they do not commute for time-like distances. Prominent examples are exponentials of a massless bose field as used in string theories (which are called vertex operators) or the Ising (dis)order fields with their noncanonical dimension. Such fields are changed by the centre

$$ZA(u,v)Z^{+} \neq A(u,v) \qquad (1)$$

even though the corresponding space-time transformation is trivial. Prominent examples of fields not effected by Z are those which fulfill Huygens' principle as currents and the energy momentum tensor in two dimensions (but <u>not</u> in higher dimensions). The decomposition theory of A results from a diagonalization of Z^4:

$$A(u,v) = \sum_{\xi,\bar{\xi}} A^{\xi,\bar{\xi}}(u,v) \qquad (2)$$

where: $$Z_L A^{\xi,\bar{\xi}}(u,v) Z_L^{+} = e^{2\pi i(d-\xi)} A^{\xi,\bar{\xi}}(u,v) \qquad (3)$$

For the right-hand light-cone belonging to the v-variable the Z_R will produce d and ξ. Here d and d are the numbers which appear in the corresponding dilations e.i.

$$U(\lambda)\ A^{\xi,\bar{\xi}}(u,v)\ U^{+}(\lambda) = \lambda^{d} A^{\xi,\bar{\xi}}(\lambda u,v) \qquad (4)$$

The physical interpretation of the phase numbers has been discussed elsewhere[22,23], there it was also shown that if one knows the dimensional spectrum of all composites of a model, then one knows the ξ-spectrum. In particular the ξ-value which appears on the vacuum is $\xi=0=\xi$. In order to encounter situations where several ξ-values of the sum (2) contribute, one must at least consider 4-point-functions[22,23]. For local fields one has asymptotically convergent Wilson expansions, their irreducible pieces $A^{\xi,\xi}$ obey in addition a global vacuum expansion[24] which has the interpretation of a Plancherel formula. With the help of the iterative application of the vacuum expansion one can show that the Wightman functions for the ξ-components factorize in u and v. Since the test functions in u and v can be varied independently, each factor fulfills positivity. Therefore it is consistent to write:

$$A^{\xi,\bar{\xi}}(u,v) = A^{\xi}(u) \; . \; A^{\bar{\xi}}(v) \qquad (5)$$

The factorization of the Hilbertspace going with the separate action of the $A^{\xi}(u)$'s is also required by considering short distance limits in v only whose leading term is the identity operator. The same global vacuum expansion for the 4-point-function of two energy momentum tensors and two fields with the help of the FQS[25] argument allows to derive the c-quantization[24]. This (for a physicist) is a somewhat simpler calculation than that based on the full Kac-determinant.

Since the local commutativity of causal fields is a bilinear relation, one may ask the question what bilinear structure of the A^{ξ}'s is required by causality. A moment of thinking reveals that it generally cannot be a commutation relation as that of a vertex operator (exponential bose field) of one light-cone. It cannot be a commutation relation at all, it must be an exchange algebra! In order to use a more suitable notation let us designate the generally infinite dimensional eigenspace of Z by i,f standing for initial, final. Consider all A-products with i and f fixed but for variable intermediate module a

$$A_{f,a}(u)A_{a,i}(u'). \tag{6}$$

An exchange algebra is a linear relation in which a product in the opposite order in u',u is written as a sum over products in the original order (always with fixed initial and final Z module).

$$A_{fa}(u')A_{ai}(u) = \sum_{a'} R^{(fi)}_{aa'} A_{fa'}(u)A_{a'i}(u') \tag{7}$$

The conformal covariant transformation property forces R to depend only on sig(u-u') and to obey a relation with a phase matrix ϕ which is formed from the eigenvalues of Z on the various modules. Associativity for a product of three component fields $A_{fa} \cdot A_{ab} \cdot A_{bi}$ yields:

$$\sum_{a'} R^{(fb)}_{aa'} R^{(a'i)}_{bb''} R^{(fb'')}_{a'a''} = \sum_{b'} R^{(ai)}_{bb'} R^{(fb')}_{aa''} R^{(a''i)}_{b'b''} \tag{8}$$

It is convenient to introduce a complex vector-space V whose basis-vectors are identified with arrows between the Z modules (eigenspaces). The possible A-components in the Z decomposition are then labelled by such arrows. The various matrices $R^{(fi)}$ may be combined to one big matrix R (with many zeros) acting as End (VxV) on pairs of consecutive arrows. The associativity of the exchange algebra then yields a defining relation for the Artin braid algebra:

$$R_{12}R_{23}R_{12} = R_{23}R_{12}R_{23} \tag{9a}$$

acting on: $V \otimes V \otimes V$

Exchanges of non-adjacent A_i's (where the index i now stands for an arrow) do not interfere and therefore one obtains the remaining Artin braid relation:

$$R_{i,i+1}R_{j,j+1} = R_{j,j+1}R_{i,i+1} \qquad (9b)$$

Since the number of fields in a product is unbounded, we are not just dealing with the braid group B_n, but rather its inductive limit B_∞. It is generated by:

$$\underline{1} \otimes \underline{1} \ldots R_{i,i+1} \otimes \underline{1} \ldots \qquad (10)$$

Einstein causality, i.e. local commutativity is equivalent to the algebraic relation:

$$R_L \cdot R_R^T = \underline{1} \qquad (11)$$

Here $R_{\substack{L\\R}}$ designate the braid matrices for the u and v light-cone exchange-algebras and the R's always refer to sign(u-u')>0, sign(v-v')>0. The commutator for time-like separation does not vanish and measures the strength of the conformally invariant interaction. A measure for this strength is the matrix:

$$R_L^2 - 1 \qquad (12)$$

Let us now look at the simplest realization of these ideas.

Consider a linear array of length q-1 of points (A-type Dynkin diagram) which represent Z_L-modules, and short arrows pointing from i to i+1 representing the action of:

$$A_i(u) \doteq A_{i+1,i}(u) \qquad (13)$$

As usual if one is dealing with operations in physics, one reads them from right to left. The conjugate operator: $A_i^+(u) = A_{i,i+1}(u)$ is represented by an arrow starting from i+1 going into the opposite direction. The general solution to the braid (Yang-Baxter) relations has been explicitly calculated by Rehren[26] and reads:

$$\begin{aligned} A_1^+(u')A_1(u) &= \eta A_1^+(u)A_1(u') \\ A_{i+1}(u')A_i(u) &= \eta\omega A_{i+1}(u)A_i(u') \\ A_{q-2}(u')A_{q-2}^+(u) &= \eta A_{q-2}(u)A_{q-2}^+(u') \end{aligned} \tag{14a}$$

$$\begin{pmatrix} A_i(u')A_i^+(u) \\ A_{i+1}^+(u')A_{i+1}(u) \end{pmatrix} = \eta M_i \begin{pmatrix} A_i(u)A_i^+(u') \\ A_{i+1}(u)A_{i+1}(u') \end{pmatrix} \tag{14b}$$

$$i = 1 \ldots q-3$$

with $$M_i = \begin{pmatrix} \alpha_i & \beta_i \\ \gamma_i & \delta_i \end{pmatrix} = \frac{1}{S_i} \begin{pmatrix} -(-\omega)^{i+1}, & t_i^{-1} & -\omega S_{i-1}S_{i+1} \\ t_i & -\omega S_{i-1}S_{i+1}, & 1 \end{pmatrix} \tag{15}$$

and $$\omega = -e^{-2\pi ip/q} \tag{16}$$

$$s_i = (-\omega)^{i/2} \frac{\sin\pi(i+1)p/q}{\sin\pi p/q} \tag{17}$$

η is a phase which is determined by the two point function of the field and t_i is as yet undetermined. Assuming that the local field A(x) is a parity symmetric scalar field one obtains[26]:

$$t_i = 1, \; p = \pm 1 \bmod q \tag{18}$$

The global conformal invariance of the exchange algebra (14) requires

$$M_i^{-1} = \eta^2 \phi_i M_i \phi_i \qquad (19)$$

$$\phi_i = \text{diag}\left(e^{-2\pi i(d_{i+2}-d_{i+1})}, e^{-2\pi i(d_i-d_{i+1})}\right)$$

This fixes the differences of dimensions:

$$d_i - d_1 = \frac{(i-1)(i-2)}{2}(d_3-d_1)-(i-1)(i-3)(d_2-d_1) \quad \text{mod } Z$$

(20)

Using $d_1 = 0$ (vacuum) and the relation between d_2 and the phase of the two point function one realizes that the subset of the minimal model spectrum

$$d_i = \frac{[q-i(q-1)]^2-1}{4q(q-1)} \qquad (21)$$

is a solution of (18),(19) and (20) with

$$p=q-1 \quad \text{and} \quad \eta = e^{2\pi i d_2}$$

The SU(2) loop group trajectory to the level $k = q-2$:

$$d_i = \frac{j(j+1)}{q}, \qquad j = \frac{i-1}{2} \qquad (23)$$

$$\text{and} \qquad p = 1, \qquad \eta = -e^{2\pi i d_2} \qquad (24)$$

is another solution[26,23].

The "mod restriction" of all formulas allows for other d_i values for which no models are known.

We emphasize that all these statements are purely algebraic. The explicit form of the short-distance singularities, the value of the Virasoro charge, as well as the form of the correlation functions beyond the 3-point function are part of the representation theory of the exchange algebra. In certain cases, when the braid-matrices generate finite groups as for n-point functions with q=4,6 and certain 4-point functions belonging to other special situation, the problem of calculating correlation functions may be reduced to algebraic equations[1]. Another useful pure algebraic construction is the formation of "strands"[1]. By forming clusters of "threads" which remain unbraided within each cluster, the simple scheme (15) and (16) yields bigger R-matrices. This strand-formation should be distinguished from tensor-products of representations. The first possibility exists only for the Artin braid groups (and possibly similar groups). In general these new representations are reducible as with tensor-products, but their explicit reduction is somewhat more tedious. The selection rules of BPZ[5] may be derived in this way.

Strand-formation is the algebraic aspect of the fusion process for the formation of composite fields and corresponds to the construction of bound states for factorizable S-matrices[29]. The fusion of fields is, however, part of representation theory and requires the construction of short-distance limits[5].

The type A Dynkin scheme is the simplest braid scheme in order to obtain a nontrivial exchange algebra. More complicated schemes can be obtained from the Dynkin scheme by strand-formation. Such "higher branched" schemes do also occur in the construction of level k higher loop groups viz. SU(n). The schemes belonging to minimal models can be obtained by a tensor product construction from the type A schemes. In that case one is dealing with a rectangular scheme of say width q-1 and height q-2. The higher horizontal lines may be viewed as soliton sectors of the basis exchange algebra[14]. We will return to this construction in the next section.

The braid matrices of our simple scheme can be used to construct a homomorphic image of the group algebra B_∞. If one introduces the topology by a trace functional[2], the algebra turns out to be identical with the Jones algebra[2] which is a von Neumann algebra of factor type II_1. This algebra generated by the "structure constant matrices" has to be distinguished from the exchange algebra generated by the light cone fields. These fields, restricted to intervals on the light cone generate a new apparently not yet studied C* algebra (after converting the unbounded objects into bounded ones) which is not a von Neumann algebra of type II.

The work of Pasquier[27] also uses the Jones algebra, but the conceptual situation is quite different. In statistical mechanics there are certain very special models in which the dynamical variables, which are attached to the points of a two-dimensional lattice (one dimension in the transfer-matrix formulation) may be written in terms of the Jones generators. The most prominent lattice model with this property is the "restricted SOS model".

The contact of these dynamical variables with the continuous fields of the exchange algebras is only made after the very subtle and complicated scaling limit which was recently discussed by Karowski[28]. The exchange algebra (or an appropriate extension to the case of infinitely many sectors) is a general property of all QFT_2 and, as we will argue in section 4, may even be a structural property of all causal quantum field theories.

I believe that some objects which belong to B_∞, as for example the transfer-matrix will (without scaling limits!) play an important rôle in the yet unknown representation theory of exchange algebras i.e. in the construction of n-point correlation functions.

However, presently the only known explicit construction is that of 4-point functions of arbitrary composites via a detour[1]. One first forms the (reduced) strands which belong to the composite fields. The size of the braid matrices tells how many 4-point channels are coupled, i.e. the number of ξ-components in the global 4-point decomposition.

With these braid matrices one associates through the Riemann-Hilbert-problem a Fuchsian system of differential equations whose monodromy matrices are the squares of the braid matrices. The solution with the correct short distance behaviour at the points 0, 1 and ∞ are the candidates for the 4-point functions of the composite basic (primary) fields. The last step consists in applying the global operator expansion and the verification that all generalized 3-point coupling constants which appear in the expansion coefficients are real. For the special case of correlations between two arbitrary exchange fields and two stress-energy tensors this requirement[24] leads to the well known quantized Virasoro charges for c<1, i.e the positivity of the representation of the Virasoro algebra. In the general case the reality of the coupling constants between arbitrary quasiprimary fields[24] also insures the positivity of the complete quantum field theory. In principle higher point correlation-functions can be built up by successive application of the global expansion. In practice, however, this is not a realistic way to obtain closed expressions for n-point correlation function. For this reason the present method is no substitute for a systematic representation theory of exchange algebras.

It is a remarkable fact that the recognition of an underlying exchange algebra makes the non-linear bootstrap equations, expressing causality and unitarity (for example in the form of Schwinger-Dyson equations with conformal invariant boundary condition) for the first time solvable. In a way exchange algebras serve to "defuse" the "explosive nonlinear structure" on which the older bootstrap philosophy based many incorrect pictures (the existence of only one ("theory of everything") or a few solutions).

Conformal invariant theories are not realistic theories of nature. According to Wilson's, Kadanoff's and Wegner's philosophy, we should explore the regions around fixed points by thinking about "relevant perturbations". These can be classified according to their global conformal transformation properties and they carry their own "strand-braids". In general these operators wipe out the conformal invariance but maintain part of the discrete $Z_L \times Z_R$ invariance. For example a perturbation $\bar{\psi}\psi$ in the Ising model still commutes with $Z_{diag.}$. In the fermionic description of the model, the remnant of the $Z_{diag.}$ invariance is the soliton sector generated by the (dis)order-operator. Here our exchange algebra formalism makes contact with the work of Fröhlich[21] on general statistics. It also connects with Pasquier's[27] work which suggests that certain statistical mechanics models (6-vertex models, RSOS model) do not only possess a scale invariant conformal limit, but also a scaling limit with finite correlation length. However, the relation between these different observations is a special feature of two dimensions. Only here the global Z-sectors are contained in the sectors of the observable algebras. In higher dimensions the situation is different and this means that the Wilson-Kadanoff-Wegner picture cannot be expected to provide an analytic tool for the handling of realistic models like QCD. As we argue in section 4 one needs an essentially new physical idea to tackle such problems.

Although we were able to proof that the simplest nontrivial braid scheme (14), (15) has exchange algebras admitting positive definite vacuum representations, we have not been able to isolate that property of finite braid schemes which is responsible for positivity in general.

3. Demystification of Modular Mystery

From Lagrangian field theory one is accustomed to the picture of "soliton completeness". Consider for example the Sine-Gordon model. Even if one restricts one's attention to the correlation functions of the Sine-Gordon field ϕ, one cannot fail to see the particle-antiparticle states of the massive Thirring field ψ to make their appearance as intermediate states. Karowski[29] has shown how to compute the S-matrix of ψ in terms of the known ϕ-S-matrix. This computation is based on the intuitive picture:

$$\psi \cong \psi + \phi \qquad (25)$$

Presumably this is a special illustration of a generalization of a theorem by Borchers[30]. The theorem states that all fields which are relatively Einstein causal to each other have the same S-matrix if they link the one-particle state with the vacuum. In case that a field ψ is not local with respect to ϕ, but instead has a different well-defined relative space-like behaviour, one expects as a generalization of the Borchers theorem the computability of the ψ-S-matrix in terms of that for ϕ.

This picture radically changes for conformal invariant theories where all particle thresholds fall on top of each other and where the S-matrix becomes meaningless. Indeed the previously discussed local field of the minimal models which one can attach to the linear Dynkin scheme, does not obey the soliton completeness for $q > 4$. One rather has to add the "missing" modules of the discrete unitary series of the Virasoro algebra in order to obtain the well-known rectangular scheme of BPZ with width q-1 and height q-2. The local field which belongs to the exchange algebra attached to the base-line has (q-2) soliton series which are, due to symmetries of the rectangular scheme of modules, not all independent. Introducing a vertical field B belonging to an exchange algebra of a linear scheme of length q-2 one then finds[1] that with the commutative relation:

$$B_j(u')A_k(u) = iA_k(u)B_j(u') \qquad (26)$$

and similar relations involving the adjoint the resulting rectangular braid-scheme contains the complete dimensional trajectories of the minimal model:

$$d_{ij} = \frac{(j(q-1)-iq)^2-1}{4\,(q-1)q} \qquad (27)$$

One expects that the algebra generated by the two local fields A and B can also be represented in terms of only one generator field C. Indeed the field C = BA which, using the terminology of BPZ[5], would belong to a 45° arrow has exchange matrices whose smallest blocks are 4x4 matrices formed via quasi-tensor products (26) from the old 2x2 blocks. In the generic case (sufficiently large q) there exist precisely two bosonic (Einstein-causal) local fields which are attached to the same rectangular (for minimal models) respectively linear (for the SU(2) loop group) braid scheme. One is a scalar local field which amalgamates the same arrows representing the exchange fields on the u- and v-light-cone. The other is a local half integer spin field, which amalgamates the standard u-arrows with different v-arrows induced by a peculiar permutation of the modules. The permutation either takes place on the horizontal line (for q-1 even) or on the vertical (q-2 even). The permuted v-modules are different (accounting for the different dimension of the local field) but the permuted braid scheme is the same, since the braid matrices possess the symmetry j -> q - j. In this way one may understand the A-D solutions found in the study of modular invariance[10]. The case q-1 = 5 may serve as an illustration[1]:

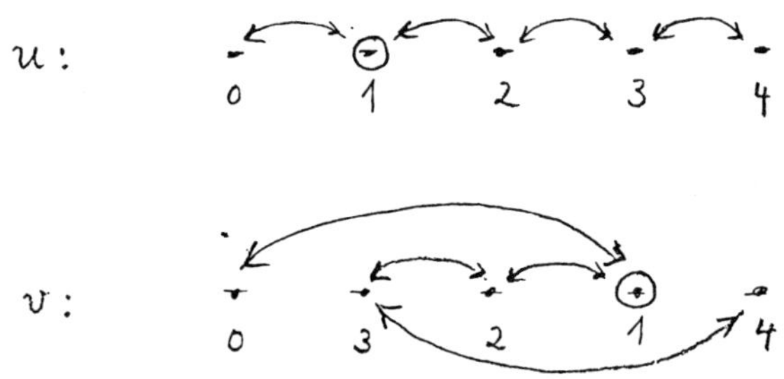

(28)

The modules belonging to the odd points (encircled) of the left half of the chain are permuted with their mirror images, and the arrows follow suite as indicated by their enumeration. The number theoretical aspects in the CIZ work[10] resides already in the braid algebra[1].

The crucial question is now: how can one recognize the "soliton completeness" of an exchange algebra scheme by using only field theoretical methods and not methods of Virasoro algebra representation theory? It turns out that for the massless situation, in which the S-matrix method does not work, the problem can be solved by constructing temperature states (KMS states) over the exchange algebra. The introduction of temperature has in some ways the same effect as a mass, but it yields a much simpler computational method. The field-theoretic way of dealing with

---

1 strictly speaking the local field belonging to the u-v unsymmetric scheme is a D-fermion whereas CIZ only treat "D-bosons".

temperature states has been introduced by Kubo, Martin and Schwinger a long time ago[8]. In algebraic terms it amounts to the introduction of a GNS state which belongs to a type I (if the "hamiltonian" is compact as in conformal field theory) factor representation of the algebra of fields. The KMS state has a remaining symmetry (sometimes referred to as the modular symmetry group) but the spectrum of the generator is unbounded from both sides (i.e. the representation is not of Feigin-Fuchs type), see the last reference in [8]. It is somewhat surprising that KMS states on conformal exchange algebras do not automatically satisfy the Nelson symmetry[9] which, in the spirit of our approach, we attribute to the "soliton incompleteness" of certain exchange algebra schemes. This symmetry then has to be imposed by requiring that the set of Z-phases agrees with the spectrum of eigenvalues which appears in the diagonalization of certain KMS matrices in which the braid matrices appear as substructures. Here we will only schematically scetch the method, for more details I refer to [1].

Consider first the simpler situation for a canonical field. Take as an example the Majorana-spinor field of the Ising model. Consider the traces:

$$\frac{1}{Z_\pm} \operatorname{tr} e^{2\pi i \tau L_0}(P_0 \pm P_2)\psi(u)(P_0 \pm P_2)\psi(0) = (\cos^2\alpha\pi)^{-1/2} F_\tau^{(\pm)}(\alpha) \tag{29}$$

$$\frac{1}{Z_1} \operatorname{tr} e^{2\pi i \tau L_0} P_1 \psi(u) P_1 \psi(0) = (\cos^2\alpha\pi)^{-1/2} F_\tau^{(1)}(\alpha) \tag{30}$$

Here $u = tg\pi\alpha$ and we have split off a conformal factor. P_i, $i = 0, 1, 2$ are the projectors on the sectors of dimension zero, 1/16 and 1/2. The conformal transformation property of ψ under the centre Z leads to:

$$\begin{aligned} F_\tau^{(\pm)}(\alpha+1) &= -F_\tau^{\pm}(\alpha) \\ F_\tau^{(1)}(\alpha+1) &= F_\tau^{(1)}(\alpha) \end{aligned} \tag{31}$$

whereas the commutation (exchange) relation and the cyclicity of the trace yield:

$$\begin{aligned} F^{(\pm)}(\alpha+\tau) \pm F^{(\pm)}(\alpha) &= f^{(\pm)}(\alpha) \\ F^{(1)}(\alpha+\tau) - F^{(1)}(\alpha) &= g(\alpha) \end{aligned} \tag{32}$$

Here the τ independent right-hand side results from the short distance singularities and is a (anti)periodic δ function (in this particular example) which vanishes for $\alpha \neq 0 \bmod 1$.

The equation can be easily solved for the KMS two-point function. Note that only if one considers all three correlation functions the quasiperiodic spectrum (--+) in the direction 1 is identical to that in the direction τ. This $1 \leftrightarrow \tau$ symmetry we take as the definition of Nelson symmetry in the operator formulation. The use of the short-distance limit definition of the energy momentum tensor:

$$T(u) = \lim_{u\to 0} : \frac{d}{du}\, \psi(u)\psi(0) : \tag{33}$$

and the representation:

$$\langle L_0 \rangle_\tau = \frac{1}{2\pi i} \quad \frac{d}{d\tau} \ln Z \tag{34}$$

then easily leads to the various well known partition-functions belonging to the different modules.

$$Z_{--} = \prod_{m \geq 1} \left(1+e^{2\pi i\tau(m+1/2)}\right) = d_{\frac{1}{2},\frac{1}{2}} \tag{35a}$$

$$Z_{-+} = \prod_{m \geq 0} \left(1-e^{2i\pi\tau(m+1/2)}\right) = d_{0,\frac{1}{2}} \tag{35b}$$

$$Z_{+-} = e^{2\pi i\tau 1/16} \prod_{m \geq 0} (1+e^{2\pi i\tau m}) = d_{\frac{1}{2},0} \tag{35c}$$

In order to facilitate a comparison with the later generalization to a non-canonical braid relation, let us use:

$$\frac{1}{Z_{0,2}} \text{ tr } e^{2\pi i L_o \tau} P_{\substack{0\\2}} \psi(u) P_{\substack{2\\0}} \psi(0) \tag{36}$$

instead of the first two traces in (35). Now the quasi-periodicity in τ direction is described in terms of a non-diagonal 2x2 matrix:

$$\begin{pmatrix} 0 & -1 \\ -1 & 0 \end{pmatrix} \tag{37}$$

Consider now the description of the Ising model in terms of the non-canonical braid field of dimension 1/16. In this case the four possible traces

$$\operatorname{tr} e^{2\pi i L_0 \tau} P_i A(u) P_j A(0), \quad ij = \begin{matrix} 0,1 \\ 2,1 \\ 1,0 \\ 1,2 \end{matrix} \qquad (38)$$

lead to a coupled problem in the KMS boundary condition. The coupling matrix turns out to be:

$$\eta \begin{pmatrix} 0 & M \\ \underline{1} & 0 \end{pmatrix} \qquad (39)$$

where the notation of the second section has been used. A straight forward diagonalization reveals that the spectrum of eigenvalues agrees with that appearing in the 1-direction. Therefore Nelson's symmetry holds and the computable Rocha-Caridi characters are three linear combinations of the previous module partition functions. The simple braid algebra based on the linear Dynkin scheme in the case of $q \geq 5$, however, leads to a violation of Nelson's symmetry. One needs the bigger rectangular braid scheme attached to C in order to rescue Nelson's symmetry and insure modular invariance of physical partition functions. It would be interesting to see how the present framework can be converted into a computational scheme for Rocha-Caridi characters. Note that our various KMS two-point functions for $\tau \to \infty$ approach the Minkowski-space 4-point functions in a special conformal frame, thus again showing the deep connection between the soliton-completeness picture and modular properties.

4. Considerations in Higher Dimensions

As already mentioned in the introduction there is no reason whatsoever why the local observables with canonical dimensions as currents or the stress-energy tensor of conformal field theories in higher dimensions obey Huygens' principle as their two dimensional counterparts do. Already in the 4 point-functions of such objects this complication is expected to make itself felt. So the stress-energy tensor itself gives rise to a rich and complicated (module dependent) dimensional spectrum. Nevertheless the global conformal decomposition theory and the global vacuum expansions still remain intact. The crucial question is what kind of exchange relations will insure the structure of Einstein causal fields? A glance at conformal two and three-point functions reveals that ξ-phases do not show up if one analytically continues along paths which stay inside the spacelike region. All spacelike points can be transformed into each other by Lorentz transformations which are continuously linked with the identity; a linear Lorentz-covariant ordering as it is typical for d=2 does not exist for space-like but only for time-like separations. The structure of the two and three point functions is therefore consistent with a time-like exchange relation:

$$A_{fm}(x')A_{mi}(x) = \sum R^{(fi)}_{mm'} A_{fm'}(x)A_{m'i}(x') \qquad (40)$$

where $R^{(fi)}$ only depends on the sign of the time-like distance $(x'-x)^2 > 0$. For spacelike separation one obtains $R_{mm} = \delta_{mm}$. The analogy with two dimensions is restored if one restricts the two-dimensional decomposition formula to sectors of $Z_L \times Z_R$ which are parity symmetric. In that case only time like exchange relations remain, and the corresponding braid matrices are the squares of the light-cone braids as in (12). It seems very improbable that higher dimensions can support braid matrices which belong to the finite Dynkin scheme or any other finite scheme. In fact not even the stress-energy tensor (which now does not commute with the centre Z of the conformal covering group) is expected to have finitely many Z sectors. A countable set of Z sectors, each belonging to a finite covering, seems to be all one can hope for. Fortunately there exists an idea which relates a conformal invariant higher dimensional theory to a (not necessarily irreducible) two-dimensional one. This idea is similar to scaling properties in the horizon discussion of the Hawking radiation[12]. One can check that the longitudinal scaling[33]:

$$\lim_{\lambda \to 0} \lambda^{(d+1-\frac{N}{2})\frac{n}{2}} \langle A(\lambda u_1, v_1, x_1^{\perp}) \ldots A(\lambda u_n, v_n, x_n^{\perp}) \rangle$$

$$= \prod_{i=1}^{n-1} \delta(\xi_i^{\perp}) W(u_1, v_1, u_2, v_2, \ldots u_n, v_n) \qquad (41)$$

relates a N-dimensional conformal theory to a two dimensional one with correlation functions W. Here u=x+t, v=x-t, where x is an arbitrary spatial coordinate selected as "longitudinal" and the remaining ones are called transversal $x^{\perp}$. For the derivation of (41) one uses the factorization:

$$\langle A(x_1) \ldots A(x_n) \rangle = \prod_{i<k} [-(x_i - x_k)^2]^{\frac{-d}{n-1}} \cdot F \qquad (42)$$

where F is a function of the conformal invariant cross-ratios. These stay finite in the limit $\lambda \to 0$ whereas the first factor splits into the transversal δ-functions and a corresponding two-dimensional product factor which belongs to the reduced dimension $\delta = d+1-\frac{N}{2}$ assuming $d > d_{can}$. (The smallest d-value allowed by positivity).

The formula (41) may either be established with the use of transversal testing functions and partial integrations, or by using spectral representation i.e. momentum-space techniques. In the latter case it takes on the form of the well known "$P_3 \to \infty$" limit used in the old light-cone manipulations.

The nature of the longitudinal scaling limit assures that the resulting distributions W are indeed those of a positive definite causal conformal two-dimensional Wightman field. There is, however, no statement that the resulting field theory belongs to only one Virasoro charge. To the contrary one would expect that the complicated relation of the global Z sectors within a sector generated by the observable stress-energy tensor for the N-dimensional theory is mirrored into a complex interweaving of many Virasoro-charge sectors in the two-dimensional companion theory. In this way the two- dimensional techniques become useful for the study of higher dimensions. However, the study of higher dimensional theories (viz. the 3-dimensional Ising model) is by no means an easy matter since the present framework is not a Lagrangian approach but rather a classification method. Without detailed investigations one can only say that assumption of a finite conformal covering will yield a rational dimensional spectrum with reduced values belonging to certain conformal QFT_2.

Conformal invariant higher dimensional models are very far away from realistic models for elementary particle physics viz. QCD. Whereas in two-dimensions the Wilson-Kadanoff-Wegner picture of relevant perturbations around fixed points is expected to be useful for analytic computations e.g. of non-scale-invariant extensions of minimal models, I am not so optimistic in case of higher dimensions. The reason is that the expected surviving part of the centre Z and the stress-energy algebra relate in a very complicated presently not understood manner. Here one needs a radical new idea. Such an idea may be related to a string picture. By this I do not mean contemporary string physics (of which a critical evaluation is presented in the last section), but rather the idea that dimensional spectra of conformal QFT_2 may be related to mass spectra and the S-matrix of higher dimensional quantum field theory in some yet essentially unknown way. We then obtain the following picture

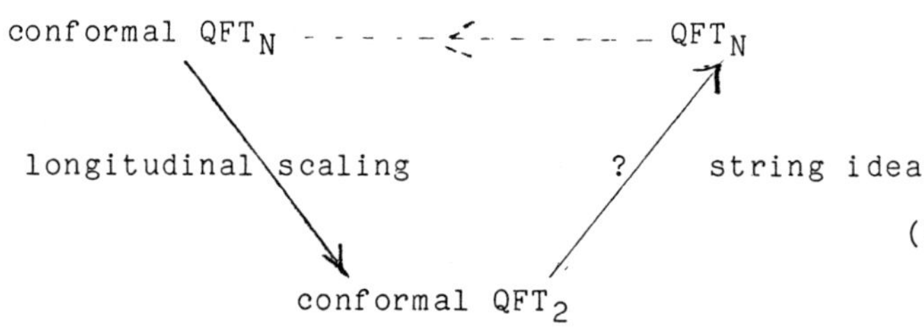

(43)

The broken arrow does not necessarily lead back to the same conformal theory. This pictures should also apply if N=2. There is some hope to lift some of the mysteries if one considers the d=2 Sine-Gordon theory at the discrete β values:

$$\beta_\nu = 8\pi \frac{\nu}{\nu+1}, \quad \nu = 1,2,... \tag{44}$$

There are indications that the "Russian Coulomb-Gas" is related to the Sine-Gordon system at β_ν by the physics of a Θ angle[35]. The Sine-Gordon system has a known S-matrix with infinitely many poles on higher Riemann sheets. The minimal models (RSOS system) may also possess a massive version with an integrable S-matrix. It would be very interesting to compare the structure of the S-matrices with that of the dimensional trajectories in order to obtain a clue about a "string" relationship.

The way in which Θ-angles lead to transmutation of additive charge quantum numbers to more subtle multiplicative charge sectors is very deep and extremely important. For example one needs such an additive charge breaking mechanism in order to understand (high-temperature) superconductivity.

5. Criticism of Contemporary String Physics and Outlook

Contemporary string physics links the set of vertex operators of dimension one, a special kind of conformal fields in two dimensional space-time, in a very specific way with the mass-spectrum and the S-matrix of a would be higher dimensional Lorentz-invariant quantum field theory. The lowest order of this S-matrix has the form of a Veneziano amplitude and describes an infinite tower of stable particles. The "stringiness" of the S-matrix lies in the appearance of this infinite tower, however, in higher loop order ($g \geq 2$) one expects, as in Feynman perturbation theory, that particles which appear in the two particle continuum of others become unstable if the internal quantum numbers permit this instability. Hence the true unitary S- matrix is indistinguishable from that of a nonperturbative quantum field theory since the principles of general quantum field theory do not forbid the occurrance of infinite towers in the analytic continuation of the S-matrix to higher Riemann sheets. Therefore there is nothing particularly "stringy" about a complete Veneziano S-matrix. Indeed this seemed to have been the original picture of the fathers of the dual model; it was a phenomenologically inspired and successful (the only phenomenological success of string theory up to now!) idea. As a result of the appearance of higher spin zero mass particles in the spectrum (the "Pomeron" in the old version) doubts began to grow about the correctness of this Ansatz for strong interactions. Later, after the string presentation had been worked out, it was suggested that a supersymmetric version[36] of this model could have something to do with quantum gravity and even constitute a "theory of everything". One of the early arguments forwarded in favour of such an interpretation was the result that the zero slope limit leads to field theoretic Lagrangians which belong to the "super-gravity" family.

The first critical question which one can ask is in what sense the quantization of classical super-gravity Lagrangians can produce a genuine "quantum gravity". The distinction "par excellence" to special relativistic quantum field theories lies in the breakdown of Einstein causality (and hence the exchange algebra structure) in favour of a new and largely unknown (see remarks below) algebraic structure. In this context it is interesting to remind oneself of the long history of failures to modify the Einstein causality of Lorentz-covariant quantum field theory via an "elementary length". As we have argued here, the perturbative picture of the standard field theories is based on a a perturbative evaluation of the causality and unitarity relations. Hence if super-gravity would fit into the standard framework this would be the safest sign that it has nothing to do with quantum gravity. Fortunately it turns out to produce non-renormalizable divergencies. Presently, however, we have no method which allows to differentiate between non-renormalizability which indicates plain inconsistency and one which reveals a new algebraic structure in the old space-like region. Therefore the zero slope limit of string theory does not tell anything decisive about quantum gravity.

The second critical question concerns the structure of the string theory S-matrix, in which, as we have argued, the "stringyness" of the lowest order is wiped out in higher loop orders. The only circumstance under which this Lorentz-covariant S-matrix may be linked with quantum gravity is if it produces infra-red divergencies (for $g \geq 2$) since otherwise, as we have argued, it would only be a candidate for an ordinary S-matrix belonging to a Lorentz-covariant local quantum field theory. Barring this possibility of infrared-divergencies, which according to my best knowledge has not yet been excluded, one should conclude that string theory is not capable to describe quantum gravity.

Can one say anything about the new algebraic structure required by quantum gravity? There is an interesting speculation of Haag[37]. According to Haag one should expect that the field algebras which one attaches to compact space-time regions are all identical (even for arbitrarily small regions). This is an extension of the Reeh-Schlieder property stating that in Einstein causal field theories, the application of the local algebra in an arbitrarily small space-time regions onto the vacuum yields a dense set of states in the Hilbert-space. The Reeh-Schlieder property is derived from the usual analytic structure of correlation function. Intuitively, there is a sort of complementarity between the "bumpyness" of the stochastic configurations appearing in the euclidean functional integral and the analytic smoothness of the obtained correlation functions. In the stochastic picture of quantum gravity, at least in some intuitive sense, one has to average in addition over all bumpy metrics which perhaps makes it palatable that now even the algebras are already determined by their "germs". It turns out that in two dimensions quantum field theories on Riemann surfaces for $g \geq 2$ (which contains all the "averaging" necessary to obtain one "conformal class") possess the Haag property. This is most conveniently seen within the recently proposed "Hawking-Unruh" construction of quantum field theories on Riemann-surfaces[38] which should be viewed as the non-commutative generalization of the KMS construction. Starting from the flat-space theory, one defines an "Unruh average" with a Fuchsian group F of first kind:

$$J^{av}(u) = \lim_{N\to\infty} \frac{1}{\sqrt{N}} \sum_{g_i \varepsilon F,\ i=1}^{N} U(g_i) J(u) U^{+}(g_i) \tag{45}$$

where for reason of simplicity we have taken a canonical free field J of dimension one.

This averaging is the precise generalization of the Unruh acceleration transformation to the case of non-commutative subgroups of the conformal symmetry group. In the same sense as the acceleration strength is a deformation of the Lorentz-boost for the 0-3 Rindler wedge, the Teichmüller parameters are deformation parameters of the Fuchsian group. In the first case the deformation parmeter is directly related to the Hawking temperature, in the second case one may speak about "Teichmüller temperatures" by analogy.

The expectation values are

$$\langle J^{av}(u) J^{av}(u')\rangle_0 = \sum_{g_i \varepsilon F_\tau} \langle J(u) U(g_i) J(u')\rangle_0$$

$$= \langle J(u) J(u')\rangle_{\tau_1 \ldots \tau_{3g-3}} \tag{46}$$

where $\tau \ldots \tau_{3g-3}$ are the deformation parameters (Teichmüller parameters) of the Fuchsian group inside the Möbius-group. They are somewhat strange distributions (they oscillate wildly but are not yet hyperfunctions) of a type never encountered before in quantum field theory[39], however, they fit perfectly into the standard reconstruction GNS framework because the positivity property is not destroyed by the averaging (45). Since the J^{av} lie in the commutant of the Fuchsian group, the generated algebras turn out to be von Neumann algebras of type II. The Haag property is best seen if one transfers the averaging to the infinitely smooth compact support test function. The application of the Fuchsian group spreads the support over the whole light cone and the space of test functions with support in a given interval spreads into a dense set on the light cone. A true two-dimensional gravity model with conformal matter also requires the integration over Teichmüller parameters with a suitable measure. It is surprising that the Haag property already holds before the τ-integrations. The use of two-dimensional models for investigating the basic problems of quantum gravity parallels the use of two-dimensional gauge theories with spinor-matter in order to discuss the issue confinement versus screening versus free charges. There the main difference to higher dimensions is that there can be no free additive charge even in the abelian case. However, multiplicative abelian charges[41] ("centre charges") of the type envisaged by Buchholz and Fredenhagen[20] may exist. The use of two-dimensional gravity models as a theoretical laboratory is very encouraging. For further details we refer to a forthcoming paper[42].

For many discussions and suggestions I am particularly indebted to K. Fredenhagen and K.H. Rehren. I also acknowledge conversations with D. Buchholz and G. Mack.

References

1.) K.H.Rehren and B.Schroer, "Einstein Causality and Artin Braids" in preparation.

2.) V.Jones, "Braid Groups, Hecke Algebras and Type II_1 factors", Japan.-U.S. Conf. Proc. (1983).

3.) B.Schroer and J.A.Swieca, Phys.Rev.D10,480 (1974).
B.Schroer, J.A.Swieca and A.H.Völkel, Phys.Rev.D11,1509 (1975).
M.Lüscher and G.Mack, Commun.Math.Phys.41,203 (1975).
J.Kupsch, W.Rühl and B.C.Yunn, Ann.Phys.(N.Y.) 89,115 (1975).

4.) M.Hortacsu, R.Seiler and B.Schroer, Phys.Rev.D2,2519 (1972).

5.) A.A.Belavin, A.M.Polyakov and A.B.Zamolodchikov, Nucl.Phys. B241,333 (1984).

6.) A.Rocha-Caridi, in: "Vertex Operators in Mathematics and Physics", eds. J.Lepowski, S.Mandelstam and I.Singer, M.S.R.I.3,451, Springer Verlag, Berlin-New York (1984).

7.) V.G.Kac and D.H.Peterson, Adv. in Mathematics 53,125 (1984).

8.) R.Kubo, J.Phys.Soc.Japan 12,570 (1957).
P.C.Martin and J.Schwinger, Phys.Rev.115,1342 (1959)
R.Haag, N.M.Hugenholtz and M.Winnink, Commun.Math.Phys. 5,215 (1967).

9.) E.Nelson, "Quantum fields and Markov fields", in: "Partial Differential Equations", D.C.Spencer, ed. Providence: American Math.Soc.

10.) A.Capelli, C.Itzykson and J.B.Zuber, Nucl.Phys.B.280 [FS18],445 (1987).

11.) S.Doplicher, R.Haag and J.E.Roberts, Commun.Math.Phys.35,49 (1979) and prior works quoted therein.

12.) R.Haag, H.Narnhofer and U.Stein, Commun.Math.Phys.94,219 (1984).

13.) N.N.Bogolinubov and D.V.Shirkov, "Introduction To The Theory of Quantized Fields", Wiley Interscience 1959.

14.) R.Picken and B.Schroer, to be published.

15.) E.C.Marino and J.A.Swieca, Nucl.Phys.B170,175 (1980). B.Schroer, Nucl.Phys.B210,103 (1982) and references therein. J.Fröhlich and P.Marchetti, "Bosonization, Topological Solitons and Fractional Charges in Two Dimensional Quantum Field Theories", ETH Zürich preprint 1987.

16.) A.Connes, Lecture Notes Cargese 1987, to be published.

17.) R.Haag and D.Kastler, J.Math.Phys. 5,848 (1964).

18.) See R.F.Streater and A.S.Wightman, "PCT, Spin, Statistics, and all That", Math.Physics Monograph Series Benjamin Inc. 1964.

19.) J.Fröhlich, Cargese Lectures 1987, to be published.

20.) D.Buchholz and K.Fredenhagen, Commun.Math.Phys. 84,1 (1982).

21.) J.Fröhlich and P.Marchetti, to be published.

22.) B.Schroer, Nucl.Phys.B, to be published.

23.) B.Schroer, Cargese Lectures 1987, to be published and references therein.

24.) K.H.Rehren and B.Schroer, Nucl.Phys.B, to be published.

25.) D.Friedan, Z.Qiu and S.Shenker, Phys.Rev.Lett.52,1575 (1984).

27.) V.Pasquier, Nucl.Phys.B285 [FS19],162 (1987).

28.) M.Karowski, "Conformal Quantum Field Theories and Integrable Systems", Brasov Summer School Lecture Notes, ETH Zürich preprint 1987, and later works.

29.) M.Karowski, Nucl.Phys.B153,244 (1979).

30.) H.-J.Borchers, Nuovo Cim.15,784 (1960).
D.Buchholz and K.Fredenhagen, Commun.Math.Phys.56,91 (1977).

31.) I.T.Todorov, Lectures presented at the Brasov Summer Bulgarian Academie of Sciences, preprint 1987.

32.) C.Itzykson and J.B.Zuber, Nucl.Phys.B275[FS17],580 (1986).

33.) Projected joint publication with K.Fredenhagen.

34.) V.S.Dotsenko and V.A.Fateev, Nucl.Phys.B240,312 (1984).

35.) In reference 28.) it is shown that the RSOS model is related to the ordinary 6 vertex model by a Θ angle construction.

36.) For the history and recent developments of these ideas see M.B.Green, J.H.Schwarz and E.Witten, "Superstring Theory", Cambridge University Press 1987.

37.) R.Haag, private communication through K.Fredenhagen.

38.) B.Schroer, Phys.Lett.199,183 (1987).

39.) W.Rühl, "Automorphic Functions", Poincare Series, and "Conformal Field Theories on Riemann Surfaces", University of Kaiserslautern, 1988.

40.) See V.Jones: "Subfactors of Type II_1 Factors and Related Topics", IHES prepring M/86/50. This holds for any discrete group whose conjugacy classes are infinite. The group algebra is of type II_1 and the commutant remains of type II.

41.) L.V.Belvedere, K.D.Rothe, B.Schroer and J.A.Swieca, Nucl.Phys.B153,112 (1979).
K.D.Rothe and B.Schroer, Nucl.Phys.B185,429 (1981).

42.) B.Schroer, in preparation.

The pairing method and bosonic anomalies

Shlomo Sternberg

Abstract. In this expository paper we use the pairing method of geometric quantization to explain the appearance of "anomalies" in the quantization of infinite dimensional bosonic systems. That is , we explain why, when quantizing quadratic Hamiltonians,one has to only pass to a $\mathbf{Z}_2$ extension of the symplectic group in the finite dimensional case, which is therefor not visible on the infinitesimal level, whereas in the infinite dimensional case one has to adjoin a whole circle. The corresponding infinitesimal cocycle is the "bosonic anomaly".

Introduction.

The "integrated version" of the Heisenberg commutation relations were formulated by Hermann Weyl in a fundamental paper in 1927 [Weyl] . In this paper he suggested that these relations have a unique irreducible unitary representation, once the value of Planck's constant is specified. This result was established by Stone and von Neumann in 1930, [von-Neumann] , and is known as the Stone-von Neumann theorem: Let V be a real vector space and (,) a non-degenerate anti-symmetric form on V. The pair consisting of V and (,) is known as a symplectic vector space. From a symplectic vector space we can construct the Lie algebra h(V) by adjoining a one dimensional space: That is, let E be a basis element in the one dimensional space $\mathbf{R}\cdot E$ and let

$$h(V) = V \oplus \mathbf{R}E \quad \text{as a vector space}$$

with the bracket relations

$$[u,v] = (u,v)E \quad \text{for } u \text{ and } v \text{ in } V$$

and

$$[E,u] = 0 \text{ , (i.e. } E \text{ is in the center of } h(V)).$$

These are just the Heisenberg commutation relations (if we choose a basis of V given by $P_1, P_2, \ldots ; Q_1, Q_2, \ldots$ with $(P_i,P_j) = 0 = (Q_i,Q_j)$ and $(P_i,Q_j) = \delta_{ij}$). The corresponding (simply connected) group is known as the Heisenberg group, H(V) . Using the exponential map, exp , we may identify H(V) with the vector space $V \oplus \mathbf{R}\cdot E$ and the multiplication law becomes

$$\exp(v + sE)\exp(w + tE) = \exp(v + w + (s + t + (v,w)/2\,)E) \;.$$

Alternatively, if we take this as the multiplication law on $V \oplus \mathbf{R}\cdot E$ we see that the corresponding Lie algebra is h(V). The Stone - von Neumann theorem asserts that if V is finite dimensional, then for any $h \neq 0$ there is a unique irreducible unitary representation, U , of H(V) which satisfies

$$U(\exp(tE)) = e^{ith}\,\mathrm{Id}.$$

K. Bleuler and M. Werner (eds.), Differential Geometrical Methods in Theoretical Physics, 261–293.

A clear and careful exposition of the Stone von Neumann theorem can be found in the book [Lion - Vergne]. Suppose that V is finite dimensional. Let g be an element of the symplectic group, Sp(V). That is, g is a real linear transformation of V such that (gu,gv) = (u,v) for all u and v in V. Since the multiplication law of H(V) is built out of the symplectic structure of V, it follows that g induces an automorphism of H(V) which we continue to denote by g; i.e. we define g(exp(v+tE)) = exp(gv +tE). This automorphism does not move E. Hence the irreducible representation U_g defined by

$$U_g(a) = U(ga)$$

must, by the Stone- von Neumann theorem be unitarily equivalent to U. So we can find a unitary operator R(g) such that

$$U(ga) = R(g)U(a)R(g)^{-1}.$$

Since U is irreducible, this equation determines R(g) up to a phase factor by Schur's lemma. So for any choice of the operators R(g) we get

$$R(g_1g_2) = c(g_1,g_2)R(g_1)R(g_2)\,.$$

In other words we get a projective representation of the group Sp(V) with "multiplier" $c(g_1,g_2)$. As far as I can determine, this important observation - that we get a projective representation of the symplectic group "free of charge" directly from the Stone - von Neumann theorem - is due to Irving Segal in the 1950's who used this fact (and, more importantly, an infinite dimensional version of it) for the construction of quantum fields. A definitive statement of the existence of this projective representation, in a form applicable to the infinite dimensional case is to be found in the fundamental paper of Shale published in 1962 [Shale].

According to general theory, a projective representation of a group is equivalent to an ordinary representation of a central extension of the group. It turns out that if V is finite dimensional all that is needed is to pass to a $\mathbf{Z}_2$ extension of Sp(V) - the projective representation we have introduced becomes an honest representation of the double cover of Sp(V), known as the metaplectic group. In other words, although the multiplier c is not trivial, its square, c^2, is a trivial multiplier. This fact was first proved by André Weil in 1964, in a breathtaking paper [Weil] in which he shows that the representation of Segal and Shale described above plays a central role in the transformation properties of the classical Jacobi θ series and its higher dimensional generalizations and that this representation and its analogues when constructed over arbitrary fields are intimately related to "quadratic reciprocity" laws. Thus constructions emanating from the study of quantum mechanics gave rise to deep new results in number theory.

The fact that all that is needed is a $\mathbf{Z}_2$ extension means that the multiplier c is not visible at the infinitesimal level. The Segal Shale procedure gives an honest representation of the Lie algebra of the symplectic group. Since this Lie algebra can be identified with the space of quadratic polynomials, this meant that there was an unambiguous and consistent way of "quantizing" quadratic Hamiltonians in terms of any given realization of the Heisenberg commutation relations. In terms of the "Fock representation" this meant that all quadratic Hamiltonians could be quantized unambiguously and consistently in terms of creation and annihilation operators. Since the physics literature was primarily concerned with infinitesimal methods, there was little mention of the need to pass to the double cover of Sp(V) to be found in physics papers of the period. In fact, as we pointed out in the opening pages of [Guillemin Sternberg] , the need for a $\mathbf{Z}_2$ extension was experimentally observed in optics experiments of the late1880's when it was shown that a photon undergoes a phase shift of $e^{\pi i/4}$ when passing through a focal point.

For infinite dimensional systems the Stone - von Neumann theorem does not hold. There are infinitely many inequivalent unitary representations of the Heisenberg group. This fact was already realized by the late 1940's [van Hove]. Indeed, to construct a suitable representation one must specify more structure on V . Thus to construct the Fock representation one assumes that V has the structure of a complex (pre) Hilbert space and that the symplectic form is (up to a sign) the imaginary part of the Hilbert scalar product , -(u,v) = Im (u,v) . Then the Fock space is the completed symmetric algebra of V . The algebraic construction of the symmetric algebra depends on the complex structure superimposed on V and the completion depends on the Hilbert space structure. As we shall recall in §1, such additional structure can be described by the choice of a complex subspace of the complexification of V . This complexification, $V^{\mathbf{C}} = V \oplus iV$, is a complex symplectic vector space. A complex subspace , L ,of $V^{\mathbf{C}}$ is called Lagrangian if (z,w) = 0 for all z and w in L and if L is not properly contained in any larger subspace with this property. We shall see in §1 that giving a certain kind of Lagrangian subspace is the same as giving the necessary additional structure. As we shall recall in § 3 each such L determines a corresponding Fock space $\mathbf{F}_L$ and an irreducible unitary representation of of H(V) on $\mathbf{F}_L$. In physics language the choice of L amounts to the choice of a "vacuum state" . It was taken for granted in the physics literature that in studying infinite dimensional systems one must specify the vacuum. Thus the non-uniqueness of the commutation relations caused barely a ripple in the physics literature of the period.

Since the representation, U_L, of H(V) on $\mathbf{F}_L$ now depends on L, and a general element of g moves L into gL, we can ask the question: For which g is the representation $U_L(g\,\cdot)$ equivalent to U_L ? This will certainly be the case if gL = L . This is the same as demanding that g belong to the unitary group of the Hilbert space structure imposed on V . For more general g the answer is: if and only if $g - g^{\dagger -1}$ is a real Hilbert Schmidt operator where $g^{\dagger}$ denotes the transpose of g relative to Re(,) , the real part of the Hilbert scalar product on V determined by L . Let us denote the group of such elements by $Sp_L(V)$. So we have $Unit_L(V) \subset Sp_L(V) \subset Sp(V)$. This fact, that $U_L(g\,\cdot)$ is equivalent to U_L if and only if $g \in Sp_L$ is the central result of the fundamental paper of Shale referred to above, together with the observation that we then get a projective representation of $Sp_L(V)$. A very clear and succinct proof of this fact together with an explicit formula for the equivalence can be found in the short note of Vergne of 1977 [Vergne] .

This then leads to the question of determining the nature of the cocycle of this projective representation in the infinite dimensional case. To the best of my knowledge, the first place where this cocycle and the corresponding central extension for $Sp_L(V)$ is worked out is in the1981paper by Graeme Segal on unitary representations of infinite dimensional groups [Segal] . It turns out that in the infinite dimensional case the central extension is by a whole circle,and that there is a corresponding infinitesimal cocycle This fact is related to the non-existence of certain determinants. In retrospect we now realize that the "Schwinger terms" and many of the "anomalies" that made their appearance in the physics literature of the 1960's and 70's are particular manifestations of this cocycle.

The present paper contains no new mathematical results. Its purpose is to explain the central extensions in the finite and infinite dimensional cases from the point of view of the pairing method that Blattner , Kostant and I introduced into geometrical quantization many years ago [Blattner] . The reader interested in a short and complete discussion of the facts should read [Vergne] and §5 of [Segal] . The pairing method in the special case at hand (and in the finite dimensional case) says that for any two choices L_1 and L_2 there is a well determined unitary operator $I_{21} : \mathbf{F}_{L1} \rightarrow \mathbf{F}_{L2}$ which intertwines the representations U_{L1} and U_{L2} of $H(V)$. We shall review this construction in §3 . If we have three L's, then we can go around the "triangle" and construct $I_{13} I_{32} I_{21}$ which intertwines the irreducible representation U_{L1} with itself and hence must be a scalar multiple of the identity: $I_{13} I_{32} I_{21} = \mu(1,2,3)\mathrm{Id}$.

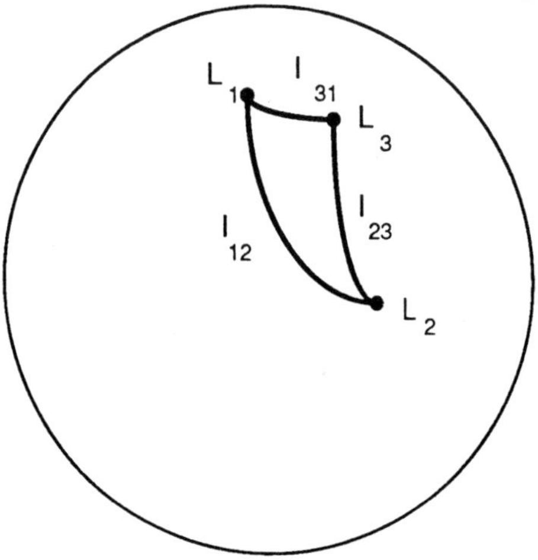

D = the space of positive Lagrangian subspaces

The intertwining operators I_{ji} continue to exist in the infinite dimensional case provided that L_i and L_j are close in the Hilbert Schmidt sense - i.e. if $L_j = gL_i$ where $g \in Sp_{L_i}(V)$.
If we consider any $g \in Sp(V)$, there is a canonical map, $C(g) : \mathbf{F}_L \rightarrow \mathbf{F}_{gL}$ which intertwines the representations U_L with U_{gL} . If $g \in Sp_L(V)$ we can then apply $I_{L,gL}$ to get back to $\mathbf{F}_L$. We thus define

$$R(g) = kI_{L,gL}C(g)$$

where k is a positive constant chosen so as to make R(g) unitary. This is our choice of representative for the projective representation, and an elementary calculation shows that the corresponding multiplier is given by

$$c(g_1,g_2) = \mu(L, g_1L, g_1g_2L) .$$

Now the evaluation of a Gaussian integral will show that the square of μ is a coboundary, i.e. that

$$\arg\ \mu^{-2}(1,2,3) = \arg[ß(1,2)ß(2,3)/ß(1,3)]$$

where $ß(i,j)$ is a certain determinant which will exist in the finite and in the infinite dimensional case. This will be explained in §3 .To complete the proof of the triviality of the square of c in the finite dimensional case we need the transformation properties of $ß$ under the action of $Sp(V)$. If we let $s = \arg ß$ this transformation property is given by eq.(2.10) below which says that

$$s(gL_1,gL_2) = s(L_1,L_2) + f(g_1,L) - f(g_2,L) \quad ,$$

where f is the argument of a certain determinental expression which does not exist in the infinite dimensional case. It is precisely the failure of the existence of this determinant which accounts for the need to extend by a whole circle in the infinite dimensional case and the concomitant nontriviality of the infinitesimal cocycle.

The application of the pairing method to the case of positive Lagrangian subspaces was worked out by Magneron in his thesis [Magneron], based in part on some notes of Vergne. In the first four sections I simply reproduce an extract from Magneron's thesis slightly reorganized to meet our current needs. I take the liberty of doing so since this thesis was apparently never published. The thesis itself contains many other interesting results and extensions to the positive semidefinite case some of which I summarize in §6. In §5 I discuss the transition to the infinite dimensional case following [Vergne] . I wish to thank Barry Simon for helpful discussions.

To repeat: in the next four sections V will be a finite dimensional real symplectic vector space. All of the mathematical arguments in these sections are to be found in [Magneron].

1. The space of positive Lagrangians.

Let V be a real symplectic vector space with symplectic form (,). We let $L(V)$ denote the set of Lagrangian subspaces of $V^{\mathbf{C}}$. We let $*$ or $\overline{}$ denote complex conjugation of $V^{\mathbf{C}}$ with respect to V. A (complex) Lagrangian subspace L of $V^{\mathbf{C}}$ is called real if $\overline{L} = L$; in this case L is the complexification of a Lagrangian subspace of the real vector space, V . A (complex) Lagrangian subspace L of $V^{\mathbf{C}}$ is called (totally) complex if $L \cap \overline{L} = \{0\}$.

Let L be a totally complex Lagrangian subspace. So $L \cap V = \{0\}$ and $L \cap iV = \{0\}$. Thus L can be considered as the graph of a map from V to iV. That is, there is a bijective linear operator J on V so that L consists of all elements of the form

$$x + iJx$$

where x ranges over all of V. Taking the real and imaginary part of the equation

$$(x+iJx, y+iJy) = 0$$

shows that

$$J \in Sp(V)$$

and

$$J^2 \doteq -Id \ .$$

We can thus think of J as defining a complex structure on V.

The map

$$\alpha: V \rightarrow V^{\mathbf{C}}, \qquad \alpha(v) = v - iJv$$

is complex linear. That is, it satisfies

$$\alpha J = i\alpha \ .$$

Furthermore

$$\alpha(V) = \bar{L} \ .$$

Similarly, the map $\bar{\alpha}$ given by

$$\bar{\alpha}(v) = (\alpha(v))^* = v + iJv$$

maps V onto L and is antilinear, i.e. satisfies

$$\bar{\alpha}J = -i\,\bar{\alpha} \ .$$

The subspace L is called <u>positive</u> if the Hermitian form $i(z, \bar{w})$, defined on all of $V^{\mathbf{C}}$ becomes positive definite when restricted to L. It is convenient to define the form $(\ ,\ )$ on V to be one half the complex conjugate of this form pulled back to V by $\bar{\alpha}$. That is,

$$(u,v) = -(i/2)(\alpha u, \bar{\alpha} v)$$

Using the properties of J it is easy to check that an alternative expression for (,) is

$$(u,v) = (u,Jv) - i(u,v).$$

The first expression for (u,v) shows that it is linear in u and antilinear in V . The second expression shows that

$$(u,v)^* = (v,u) .$$

The nondegeneracy of (,) implies that for any totally complex L the corresponding Hermitian form (,) is nondegenerate. It will be positive definite if and only if L is positive. We denote the set of all positive L's by D . We let $\bar{D}$ denote the space of Lagrangian subspaces of $V^{\mathbf{C}}$ which are positive semidefinite, i.e. to which the restriction of $i(z, \bar{w})$ is positive semidefinite. At the extreme, the real Lagrangian subspaces belong to $\bar{D}$ since the restriction of $(z, \bar{w})$ to a real Lagrangian subspace vanishes identically. We wish to parametrize D and $\bar{D}$ and describe the action of Sp(V) on them. We shall see that once we have fixed one positive Lagrangian subspace, L_0, the space D (resp. $\bar{D}$) looks like a generalization of the open (resp. closed) unit disc, and Sp(V) acts as "fractional linear transformations".

So let us fix one positive L_0 and use the subscript 0 to denote all the corresponding induced objects such as J_0, α_0 and $(,)_0$. Let L be a subspace of $V^{\mathbf{C}}$ such that $L \cap \bar{L}_0 = \{0\}$. Since $\bar{L}_0$ is negative definite, every element of $\bar{D}$ has this property. Since $L_0 \oplus \bar{L}_0 = V^{\mathbf{C}}$, we can think of L as defining the graph of a complex linear map, $S(L) : L_0 \to \bar{L}_0$. We can then define the map $T = T(L) : V \to V$ by

$$T(L) = \alpha_0^{-1} S(L)\, \bar{\alpha}_0 .$$

Thus T(L) is an antilinear map of $V \to V$ with respect to the complex structure induced by L_0, i.e. it satisfies

$$TJ_0 = -J_0T .$$

By definition, L consists of all elements of the form

$$\bar{\alpha}_0 x + \alpha_0 Tx \quad , \qquad x \in V \tag{1.1}$$

Notice that

$$(\bar{\alpha}_0 x + \alpha_0 Tx,\ \bar{\alpha}_0 y + \alpha_0 Ty) = (\alpha_0 Tx,\ \bar{\alpha}_0 y) + (\bar{\alpha}_0 x, \alpha_0 Ty) = 2i\,[(Tx,y)_0 - (Ty,x)_0]\ .$$

Thus L is Lagrangian if and only if T is symmetric, i.e. satisfies

$$(Tx,y)_0 = (Ty,x)_0 \quad \text{for all } x \text{ and } y \text{ in } V.$$

Notice that since T is antilinear, this will hold if and only if T is symmetric with respect to the orthogonal form $\mathrm{Re}\,(\ ,\ )_0$. That is, if and ony if

$$\mathrm{Re}\,(Tx,y)_0 = \mathrm{Re}(Ty,x)_0 \quad \text{for all } x \text{ and } y \text{ in } V. \tag{1.2}$$

We have thus shown

Prop.1.1. The set of Lagrangian subspaces which satisfy $L \cap \bar{L}_0 = \{0\}$ is in one to one correspondence with the set of all antilinear maps $T: V \to V$ which satisfy (1.2) .

Let us now examine which of these Lagrangian subspaces are positive. The requirement is that

$$i(z, \bar{z}) = -i(\bar{z}, z)$$

be positive for all z in L. Writing $z = \bar{\alpha}_0 x + \alpha_0 Tx$, this becomes

$$-i(\alpha_0 x + \bar{\alpha}_0 Tx,\ \bar{\alpha}_0 x + \alpha_0 Tx) > 0.$$

But since L_0 and $\bar{L}_0$ are Lagrangian, we have

$$-(i/2)(\alpha_0 x + \bar{\alpha}_0 Tx,\ \bar{\alpha}_0 x + \alpha_0 Tx) = (x,x)_0 - (Tx,Tx)_0\ .$$

Thus we have proved

Prop1.2. The space D is in one to one correspondence with the set of all antilinear maps T satisfying (1.2) and

$$(Tx,Tx)_0 < (x,x)_0 \tag{1.3}$$

while the space $\bar{D}$ is in one to one correspondence with the set of all antilinear maps T satisfying (1.2) and

$$(Tx,Tx)_0 \leq (x,x)_0 \tag{1.4}$$

The set of antilinear maps satisfying (1.2) is clearly a real vector space. If T is such a map, then a direct check shows that J_0T is again antilinear and satisfies (1.2). In other words, the map $T \dashrightarrow J_0T$ is a real linear transformation of the space of all T into itself and this map clearly has square $T \dashrightarrow -T$. Thus the map $T \dashrightarrow J_0T$ makes the space of all T into a complex vector space. In particular, the space of all T satisfying (1.3) is a complex manifold. We have thus made D into a complex manifold.

Finally, let us determine the complex structure, J_L, and the Hermitian form, $(\, , \,)_L$, induced by L on V in terms of the parametrization $T = T(L)$. If we substitute the explicit expressions for α_0 and $\bar{\alpha}_0$ into (1.1) we see that L consists of all elements of the form

$$(I+T)x + i(I+T)J_0x \quad , \quad x \in V .$$

Writing $(I+T)x = y$, this is the same as the set of all

$$y + i[(I+T)J_0(I+T)^{-1}]y \quad , \quad y \in V .$$

Thus

$$J_L = (I+T)J_0(I+T)^{-1} = J_0(I-T)(I+T)^{-1} . \tag{1.5}$$

Therefor

$$\mathrm{Re}(u,v)_L = (u, J_0(I-T)(I+T)^{-1} v) ,$$

or

$$Re(u,v)_L = Re(u, (I-T)(I+T)^{-1} v)_0 \tag{1.6}$$

2. the action of the symplectic group on D.

It is clear that Sp(V) acts transitively on D and that the isotropy group of a given positive definite subspace L_0 is the unitary group U(V) (relative to the complex structure and Hermitian form, $(\, , \,)_0$ induced by L_0 on V).

Let L be a subspace of V^C such that $L \cap \bar{L}_0 = \{0\}$ and let $g \in Sp(V)$. Then

$$\bar{\alpha} + \alpha T(L) \; : V \rightarrow L \, .$$

Hence

$$g[\, \bar{\alpha} + \alpha T(L)]$$

and

$$\bar{\alpha} + \alpha T(gL)$$

are both bijective maps of $V \rightarrow gL$. Thus the map

$$v \longrightarrow (g[\, \bar{\alpha} + \alpha T(L)])^{-1}[\, \bar{\alpha} + \alpha T(gL)\,]v = u$$

is an automorphism of V . We can write the relation between u and v as

$$g[\, \bar{\alpha} + \alpha T(L)]\,u = [\, \bar{\alpha} + \alpha T(gL)\,)]v \, .$$

Recalling that $\alpha = I - iJ_0$, the real and imaginary part of this equation become

$$g[I+T(L)]u = [I+T(gL)]v$$

and

$$gJ_0[I-T(L)]u = J_0[I-T(gL)]v \, .$$

For any linear transformation, g , of V, let $g^\dagger$ denote its transpose with respect to the orthogonal form $Re(\, , \,)_0$. Since the symplectic form $(\, , \,) = Re(J_0 \cdot, \cdot)_0$ we see that $g \in Sp(V)$ if and only if

$$gJ_0 g^\dagger = J_0 \, .$$

Thus if we define $g^{\#} = (g^{\dagger})^{-1}$ for invertible g, the condition that an invertible linear transformation belong to Sp(V) is that

$$gJ_0 = J_0 g^{\#} .$$

So replacing gJ_0 by $J_0 g^{\#}$ in the second of the equations above we get

$$g[I+T(L)]u = [I+T(gL)]v$$

and

$$g^{\#}[I-T(L)]u = [I-T(gL)]v .$$

Adding these equations we get

$$v = [(1/2)(g-g^{\#})T(L) + (1/2)(g+g^{\#})]u .$$

In particular, the operator

$$\left[\frac{1}{2}(g-g^{\#})T(L) + \frac{1}{2}(g+g^{\#})\right]$$

is invertible. Subtracting the second equation from the first gives

$$T(gL)v = \left[\frac{1}{2}(g+g^{\#})T(L) + \frac{1}{2}(g-g^{\#})\right]u$$

$$= \left[\frac{1}{2}(g+g^{\#})T(L) + \frac{1}{2}(g-g^{\#})\right]\left[\frac{1}{2}(g-g^{\#})T(L) + \frac{1}{2}(g+g^{\#})\right]^{-1} v .$$

So let us set

$$a(g) = \frac{1}{2}(g+g^{\#}) \quad \text{and} \quad b(g) = \frac{1}{2}(g-g^{\#}) \tag{2.1}$$

Then

$$T(gL) = (aT(L) + b)(b(T(L) + a)^{-1} \quad .$$

To summarize: (2.1) sets up a one to one correspondence between Sp(V) and the set of pairs of endomorphisms (a,b) of V such that

$$a \text{ is } J_0 \text{ linear, } b \text{ is } J_0 \text{ antilinear} \tag{2.2}$$

$$ab^{\dagger} = ba^{\dagger} \text{ and } a^{\dagger}b = b^{\dagger}a \tag{2.3}$$

and

$$aa^{\dagger} - bb^{\dagger} = (a\text{-}b)(a^{\dagger} + b^{\dagger}) = a^{\dagger}a - b^{\dagger}b = I \; . \tag{2.4}$$

The multiplication law is clearly given by

$$(a_1,b_1)(a_2,b_2) = (a_1a_2+b_1b_2 \, , a_1b_2+b_1a_2) \; . \tag{2.5}$$

In terms of this parametrization, the action of Sp(V) on D (where L is parametrized by T(L)) is given by the "fractional linear transformation"

$$T \longrightarrow (aT+b)(bT+a)^{-1} \quad . \tag{2.6}$$

Notice that $T = 0$ corresponds to the base point L_0 and the subgroup of Sp(V) fixing L_0 is therefor characterized by $b = 0$. Then (2.2) and (2.4) imply that the operators a for the subgroup must be unitary. In other words, the stabilizer of L_0 is just U(V) , the unitary group relative to the Hermitian form $(\, ,\,)_0$. If we identify the tangent space to D at L_0 with the vector space of all T, we see from (2.2) that the elements of U(V) commute with multiplication by J_0 , that is preserve the complex structure on the tangent space. But this then implies that Sp(V) preserves the complex structure on D . In other words, Sp(V) acts as holomorphic transformations.

Let us write B(g,L) for the denominator occurring in (2.6) That is, let us define

$$B(g,L) = bT + a \qquad \text{if } g = (a,b) \text{ and } T = T(L) \quad . \tag{2.7}$$

If $g_1 = (a_1,b_1)$ and $g_2 = (a_2,b_2)$ are two elements of Sp(V) then

$$(a_1b_2+b_1a_2)T + a_1a_2+b_1b_2 = b_1(a_2T+b_2) + a_1(b_2T+a_2) \; .$$

So using the multiplication law (2.5) we see that

$$B(g_1g_2,L) = B(g_1, g_2L)\,B(g_2,L) \; . \tag{2.8}$$

For future use we want to define several functions related to the actionof Sp(V) on D. Let us set

$$\mathbf{T} = \mathbf{R}/2\pi\mathbf{Z}$$

so that the arguments of non-zero complex numbers , denoted by arg z , $z \in \mathbf{C}^*$, will take values in $\mathbf{T}$. Let the function f: Sp(V)$\times$ D $\rightarrow$ $\mathbf{T}$ be defined by

$$f(g,L) = \arg \det_{\mathbf{C}} B(g,L) = \arg \det_{\mathbf{C}}(bT(L)+a) \quad \text{where} \quad g = (a,b) \; . \tag{2.9}$$

In this expression the operator bT+a is complex linear, and the subscript **C** means that we are computing the determinant of the linear transformation bT+a as a linear transformation over the complex numbers. So if V is 2n dimensional this is an n by n complex determinant. Taking the complex determinant of (2.8) gives

$$f(g_1g_2,L) = f(g_1, g_2L) + f(g_2,L) \quad . \tag{2.10}$$

In other words f is a cocycle of the Sp(V) action on D with values in $\mathbf{T}$.

Next define the function s: D $\times$ D $\rightarrow$ $\mathbf{T}$ by

$$s(L_1,L_2) = \arg\det_{\mathbf{C}}(I - T_1T_2) \tag{2.11}$$

We claim that

$$s(gL_1,gL_2) = s(L_1,L_2) + f(g_1,L) - f(g_2,L) \tag{2.12}$$

Indeed, let us write $g = (a,b)$. Then, since the T's are symmetric with respect to Re $(\ ,\ )_0$, i.e. $T = T^{\dagger}$, we have

$$I - (aT_1+b)(bT_1+a)^{-1}(aT_2+b)(bT_2+a)^{-1} =$$

$$I - [(aT_1+b)(bT_1+a)^{-1}]^{\dagger}(aT_2+b)(bT_2+a)^{-1} =$$

$$I - (bT_1+a)^{-1\dagger}(aT_1+b)^{\dagger}(aT_2+b)(bT_2+a)^{-1} =$$

$$(bT_1+a)^{-1\dagger}[(bT_1+a)^{\dagger}(bT_2+a) - (aT_1+b)^{\dagger}(aT_2+b)](bT_2+a)^{-1} =$$

$$(bT_1+a)^{-1\dagger}[(T_1b^{\dagger}+a^{\dagger})(bT_2+a) - (T_1a^{\dagger}+b^{\dagger})(aT_2+b)](bT_2+a)^{-1} =$$

$$(bT_1+a)^{-1\dagger}[I - T_1T_2](bT_2+a)^{-1}$$

in view of the identities satisfied by (a,b). Now for a complex linear map (with respect to J_0) the adjoint ,†, with respect to the orthogonal form Re $(\ ,\ )_0$ is the same as the adjoint , * , with respect to the Hermitian form , $(\ ,\ )_0$. We can therefor summarize the preceding computation by saying that if $T_1 = T(L_1)$ and $T_2 = T(L_2)$ then

$$I - T(gL_1)T(gL_2) = (bT_1+a)^{-1*}[I - T_1T_2](bT_2+a)^{-1} . \tag{2.13}$$

If we define

$$A(L_1,L_2) = (I - T_1T_2) \tag{2.14}$$

then we can write (2.13) as

$$A(gL_1,gL_2) = B(g,L_1)^{*-1} A(L_1,L_2) B(g,L_2)^{-1} \tag{2.15}$$

Taking the argument of the complex determinant of both sides of this equation proves (2.12) . We shall see in §5 that it is the failure of $\det_{\mathbf{C}}B(g,L)$ to exist in the infinite dimensional case which accounts for the nontriviality of the cocycle.

3.pairings and determinants

Let V be a real symplectic vector space . Let H(V) denote the Heisenberg group . Let $\mathbf{F}(H(V))$ denote the space of smooth functions on H(V). We have a natural left and right action of H(V) on $\mathbf{F}(H(V))$ given by

$$[\pounds(a)f](n) = f(a^{-1}n)$$

and

$$[R(a)f](n) = f(na) .$$

We shall denote the corresponding actions of the Lie algebra, h(V) of H(V) on $\mathbf{F}(H(V))$ by £ and r so that

$$[\pounds(X)f](n) = (d/dt)f((\exp\text{-}tX)n)_{|t=0}$$

and

$$[r(X)f](n) = (d/dt)f(n(\exp tX))_{|t=0} .$$

We extend the action of h(V) on $\mathbf{F}(H(V))$ to an action of the complexified Lie algebra $H(V)^{\mathbf{C}}$ on $\mathbf{F}(H(V))$ in the usual way by

$$\pounds(X_1+iX_2) = \pounds(X_1) + i\pounds(X_2)$$

and

$$r(X_1+iX_2) = r(X_1) + ir(X_2).$$

Let L be a positive (definite) Lagrangian subspace of $V^{\mathbf{C}}$. The Fock space, F_L , associated to L consists of (the Hilbert space completion of) those elements of $\mathbf{F}(H(V))$ which satisfy

(i) $f(ue^t) = e^{-it}f(u)$

(ii) $r(X)f = 0$ for $X \in L$

and

(iii) $\int_V |f|^2 dx < \infty$.

Condition (iii) is to be understood as follows: in view of condition (i) $|f|$ is a well defined function on $H(V)/\mathbf{R}^+$ which we may identify with V. We choose a Lebesgue measure on V. (This is unique up to a multiplicative positive constant.) This then determines, via (iii), a scalar product on the space of all elements of $\mathbf{F}(H(V))$ satisfying (i), (ii) and (iii). The completion of this space in the norm provided by (iii) is F_L. We get a representation of $H(V)$ on F_L, via the restriction of the left action, $\pounds$, on $\mathbf{F}(H(V))$ to F_L. This is, as we shall see, the Fock representation associated to L.

Let F denote the completion of the space of functions which satisfy just conditions (i) and (iii). Thus each F_L is a subspace of F and we get a pairing between two such subspaces given by

$$(f_1,f_2)_F = \int f_1 \overline{f_2} dx \qquad f_1 \in F_{L1},\ f_2 \in F_{L2} .$$

This pairing induces a linear map from F_{L1} to F_{L2} given by

$$P_2 \circ I_1$$

where I_1 denotes the injection of F_{L1} into F and where P_2 denotes orthogonal projection of F onto F_{L2}. Since (as we shall see) the representation of $H(V)$ on F_L is irreducible for each L, this map can be multiplied by some positive scalar so as to become unitary. We have thus defined a canonical unitary intertwining operator

$$I_{21}\colon F_{L1} \to F_{L2}, \qquad I_{21} = c P_2 \circ I_1, \qquad \text{where} \qquad c \in \mathbf{R}^+ .$$

Let L_0 be a fixed positive complex Lagrangian subspace, which then allows us to parametrize all other positive Lagrangian subspaces by antilinear symmetric operators $T = T(L)$ satisfying $T^2 < 1$. We now want to describe the notion of the relative vacuum, that is, the vacuum vector for L_0 inside F_L for a positive Lagrangian subspace, L. It will be the vector (characterized up to scalar multiple) in F_L which satisfies $\pounds(X)f = 0$ for all $X \in \overline{L}_0$. We shall normalize it by the requirement that $f(e) = 1$ where e is the identity element of $H(V)$.

<u>Proposition 3.1. There is a unique function</u> f_L <u>in</u> $\mathbf{F}(H(V))$ <u>which satisfies</u>

(a) $f_L(e) = 1$,

(b) $f_L(ue^t) = e^{-it}f(u)$

(c) $r(y)f_L = 0$ for all $Y \in L$,

and

(d) $\pounds(X)f_L = 0$ for all $X \in \overline{L_0}$.

In fact, the function f_L, which, by (b), is determined by its values on V, is given by

$$f_L(\exp v) = e^{-(1/4)((I-T)v,v)_0}$$

and so $f_L \in \mathbf{F}_L$.

For the proof of this proposition let us consider the group $H^{\mathbf{C}}(v)$ which is defined to be the connected and simply connected group whose Lie algebra is $h(V)^{\mathbf{C}}$, the complexification of $h(V)$. The Heisenberg group $H(V)$ is a subgroup of $H^{\mathbf{C}}(V)$ and we can think of $H^{\mathbf{C}}(V)$ as the complexification of $H(V)$.We claim that the following lemma holds:

Lemma. There exists a unique C^∞ function, ϕ ,defined on $H^{\mathbf{C}}(V)$ such that

1) $\phi(e) = 1$

2) $\phi((\exp X)u) = \phi(u)$ for all $X \in \overline{L_0}$ and all $u \in H^{\mathbf{C}}(V)$

3) $\phi(u(\exp Y)) = \phi(u)$ for all $Y \in L$ and all $u \in H^{\mathbf{C}}(V)$

and

4) $\phi(e^t u) = e^{-it}\phi(u)$ for all $u \in H^{\mathbf{C}}(V)$.

Furthermore this unique function ϕ is holomorphic on the complex group $H^{\mathbf{C}}(V)$.

Proof of the lemma. Observe that

$$V^{\mathbf{C}} = L \oplus \overline{L_0}$$

and we shall let p_L and p^* denote the corresponding projections. Any $u \in H^{\mathbf{C}}(V)$ can be written as $u = \exp(v)e^z$ where $v \in V^{\mathbf{C}}$ and $z \in \mathbf{C}$. Writing $v = p^*v + p_L v$ we see that

$$u = (\exp p^*v)(\exp p_L v)e^{-\frac{1}{2}(p^*v, p_L v) + z}$$

so conditions a)-d) uniquely determine ϕ to be the function

$$\phi(u) = e^{i\frac{1}{2}(p^*v, p_L v) - iz} .$$

Since p^* and p_L are complex linear maps, it is evident that ϕ is holomorphic. This proves the lemma. Now the map

$$X \longrightarrow \phi(u(\exp X)) \qquad X \in V^C$$

is holomorphic and hence

$$\frac{d}{dt}\phi(u\ (\exp tiX))|_{t=0} = i\frac{d}{dt}\phi(u\ (\exp tiX))|_{t=0}$$

with a similar assertion for left multiplication. Hence the restriction of the function ϕ to $H(V)$ satisfies all the requirements of the proposition. To prove the uniqueness observe that if we had two such functions, their quotient g would satisfy $g(ue^t) = g(u)$ and hence $g(uu') = g(u'u)$ for all $u, u' \in H^C(V)$. This implies that $r(X)g = \pounds(X)g$ for all $X \in V^C$. Since $V^C = L \oplus \overline{L_0}$ this implies the differential of g vanishes at all points of $H(V)$ and hence that g is a constant, and must be the constant 1 by condition a) of the proposition. To get the explicit expression for f_L given in the proposition observe that in terms of the maps α and $\bar{\alpha}$ and T we have

$$L = (\bar{\alpha} + \alpha T)V \text{ and } \overline{L_0} = \alpha V$$

so every $v \in V$ can be written as

$$v = \tfrac{1}{2}(\alpha + \bar{\alpha})v = \tfrac{1}{2}(\bar{\alpha} + \alpha T)v + \tfrac{1}{2}\alpha(I-T)v.$$

Hence the restiction of p_L and p^* to V are given by

$$p_L v = (1/2)(\bar{\alpha} + \alpha T)v \text{ and } p^*v = (1/2)\alpha(I-T)v.$$

Plugging these two expressions into the restriction of ϕ to $H(V)$ and using the fact that L and $\overline{L_0}$ are Lagrangian gives the expression for f_L, completing the proof of the proposition.

Notice the special case when we take $L = L_0$. Then $T = 0$ and, writing f_0 instead of f_{L0} we see that

$$\phi_0(\exp v) = e^{-(1/4)(v,v)_0}.$$

Let us write the most general element of $\mathbf{F}_{L0}$ as

$$f(\exp v) = g(v) f_0(\exp v).$$

Notice that $g = (f/f_0)(\exp v)$ being the quotient of two functions which satisfy condition a) of the proposition is well defined as a function of v. Furthermore, the condition $r(X)f = 0$ for all $X \in L_0$, becomes the requirement that g be holomorphic. Also,for $X \in L_0$, we have, writing $X = X_1 + i X_2$ with X_1 and $X_2 \in V$,

$$[\mathcal{L}(X)f](\exp v) =$$

$$\frac{d}{dt}[e^{it(X_1 v)} f((\exp v)(\exp -tX_1)) + ie^{it(X_2 v)} f((\exp v)(\exp -tX_2))]|_{t=0}$$

$$= i(X,v) f(\exp v).$$

Furthermore, for $X \in \overline{L}_0$, we have, since $\mathcal{L}(X) f_0 = 0$,

$$[\mathcal{L}(X)f](\exp v) = (D_X g)(v) f_0(\exp v).$$

In short, we have identified $\mathbf{F}_{L0}$ with the space of all holomorphic functions g which have finite norm relative to the scalar product

$$\langle g_1, g_2 \rangle = \int g_1 \, \overline{g}_2 e^{-\frac{1}{2}(v,v)_0} dv$$

and where the action of $V^{\mathbf{C}}$ can be described by

$$\mathcal{L}(X)g = D_X g \qquad \text{for } X \in \overline{L}_0 \qquad \text{(annihilation operators)}$$

and

$$\mathcal{L}(X)g = i(X,\cdot)g \qquad \text{for } X \in L_0 \qquad \text{(creation operators).}$$

This gives the explicit identification of $\mathbf{F}_{L0}$ with Fock space, and proves the irreducibility of the representation of H(V) on $\mathbf{F}_{L0}$ and hence on any $\mathbf{F}_L$.

Now suppose that we have several positive complex Lagrangian subspaces L_1, L_2, L_3, etc. and let us write f_i instead of f_{L_i} . It follows from the uniqueness properties of the f_i that

$$l_{ji}f_i \text{ is a multiple of } f_j .$$

If we write

$$l_{ji}f_i = \beta(j,i)\, f_j$$

then , by definition,

$$(l_{ji}f_i,f_j)_{L_j} = \beta(j,i)(f_j,f_j)_{L_j}$$

where $(f_j,f_j)_{L_j}$ is a positive number and $(l_{ji}f_i,f_j)_{L_j}$ is a positive multiple of $\int f_i\, \bar{f}_j dv$ by the definition of l_{ji} . Thus

$$\arg \beta(j,i) = \arg \int_V f_i\, \bar{f}_j dv \quad \text{as an element of} \quad \mathbf{T} = \mathbf{R}/2\pi\mathbf{Z} .$$

Now

$$\int_V f_1 \bar{f}_2 dv = \int_V e^{-\frac{1}{2}(v,v)_0 + \frac{1}{4}[(T_i v, v)_0 + (v, T_j v)_0]} dv \tag{3.1}$$

and the integrand (and hence the integral) is holomorphic in T_i and antiholomorphic in T_j . Let us denote this integral by $\phi(T_i,T_j)$. We claim that

$$\arg\{\phi(T_i,T_j)^2\} = \arg \det_{\mathbf{C}}(I - T_iT_j) \quad . \tag{3.2}$$

In this equation, recall that T_i and T_j are antilinear and hence T_iT_j is complex linear so the determinant as a complex linear transformation makes sense. Furthermore, as a function of T_i it is holomorphic and as a function of T_j it is antiholomorphic. Expansion in power series shows that any function of (T_i,T_j) which is holomorphic in the first and antiholomorphic in the second is completely determined by its values along the diagonal, $T_i = T_j$. So it is enough to prove the formula for

$$T_i = T_j = T.$$

Now

$$\phi(T,T) = \int e^{-\frac{1}{2}\mathrm{Re}((I-T)v,v)_0}\, dv$$

and $I - T$ is a symmetric operator with respect to the orthogonal form $\mathrm{Re}\ (\ ,\ )_0$ with strictly positive eigenvalues. Hence it has a unique positive symmetric square root and we can write

$$\mathrm{Re}((I-T)v,v)_0 = \mathrm{Re}((I-T)^{\frac{1}{2}}v,(I-T)^{\frac{1}{2}}v)_0$$

and a change of variables in the expression for $\phi(T,T)$ gives

$$\phi(T,T) = \det(I-T)^{-\frac{1}{2}} \int e^{-\frac{1}{2}(v,v)_0} dv \ .$$

In this formula the determinant is the real determinant of the real operator T . Now T is a symmetric operator and hence diagonizable. Suppose that e is an eigenvector of T with eigenvalue μ . Then

$$TJ_0e = -J_0Te = -\mu J_0 e \ .$$

In other words J_0e is an eigenvector of T with eigenvalue $-\mu$ .So the eigenvalues of T occur in pairs $\pm\mu$. Furthermore, the two dimensional real space spanned e and $J_0 e$ is a complex one dimensional space which is an eigenspace for the complex linear transformation T^2 with eigenvalue μ^2 . Hence

$$\det_C(I - T^2) = \prod_{\mu > 0}(1 - \mu^2) = \prod_{\mu > 0}(1 + \mu)(1 - \mu) = \det(I - T).$$

This proves that

$$\phi(T,T) = K\det_C^{-1}(I - T^2) \quad \text{with} \quad K = \left[\int e^{-\frac{1}{2}(v,v)_0}dv\right]^2 ,$$

completing the proof of (3.2).

Now let L_1, L_2, and L_3 be three positive Lagrangian subspaces. Then

$$I_{13}I_{32}I_{21}$$

intertwines the irreducible representation F_{L_1} with itself and hence , by Schur's lemma, is some multiple of the identity by a complex number of modulus one, so we may write

$$I_{13}I_{32}I_{21} = \mu(1,2,3)\,Id$$

where $|\mu(1,2,3)| = 1$. Applying both sides of this equation to the element f_1 shows that

$$\mu(1,2,3) = \beta(1,3)\beta(3,2)\beta(2,1)$$

so, in the terminology of section 2 where we defined $s(1,2) = \arg\det_C(I - T_1T_2)$ we see that

$$-2\arg\mu(1,2,3) \quad = \quad s(1,2)+s(2,3)+s(3,1) \quad , \tag{3.3}$$

which we can write more succinctly as

$$-2\arg\mu = \delta s. \tag{3.4}$$

4. the metaplectic representation

The group $Sp(V)$ acts on $H(V)$ and hence on the space of functions on $H(V)$. We will denote this action by C , so that

$$[C(g)f](n) = f(g^{-1}n) \text{ where } g\in Sp(V) \text{ and } n\in H(V).$$

Since $g^{-1}(n_1gn_2) = (g^{-1}n_1)n_2$ we see that $C(g)$ carries $\mathbf{F}_L$ into $\mathbf{F}_{gL}$. If we let $£_L$ denote the irreducible representation of $H(V)$ on $\mathbf{F}_L$ given by the restriction of the left action, $£$, to $\mathbf{F}_L$, then up to a positive scalar, $C(g) : \mathbf{F}_L\rightarrow\mathbf{F}_{gL}$ is a unitary operator which intertwines $£_L$ with $£_{gL}(gn)$, i.e.,

$$C(g)£_L(n)C(g)^{-1} = £_{gL}(gn) \quad .$$

Now $I_{L,gL}$ intertwines $£_{gL}$ with $£_L$. Thus let us define

$$R_L(g) = kI_{L,gL}\, C(g) \tag{4.1}$$

where the positive constant k is chosen so as to make $R(g)$ unitary. Then

$$R_L(g)\, £_L(n)R_L(g)^{-1} = £_L(gn) \quad . \tag{4.2}$$

Since $£_L$ is irreducible, equation (8.2) determines $R_L(g)$ up to multiplication by a scalar of absolute value one. Hence R gives a projective representation of $Sp(V)$. To determine the multiplier for this representation observe that Sp(V) invariance implies that

$$C(g)\, I_{L1,L2} = I_{gL1\,,\,gL2}\, C(g)$$

so, up to positive constants,

$$R_L(g_1)R_L(g_2) = I_{L,g1L}C(g_1)I_{L,g2L}C(g_2) = I_{L,g1L}\; I_{g1L,g1g2L}C(g_1g_2) =$$

$$\mu(L,g_1g_2L,g_1L)I_{L,g1g2L}C(g_1g_2) \;.$$

Since the positive constants have been chosen so as to make $R_L(g)$ unitary, we see that

$$R_L(g_1g_2) = c_L(g_1,g_2)R_L(g_1,g_2)$$

with (4.3)

$$c_L(g_1,g_2) = \mu(L,g_1L,g_1g_2L) \;.$$

Now

$$-2 \arg \mu(L,g_1L,g_1g_2L) = s(L,g_1L) + s(g_1L,g_1g_2L) + s(g_1g_2L,L) \qquad \text{by} \quad (3.3)$$

$$= s(L,g_1L) + f(g_1,L) - f(g_1,g_2L) + s((L,g_2L) + s(g_1g_2L,L) \qquad \text{by} \quad (2.8)$$

$$= [s(L,g_1L) + f(g_1,L)] + [s(L,g_2L) + f(g_2,L)] - [s(L,g_1g_2L) + f(g_1g_2,L)] \qquad \text{by} \quad (2.10).$$

Let us define

$$s_L(g) = e^{i\,[s(L,g1L) + f(g1,L)]} \;.$$

Then

$$\mu(L,g_1L,g_1g_2L)^{-2} = s_L(g_1)\, s_L(g_2)/ s_L(g_1g_2) \quad . \tag{4.4}$$

Define the group G_L to consist of all pairs (g,u) where $g \in Sp(V)$ and where u is a complex number with $|u| = 1$ satisfying

$$u^2 = s_L(g)^{-1}, \tag{4.5}$$

and with the multiplication law

$$(g_1,u_1)(g_2,u_2) = (g_1g_2,u_1u_2c_L(g_1g_2)^{-1}) . \tag{4.6}$$

By (4.4) and (4.5) this multiplication is well defined.

Define

$$\mathcal{R}_L(g,u) = uR_L(g) \ .$$

Then

$$\mathcal{R}_L[(g_1,u_1)(g_2,u_2)] = u_1u_2c_L(g_1g_2)^{-1}R_L(g_1g_2) = u_1R_L(g_1)u_2R_L(g_2)$$

by (4.3) . So $\mathcal{R}_L$ is an ordinary unitary representation of the group G_L . In view of (4.5) and (4.6) G_L is a central extension of Sp(V) by $\mathbf{Z}_2$. It is not difficult to see that all the groups G_L are isomorphic and that they all have a natural topology which makes them a double cover of Sp(V) and makes the representation $\mathcal{R}_L$ continuous.

5. the infinite dimensional case.

In the infinite dimensional case the first thing we have to consider is the topology on V . A choice of a positive Lagrangian subspace of $V^{\mathbf{C}}$ puts a Hermitian scalar product on V, making V into a preHilbert space. The simplest, although perhaps not the most natural thing to do, is to complete V with respect to this Hilbert norm. According to (1.6), all T with $\|T\| < 1$ will then put Hilbert space structures on V with equivalent topologies. So, for simplicity, this is what we shall do . So we will assume that V is a topological vector space which, topologically, is a separable complex Hilbert space and that the symplectic form is given as the imaginary part of one of a family of topologically equivalent Hilbert space scalar products on V, all differing from one another by the formulas of §1 with $\|T\| < 1$.

The most serious obstacle to overcome in the infinite dimensional theory is, of course, the fact that Liouville (i.e. Lebesgue) measure loses all meaning. Thus the space $L^2(V)$ which we used in the definition of the Fock spaces and the pairing in §3 is not available In fact, the thrust of the pioneering work of Irving Segal in the 1950's was to show that we may replace a small piece of the missing $L^2(V)$ by a stochastic process indexed by V . For the study of interacting quantum fields the full force of this construction seems to be necessary. A very readable treatment is available in the book [Simon] . For the case of free bosonic fields that we are studying the situation simplifies because one deals only with Gaussian processes. In any event, from this point of view the content of Shale's theorem is as follows: Each choice of L determines a probability space, X_L , with probability measure ν_L .The Fock space $\mathbf{F}_L$ becomes $L^2(X_L, \nu_L)$. The pair (X_L, ν_L) and the corresponding L^2 (X_L, ν_L) is uniquely determined (in the measure theoretical sense) as being the unique Gaussian process indexed by the Hilbert space structure (V,L) . If we fix one space L_0 as in §1, then we can identify X_L with X_{L0} in such a way that the probability measures ν_L and ν_{L0} become absolutely continuous with respect to one another if and only if T = T(L) is Hilbert Schmidt. Here is a sketch of the construction of an individual Fock space taken from [Vergne]. It is, of course, a special case of the general construction: Having chosen a complex Hilbert space structure on V, a complex valued function P is called a homogeneous polynomial of degree if there exists a continuous (complex) multilinear map E: V $\to$ **C** such that P(v) = E(v,...,v) . A finite sum of homogenous polynomials is simply called a polynomial. A function f is called holomorphic if

$$f(v) = \Sigma_m P_m(v)$$

where the P_m are homogeneous functions of degree m, and where the series converges absolutely and uniformly on every ball. Let $\mathbf{A}_L$ denote the space of holomorphic functions. For each finite dimensional subspace, M of V, let dm(x) denote the normalized Gaussian measure (with variance given by $\mathrm{Re}(\, ,)_L$) so

$$dm(x) = (2\pi)^{-n/2} e^{-(x,x)_L/2} dx_1 \cdots dx_n$$

in terms of a real orthonormal basis of (the underlying real space structure of) M. A function f on V is said to be based on M if f(v) = $F(P_M v)$ where P_M denotes orthogonal projection onto M and where F is a polynomial on M . If f_1 and f_2 are based on M , define

$$<f_1, f_2> = \int_M f_1(x)\, \overline{f_2(x)} dm(x) \ .$$

It is easy to check that this is independent of M, i.e. if the functions f are also based on some other space they have the same inner product. Let $\mathbf{B}_L$ denote the space of functions which are basedon some finite dimensional subspace.So $\mathbf{B}_L$ is a subspace of $\mathbf{A}_L$. Then $\mathbf{F}_L$ is defined to be the completion of $\mathbf{B}_L$ in the norm just defined and one can identify $\mathbf{B}_L$ with the subspace of $\mathbf{A}_L$ consisting of those holomorphic functions f for which

$$||f||^2 = \sup \int_M |f(v)|^2 dm(v) < \infty$$

where the supremum varies over all finite dimensional subspaces of M . One then gets a projective representation of the additive group of V in the usual fashion and hence a representation, $\pounds_L$, of H(V) on

F_L .The constant function 1 is the vacuum state for L, we shall denote it by Ω_L . The vacuum Ω_L satisfies

$$\pounds(X)\,\Omega_L = 0 \text{ for all } X \in \bar{L}$$

but is only determined up to phase by this property. Two Fock spaces F_{L1} and F_{L2} give unitarily equivalent representations of H(V) if and only if they are Hilbert-Schmidt related. If I_{21} is a unitary operator implementing the equivalence, then the "vacuum overlap theorem" asserts that

$$<I_{21}\,\Omega_{L1}, \Omega_{L2}> \neq 0 \; .$$

We can then fix the arbitrary phase in I_{21} by requiring that

$$<I_{21}\,\Omega_{L1}, \Omega_{L2}> \quad > 0 \; .$$

We have thus constructed the pairing between any two Hilbert-Schmidt related Fock spaces.

Having fixed one L_0 , the element f_L in F_L given in Prop.3.1 is (up to positive multiple) just $I_{L,L0}\Omega_{L0}$. The complex number ß(i,j) is then defined just as in §3, and the content of eq. (3.2) is the assertion that

$$2 \arg ß(i,j) = \arg \det_{\mathbf{C}} (I - T_i T_j) \; .$$

Now the right hand side of this equation is well defined since T_i and T_j are Hilbert-Schmidt, and is a continuous function of the two variables in the Hilbert-Schmidt norm. The continuity of the metaplectic representations[Shale] implies that the left hand side is also continuous. Hence this equation is true for arbitrary Hilbert-Schmidt operators T_i and T_j since we have verified it for the case of finite rank operators.

Thus (3.3) holds in the infinite dimensional space, as does the discussion in §4 up through eq.(4.3) (where the elements g are restricted to lie in $Sp(V)_{L0}$) . The trouble comes in the passage from (4.3) to (4.4) since the function f , and hence the function s_L is not defined because $\det_{\mathbf{C}}$ B(g,L) (see §2) makes no sense. So let us see what we can do with the operators themselves instead of their determinants. Using the definition (2.14) we can rewrite (4.3)and (3.3) as

$$\arg c_L(g_1,g_2)^{-2} = \arg \det[A(L,g_1L)A(g_1L,g_1g_2L)A(g_1g_2L,L)] \; .$$

Now, by (2.15) we have

$$A(g_1L,g_1g_2L) = B(g_1,L)^{*-1}A(L,g_2L)B(g_1,g_2L)^{-1}$$

while

$$A(g_1g_2L,L) = A(L,g_1g_2L)^* .$$

So if we use (2.8) , and write A_1 for $A(g_1,L)$ etc. we see that

$$\arg c_L(g_1,g_2)^2 = \arg \det_{\mathbf{C}}[(A_{12}^{*-1}B_{12})(B_2^{-1}A_2^{-1})(B_1^*A_1^{-1})] . \tag{5.1}$$

In finite dimensions each of the factors has a determinant and (5.1) reduces to (4.4) . In infinite dimensions these separate determinants do not exist. Let us evaluate (5.1) at $L = L_0$ where all the A's become the identity and $B = a$ so the right hand side of (5.1) becomes

$$\arg \det_{\mathbf{C}}[a_{12}a_2^{-1}a_1^*] .$$

Since $a_{12} = a_1a_2 + b_1b_2$ this becomes

$$\arg \det_{\mathbf{C}}[(a_1 + b_1b_2a_2^{-1})a_1^*] . \tag{5.2}$$

Let us compute the infinitesimal version of this cocycle. For this observe that the infinitesimal version of the parametrization

$$g = \begin{pmatrix} a & b \\ \bar{b} & \bar{a} \end{pmatrix}$$

given by (2.2) - (2.5) is that elements of the Lie algebra of $Sp(V)_{L_0}$ will be of the form

$$\xi = \begin{pmatrix} X & Y \\ \bar{Y} & \bar{X} \end{pmatrix}$$

where Y is antilinear and Hilbert-Schmidt. Now if c is a group cocycle, the corresponding Lie algebra cocycle is given by

$$\gamma(\xi_1,\xi_2) = \frac{d}{dt}\frac{d}{ds}c(\exp t\,\xi_1, \exp t\,\xi_2)_{|t=s=0} - \frac{d}{dt}\frac{d}{ds}c(\exp t\,\xi_2, \exp t\,\xi_1)_{|t=s=0}$$

and we will take the ξ in this formula as given above. Doing the differentiations in (5.2) gives

$$\gamma(\xi_1,\xi_2) = -i\,\mathrm{tr}_{\mathbf{C}}(Y_1Y_2 - Y_2Y_1) \quad . \tag{5.3}$$

In this formula observe that the trace exists since the Y's are Hilbert-Schmidt. Since the Y's are antilinear, the complex trace of the commutator $[Y_1,Y_2]$ which is a complex linear operator will not vanish , in general. Let us check whether this cocycle is a coboundary. So we are looking for a linear function b such that

$$\gamma(\xi_1,\xi_2) = b([\xi_1,\xi_2]) \ .$$

Clearly the only possibility is for b to be given by $b(\xi) = -i\,\mathrm{tr}_{\mathbf{C}}(X)$. In finite dimensions this trace will exist, and so γ is a coboundary as we already know. But in infinite dimensions $\mathrm{tr}_{\mathbf{C}}(X)$ will not exist. Thus γ defines a non-trivial cohomology class, showing that we must pass to a central extension by a whole circle to get the metaplectic representation in the infinite dimensional case. Thus the cocycle defined by (5.3) gives the bosonic anomaly in its general form.

6. additional comments.

1. In [Kostant Sternberg] we discussed the Fermionic cocycle in great detail from an algebraic point of view. A more analytic and comprehensive treatment can be found in [Araki] . See also [Shale and Stinespring] . There are many points of similarity between the bosonic and fermionic situations, especially in the form of the cocycle, see also [Kac-Petersen]. The analogy is between the metaplectic representation on the one hand, and the "spin" representation of a totally split orthogonal group on the other. The role of a Lagrangian subspace (maximally totally isotropic with respect to the symplectic form) is a maximally totally isotropic subspace with respect to the orthogonal form. However there are some important differences which are worth emphasizing:

a) In finite dimensions the Clifford algebra has a unique irreducible module up to isomorphism. This is an algebraic assertion. The symplectic analogue of the Clifford algebra in the symplectic case is the Weyl algebra . It does not have a unique irreducible module. It is only at the level of the unitary theory, when we demand a unitary representation of the Heisenberg group that the Stone von Neumann theorem gives a unique representation.

b) In finite dimensions the spin module is finite dimensional while the representation of the Heisenberg group is infinite dimensional.

c) A choice of a totally complex maximally isotropic subspace of $V^{\mathbf{C}}$ determines a $J \in O(V)$ satisfying

$J^2 = -I$ just as in the symplectic case, the proof being identical. But such a J need not exist. For example it won't exist if the space V has signature (n,n) with n odd. More to the point, the Hermitian form induced on V will have the original scalar product as its real part, and its signature is thus fixed independent of the choice of J . So in the orthogonal case it is best to start with a complex orthogonal vector space , W, which is totally split (e.g. even dimensional if the dimension is finite). One can then choose a complex conjugation, σ , so that the Hermitian form $(z,\sigma w)$ is positive definite. In other words, the positivity is coded into the complex conjugation. Conversely, starting with a Hilbert space, W with Hilbert space scalar product (,) and a complex conjugation, one gets an orthogonal scalar product $(z,w) = (z,\sigma w)$. Then a subspace L will be maximally isotropic for (,) if and only if L is perpendicular to σL with respect to (,) and $W = L \oplus \sigma L$. Each such L then determines a "Fermionic Fock space", cf. [Shale and Stinespring] and [Araki].

2. The space D is not only a complex manifold, it is a Kaehler manifold with a Kaehler metric which is invariant under Sp(V) .(This in the finite dimensional case; in the infinite dimensional case one must consider the restricted D and Sp(V).) Indeed, there is a holomorphic line bundle over D , the one which assigns to each L the complex line in $\mathbf{F}_L$ generated by its vacuum state, and the curvature of its canonical connection will be the Kaehler form which we denote by ω . Let L_1, L_2 and L_3 be three points in D and let

$$\tau(1,2,3) = \int_\Delta \omega$$

where Δ is any oriented surface whose boundary is the geodesic triangle joining L_1, L_2 and L_3. Since ω is closed, this integral is independent of the choice of the surface Δ . Bu definition, τ is cyclically symmetric in L_1, L_2 and L_3 and antisymmetric under the interchange of two of the three points. Also, τ is a cocycle in the sense that evaluating on the four triangles of a tetrahedron gives zero by Stokes' theorem. Furthermore, τ is Sp(V) invariant. Magneron shows in his thesis (in the finite dimensional case) that

$$\mu(1,2,3) = e^{-(\pi i/4)\tau(1,2,3)}$$

and that the $\mathbf{R}$ valued function τ extends to the boundary, $\overline{D}$. (In fact this is almost an immediate consequence of the definition of the Kaehler form as the curvature of the line bundle.)Furthermore, when restricted to the totally real subspaces , τ takes its values in $\mathbf{Z}$ and coincides with the usual Maslov index, thus extending ideas of [Souriau] to the positive definite case . The Kaehler structure and the line bundle play an important role in the infinite dimensional theory, cf. [Pressley and Segal], chapter 7.

3. In many of the physical and number theoretical applications of the metaplectic representation for finite dimensional V , cf. [Sternberg and Wolf] and the second half of [Lion - Vergne] ,it is useful to consider complex structures on V giving Hermitian forms which are not positive definite. This is within the framework of the theory of "Howe Pairs" [Howe] : Let G and H be subgroups of Sp(V) which mutually centralize one another . That is, G consists of all elements of Sp(V) which commute with all elements of H and vice versa. Then under suitable hypotheses about G or H the metaplectic representation decomposes into a discrete direct sum of irreducibles of $G \times H$. Thus there will be a discrete family of types of irreducibles of G occurring in the decomposition of the restriction of the metaplectic representation , each parametrized by an irreducible representation of H . The case G = U(p,q) and H = U(1) inside Sp(2(p+q)) was worked out in [Sternberg-Wolf] and the more general cases G = U(p,q) H = U(k) inside Sp(2k(p+q)) and G = Sp(2n) , H = O(k) inside Sp(2kn) were worked out geometrically in {Kashiwara Vergne] where a general formula for the parametrizing representations of H was provided. In [Kazhdan Kostant Sternberg] a recipe for evaluating the the representations using the moment map and geometric quantization was put forth, and this program was carried out in [Adams]. In some sense, the Howe pair method can be considered as a sort of "quantum gauge principle" where the "gauge group" H, serves to determine and to parametrize the "physical group", G . Here are some interesting physical examples: Take V = $\mathbf{R}^8$ = $\mathbf{C}^{2,2}$, H = U(1) and G = U(2,2) . Since SU(2,2) is the (fourfold covering of the) conformal group of (the completion of)Minkowski space time, we see that the "gauge group" , H , "picks out" the conformal geometry of space time. Since H is abelian, its irreducibles are all one dimensional and hence the metaplectic representation , when restricted to SU(2,2) becomes a multiplicity free discrete direct sum of irreducible representations. These representations have the remarkable property that they remain irreducible when restricted to the (ten dimensional) Poincaré group (the isometries of Minkowski space time) sitting inside SU(2,2) , cf. [Sternberg and Wolf].Furthermore each massless discrete spin representation of the Poincaré group appears exactly once. Thus H "picks out" the massless elementary particles of the Poincaré group , which are, as is well known, conformally invariant. This is the Howe pair U(2,2) × U(1) in Sp(8) .

Now suppose that we modify U(1) slightly, changing it to O(2) (by adding the disconnected component). The centralizer of O(2) is Sp(4) , so our Howe pair is Sp(4) × O(2) inside Sp(8) . The general theory of [Kashiwara Vergne] applies . In this case it tells us how to decompose the tensor product of two copies of the metaplectic representation of Sp(4) - we are identifying this tensor product with the metaplectic representation of Sp(8) . But Sp(4) is locally isomorphic to the anti de Sitter group SO(3,2) , and the representation we get was discovered by Dirac. Thus this time the "gauge group" O(2) "picks out" anti de Sitter geometry. And if we are willing to pass from Minkowski space to anti- de Sitter space, all the mass zero particles can be realized as composites of two copies of a single elementary particle as was pointed out by Flato and Fronsdal, who developed a field theory for such a particle [Flato and Fronsdal].

In string theory one wants to take V to be the symplectic vector space consisting of all maps of the cylinder into $\mathbf{R}^{1,d-1}$ which satisfy the wave equation. (This space has a natural symplectic structure.) The group H (which we would like to "eliminate') is the group of conformal transformations of the cylinder, and its centralizer is the d dimensional Poincaré group. This time, the natural Hermitian form is not positive definite. As far as the group H is concerned this is relatively harmless since H embeds diagonally into d copies of the symplectic group of the space of **R** valued solutions of the wave equation,and the induced cocycle on H is the straightforward to compute. (The fact that one needs to cancel this cocycle with a corresponding fermionic cocycle coming from the adjoint representation of h accounts for the requirement that d = 26. The general theory underlying this construction is explained in [Kostant-Sternberg] .) These examples suggest, however, that one would want to consider non positive definite situations in the infinite dimensional case. Within the framework of a purely infinitesimal approach one can carry out such a program in a category of highest weight representations [Kac]. However even for finite dimensional situations this approach is unsatisfactory for bosons as was pointed out in 1a (an observation that goes back to [Weyl]). Also, in the actual applications to string theory the (d -dimensional) Poincaré content is evaluated by the Wigner - Mackey little group method which is group theoretical in character. So, from the mathematical point of view, the correct category in which to operate remains an open and important question. Experience in finite dimensions shows that passing to the non-positive definite realm can involve a huge amount of technical machinery, cf. [Rawnsley-Schmid-Wolf].

REFERENCES

Adams, J. "Coadjoint orbits and reductive dual pairs" Adv. in Math. 63 138-151 (1987)

Araki,H. "Bogoliubov automorphisms and Fock representations of canonical anticommutation relations" Contemporary Mathematics 62 24-141 (1987)

Blattner, R.J. "Quantization and representation theory" Proc. Sym. Pure and App. Math. 26 A.M.S. Prov. R.I. 147-165 (1973)

Dirac,P. A . M. "A remarkable representation of the 3+2 deSitter group" J. Math. Phy. 4 901-909 (1963)

Flato, M. and Fronsdal , C. "Quantum field theory of singletons" J. Math. Phy. 22 1100-1105 (1981)

Guillemin,V. and Sternberg,S . "Geometric Asymptotics" Math. Surveys 14 A.M.S. , Prov., R.I. (1977)

Howe, R. "Remarks on classical invariant theory" preprint circa1977

Kac, V. "Infinite dimensional Lie algebras" Camb. Univ. Press. 1985

Kac,V. and Peterson, D. Proc. Nat. Acad. Sci. 78 3308 (1981)

Kashiwara, M. and Vergne, M. "On the Segal - Shale - Weil representation and harmonic polynomials" Inv.Math. 44 1-47 (1978)

Kazhdan,D., Kostant,B. , and Sternberg, S. "Hamiltonian group actions and dynamical systems of Calogero type" Comm. Pure and App. Math. 31 481-507 (1978)

Kostant, B. and Sternberg,S. "Symplectic reduction, BRS cohomology, and infinite dimensional Cliffordalgebras" Ann. of Phy. 176 40-113 (1987)

Lion,G. and Vergne, M ."The Weil representation, Maslov index and Theta series" Birkhäuser Boston (1980)

Magneron, B. "Une généralisation de la notion d'indices de Maslov aux plans lagrangiens positives" Thesis, Université Marie et Pierre Curie , Paris 6 (1979)

Pressley and Segal, G . "Loop Groups" Oxford Univ. Press 1986

Segal, G. "Unitary Representations of some Infinite Dimensional Groups" Comm. Math. Phys.80 301-342 (1981)

Shale, D. "Linear symmetries of free boson fields" Trans Amer. Math. Soc. 103 149-167 (1962)

Shale, D. and Stinespring,W.F. J. Math. Phys 14 315-322 (1965)

Simon, B. "The $P_2(\Phi)$ Euclidean (Quantum) Field Theory" Princeton Series in Physics , Princeton University Press (1974)

Souriau,J.M. "Construction explicite de l'indice de Maslov et applications" Fourth int. coll. on group theoretical methods in physics, Nijmegen (1975)

Sternberg, S. "Some recent results on the metaplectic representation" Group theoretical methods in physics, Proceedings , Tübingen. (1977) Springer Lecture notes i Physics, Springer Verlag Berlin, Heidelberg, New York.

Sternberg, S. and Wolf,J.A. "Hermitian Lie algebras and metaplectic representations" Trans. A.M.S. 238 1-43 (1978)

van Hove,L. "Sur certaines représentations unitaires d'un group infini de transformations" Acad. Roy. Belgique , Cl. Sci. Mém. Coll. in 8° 26 102 pp. (1951)

Vergne, M. "Seconde quantification et groupe symplectique" C.R. Ac.Sci. (Paris) 285 A 191-194 (1977)

von Neumann, J. " Die Eindeutigket der Schrödingershen Operatoren" Ann. Math. Pure Appl. 104 570-578 (1931)

Weil, A. " Sur certains groupes d'operateurs unitaires", Acta Math. 111 143-211 (1964)

Weyl, H. "Quantenmechanik und Gruppentheorie" Zeit. f. Phy. 46 1-46 (1927)

A Multisymplectic Approach to the KdV Equation

MARK J. GOTAY

Mathematics Department
United States Naval Academy
Annapolis, MD 21402-5000 USA

Abstract

A canonical multisymplectic formulation of the KdV equation is presented which, when space + time decomposed, gives rise to the usual symplectic description of dynamics on the appropriate space of Cauchy data. In addition to allowing one to treat the KdV equation covariantly, this formalism enables one to derive the Gardner symplectic structure for the KdV equation in a completely systematic way.

K. Bleuler and M. Werner (eds.), Differential Geometrical Methods in Theoretical Physics, 295–305.

I. Introduction

Oftentimes the study of evolution equations begins with these equations presented in the specialized form

$$\underline{u}_t = \mathcal{D}\cdot\delta\mathcal{H}[\underline{u}], \tag{1.1}$$

where $\underline{u}$ is a matrix of fields, $\mathcal{D}$ is a matrix differential operator, δ is the variational derivative with respect to $\underline{u}$ and $\mathcal{H}$ is the Hamiltonian density [O]. While ultimately very useful, this form is somewhat unsatifactory from the point of view of classical field theory, since (*i*) it essentially ignores the variational aspects of the theory, (*ii*) it is intrinsically noncovariant, (*iii*) it presupposes that an initial value analysis has already been carried out (at least formally), and (*iv*) the underlying symplectic geometry of the theory is not readily apparent. Ideally, one would like to be able to *derive* (1.1) in a "canonical" way, that is, from a Lagrangian variational principle followed by a space + time decomposition of the corresponding Hamiltonian formalism. But this program has been hampered by the lack of an adequate *covariant* Hamiltonian formalism in the calculus of variations [dLR]. Such a theory has now been developed [Go]; the key ingredient is the notion of a "multisymplectic" structure [Ki].

The principal advantage of this canonical approach is that it is both general and systematic. Consequently, one can deal *directly* with higher order Lagrangians as well as multiple integrals in the calculus of variations in much the same way as one treats ordinary mechanics. It is also straightforward to handle systems which have initial value constraints. Moreover, after a space + time decomposition, this multisymplectic framework gives rise to the standard symplectic description of dynamics on the appropriate space of Cauchy data, and (1.1) effectively reappears as the complete set of dynamic field equations thereupon.

Here I apply this formalism to the KdV equation

$$u_t = 6uu_x - u_{xxx}. \tag{1.2}$$

The analysis has two interesting features. First, the standard KdV Lagrangian is second order, so this provides a simple illustration of how to canonically analyze higher order field theories *without* having to reduce the order of the Lagrangian. Second, it enables one to systematically *derive* the somewhat "mysterious" Gardner symplectic structure [Ga] for the KdV equation from first principles.

The reader is referred to [Go] for the relevant definitions and the details of the formalism; notation is explained in the Appendix. Throughout, I work formally in the C^∞ category.

II. Covariant Formulation

To obtain a Lagrangian for the KdV equation it is necessary to introduce the potential $v_x = u$. The corresponding covariant configuration space Y is the trivial bundle $\mathbf{R}^2 \times \mathbf{R} \to \mathbf{R}^2$ with coordinates $x^\mu = (x, t)$ on "spacetime" $\mathbf{R}^2$ and v on the fiber. Then the Lagrangian density, viewed as a map $\mathcal{L}$: $J^2Y \to \Lambda^2(\mathbf{R}^2)$, is

$$\mathcal{L}[v] = \left(\tfrac{1}{2}v_t v_x - v_x^3 - \tfrac{1}{2}v_{xx}^2\right) dx \wedge dt.$$

The first step is to construct a Cartan form $\Theta_{\mathcal{L}}$ on J^3Y; the formalism requires it to be "quasi-symmetric" [Ko]. Since $\mathcal{L}$ is only second order, there a unique such $\Theta_{\mathcal{L}}$; it is

$$\Theta_{\mathcal{L}} = -v_{xx}dv_x \wedge dt - \tfrac{1}{2}v_x\, dv \wedge dx$$

$$+ \left[\tfrac{1}{2}v_t - 3v_x^2 + v_{xxx}\right] dv \wedge dt \qquad (2.1)$$

$$+ \left[2v_x^3 + \tfrac{1}{2}v_{xx}^2 - \tfrac{1}{2}v_t v_x - v_x v_{xxx}\right] dx \wedge dt.$$

In terms of the Cartan form, a section γ of Y satisfies the KdV equation iff

$$(j^3\gamma)^*\left[i(\xi)\, d\Theta_{\mathcal{L}}\right] = 0 \qquad (2.2)$$

for all vector fields ξ on J^3Y.

The covariant phase space for the KdV equation is $J^{1\bigstar}(J^1Y)$, where the star denotes an affine $\Lambda^2(\mathbb{R}^2)$-valued dual. This space can be canonically identified with the subbundle

$$Z = \left\{z \in \Lambda^2(J^1Y) \mid i(\xi)i(\eta)z = 0 \text{ for all } \xi,\eta \in V\pi^1\right\}.$$

of $\Lambda^2(J^1Y)$. Coordinates on Z are $(x^\mu, v, v_\mu; p^\nu, p^{\mu\nu}, p)$. The space Z carries a canonical 2-form given in these coordinates by

$$\Theta = \left[p^{\mu\nu}\, dv_\mu + p^\nu\, dv\right] \wedge dx_\nu + p d^2x. \qquad (2.3)$$

The *multisymplectic form* is the 3-form $\Omega = -d\Theta$.

The connection between the variational and multisymplectic formalisms is provided by the covariant Legendre transformation $\sigma_{\mathcal{L}}: J^3Y \to Z$. It is the the J^1Y-bundle map defined by the factorization

$$\Theta_{\mathcal{L}} = \sigma_{\mathcal{L}} \circ \wedge^2 \left[T\pi_1^3 \right]$$

of the Cartan form. Using (2.1) this yields the coordinate expression for $\sigma_{\mathcal{L}}$:

$$p^{tt} = 0 \ , \ p^{tx} = p^{xt} = 0 \ , \ p^{xx} = -v_{xx}$$

$$p^t = \tfrac{1}{2} v_x \ , \ p^x = \tfrac{1}{2} v_t - 3v_x^2 + v_{xxx} \qquad (2.4)$$

$$p = 2v_x^3 + \tfrac{1}{2} v_{xx}^2 - \tfrac{1}{2} v_t v_x - v_x v_{xxx}.$$

This transformation is clearly degenerate; its image is the "covariant primary constraint set" $P \subset Z$ defined by equations (2.4). However, $\sigma_{\mathcal{L}}$ *is* a submersion onto its image P.

Noting that $\sigma_{\mathcal{L}}{}^*\Theta = \Theta_{\mathcal{L}}$, it then follows from (2.2) that $\gamma \in \mathcal{Y}$ is a solution of the KdV equation iff $\rho = \sigma_{\mathcal{L}} \circ j^3\gamma$ satisfies

$$\rho^* \left[i(\xi)\Omega \right] = 0$$

for all vector fields ξ on P. This is the multisymplectic version of the KdV equation.

III. Space + Time Decomposition

I now 1+1 decompose the covariant formalism of the previous section. Fix a Cauchy surface $\Sigma \subset \mathbb{R}^2$ and choose spacetime coordinates such that Σ is the level set $t = 0$. Take the evolution direction ζ on Y to be ∂_t. Assume all fields vanish sufficiently rapidly at spatial infinity that boundary terms may be ignored. The decomposition will proceed in two stages.

Consider the space $\mathcal{Y}_\Sigma$ of sections of $Z|\Sigma$. The canonical 2-form Θ on Z induces a 1-form Θ_Σ on $\mathcal{Y}_\Sigma$ by integration:

$$\Theta_\Sigma(\rho)\cdot\xi = \int_\Sigma \rho^*\left[i(\xi)\Theta\right]$$

where $\rho \in \mathcal{Y}_\Sigma$ and $\xi \in T_\rho\mathcal{Y}_\Sigma$. Set $\Omega_\Sigma = -d\Theta_\Sigma$; this defines a closed 2-form on $\mathcal{Y}_\Sigma$. However, $(\mathcal{Y}_\Sigma,\Omega_\Sigma)$ is not symplectic; using (2.3) it is straightforward to check that

$$ker\ \Omega_\Sigma = span\left\{\frac{\delta}{\delta p^{tx}}, \frac{\delta}{\delta p^{xx}}, \frac{\delta}{\delta p^{x}}, \frac{\delta}{\delta p}\right\}.$$

Then reduction yields $(\mathcal{Y}_\Sigma,\Omega_\Sigma)/(ker\ \Omega_\Sigma) \approx (T^*(\Gamma(J^1Y)_\Sigma),\bar{\Omega}_\Sigma)$, where $\bar{\Omega}_\Sigma = -d\bar{\Theta}_\Sigma$ is the canonical symplectic form on the cotangent bundle. Canonical "coordinates" on $T^*(\Gamma(J^1Y)_\Sigma)$ are $(v, v_x, v_t; p^t, p^{xt}, p^{tt})$ and the canonical 1-form is

$$\bar{\Theta}_\Sigma(\bar{\rho}) = \int_\Sigma \left[\bar{\rho}^*\left[p^{tt}\delta v_t + p^{xt}\delta v_x + p^t\delta v\right]\right] \otimes dx \qquad (3.1)$$

where $\bar{\rho} \in T^*(\Gamma(J^1Y)_\Sigma)$. Note that this reduction essentially amounts to eliminating the *spatial* "multimomenta" p^{tx}, p^{xx}, p^x and p from the formalism, these being irrelevant for the initial value problem.

The second stage of the decomposition is to eliminate redundant *spatial* derivatives (in this case, just v_x). This is accomplished by requiring sections of $(J^1Y)_\Sigma$ to be *spatially* holonomic, i.e., to lie in the space $(j^1\mathcal{Y})_\Sigma \subset \Gamma(J^1Y)_\Sigma$. The corresponding inclusion i induces a symplectic projection i^*: $(T^*(\Gamma(J^1Y)_\Sigma),\bar{\Omega}_\Sigma) \to (T^*(j^1\mathcal{Y})_\Sigma,\omega_\Sigma)$, where again $\omega_\Sigma = -d\theta_\Sigma$ is the canonical symplectic form on the cotangent bundle.

Moreover, one may use the evolution direction ζ to identify $(j^1\mathcal{Y})_\Sigma$ with $T\mathcal{Y}_\Sigma$. Thus, finally, the 1+1 decomposed phase space is $(T^*T\mathcal{Y}_\Sigma, \omega_\Sigma)$. In canonical "coordinates" $(v, v_t; \pi, \pi^t)$ on this space the canonical 1-form is

$$\theta_\Sigma(\varphi) = \int_\Sigma \varphi^* \left[\pi\delta v + \pi^t \delta v_t\right] \otimes dx \tag{3.2}$$

for $\varphi \in T^* T\mathcal{Y}_\Sigma$, and the projection i^* is given by $(v, v_x, v_t; p^t, p^{xt}, p^{tt}) \to (v, v_t; \pi, \pi^t)$, where

$$\pi = \left[p^t - D_x p^{xt}\right] \ , \ \pi^t = p^{tt}, \tag{3.3}$$

cf. (3.1) and (3.2).

IV. Initial Value Analysis

Turn next to a (formal) initial value analysis of the KdV equation. According to general principles, the Hamiltonian - defined as a functional on $\mathcal{P}_\Sigma$ - is given by

$$H_\Sigma(\rho) = \int_\Sigma \rho^* \left[i(j^1\zeta)\Theta\right],$$

where $j^1\zeta$ is the first jet prolongation of the evolution direction ζ. Taking (2.3) and (2.4) into account, this becomes

$$H_\Sigma(\rho) = -\int_\Sigma \rho^* \left[p^{xx} D_x v_x + p^x D_x v + p\right] dx. \tag{4.1}$$

Now $\mathcal{P}_\Sigma$ projects through the two redutions of the previous section to a submanifold $\mathcal{C}_\Sigma$ of $T^*T\mathcal{Y}_\Sigma$, the "1+1 primary constraint set." From (2.4) and (3.3), $\mathcal{C}_\Sigma$ is defined by the constraints

$$\pi^t = 0 \ , \ \pi = \tfrac{1}{2} D_X v. \tag{4.2}$$

It turns out that the first of these constraints is first class in the sense of Dirac, while the second is second class [N,Ke].

It is possible to show that the Hamiltonian (4.1) on $\mathcal{P}_\Sigma$ induces the function h_Σ on $\mathcal{C}_\Sigma$ given by

$$h_\Sigma(\varphi) = \int_\Sigma \varphi^* \left[(D_X v)^3 + \tfrac{1}{2}(D_X D_X v)^2 \right] dx, \tag{4.3}$$

while ω_Σ pulls back to a closed 2-form η_Σ on $\mathcal{C}_\Sigma$ which, in view of (3.2) and (4.2), is

$$\eta_\Sigma(\varphi) = \tfrac{1}{2} \int_\Sigma \varphi^* \left[\delta v \wedge \delta D_X v \right] \otimes dx. \tag{4.4}$$

Hamilton's equations on $\mathcal{C}_\Sigma$ are

$$i(\xi)\eta_\Sigma = dh_\Sigma, \tag{4.5}$$

which are to be solved for the evolution vector field ξ. One checks that there are no initial value constraints other than (4.2) [N,Ke], so $\mathcal{C}_\Sigma$ is in fact the *final* constraint set. Thus, writing $\xi = v\delta/\delta v + v_t \delta/\delta v_t$, (4.5) may be solved in the form

$$v = 3(D_X v)^2 - (D_X^3 v) \ , \quad v_t = arbitrary. \tag{4.6}$$

Differentiating the first of these and eliminating the potential v yields (1.2). On the other hand, (4.6)(*ii*) reflects the gauge freedom of the system, corresponding to the presence of the first class constraint (4.2)(*i*).

One may eliminate this trivial ambiguity by reducing $(\mathcal{E}_\Sigma, \eta_\Sigma)$ by $ker\ \eta_\Sigma = span\{\delta/\delta v_t\}$. The reduced space can be identified with $\mathcal{Y}_\Sigma$ (now with "coordinate" $u = D_X v$), and the reduced symplectic form is, from (4.4),

$$\bar{\eta}_\Sigma(\bar{\varphi}) = \tfrac{1}{2}\int_\Sigma \int_{-\infty}^{X} \bar{\varphi}^* \left[\delta u(y) \wedge \delta u(x)\right] \otimes dydx \tag{4.7}$$

where $\bar{\varphi} \in \mathcal{Y}_\Sigma$. This is the KdV symplectic structure as presented in [ZF].

Thus, ultimately one has the KdV equation realized as a Hamiltonian system on $\mathcal{Y}_\Sigma$:

$$i(\bar{\xi})\bar{\eta}_\Sigma = d\bar{h}_\Sigma, \tag{4.8}$$

where the reduced Hamiltonian is

$$\bar{h}_\Sigma(\bar{\varphi}) = \int_\Sigma \bar{\varphi}^* \left[u^3 + \tfrac{1}{2}(D_X u)^2\right] dx, \tag{4.9}$$

cf. (4.3). Solving (4.8), one obtains immediately from (4.7) the KdV equation in the form (1.1), where $\mathcal{D} = D_X$ and $\mathcal{H}[u]$ is the Hamiltonian density corresponding to (4.9).

Remarks: 1. This treatment of the KdV equation should be compared with those given by Nutku [N] and Kentwell [Ke], who reduce the Lagrangian to first order. Their method has the disadvantage of introducing spurious *secondary* constraints. Yet other approaches can be found in [BdK] and [S].

2. Actually, Gardner presented not the symplectic form (4.7), but rather its associated Poisson bracket [Ga]. The connection between these is given in [BdK]; this paper also examines in some detail symplectic structures which generalize (4.7).

Acknowledgements

It is a pleasure to thank Ian Anderson, Peter Olver and David Sattinger for several very useful discussions, and Utah State University for its hospitality while this research was carried out. This work was supported by a grant from the U.S. Naval Academy Research Council.

Appendix

Let $\pi: K \to X$ be a fibration. Adapted coordinates on K are v^A along the fibers and x^μ on the base. The space of sections of π will be denoted by either ΓK or the script letter $\mathcal{K}$. Similarly, for $\Sigma \subset X$, the space of sections of $K|\Sigma \to \Sigma$ will be denoted by either ΓK_Σ or $\mathcal{K}_\Sigma$.

The volume form on X is $d^n x = dx^1 \wedge \cdots \wedge dx^n$, where $n = dim\ X$. Then $d^{n-1}x_\mu := i(\partial_\mu)\, d^n x$.

For $f \in C^\infty(K)$, the *formal partial derivative* $D_\mu f$ of f in the direction of x^μ is defined by $\varphi^*(D_\mu f) = \partial_\mu(f \circ \varphi)$ for all $\varphi \in \mathcal{K}$.

The *vertical bundle* $V\pi$ is the subbundle $ker\ T\pi$ of TK.

The rth *jet bundle* of K is J^rK with coordinates $x^\mu, v^A, v^A{}_{\mu_1}, \ldots, v^A{}_{\mu_1 \ldots \mu_r}$. The projections $J^rK \to J^sK$ for $r \geq s$ and $J^rK \to X$ are denoted π^r_s and π^r, respectively. Set $J^0K = K$ and $\pi^0 = \pi$. The rth jet prolongation is denoted j^r.

References

[BdK] M.J. Bergvelt and E.A. de Kerf, *Poisson Brackets for Lagrangians Linear in the Velocity*, Lett. Math. Phys., 10, 13-19 (1985).

[dLR] M. de León and P.R. Rodrigues, *Generalized Classical Mechanics and Field Theory* (North Holland, 1985).

[Ga] C.S. Gardner, *Korteweg-de Vries Equation and Generalizations. IV. The Korteweg-de Vries Equation as a Hamiltonian System*, J. Math. Phys., 12, 1548-1551 (1971).

[Go] M.J. Gotay, *Multisymplectic Structures and the Hamiltonian Formalism in the Calculus of Variations*, in preparation (1988).

[Ke] G.W. Kentwell, *The Hamiltonian Structure of the Korteweg-de Vries Equation*, preprint (1987).

[Ki] J. Kijowski, *A Finite-dimensional Canonical Formalism in the Classical Field Theory*, Commun. Math. Phys., 30, 99-128 (1973).

[Ko] I. Kolář, *A Geometrical Version of the Higher Order Hamilton Formalism in Fibered Manifolds*, J. Geom. Phys., 1, 127-137 (1984).

[N] Y. Nutku, *Hamiltonian Formulation of the KdV Equation*, J. Math. Phys., 25, 2007-2008 (1984).

[O] P.J. Olver, *Applications of Lie Groups of Differential Equations*, Graduate Texts in Mathematics, 107 (Springer,1986).

[S] W.F. Shadwick, *The Hamiltonian Structure Associated to Evolution-type Lagrangians*, Lett. Math. Phys., 6, 271-276 (1982).

[ZF] V.E. Zakharov and L.D. Faddeev, *Korteweg-de Vries Equation: A Completely Integrable Hamiltonian System*, Funct. Anal. Appl., 5, 280-287 (1971).

SINGULAR POINTS IN LEVEL SETS OF THE MOMENTUM MAP AND QUANTUM THEORY

H. Römer
Fakultät für Physik
Universität Freiburg
Hermann-Herder-Str. 3
D-7800 Freiburg i.Br./FRG

The phase space of Hamiltonean systems is a symplectic manifold (M,ω). Let a Lie group G act on (M,ω) by symplectic automorphisms. The momentum map [1] provides a geometric formulation of the relationship between symmetries and conserved quantities. It is a map

$$J: \; M \rightarrow \mathfrak{g}^*$$

where $\mathfrak{g}^*$ is the dual of the Lie algebra $\mathfrak{g}$ of G, defined in the following way:

$$d < J, \xi > \; = \; i_{\xi_M} \omega \quad .$$

Here ξ_M is the fundamental vector field of the symplectic action associated to $\xi \in \mathfrak{g}$

We, furthermore, assume that J is equivariant, i.e. that for every $g \in G$ the diagram

$$\begin{array}{ccc} M & \xrightarrow{\;g\;} & M \\ J \downarrow & & \downarrow J \\ \mathfrak{g}^* & \xrightarrow{\;Ad^*_g\;} & \mathfrak{g}^* \end{array}$$

is commutative. Equivariant momentum maps exist, for instance, if (M,ω) is exact symplectic: $\omega = -d\theta$ and if the symplectic action leaves θ invariant. This is the case of principal interest in the sequel. Under these circumstances we have

$$< J, \xi > \; = \; \theta(\xi_M) \quad .$$

The critical points m of J are just the points with nontrivial continuous isotropy group G_m.

Example: For rotations of an N particle system in Euclidean space one finds

$$J(q, p) = \sum_{\alpha=1}^{N} \vec{q}_\alpha \times \vec{p}_\alpha \quad .$$

K. Bleuler and M. Werner (eds.), Differential Geometrical Methods in Theoretical Physics, 307–315.

$0 \in \mathfrak{g}^*$ is the only critical value of J and the level set $J^{-1}(0)$ is a quadratic cone.

This structure of $J^{-1}(0)$ is typical. In fact, it has been shown by J. Arms, J. Marsden and V. Moncrief[2], that $J^{-1}(0)$ has at most quadratic singularities, if (M,ω) is an (almost) Kähler manifold and if G acts by (almost) Kähler automorphisms and if an equivariant momentum map exists.

Here we intend to point out the possible relevance of such quadratic singularities in phase (and configuration) space for the associated quantized field theories[3,4]. Before coming to this we briefly mention some results about the opposite problem of avoiding singularities of this type.

The Marsden-Weinstein-Reduction[1] allows the construction of reduced phase spaces by dividing out symmetry degrees of freedom:

Thm Let (M,ω) be a symplectic manifold and let a Lie Group G act on (M,ω) with an equivariant momentum map J. If $\mu \in \mathfrak{g}^*$ is a weakly regular value of J and if the isotropy group G_μ of μ acts properly with conjugate stability groups on $J^{-1}(\mu)$, then

$$M_\mu = J^{-1}(\mu)/G_\mu$$

is a manifold with a symplectic structure ω_μ, uniquely defined by

$$\pi_\mu^* \omega_\mu = i_\mu^* \omega$$

$$(\pi_\mu : J^{-1}(\mu) \to M_\mu \quad \text{projection}, \quad i_\mu : J^{-1}(\mu) \hookrightarrow M) \quad .$$

This reduction procedure has two main applications:

- Reducing Hamiltonean systems to smaller ones
- Demonstrating complete integrability of Hamiltonean systems by identifying them as reduced systems of larger Hamiltonean systems, which are trivially integrable.

The conditions under which M_μ is really a manifold are rather restrictive. M. Otto has shown[5] that one can do better if (M,ω) is an (almost) Kähler manifold, i.e. if there is an (almost) complex structure I on M such that $\omega(IX,IY) = \omega(X,Y)$ and $g(X,Y) =: -\omega(IX,Y)$ is a Riemannian metric on M.

For a (closed and connected) subgroup $H \subset G$ let us denote by $M_{(H)}$ the (not necessarily symplectic) submanifold of points $m \in M$, such that the identity component of the isotropy group G_m is conjugate to H. Then

Thm (M. Otto, 1986)[5]
Let G act properly on (M,ω,I) by almost Kähler automorphisms with equivariant momentum map J. Then for every $\mu \in \mathfrak{g}^*$

$$M_{\mu(H)} = (M_{(H)} \cap J^{-1}(\mu))/G_\mu$$

is a symplectic manifold, whose symplectic form $\tilde{\omega}_\mu$ is uniquely determined by

$$i_{\mu}^{*}\omega = p^{*}\tilde{\omega}_{\mu} \quad .$$

This time, reduction of phase space proceeds in two steps

α) Decomposition into orbit types $M = \bigcup_{(H)} M_{(H)}$

β) Reduction of $J^{-1}(\mu) \wedge M_{(H)}$

New reduced phase spaces and Hamiltonean systems not obtainable by a reduction in one step (because $M_{(H)}$ need not be symplectic) can be constructed this way.

The hypothesis of Otto's theorem in particular applies[5)] to the most important case $M = T^{*}N$, where N is a Riemannian manifold; If G acts on N by isometries, then $T^{*}N$ admits a natural almost Kähler structure, such that the action on N can be lifted to an action on $T^{*}N$ respecting this structure.

Let us now come to a discussion of the possible role of quadratic singularities. The principal application we have in mind are theories with symmetries of gauge type like Yang-Mills theories[2,6)] and Einsteinian gravity[2,7)]. Both can be formulated as infinite dimensional Hamiltonean systems with an infinite dimensional invariance group: The group of gauge transformations (= principal bundle isomorphisms) for Yang-Mills theories and the diffeomorphism group for Einsteinian gravity.

For a Yang-Mills theory with group G on the space-time manifold $M \times \mathbb{R}$ (M compact, orientable, D-dimensional, Riemannian) the Langrangean in the temporal gauge $A_0 \equiv 0$ is given by

$$\mathcal{L} = -\frac{1}{4} F^{\mu\nu} \cdot F_{\mu\nu} = \frac{1}{2} \dot{A}_i \cdot \dot{A}_i - \frac{1}{4} F^{ij} \cdot F_{ij}$$

$$(\mu, \nu = 0, 1, \ldots, D \quad ; \quad i, j = 1, \ldots, D)$$

Thus, the momentum conjugate to A_i on M is

$$\pi_i = \dot{A}_i =: E_i \quad .$$

The phase space $P = \{(A_i; E_i)\}$ and its tangent spaces

$$T_{(A;E)} P = \{(\alpha_i; \varepsilon_i)\}$$

are function spaces. An exact symplectic form $\omega = -d\theta$ on P is defined by

$$\theta(\alpha;\varepsilon) = \int_M E_i \cdot \alpha_i = (E, \alpha)$$

where (.,.) is the Hodge scalar product. The generators of gauge transformations are given by $\mathfrak{g}$-valued functions χ, and the fundamental vector field χ_P associated to χ is

$$\chi_P(A, E) = (\partial_i \chi + A_i \chi ; [E_i, \chi]) = (d_A \chi, [E, \chi]) \quad .$$

This gives the momentum map[2,6)]

$$<J(A;E),\chi> = \theta(\chi_P) = (E, d_A\chi) = (d_A^* E, \chi).$$

The level set $J^{-1}(0)$ consists of pairs (A;E) with

$$d_A^* E = 0 \quad , \qquad \text{in coordinates}$$

$$\partial_i E_i + [A_i, E_i] = 0 .$$

This is just Gauß's law, the constraint which every admissible configuration of the Yang-Mills field has to obey.

The critical points of J are the symmetric connections, invariant under some subgroup of the group of gauge transformations and reducible to a smaller group $H < G$.

The Hamiltonean formulation of Einsteinian gravity is provided by the ADM-formalism[8], also called 3+1-decomposition. The space-time manifold $M^{(4)}$ is assumed to be of the form

$$M^{(4)} = M^{(3)} \times \mathbb{R} .$$

In local coordinates the metric tensor is parametrised in the form

$$ds^2 = (N^2 - N^i N_i)\, dt^2 - 2 N_i dx^i dt + \gamma_{ij}\, dx^i dx^j ,$$

where N is the lapse function, N_i the shift vector and γ_{ij} a Riemannian metric on $M^{(3)}$. The Einstein-Hilbert action is

$$S = \int_{M^{(4)}} d^4x \, \{ g^{1/2} (R - 2\Lambda) \} + 2 \int_{\partial M^{(4)}} \gamma^{1/2} K$$

$$= \int_{M^{(4)}} d^4x \, \gamma^{1/2} N \{ K_{ij} K^{ij} - K^2 R\{\gamma\} \}$$

where

$$K_{ij} = \frac{1}{2N} \{ N_{i|j} + N_{j|i} - \dot{\gamma}_{ij} \}$$

is the second fundamental form on $M^{(3)}$ and $R\{\gamma\}$ the curvature scalar of the 3-metric γ_{ij} ; Λ is the cosmological constant.
The momenta conjugate to γ_{ij} are

$$\pi_{ij} = -\gamma^{1/2} (K_{ij} - \gamma_{ij} K) ,$$

and the constraint equations, which all admissible initial configurations have to satisfy are

$$\mathcal{H}(\gamma, \pi) = \gamma^{-1/2} \{ \pi^{ij} \pi_{ij} - \tfrac{1}{2} (\pi^i_i)^2 \} - \gamma^{1/2} R\{\gamma\} = 0$$

$$\mathcal{H}_i(\gamma, \pi) = -2 \pi_{ij}{}^{|j} = 0 .$$

These constraint equations can be interpreted as the vanishing of the momentum map associated to the action of diffeomorphisms on the metric. The solution space of the constraint equations was shown to have at most quadratic conical singularities[2].

These singular points are critical points of J and correspond to metrics which admit a Killing vector field generating an isometry.

The common feature of the Yang-Mills and gravity case is the fact that the set of admissible initial configurations is $J^{-1}(0)$ and is not a manifold but rather an "orbifold" with possible quadratic conical singularities located at admissible configurations with a certain symmetry.

We shall sketch three possible consequences of the emergence of conical singularities in the state space of classical theories for their associated quantum theories:

a) Nonequivalence of extrinsic and intrinsic quantization[3]
b) Linearization instability[2,7,9]
c) Schrödinger equations on orbifolds[3]

a) Consider a classical theory with a classical field φ, subject to a constraint $F(\varphi) = 0$. Quantization is a procedure of associating a quantum field Φ to the classical field φ. Two different strategies are conceivable:[3]

i) Extrinsic quantization
First one converts φ into an operator field Φ and then one takes care of the constraint by demanding that the operator $F(\Phi)$ has vanishing matrix elements for all vectors in the Hilbert space of physical states.

ii) Intrinsic quantization
First one solves for the constraint $F(\varphi) = 0$ on the classical level then one quantizes only the remaining degrees of freedom.

The second procedure looks more attractive from a conceptual point of view, it is, however, usually hard to implement because solving for the constraints will in general be possible only locally and will, moreover, normally result in theories with non polynomial Lagrangeans. The equivalence of the intrinsic and extrinsic procedure is by no means clear. The usual strategy followed in path integral quantization consists in manipulating the constraint into the classical action $S(\varphi)$ in the exponent of the integrand of the path integral:

$$Z = \int \mathcal{D}\varphi \, \delta(F(\varphi)) \, e^{iS(\varphi)}$$

$$\sim \int \mathcal{D}\varphi \mathcal{D}\sigma \, e^{iS(\varphi) + i\sigma \cdot F(\varphi)}$$

This strategy has to be classified as extrinsic because it admits fluctuations about the allowed classical configurations φ with $F(\varphi) = 0$ at least in intermediate steps.

Extrinsic and intrinsic quantization are expected to be inequivalent if the constraint set $F^{-1}(0)$ contains conical singularities. In fact, such a singularity is wiped out if fluctuations into a neighbourhood of the constraint set are admitted. Let us illustrate this by a simple example[3]. We compare intrinsic and extrinsic quantization for σ -models, where the classical field assumes its values

α) on a 1-sphere $\quad F_S(\varphi_1,\varphi_2) = \varphi_1^2 + \varphi_2^2 - a^2 = 0$

β) on a 1-cone $\quad F_C(\varphi_1,\varphi_2) = \varphi_1^2 - \varphi_2^2 = 0$.

For the generating functional one obtains

(i) in the extrinsic formulation

$$Z_S = \int \mathcal{D}\varphi_1' \mathcal{D}\varphi_2' \mathcal{D}\sigma \exp \int dx \{ -\tfrac{1}{2}\varphi_i' \Delta_S^{(i)} \varphi_i' + \tfrac{1}{2} J^i (\Delta_S^{(i)})^{-1} J^i - \tfrac{1}{2}\sigma a^2 \}$$

$$Z_C = \int \mathcal{D}\varphi_1' \mathcal{D}\varphi_2' \mathcal{D}\sigma \exp \int dx \{ -\tfrac{1}{2}\varphi_i' \Delta_C^{(i)} \varphi_i' + \tfrac{1}{2} J^i (\Delta_C^{(i)})^{-1} J_i \}$$

with

$$\Delta_S^{(1)} = \Delta_S^{(2)} = \Delta_C^{(1)} = -\nabla^2 + m^2 - i\sigma$$

$$\Delta_C^{(2)} = -\nabla^2 + m^2 + i\sigma$$

$$\varphi_i = \varphi_i' + (\Delta_{S,C}^{(i)})^{-1} J_i \quad ;$$

(ii) in the intrinsic formulation

$$\tilde{Z}_S = \tfrac{1}{2} \int \mathcal{D}\varphi \exp \{ -\int dx \, [\tfrac{1}{2} (1 + \frac{\varphi^2}{a^2-\varphi^2}) \partial_\mu \varphi \partial^\mu \varphi + \tfrac{1}{2} m^2 a^2 - J^2 \varphi - J^1 \sqrt{a^2-\varphi^2} + \ln |(a^2-\varphi^2)^{-1/2}|] \} + (J^1 \leftrightarrow J^2)$$

$$\tilde{Z}_C = \tfrac{1}{2} \int \mathcal{D}\varphi \exp \{ -\int dx \, [\partial_\mu \varphi \partial^\mu \varphi + m^2 \varphi^2 + (J^1 + J^2)\varphi - \ln|\varphi| \} + (J^1 \rightarrow -J^1)$$

No essential difference between the spherical and the conical case is visible in the extrinsic formulation, in contrast to the intrinsic approach, where one seriously has to face the problem of the singularity at $\varphi = 0$ for the cone.

b) Consider again the constraint equation $F(\varphi) = 0$ and its linearization about φ_0 : $DF(\varphi_0)\eta = 0$.

As long as φ_0 is a regular point of F, linearization stability holds: every solution of the linearized equation is tangent to an curve of solutions of the full equation $F(\varphi) = 0$ through φ_0.

If φ_0 is a critical point of F, the phenomenon of linearization instability is observed: The linearized equation has spurious solutions not tangent to any curve of solutions of the original constraint equation. These spurious solutions have to be eliminated by subsidiary conditions, the lowest of which are quadratic in η of the type

$$\langle v, D^2 F_0(\varphi_0)(\eta,\eta) \rangle = 0$$

for every v orthogonal to the range of DF(φ_0).
Conical singularities of the level set $J^{-1}(0)$ as considered above are an example of this situation. We mentioned that for Young Mills and gravity theories they are quadratic and correspond to configurations with some symmetry. For singularities of quadratic type quadratic subsidiary conditions on the solutions of the linearized equation are sufficient. In ref.[7], it is shown for Einstein gravity how the spurious solutions of the linearized equations have to be eliminated by imposing quadratic subsidiary conditions. In addition, a consistent quantization procedure for the linearized classical equation is given, in which the quadratic constraints are converted into operators, which have to vanish on physical Hilbert space states[7]. (Fortunately, for technical reasons, this kind of linearization instability does not occur for linearization about a flat Minkowski background metric). If linearization instability occurs, the quadratic subsidiary conditions imply that the linearization has to have (at least) the same amount of symmetry as the configuration about which linearization is performed. In ref.[7] the Everett-Wheeler interpretation[10] is proposed as a possible explanation for this inforced symmetry: it arises by incorporating the observer into the system.

In a more conventional interpretation the additional symmetry of the admissible linearized solutions means that quantum fluctuations from a symmetric configuration into a less symmetric one are suppressed to linear order.

c) If one wants to go beyond quantizing the linearized version of Yang-Mills or gravity theory, the Schrödinger picture provides the correct conceptual framework. The state vectors are wave functionals defined on the set of admissible classical configurations. If, as it is the case in the situations under consideration, the configuration space contains quadratic conical singularities and, hence, is not a manifold everywhere, one immediate difficulty arises:
The functional derivatives $\delta/\delta\varphi$ which occur in the Schrödinger equation are undefined and a regularization procedure has to be found. Leaving out the singular points in a first stage on arrives at a problem of self adjoint extensions of symmetric linear operators.

Physically one expects that the conical singularities will play an important part in the discussion of the behaviour of the Schrödinger amplitude. Remembering that fluctuations into configurations of lower degree of symmetry are suppressed in the linear approximation one expects [3] either a particular suppression or an enhancement of the Schrödinger amplitude in the vicinity of a conical singularity. In the latter case this would be a mechanism for enhancing symmetric configurations by quantum effects.

This expectation is confirmed[3)] in the most simple model one can imagine:
We drop an infinity of nonsingular degrees of freedom and consider a free Schrödinger equation on a two dimensional circular cone. The singular tip of the cone is the only point invariant under rotations about the axis of the cone. We first puncture the cone by leaving out the tip.

The stationary Schrödinger equation is

$$H\psi = -\frac{1}{2m}\Delta\psi = E\psi \qquad (\hbar = 1) \quad ,$$

where the Laplace operator in polar coordinates centered at the tip is given by

$$\Delta = \frac{\partial^2}{\partial r^2} + \frac{1}{r}\frac{\partial}{\partial r} + \frac{1}{r^2 \sin^2\alpha}\frac{\partial^2}{\partial \varphi^2} \quad .$$

Here α is the half opening angle of the cone, r is the distance from the tip and φ the angle about the axis. The solutions of the Schrödinger equation for $E = k^2/2m$ are

$$\psi(r,\varphi) = A_n J_{n/\sin\alpha}(kr)e^{\pm in\varphi} + B_n Y_{n/\sin\alpha}(kr)e^{\pm in\varphi} \qquad (n=0,1,\ldots) \quad .$$

The requirement of local squareintegrability around the tip leads to the following behaviour of the total probability of the region $r \leq a$ for small a for the cone ($\alpha < \frac{\pi}{2}$) and the plane ($\alpha = \frac{\pi}{2}$):

	cone	plane
$n = 0$	a^2	a^2
$n > 0$	$a^{2+\frac{2n}{\sin\alpha}}$	a^{2+2n}

The only rotationally invariant solution corresponds to n = 0. Comparing the results for the cone and the plane we see that $n > 0$ is more strongly suppressed around the origin for the cone than for the plane. In other words, on the cone the symmetric solution with n = 0 shows an extra enhancement as compared to the non symmetric ones.

The same result is obtained by an alternative regularisation procedure, where the singularity is smoothened out by fitting a sphere into the cone and letting the radius of the sphere shrink to zero.

So, already in this simple model example we see the special relevance of conical singularities in the classical state space, which can be expected to hold for quantum Gravity and quantum Yang Mills theories.

1) A standard reference for Mechanics on symplectic manifolds is R. Abraham, J.E. Marsden: Foundations of Mechanics 2nd edition 1978 The Benjamin/Cummings Publishing Comp., Inc.

2) Arms, J.M., Marsden, J.E., and Moncrief, V., Comm.Math.Phys. 78, 455 (1981);
Arms, J.M., Marsden, J.E., and Moncrief, V., Ann.Phys. 144, 81 (1982);
Marsden, J.E., "Lectures on Geometric Methods in Mathematical Physics" 37, Society for Industrial and Applied Mathematics, Philadelphia, Pennsylvania, 1982;
Marsden, J.E., "Applications of Hamiltonian Structures", talk Clausthal, 4th Summer workshop on Math.Phys., 1984

3) B. de Barros Cobra Damgaard, H. Römer: "Quantum gravity and Schrödinger Equations on Orbifolds", Lett.Math.Phys. 13 (1987) 189.
B. de Barros Cobra Damgaard: "Kegelartige Singularitäten in der Quantengravitation", Doctoral dissertation, University of Freiburg, 1986

4) Singularities in phase space are discussed from different points of view e.g. in B. Kostant, S. Sternberg Ann.Phys. 176 (1987) 49
J. Śniatycki, A. Weinstein, Lett.Math.Phys. 7 (1983) 155
M. J. Gotay, L. Bos, J.Differential Geometry 24 (1986) 181 and references therein

5) M. Otto: "A reduction Scheme for Phase Spaces with Almost Kähler Symmetry. Regularity Results for Momentum Level Sets" to be published in Journal of Geometry and Physics
M. Otto: Phasenraumreduktion bei Fast-Kähler-Symmetrie, MSc Thesis. Freiburg University 1986

6) Moncrief, V., Ann.Phys. 108, 387 (1977);
Arms, J.M., Math.Proc.Camb.Phil.Soc. 90, 361 (1981)

7) Moncrief, V., Phys.Rev. D18, 983 (1987)

8) R. Arnowitt, S. Deser, C.W. Misner (1962) in "Gravitation: An introduction "to Current Research" ed. L. Witten (J. Wiley & Sons, Inc. N.Y.)

9) Geroch, R. and Lindblom L., J.Math.Phys. 26, 361 (1981)

10) Wheeler, J., "Superspace and the Nature of Quantum Geometrodynamics", in C.M. De Witt, J.A. Wheeler (eds.), Battelles Rencontres, W.A. Benjamin, New York, 1968

KILLING SPINORS AND UNIVERSALITY OF THE HIJAZI INEQUALITY

André LICHENROWICZ
Collège de France
11, Place Marcelin Berthelot
75231 Paris Cedex 05 - FRANCE

ABSTRACT - Introduction of a first order operator D on the spinors, in terms of the Dirac operator; D is connected with the equation of the global twistors and conformally covariant. For a compact spin manifold of dimension $n \geq 3$, the operator $(\Delta-\rho)$ on the spinors ($\rho = (n/4(n-1)R)$ is positive; Hijazi inequality holds for any spinor ψ such that $(P\psi, P\psi) = \lambda(\psi,\psi)$ with λ = const. In the limiting case, the manifold admits a Killing spinor.

Introduction

Killing spinors were fist introduced in Mathematical Physics : general relativity, 11-dimensional (resp. 10-dimensional) supergravity theory, supersymmetry, matter fields (see, for example [1] to [5]). The context is often the following : the main space is a fiber bundle on a space-time; the fibers are properly Riemannian manifolds (for example, spheres or homogeneous spaces) admitting a spinor structure. In many cases Killing spinors appear on these manifolds.

The notion also appears in a purely geometric way in direct relation with the Dirac operator P on a spin manifold. A Killing spinor is automatically an eigenspinor of P and is a generalization of the notion of parallel spinor. Many years ago, I studied same properties of parallel spinors in the context of harmonic spinors [6]. Some more interesting steps in this direction have been recently taken by Friedrich [7] and

K. Bleuler and M. Werner (eds.), Differential Geometrical Methods in Theoretical Physics, 317–332.

Hijazi [8]. For a compact spin manifold, Hijazi has given a lower bound for the square of the eigenvalues of P in terms of conformal geometry. The limiting case is precisely the case when the manifold admits Killing spinors.

I will give here new properties of the Killing spinors (Theorems 1 and 2) and, introducing a first order operator D on the spinors, connected with the so-called twistor operator, I shall place the Hijazi inequality in its true context, showing that, in a suitable sense, the inequality is *universal*. My approach differs from that of Hijazi (Theorem 3 and 4).

The physical philosophy which appears can be the following : consider a compact spin manifold of a basis for a Euclidean model for fermionic fields ψ. If the manifold admits Killing spinors, we can consider the corresponding eigenvalue ν_1 of P as associate with a ground state : if ψ satisfies $(P\psi, P\psi) = \lambda^2(\psi,\psi)$, we have $\lambda^2 \geqslant \nu_1^2$.

I - Definitions and general formulas.

1 - SPIN MANIFOLDS AND CORRESPONDING CONNECTIONS.

a) Let (W,g) be an oriented Riemannian manifold of dimension $n \geqslant 2$ (with $n = 2\nu$ or $n = 2\nu+1$) endowed with a positive definite metric g. If (W,g) has a *spin structure* with *fundamental vector-spinor* γ, there is a principal Spin(n)-bundle $\mathcal{E}$ on W which is a two-fold covering $p : \mathcal{E} \to E$ of the SO(n)-bundle of the orthogonal frames; a point z of $\mathcal{E}$ is a *spin frame*; (W,g) being referred to the orthonormal frames, the components of γ are given by the $2^\nu \times 2^\nu$ Dirac matrices satisfying

$$\gamma_\alpha \gamma_\beta + \gamma_\beta \gamma_\alpha = -2 g_{\alpha\beta} e \qquad (\alpha,\beta = 1,\dots,n) \qquad (1\text{-}1)$$

It is well-known that the γ_α can be chosen *anti Hermitian* ($\tilde{\gamma}_\alpha = -\gamma_\alpha$, where ~ denotes the adjunction). The group Spin(n) can be considered as a group of $2^\nu \times 2^\nu$ unitary matrices Λ (with $\tilde{\Lambda} = \Lambda^{-1}$) satisfying

$$\Lambda \gamma_\alpha \Lambda^{-1} = A_\alpha^{\lambda'} \gamma_{\lambda'} \tag{1-2}$$

where $A = \left(A_\alpha^{\lambda'}\right)$ is the projection of Λ on So(n).

If ψ is a contravariant 1-spinor ($\psi(z\Lambda^{-1}) = \Lambda\psi(z)$, where z is a spinor frame), $\tilde{\psi}$ is a covariant 1-spinor and the space Σ_x ($x \in W$) of the contravariant 1-spinors in x admits a canonical structure of Hermitian vector space given by the scalar product $(\psi^{(1)}, \psi^{(2)}) = \tilde{\psi}^{(2)} \psi^{(1)}$; Σ is the vector bundle of the contravariant 1-spinors of (W,g).

b) The Riemannian connection of (W,g) induces a connection on the bundle $\mathcal{E}$ by the isomorphism between the Lie algebras of Spin(n) and SO(n).

Introduce local sections of the bundle E; if $\omega_U = (\omega_{\alpha\beta})$ (with $\omega_{\alpha\beta} + \omega_{\beta\alpha} = 0$) is, for a domain U of W, the 1-form defining the Riemannian connection, the corresponding spin connection is given by the 1-form $\sigma_U = -(1/4)\,\omega_{\alpha\beta}\,\gamma^\alpha\gamma^\beta$. We denote by ∇ the corresponding covariant differentiation. We have $\nabla\gamma = 0$ and the adjunction commutes with ∇. Let RC $= (R_{\lambda\mu,\alpha\beta})$ be the Riemannian curvature tensor of (W,g), (Ri) $= (R_{\alpha\beta})$ its Ricci tensor and R its scalar curvature. We have on U the classical formulas :

$$R_{\alpha\beta,\lambda\mu}\,\gamma^\beta\gamma^\lambda\gamma^\mu = 2\,R_{\alpha\beta}\,\gamma^\beta \qquad R_{\alpha\beta}\,\gamma^\alpha\gamma^\beta = -R$$

It follows from the Ricci identity that for a spinor field ψ :

$$\gamma^\beta(\nabla_\alpha\nabla_\beta - \nabla_\beta\nabla_\alpha)\,\psi = -(1/2)\,R_{\alpha\beta}\,\gamma^\beta\,\psi \tag{1-3}$$

c) Let P be the Dirac operator of (W,g), $\tilde{P}$ its adjoint. We have :

$$P\psi = \gamma^\alpha\nabla_\alpha\psi \qquad \tilde{P}\tilde{\psi} = -\nabla_\alpha\tilde{\psi}\gamma^\alpha$$

We set $\Delta\psi = P^2\psi$. An elementary calculus gives

$$\Delta\psi = -\nabla^\alpha\nabla_\alpha\psi + (R/4)\,\psi \tag{1-4}$$

If W is *compact*, introduce the global scalar product $\langle \psi^{(1)}, \psi^{(2)} \rangle = \int_W (\psi^{(1)}, \psi^{(2)})\,\eta$, where η is the Riemann volume element. We have

$$\langle \Delta \psi^{(1)}, \psi^{(2)} \rangle = \langle P \psi^{(1)}, P \psi^{(2)} \rangle = \langle \psi^{(1)}, \Delta \psi^{(2)} \rangle$$

We see that *the spectrum of P is real* and that Δ is *positive* selfadjoint.

d) Let D be the first-order differential operator defined on the spinor by :

$$D \psi = \nabla \psi + (1/n) \gamma P \psi \tag{1-5}$$

D satisfies $\gamma^{\alpha} D_{\alpha} \psi = 0$. By differentiation of (1-5) we obtain

$$- \nabla^{\alpha} D_{\alpha} \psi = ((n-1)/n) (\Delta - \rho) \psi \qquad (\rho = (n/4(n-1))R) \tag{1-6}$$

If $D \psi = 0$, we have $(\Delta - \rho) \psi = 0$

If there is on (W,g) a field ψ without zero, set

$$(P \psi, P \psi) = r^2 (\psi,\psi) \qquad (r \geqslant 0) \tag{1-7}$$

If r is > 0, the spinor $r^{-1} P \psi$ admits same square as ψ. There exists $\Lambda : x \in W \to \Lambda(x) \in \mathrm{Spin}(x)$ such that

$$P \psi = r \Lambda \psi \tag{1-8}$$

We will study spinor fields (having or not zeroes) satisfying (1-7) or (1-8), r being $\geqslant 0$. For r and Λ, introduce a connection on the vector bundle Σ given by $\sigma_U^{(x,\Lambda)} = \sigma_U + (r/n) \gamma \Lambda I_U$. We denote by $\nabla^{(r,\Lambda)}$, the corresponding covariant differentiation. We have :

$$\nabla^{(r,\Lambda)} \psi = \nabla \psi + (r/n) \gamma \Lambda \psi$$

If ψ satisfies (1-8), $\nabla^{(r,\Lambda)} \psi = D \psi$; a field such that $\nabla^{(r,\Lambda)} \psi = 0$ satisfies (1-8). It is easy to prove

Proposition 1 - Any field $\psi \not\equiv 0$ *such that* $\nabla^{(r,\Lambda)} \psi = 0$ *has no zero on W.*

e) Take r = const., $\Lambda = e^{i\theta}$ Id where θ = const. and set $\nu_1 = r\, e^{i\theta} \in \mathbb{C}$. *A Killing spinor of (W,g) is a spinor $\psi \not\equiv 0$ such that*

$$\nabla^{\nu_1} \psi = \nabla \psi + (\nu_1/n)\, \gamma\, \psi = 0 \qquad (1\text{-}9)$$

where ν_1 is a complex constant. We have then $P\,\psi = \nu_1\, \psi$ and ν_1 is an eigenvalue of P. If $\nu_1 = 0$, we have a parallel spinor. We say that a Killing spinor is non trivial if $\nu_1 \neq 0$.

2 - AN UNIVERSAL FORMULA.

a) Let ψ be an arbitrary spinor field. We associate with ψ a real scalar and a real 1-form :

$$u(\psi) = u = (\psi,\psi) = \tilde{\psi}\, \psi \geqslant 0 \qquad T(\psi) = \tilde{\psi}\, \gamma\, P\, \psi - \tilde{P}\, \tilde{\psi}\, \gamma\, \psi \qquad (2\text{-}1)$$

If we evaluate the positive Laplacian of u in terms of $P\,\psi$ and $D\,\psi$, we obtain *the universal formula*, valid for an arbitrary vector field ψ (δ codifferentiation operator)

$$(1/2)\, \Delta u + (1/2)\delta\, T(\psi) = -\, (R/4)u + ((n-1)/n)(P\psi,P\psi) - (D\psi,D\psi) \qquad (2\text{-}2)$$

Suppose W is *compact*. We obtain by integration of (2-2)

$$\frac{n-1}{n} \langle\, \Delta\, \psi - \rho\, \psi,\ \psi\, \rangle = \langle\, D\psi,\ D\psi\, \rangle \qquad (2\text{-}3)$$

It follows from (1-6) and (2-3) :

Proposition 2 - On a spin manifold (W,q), every field ψ such that $D\psi = 0$ satisfies

$$\Delta\, \psi = \rho\, \psi \qquad \rho = (n/4)(n-1))\ R \qquad (2\text{-}4)$$

Conversely, if W is compact, every solution of (2-4) is such that $D\psi = 0$.

b) *Suppose now $D\psi = 0$*, W being compact or not. An elementary computation gives :

$$(1/2)\ \delta\ T(\psi) = (P\psi, P\psi) - \rho u$$

and, in this case, (2-2) can be written :

$$(n/2)\ \Delta u = \rho u - (P\psi, P\psi) \qquad (\text{for } D\psi = 0) \tag{2-5}$$

We note that we have for $D\psi = 0$

$$n\ d u = -\ T(\psi) \tag{2-6}$$

Moreover from (1-3) and (2-4), we deduce the formula

$$(n-2/n)\ \nabla_\alpha P\psi - (1/n)\ \rho\ \gamma_\alpha \psi + (1/2)\ R_{\alpha\beta}\ \gamma^\beta \psi = 0 \tag{2-7}$$

Introducce the real vector

$$J(\psi) = i\ \tilde{\psi}\ \gamma\ \psi$$

Taking the product of (2-7) by $\tilde{\psi}$, we obtain

$$(n-2)\ \nabla_\alpha(\tilde{\psi}\ P\psi) - (n-2/n) i\ J_\alpha(P\psi) + i\ \rho\ J_\alpha(\psi) - i(n/2)\ R_{\alpha\beta}\ J^\alpha(\psi) = 0$$

It follows by adjunction and addition

$$(n-2)\ \nabla_\alpha \left\{\tilde{\psi}(P\psi) + (\tilde{P}\tilde{\psi})\ \psi\right\} = 0$$

Proposition 3 - Let (W,g) be a spin manifold of dimension $n \geqslant 3$. Every field ψ such that $D\psi = 0$ satisfies :

$$\tilde{\psi}(P\psi) + (\tilde{P}\tilde{\psi})\ \psi = \text{const.} \tag{2-8}$$

II - Killing spinors and connection $\nabla^{(f)}$.

3 - SPINOR ψ SUCH THAT $\nabla^{(f)}\psi = 0$.

a) Let f be a complex-valued function on W; we set $f = r\,e^{i\theta} = a + ib$, where a, b are real-valued functions. For $\Lambda = e^{i\theta}$ Id, the covariant derivative $\nabla^{(r,\Lambda)}\psi$ can be written :

$$\nabla^{(f)}\psi = \nabla\,\psi + (f/n)\,\gamma\,\psi \tag{3-1}$$

Suppose $n \geqslant 2$ and $\nabla^{(f)}\psi = 0$ that implies $P\,\psi = f\,\psi$. If $\psi \not\equiv 0$, ψ is without zero and $u(\psi) > 0$. It follows from (1-3) that we have

$$\nabla_\alpha f\,\psi + (1/n)\nabla_\alpha f\,\gamma^\beta\gamma_\alpha\psi + (1/2)\left(R_{\alpha\beta} - (4(n-1)/n^2)f^2 g_{\alpha\beta}\right)\gamma^\beta\psi = 0 \tag{3-2}$$

Taking the product by γ^α we have :

$$\nabla_\alpha\,f\,\gamma^\alpha\,\psi - (\rho - f^2)\,\psi = 0 \tag{3-3}$$

b) In this case, we have, according to (3-1) :

$$n\,du(\psi) = -\,2b\,J(\psi) \qquad \nabla_\alpha\,J_\beta(\psi) + \nabla_\beta\,J_\alpha(\psi) = -\,(4b/n)u\,g_{\alpha\beta} \tag{3-4}$$

Now multiplying (3-3) by $\tilde{\psi}$ and separating real and purely imaginary parts, we get :

$$J^\alpha(\psi)\,\nabla_\alpha\,a - 2\,ab\,u = 0 \tag{3-5}$$

and

$$J^\alpha(\psi)\,\nabla_\alpha\,b - (\rho - a^2 + b^2)\,u = 0 \tag{3-6}$$

If n is $\geqslant 3$, (2-8) takes the simple form

$$a\,u = C = \text{const.} \tag{3-7}$$

4 - KILLING SPINORS.

a) Let $\psi \not\equiv 0$ be a Killing spinor on (W,g) of dimension $n \geqslant 2$. We can take $f = \nu_1 \in \mathbb{C}$. It follows from (3-5) that ν_1 is either real or purely imaginary and we deduce from (3-2), (3-3)

$$R_{\alpha\beta} = (R/n)\ g_{\alpha\beta} \qquad \nu_1^2 = \rho = (n/4(n-1))\ R \tag{4-1}$$

In particular, if ψ is a parallel spinor ($\nu_1 = 0$), the Ricci tensor is null. We consider in the following non-trivial Killing spinors ($\nu_1 \neq 0$).

If W is compact, ν_1 is real since the spectrum of P is real. Conversely if $\nu_1 \neq 0$ is real, the Ricci tensor is positive definite and W is compact according to a classical theorem of Myers. We have the following well-known proposition [9] :

Proposition 4 - If (W,g) admits a non-zero Killing spinor, it is an Einstein space and $\nu_1^2 = \rho$*. If* $\nu_1 \neq 0$ *is real, W is compact, if* $\nu_1 \neq 0$ *is purely imaginary W is non compact.*

If ν_1 is real, it follows from (3-4) that u = const and if $J(\psi) \not\equiv 0$ it defines an infinitesimal isometry, that is a Killing vector. It is the origin of the name of Killing spinor.

I have proved the following :

Proposition 5 - Let (W,g) be spin manifold admitting a non trivial Killing spinor. There are not non trivial parallel k-forms ($k \neq 0, n$) *on (W,g). In particular such a manifold is necessarily irreducible and non-Kählerian.*

b) Let (W,g) be a compact spin manifold with a non trivial Killing spinor ψ and an infinitesimal isometry $X \not\equiv 0$. Using harmonic considerations, we can prove

Theorem 1 - Let (W,g) be a compact spin manifold with a non trivial Killing spinor ψ *(that is* $\overset{\nu_1}{\nabla}\psi = 0$*) and an infinitesimal isometry* $X \not\equiv 0$ *having a zero on W. The Dirac operator admits also the eigenvalue* $\nu_2 = -\nu_1(1 + 2/n)$*. It is in particular the case if W has a Euler* *characteristic* $\neq 0$ *or is a Riemannian homogeneous space.*

5 - CHARACTERIZATION OF KILLING SPINORS.

a) Consider a spinor $\psi \not\equiv 0$ such that $\nabla^{(f)}\psi = 0$. Multiplying (3-3) by γ_β, we have

$$\nabla_\beta f \psi + (\rho - f^2) \gamma_\beta \psi + \frac{1}{2} \nabla_\alpha f(\gamma^\alpha \gamma_\beta - \gamma_\beta \gamma^\alpha) \psi = 0$$

and multiplying by $\tilde{\psi}$

$$u \nabla_\beta f - i(\rho - f^2) J_\beta(\psi) + \frac{1}{2} \nabla_\alpha f \tilde{\psi}(\gamma^\alpha \gamma_\beta - \gamma_\beta \gamma^\alpha) \psi = 0 \tag{5-1}$$

where the factor $\tilde{\psi} (\gamma^\alpha \gamma_\beta - \gamma_\beta \gamma^\alpha)\psi$ is purely imaginary. Separating the real and purely imaginary parts of (5-1) we have :

$$u \nabla_\beta a - 2 ab J_\beta(\psi) + \frac{i}{2} \nabla_\alpha b \tilde{\psi}(\gamma^\alpha \gamma_\beta - \gamma_\beta \gamma^\alpha)\psi = 0 \tag{5-2}$$

and

$$u \nabla_\beta b - (\rho - a^2 + b^2) J_\beta(\psi) - \frac{i}{2} \nabla_\alpha a \tilde{\psi}(\gamma^\alpha \gamma_\beta - \gamma_\beta \gamma^\alpha)\psi = 0 \tag{5-3}$$

Suppose that f is a real-valued function ($f = a$). It follows from (5-2) that $a = \text{const}$. Therefore ψ is a Killing spinor corresponding to a real ν (result of Hijazi by another method).

b) *Suppose now* $n \geqslant 3$. We have $a u = C$ and $a \nabla_\alpha u + u \nabla_\alpha a = 0$. According to (3-3), (5-2) can be written :

$$u \nabla_\beta a + n a \nabla_\beta u + \frac{i}{2} \nabla_\alpha b \tilde{\psi}(\gamma^\alpha \gamma_\beta - \gamma_\beta \gamma^\alpha)\psi = 0$$

let

$$(n-1) u \nabla_\beta a - \frac{i}{2} \nabla_\alpha b \tilde{\psi}\left(\gamma^\alpha \gamma_\beta - \gamma_\beta \gamma^\alpha\right)\psi = 0 \tag{5-4}$$

Suppose $C \neq 0$; a is without zero. Multiplying (5-1) by $\nabla^\beta b$ we have $\nabla^\beta a \nabla_\beta b = 0$ and so $\nabla^\beta u \nabla_\beta b = 0$.

It follows from (3-4) $b (J^\alpha(\psi))\nabla_\alpha b = 0$, that is, according to (3-6)

$$b (\rho - a^2 + b^2) = 0$$

It is easy to see that if $a \not\equiv 0$, we have $(\rho - a^2 + b^2) = 0$. It follows from (5-4)

$$(n-1)\ u\ \nabla^\beta a\ \nabla_\beta a - \frac{i}{2} \nabla_\alpha b\ \nabla_\beta a\ \tilde{\psi}(\gamma^\alpha \gamma^\beta - \gamma^\beta \gamma^\alpha)\psi = 0$$

But we deduce from (5-3) :

$$u\ \nabla^\beta b\ \nabla_\beta b - \frac{i}{2} \nabla_\alpha a\ \nabla_\beta b\ \tilde{\psi}(\gamma^\alpha \gamma^\beta - \gamma^\beta \gamma^\alpha))\psi = 0$$

We obtain by addition

$$(n-1)\ \nabla^\beta a\ \nabla_\beta a + \nabla^\beta b\ \nabla_\beta b = 0$$

It follows $a = \text{const} \neq 0$, $b = \text{const}$ and ψ is a Killing spinor. We have :

Theorem 2 - On the spin manifold (W,g) of dimension $n \geqslant 3$, let $\psi \not\equiv 0$ be a spinor such that $\nabla^{(f)}\psi = 0$, where f is a complex-valued function with a real part $a \not\equiv 0$. Then f is a constant and ψ is a Killing spinor on (W,g) which is compact.

For $n \geqslant 3$, we obtain an extension of the above result of Hijazi.

III - The Hijazi Inequality and its Universality.

6 - CONFORMAL CHANGE OF METRIC.

In this section, we suppose that *W is compact and $n \geqslant 3$.*
a) Consider a metric $\bar{g} = \exp(2c)g$ for some real-valued function c. The quantities corresponding to $(W,\bar{g})$ are covered by a bar. It is covenient to write the conformal factor under the form $\exp(2c) = h^{4/n-2}$, where h is > 0. The scalar curvature $\bar{R}$ of $(W,\bar{g})$ is given by

$$\bar{R} \exp(2c) = (4(n-1)/n-2)\ h^{-1}\ \Delta h + R \tag{6-1}$$

where Δ is the positive Laplacian. We are led to introduce the Yamabe operator :

$$L\ h = (4(n-1)/(n-2))\ \Delta h + R\ h \tag{6-2}$$

It is known that if μ_1 is the first eigenvalue of L, there is a function $h_1 > 0$ such that $L\ h_1 = \mu\ h_1$ and so

$$\mu_1 = (4(n-1)/(n-2))\ h_1^{-1}\ \Delta h_1 + R \tag{6-3}$$

If $\mu_1 \neq 0$, $\bar{R}_1$ corresponding to h_1 has everywhere the sign of μ_1, or is null if $\mu_1 = 0$. Study the function

$$Q(h) = (4(n-1)/n-2)\ h^{-1}\Delta h + R - \mu_1 = (4(n-1)/n-2)(h^{-1}\Delta h - h_1^{-1}\ \Delta h_1)$$

where h is > 0. Trivially $\langle \Delta h, h_1 \rangle = \langle h, \Delta h_1 \rangle$; therefore $(h_1\ \Delta h - h\ \Delta h_1)$ take on W values of arbitrary signs. It follows that for any $h > 0$

$$\inf_W\ [4(n-1)(n-2)\ h^{-1}\ \Delta h + R] \leqslant \mu_1 \leqslant \sup_W\ [(4(n-1)/(n-2))\ h^{-1}\ \Delta h + R]$$

For h = const. $\inf R \leqslant \mu_1 \leqslant \sup R$. If R = const. $\mu_1 = R$.

b) A spin structure $\bar{\gamma}$ of $(W,\bar{g})$ corresponds to a spin structure of (W,g) and a spinor $\bar{\psi}$ of $(W,\bar{g})$ to a spinor of (W,g); in this correspondance $\gamma^\alpha \to \exp(-\ c)\ \gamma^\alpha$. It is known (see for example [13]) that, for any spinor ψ, we have :

$$\bar{\nabla}_\alpha\ \bar{\psi} = \overline{\nabla_\alpha \psi} - (1/2)\ \partial_\beta\ c\ \overline{\gamma_\alpha\ \gamma^\beta\ \psi} - \frac{1}{2}\ \partial_\alpha\ c\ \bar{\psi} \tag{6-4}$$

It follows that if $\varphi = \exp(-(n-1)/2)c)\ \psi$, we have

$$\bar{P}\ \bar{\varphi} = \exp(-c)\ \exp(-(n-1)2)c)\ \overline{P\ \psi} \tag{6-5}$$

7 - GENERALIZATION OF THE HIJAZI INEQUALITY

a) According to (6-5), we have the following

Proposition 6 - Let $\psi \not\equiv 0$ *be a spinor of* (W,g) *such that* $(P\psi, P\psi) = \lambda^2(\psi,\psi)$ $(\lambda = \text{const.} > 0)$. *For any real-valued function c, the spinor* $\bar{\varphi}$ *of* $(W,\bar{g} = \exp(2c)g)$ *corresponding to* φ *satisfies* $(\bar{P}\bar{\varphi},\bar{P}\bar{\varphi}) = r^2(\bar{\varphi},\bar{\varphi})$ *where* $r = \lambda\exp(-c) > 0$. Conversely let $\bar{\varphi}$ be a spinor of $(W,\bar{g})$ such *that* $(\bar{P}\bar{\varphi}, \bar{P}\bar{\varphi}) = r^2(\bar{\varphi},\bar{\varphi})$ *where r is* > 0; *for c such that* $r = \lambda\exp(-c)$, *the spinor* $\psi = \exp\left[\left(\frac{n-1}{2}\right)c\right]\psi$ *of* (W,g) *satisfies* $(P\psi,P\psi) = \lambda^2(\psi,\psi)$.

b) If μ_1 is the first eigenvalue of (6-2), there is a factor $\exp(2c_1)$ such that $\mu_1 = \bar{R}_1\exp(2c_1)$, where $\bar{R}_1$ is the scalar curvature of $\bar{g} = \exp(2c_1)g$. Let ψ be an arbitrary spinor of (W,g), $\bar{\varphi}$ be the conformal spinor of $(W,\bar{g})$ associated with ψ. Apply (2-2) to $(W,\bar{g})$, and to $\bar{\varphi}$. We obtain by integration on $(W,\bar{g})$:

$$\int_W \left[\left(\bar{R}_1/4\right)\bar{u} - ((n-1)/n)\,(\bar{P}\bar{\varphi},\bar{P}\bar{\varphi})\right]\bar{\eta} + \int_W (\bar{D}\bar{\varphi},\bar{D}\bar{\varphi})\,\bar{\eta} = 0 \tag{7-1}$$

where $\bar{\eta}$ is the volume element of $(W,\bar{g})$ and $\bar{u}$ the square of $\bar{\varphi}$. The first integral is $\leqslant 0$ and the equality corresponds to the case where $\bar{D}\bar{\varphi} = 0$. We now have :

$$\bar{\eta} = \exp(n\,c_1)\,\eta \qquad \bar{u} = \exp(-(n-1)c_1)\,u$$

It follows from (7-1) :

$$\int_W [(\mu_1/4)u - (n-1)/n(P\psi,P\psi)]\exp(-\,c_1)\,\eta \leqslant 0 \tag{7-2}$$

Theorem 3 - Let (W,g) *be a compact spin manifold of dimension* $n \geqslant 3$, μ_1 *the first eigenvalue of the Yamabe operator. For any spinor* ψ, *there is a domain of W on which*

$$(P\psi,P\psi) \geqslant (n/4(n-1)\,\mu_1(\psi,\psi) \tag{7-3}$$

If $\psi \not\equiv 0$ is such that $(P\psi, P\psi) = \lambda^2 (\psi,\psi)$ (λ = const > 0), we have the "Hijazi inequality"

$$\lambda^2 \geqslant (n/4(n-1))\ \mu_1 \tag{7-4}$$

The theorem is of interest only if $\mu_1 > 0$. The inequality (7-4) has been obtained by Hijazi for the spectrum of the Dirac operator by a different method [8].

8 - THE LIMITING CASE.

Suppose $\mu_1 > 0$ and set $\lambda_1^2 = (n/4(n-1))\ \mu_1$ (λ_1 = const > 0).
a) Suppose that we have a spinor $\psi \not\equiv 0$ such that $(P\psi, P\psi) = \lambda_1^2 (\psi,\psi)$. We have $(\overline{P}\ \overline{\varphi}, \overline{P}\ \overline{\varphi}) = r_1^2\ (\overline{\varphi},\overline{\varphi})$ where $r_1 = \lambda_1 \exp(-\ c_1)$ satisfies

$$\left(\overline{R}_1/4\right) - ((n-1)/n)\ r_1^2 = 0 \tag{8-1}$$

It follows from (7-1) that $\overline{D}\ \overline{\varphi} = 0$. We have $\overline{u}$ = const according to (2-5) and according to (2-4) :

$$\overline{\Delta}\ \overline{\varphi} = r_1^2\ \overline{\varphi} \tag{8-2}$$

Introduce on $(W,\overline{g})$ the spinor $\overline{\chi} = \overline{P}\ \overline{\varphi}$ such that $(\overline{\chi},\overline{\chi}) = r_1^2\ \overline{u}$. It follows from (8-2) that $(\overline{P}\ \overline{\chi}, \overline{P}\ \overline{\chi}) = r_1^2\ (\overline{\chi},\overline{\chi})$. We can apply (2-2) to $(W,\overline{g})$ and to $\overline{\chi}$. We obtain according to (8-1) :

$$(1/2)\ \overline{\Delta}\left(r_1^2\ \overline{u}\right) + (1/2)\ \overline{\delta}\ \overline{T}(\overline{\chi}) = -\ (\overline{D}\ \overline{\chi}, \overline{D}\ \overline{\chi})$$

and we have by integration $\overline{D}\ \overline{\chi} = 0$. It follows from (8-1) and (2-5) that $r_1^2\ \overline{u}$ = const and so *r_1 = const, c_1 = const*. By the homothety corresponding to c_1 we deduce from $\overline{u}$ = const that u = const, from $\overline{D}\ \overline{\varphi} = 0$ that $D\psi = 0$. According to (2-5) and (2-4) :

$$\Delta \psi - \lambda_1^2\ \psi = (P - \epsilon\ \lambda_1)(P + \epsilon\ \lambda_1)\ \psi = 0 \tag{8-3}$$

where $\epsilon = \pm 1$. Choose ϵ so that $\alpha = (P + \epsilon\lambda_1)\psi \not\equiv 0$ and set $\nu_1 = \epsilon\lambda_1$. We have $(P - \nu_1)\alpha = 0$.

b) Study the spinors α such that $P\alpha = \nu_1\alpha$. It follows from a that $D\alpha \neq 0$, that is

$$\nabla\alpha + (\nu_1/n)\,\gamma\,\alpha = \nabla^{(\nu_1)}\alpha = 0$$

Lemma (Hijazi). On a compact spin manifold of dimension $n \geqslant 3$, every spinor α satisfying $P\alpha = \nu_1\alpha$ where $\nu_1^2 = (n/4(n-1))\,\mu_1 \neq 0$ is a non trivial Killing spinor [8].

It follows for the limiting case of the inequality (7-4) :

Theorem 4 - Let (W,g) be a compact spin manifold of dimension $n \geqslant 3$. If $\psi \not\equiv 0$ is a spinor field of (W,g) such that $(P\psi, P\psi) = \lambda_1^2(\psi,\psi)$, where $\lambda_1^2 = n/4(n-1)\,\mu_1 \neq 0$, (W,g) admits a Killing spinor of the form $(P + \epsilon\lambda_1)\psi$ $(\epsilon = \pm 1)$ corresponding to the eigenvalue $\nu_1 = \epsilon\lambda_1$. In particular (W,g) is an Einstein space.

If $P\psi = \lambda_1 e^{i\theta}\psi$, ψ itself is a Killing spinor.

9 - STUDY OF CASES WHERE $D\psi = 0$.

a) Let (W,g) be a compact spin manifold of *constant scalar curvature $R \neq 0$*. If $\psi \not\equiv 0$ is such that $D\psi = 0$, we have $\Delta\psi = \rho\psi$ and ρ is > 0. We have $R = \mu_1$ and so $\rho = \lambda_1^2$. Therefore ψ satisfies (8-3).

Proposition 7 - Let (W,g) be a compact spin manifold with a constant scalar curvature $R \neq 0$. If $\psi \not\equiv 0$ satisfies $D\psi = 0$, (W,g) admits a Killing spinor.

b) Let (W,g) be a compact spin manifold admitting a spinor $\psi \not\equiv 0$ such that

$$\nabla^{(\lambda,\Lambda)}\psi = \nabla\psi + (\lambda/n)\,\gamma\,\Lambda\,\psi = 0 \qquad (9\text{-}1)$$

which implies $P\psi = \lambda\Lambda\psi$. We have

$$(n/2)\ \Delta u = (\rho - \lambda^2)\ u \tag{9-2}$$

Applying the formula (2-2) to $\chi = P\ \psi$ and evaluating $T(\chi)$ in terms of $T(\psi)$, we have

$$(1/2)(\lambda^2 + \rho)\ \Delta u - (n/2)\ \nabla^\alpha \rho\ \nabla_\alpha u = -\ (D\chi, D\chi) \tag{9-3}$$

Let M be the maximum of u on W and S the largest set where u = M. At $x \in S$ we have $\Delta u \geqslant 0$ and so $\rho \geqslant \lambda^2$. At x we have $\lambda^2 + \rho \geqslant 2\ \lambda^2 > 0$. The left hand side of (9-3) defines on W an operator on the functions which is elliptic on an open domain U of W containing x; it follows from (9-3) that u = M on U and S is an open subset for W; S being both open and closed in W coincides with W. We get from (9-2) that $\rho \doteq \lambda^2 = \text{const}$ and $\Delta\psi = \lambda_1^2\ \psi$.

Proposition 8 - Let (W,g) be a compact spin manifold of dimension $n \geqslant 3$. If $\psi \not\equiv 0$ is such that $\nabla^{(\lambda,\wedge)}\ \psi = 0$ we have $\lambda = \lambda_1$ and (W,g) admits a Killing spinor.

10 - EXAMPLE

Introduce on (W,g) a real 1-form A and a real scalar m. Suppose that there is spinor $\psi \not\equiv 0$ satisfying

$$P\ \psi = i\ A_\alpha\ \gamma^\alpha \wedge \psi + i\ m \wedge \psi \tag{10-1}$$

An elementary computation gives $(P\psi, P\psi) = (|A|^2 + m^2)(\psi,\psi)$, where $|A|^2 = A^\alpha .\ A_\alpha$. From Theorems 3 and 4, we deduce :

Corollary - Let (W,g) be a compact spin manifold of dimension $\geqslant 3$ admitting a spinor field $\psi \not\equiv 0$ satisfying (10-1). Then

1) Either there is a point of W such that $(|A|^2 + m^2) > \lambda_1^2$ at this point.

2) Or we have everywhere $(|A|^2 + m^2) = \lambda_1^2 = \text{const}$ and (W,g) admits a Killing spinor.

REFERENCES

[1] *E. Cremmer, B. Julia, J. Scherek*, Phys. Lett 76 B, p. 409 (1978).

[2] *M.J. Duff, C.N. Pope* in *Supergravity 81* (Ed. S. Ferrara, J.G. Taylor) Cambridge Univ. Press London 1982.

[3] *L.P. Hughston, R. Penrose, P. Sommers, M. Walker*, Comm. Math. Phys. 27, p.303 (1982).

[4] *B. Biran, F. Englert, B. de Wit, H. Nicolai*, Phys. Lett. 124 B, p. 45 (1983).

[5] *P. Candelas, G.T. Horowitz, A. Strominger, E. Witten*, Nucl. Phys. B 258, p.46 (1985).

[6] *A. Lichnerowicz*, C.R. Acad. Sci. Paris A-B, 257, p.7 (1963).

[7] *T. Friedrich*, Math. Nach 97, p.117 (1980).

[8] *O. Hijazi*, A conformal lower bound for the smallest eigenvalue of the Dirac operator and Killing spinors. Comm. Math. Phys. 104, 151, (1986).

[9] *M. Cahen, S. Gutt, L. Lemaire, P. Spindel*, Killing spinors, Bull. Soc. Math. Belg. 38 A (1986) (to appear).

[10] *J.L. Kazdan, F.W. Warner*, J. of Diff. Geom. 10, p.113 (1975).

[11] *T. Aubin*, The scalar curvature in *Differential Geometry and Relativity* (Ed. M. Cahen and M. Flato) Reidel Dordrecht p.5-18 (1976).

[12] *A. Lichnerowicz*, C.R. Acad. Sci. Paris 303, série I, p. 811 (1986) and Variétés spinorielles et universalité de l'inégalité de Hijazi, C. R. Acad. Sci. Paris 304, série I, p.227 (1987).

[13] *H. Baum*, Spin Struktures, p.138 Teubner, Leipzig (1981).
T. Branson and Y. Kossmann-Schwarzbach, Lett. Math. Phys. 7, p.138, (1983).

TOWARDS A RENORMALIZABLE THEORY OF QUANTUM GRAVITY*

Yuval Ne'eman[1,2]
Sackler Faculty of Exact Sciences, Tel Aviv University,
Tel Aviv, Israel
and
Djordje Šijački
Institute of Physics, P O Box 57, Belgrade, Yugoslavia

ABSTRACT. The difficulties in quantizing gravity are due to the non-Lie-algebra nature of the gauged translation operators and to the dimensional nature of the Newton constant. The first issue may be dealt with by gauging the infinite Lie algebra of the diffeomorphisms. The second is solved if gravity is represented by a gl(4,R) gauge with dimensionless coupling, with Einstein's theory and Newton's coupling resulting from spontaneous symmetry breakdown (i.e. a low energy effective theory).

1. Gauging Diffeomorphisms as Parallel-transport Translations

Renormalizable relativistic quantum field theories (RQFT) consist in (a) scalar or pseudoscalar interactions and (b) local gauge theories, either abelian (QED) or Yang-Mills (YM) type (QCD). A combination between the two classes results in the spontaneously-broken YM model with massive gauge bosons, as exemplified by the electroweak U(2) theory. Einstein's gravity though twice in the nature of a local gauge theory ((a) Lorentz SO(1,3) on the tetrad or anholonomic frames; (b) holonomic covariance Diff(4,**R**)) defeats these treatments [1]. One way of understanding the difficulties is to follow the soft-group-manifold (SGM) approach [2]. The effective gauge resulting from spontaneous fibrations [3] does include a conventionally fibered

* Supported in part by the US DOE Grant DE-FG05-85ER40200 and by RZNS (Belgrade)
1 Wolfson Chair Extraordinary in Theoretical Physics
2 Also on leave from University of Texas, Austin, TX 78712, USA

K. Bleuler and M. Werner (eds.), Differential Geometrical Methods in Theoretical Physics, 333–343.

homogeneous part $\overline{SO}(1,3)$, a Lie algebra; however, it also includes the parallel-transport curved-space translations as given by the anholonomic covariant derivatives,

$$\partial_\mu \to D_a = b^\mu_a D_\mu = b^\mu_a (\partial_\mu - \Gamma_\mu^{[c}{}_{d]}\{\Lambda_c{}^d\}) \tag{1}$$

($\{\Lambda_c{}^d\}$ are the matrix representations of $J_c{}^d$, the $\overline{SO}(1,3)$ generators, $b_a{}^\mu(x)$ is the inverse-tetrad). The commutation relations [4] are the same as those of the Poincaré algebra, except for those between "translations",

98789 $$[D_a, D_b] = R_{ab}{}^{cd}\{\Lambda_{cd}\} + R_{ab}{}^c D_c \tag{2}$$

These are the "anholonomized general co-ordinate transformations" (AGCT) of ref. [2] . Mathematically, these are Lie derivatives [3]. The overall curved-modified (non-Lie) algebra will be denoted mso(1,3) and it goes over into the Poincaré algebra iso(1,3) in the limit when (1) is ∂_μ i.e. in the flat limit of the Equivalence Principle. Equation (2) represents the key obstacle to renormalizability in the Hamiltonian approach, for instance [5] . When Gravity is embedded in Supergravity, we have [2] in addition to eq. (2) non-vanishing curvature terms for $[D_\alpha, D_\beta]_+$ and $[D_\alpha, D_b]_-$ where D_α are the "super"parallel transports. In this case, the method of "auxiliary fields" [6] redefines the structure function component on the r.h.s. of (2) as additional fields in the superalgebra multiplet. One thus recovers a Lie superalgebra structure, with a somewhat more complicated Lagrangian. An alternative approach to the replacement of eq. (1) by a Lie algebra could consist in gauging the infinite Lie algebra of $\overline{\mathrm{Diff}}(4,R)$. The method has been applied in recent work on the string and superstring [7]. However, for D>2, new difficulties arise, but the method may still be useful. It would amount to gauging $\overline{\mathrm{Diff}}(4,R)$ anholonomically and then breaking the symmetry first to $\overline{\mathrm{MSL}}(4,R)$ and finally to $\overline{\mathrm{MSO}}(1,3)$. The linear representations of $\overline{\mathrm{Diff}}(D,R)$ are not known for D>2, but Ogievetsky has shown (8) that the $\overline{\mathrm{Diff}}(4,R)$ algebra is given by the algebraic closure of [conformal (1,3), $s\ell(4,R)$], where conformal (1,3) = $so(2,4)$.

2. Difficulties with the Dimensionality of Newton's Constant

For a Yang-Mills or other Lie-algebra-valued connection $A = A_\mu{}^i\{\lambda_i\}dx^\mu$ (such as the spin-connection) with dimensionless algebraic generators λ_i, the dimension $\dim(A_\mu{}^i) = 1$ so as to match that of the spatial derivative in the expression for the covariant derivative $D_\mu = \partial_\mu - A_\mu{}^i\{\lambda_i\}$. However, for diffeomorphisms, or in gauging translations with the metric $g_{\mu\nu}(x)$ or tetrad $e_\mu{}^a(x)$ as gauge fields, $\{\lambda_i\}$ is itself replaced by ∂_μ or D_μ. Thus the inverse tetrad $b_a{}^\mu(x)$ acts as a dimensionless multiplier in (1). The dimensionless nature of the metric field $g_{\mu\nu}(x)$, essential to its kinematical role via the Equivalence Principle, is at the bottom of most of the difficulties in the quantization procedure. On the one hand, although the Einstein (or Einstein-Cartan) Lagrangian $\varepsilon_{abcd} R^{ab} \wedge e^c \wedge e^d$ involves a 4-form, its dimension is 2, forcing the Newtonian constant to have dimension 2, thus causing the need for an infinite number of counter-terms for that Lagrangian. Moreover, this scale dependence imposes on the theory a requirement of finiteness, rather than renormalizability, an extremely strong condition. On the other hand, when terms quadratic in the curvature are added, since the connection is not an indpependent field and has the form ∂g, we get contributions $\sim (\partial g)^4$. $(\partial^2 g)^2$ or $(\partial g)^2(\partial^2 g)$. Normally, since renormalizability implies cut-off independence (= scale invariance) and a dimensionless action, terms of order 4 in the Lagrangian should dominate the high-energy region. We indeed revert, hereforth, to renormalizability rather than finiteness in the solution we offer. We assume that terms of order 4 dominate the high-energy region and provide the key to a renormalizable theory, with the terms of dimension 2 dominating the low-energy and macroscopic regime; this scheme [9] is similar to YM with spontaneous symmetry breakdown, where the dim.4, F^2 and ϕ^4 terms dominate at high energy, and the $\mu^2 \phi^2$ at low-energy. However, the R^2 terms in gravity involve as we noted p^4 terms, of the form $\{p^2(p^2-m^2)\}^{-1} = m^{-2}\{(p^2-m^2)^{-1} - (p^2)^{-1}\}$ i.e. a wrong sign for one of the poles, a ghost. This may be remedied by adding terms that would make the connection Γ independent of the metric and produce an analog to YM. A recent calculation has, however, shown that the most general Lagrangian quadratic in Poincaré curvature and torsions, when linearized and having selected the coefficients of the various Poincare irreducible pieces so as to remove all ghosts, tachyons, frozen states and p^4 ghosts, ends up containing nothing but the original massless graviton [10]. (Torsion)^2 terms are of dimension 2 and no help in this matter. Moreover, as long as the theory stays Riemannian, the metricity condition

$$D_\lambda\, g_{\mu\nu} = 0 \tag{3}$$

makes the spin-connection depend on the metric (and matter fields).

The solution we suggest is to gauge the entire GL(4,R) instead of the Lorentz group, for the homogeneous part. The theory is then affine and non-metric at high energy, with eq. (3) emerging as a a low-energy result [9]. Similar answers seem indicated [11] in the Hamiltonian method of quantization of Gravity and in yet another approach [12].

3. General Affine Gauge Theory

We gauge [13-15] the general affine group $\overline{GA}(4,R) = T_4 \circledS \overline{GL}(4,R)$, a semidirect product of translations and the double-covering $\overline{GL}(4,R)$ of the general linear group $GL(4,R)$, generated by Q_{ab}. Here $\overline{GL}(4,R) = R_+ \otimes \overline{SL}(4,R) \supset R_+ \otimes SO(1,3)$, where R_+ is the dilation subgroup. A general affine metric g_{ab} transforms as a $\underset{\sim}{10}$ under global $GL(4,R)$: $\{g_{ab}\} = \{A_a{}^c A_b{}^d \eta_{cd} \mid , A_a{}^b \in GL(4,R)\}$; the Minkowski metric η is defined in a given "flat" $\overline{GL}(4,R)$ gauge and is a Lorentz-subgroup invariant only. Raising and lowering $\overline{GL}(4,R)$ indices with g_{ab} thus corresponds to the usual result when one goes to the flat gauge. The antisymmetric operators $Q_{[ab]} = (Q_{ab}-Q_{ba})$ generate the Lorentz subgroup $SO(1,3)$, the symmetric traceless operators (shears) $Q_{ab} = (Q_{ab} + Q_{ba}) - g_{ab} Q_c{}^c$ generate the proper 4-volume-preserving deformations, while the trace $Q = Q^a{}_a$ generates scale-invariance R_+. $Q_{[ab]}$ and $Q_{(ab)}$ generate together the $\overline{SL}(4,R)$ group. The gauge potentials are the tetrads $e^a{}_\mu$ and connections $\Gamma^a{}_{b\mu}$. The antisymmetric $\Gamma_{[ab]\mu}$, traceless symmetric $\Gamma_{(ab)\mu}$ and the trace $\Gamma^a{}_{a\mu}$, parts correspond, respectively, to the local Lorentz, shear and dilation transformations. The corresponding field strengths are the torsion $R^a{}_{\mu\nu} = \partial_\mu e^a{}_\nu + \Gamma^a{}_{b\mu} e^b{}_\nu - (\mu \leftrightarrow \nu)$, and generalized "curvature"

$R^a{}_{b\mu\nu} = \partial_\mu \Gamma^a{}_{b\nu} + \Gamma^c{}_{b\nu} \Gamma^a{}_{c\nu} - (\mu \leftrightarrow \nu)$. Let ϕ be a generic matter field with a global-general-affine matter Lagrangian $L_m(\phi,\partial\phi)$. The total gauge-invariant Lagrangian reads $L = L_m(\phi,\partial\phi,e,\Gamma) + L_g(e,\partial e,\Gamma,\partial\Gamma)$, where Γ enters L_m through the covariant derivative $D_\mu = \partial_\mu - i\Gamma^c{}_{d\mu} Q^d{}_c$. Variation with respect to ϕ yields the matter-field equations $\delta L_m/\delta\phi = 0$. Variation with respect to $e^a{}_\mu$ and $\Gamma^a{}_{b\mu}$ implies two (gravity) gauge field equations with the corresponding momentum and hypermomentum (angular momentum and deformation) currents as sources

$$-\delta L_g/\delta e^a{}_\mu = e\theta^\mu{}_a = \delta L_m/\delta e^a{}_\mu; \; -\delta L_g/\delta \Gamma^a{}_{b\mu} = e\Upsilon^{b\mu}{}_a = \delta L_m/\delta \Gamma^a{}_{b\mu} \tag{4}$$

where $e = \det(e^a{}_\mu)$ and (quasi) conservation laws:

$$D_\mu(e\theta_a{}^\mu) = e\theta_b{}^\mu R^b_{a\mu} + e\Upsilon^{d\mu}_c R^c_{da\mu} \; ; \; D_\mu(e\Upsilon^b{}_a{}^\mu) = - e^b_\mu \, e\theta^\mu{}_a \tag{5}$$

The two gravity equations can be rewritten in the form

$$D_\nu \pi^{\mu\nu}_a - \varepsilon^\mu{}_a = \varepsilon\theta^\mu{}_a; \; D_\nu \pi^a{}_b{}^{\mu\nu} - \varepsilon^a{}_b{}^\mu = e\Upsilon^a{}_b{}^\mu \tag{6}$$

where

$$\pi_a^{\ \mu\nu} = \partial L_g/\partial\partial_\nu e^a_{\ \mu} = 2\partial L_g/\partial R^a_{\ \nu\mu} \ , \ \pi^a_{\ b}{}^{\mu\nu} =$$

$$= \partial L_g/\partial\partial_\mu \Gamma^b_{\ a\mu} = 2\partial L_g/\partial R^b_{\ a\mu\nu} \ , \tag{7}$$

$$\varepsilon_a^{\ \mu} = e_a^{\ \mu} L_g - R^b_{\ a\nu} \pi_b^{\ \nu\mu} - R^b_{\ ca\nu} \pi^c_{\ b}{}^{\nu\mu} \ ,$$

$$\varepsilon^a_{\ b}{}^{\mu} = e^a_{\ \nu} \pi_b^{\ \nu\mu} \ . \tag{8}$$

4. $\overline{GA}(4,R)$ or $\overline{SA}(4,R)$ Matter Fields

The $\overline{SA}(4,R)$ unirreps [16-18] are induced from the corresponding little group unirreps. The little group turns out to be $SA(3,R)^{\sim} = \tilde{T}_3 \circledS SL(3,R)$ and thus we have the following possibilities:
(i) The whole little group is represented trivially, and we have a scalar state, which corresponds to a scalar field $\varphi(x)$, invariant under both $\overline{SL}(4,R)$ and $SL(2,C)$. We exclude scale invariance so that $\varphi(x)$ is invariant only under $\overline{SA}(4,R)$.
(ii) $\tilde{T}_3$ is represented trivially, and the corresponding states are described by the SL(3,R) unirreps, which are infinite-dimensional owing to the $\overline{SL}(3,R)$ noncompactness. The corresponding $\overline{SL}(4,R)$ matter fields are, therefore, necessarily infinite-dimensional and when reduced w.r.t. the $\overline{SL}(3,R)$ subgroup should transform w.r.t. its unirreps.
(iii) The little group $\overline{SA}(3,R)^{\sim}$ is represented nontrivially, and we find the states which are characterized "effectively" by three real numbers in addition to the $\overline{SA}(2,R)^{\sim}$ unirreps.

Finally, for quarks or leptons, we make use of the $\overline{GA}(4,R)$ nonlinear representations [19] which are realized through g_{ab}. The stability subgroup is SL(2,C), and the representations are linear for the Poincaré subgroup.

5. Deunitarizing Automorphism

Had the whole $\overline{SL}(4,R)$ been represented unitarily, the Lorentz boost generators would have a hermitian intrinsic part; as a result, when boosting a particle, one would obtain another particle - contrary to experience. There exists, however, a remarkable inner "deunitarizing" automorphism A [17,20], which leaves the $R_+ \otimes \overline{SL}(3,R)$ subgroup intact, and which maps the $Q_{(ok)}$, $Q_{[ok]}$ generators into $iQ_{[ok]}$, $iQ_{(ok)}$ respectively (k=1,2,3). The deunitarizing automorphism allows us to start with the unitary representations of the $\overline{SL}(4,R)$ group, and upon its application, to identify the finite (unitary) representations of the abstract $\overline{SO}(4,R)$ compact subgroup with nonunitary representations of the physical Lorentz group, while the infinite (unitary) representations of the abstract SO(1,3) group now represent (non-unitarily) the compact $\overline{SO}(4)/\overline{SO}(3)$ generators. The non-hermiticity of the intrinsic boost operators cancels their "intrinsic" physical action precisely as in finite tensors or spinors, the boosts thus acting kinetically only. In this way, we avoid a disease common to infinite-component wave equations.

6. $\overline{SL}(4,R)$ Wave Equations

In previous applications of $\overline{SL}(4,R)$ manifields [21] we have used manifields transforming with respect to $\overline{SL}(4,R)$ multiplicity-free representations and satisfying Lorentz-invariant wave equations. It has been shown [22] that such $\overline{SL}(4,R)$ representations cannot satisfy $\overline{SL}(4,R)$ invariant equations.

For $\overline{SL}(4,R)$-invariant wave equations [23] and the corresponding Lagrangians we have to make use of $\overline{SL}(4,R)$ unirreps with non-trivial multiplicity. All these spinorial and tensorial representations have been classified [18]. Let us denote a generic $\overline{SL}(4,R)$ unirrep by $D(c,d,e;(j_1,j_2))$ where c,d,e are the representation labels, and (j_1,j_2) denote the lowest $\overline{SO}(4)=SU(2)\otimes SU(2)$ representation contained in the given $\overline{SL}(4,R)$ representation.

For the $\overline{SL}(4,R)$ tensorial field we take an infinite-component field ϕ which transforms w.r.t. an A-deunitarized unirrep belonging to the Principal series of representations: $D^{pr.}_{SL4R}(c_2,d_2,e_2:(00))$, $c_2,d_2,e_2\varepsilon R$. The manifield ϕ obeys a Klein-Gordon-like equation

$$(g^{ab}\ell_a\partial_b + M^2)\phi(x) = 0 \tag{9}$$

For the $\overline{SL}(4,R)$ spinorial fields we take an infinite-component field ψ which transforms w.r.t. an A-deunitarized unirrep belonging to the Principal series of representations:

$$D^{pr.}_{\overline{SL}(4,R)}(c_2,d_2,e_2;(\tfrac{1}{2},0)) + D^{pr.}_{\overline{SL}(4,R)}(c_2,d_2,e_2;(0,\tfrac{1}{2})),$$

$c_2,d_2,e_2\varepsilon R$,

while $(\frac{1}{2},0)$ and $(0,\frac{1}{2})$ denote parity conjugated spinorial representations. The manifield ψ satisfies a Dirac-like equation

$$(ig^{ab}X_a\partial_b - M)\ \psi\ (x) = 0, \tag{10}$$

where X_a is an $\overline{SL}(4,R)$ 4-vector acting in the space of our spinorial manifield. We construct X_a in the following way: first we embed SL(4,R) into SL(5,R), and then select a pair of (mutually conjugate) Principal series representations which contain in the $\overline{SL}(4,R)$ reduction our spinorial representations. Let the $\overline{SL}(5,R)$ generators be $Q_{\hat{a}\hat{b}}$, $\hat{a}\hat{b} = 0,1,2,3,5$. We define $X_a = O_{[5a]}$, $a = 0,1,2,3$ and thus arrive at the sought-for $\overline{SL}(4,R)$ 4-vector. The transition to the $\overline{GL}(4,R)$ symmetry is straightforward. Note that the X_a operators constructed in this way yield upon commutaiton the SL(2,C) generators (generalizing a Dirac γ-matrices property).

7. General Affine Lagrangian

As basic fields we take $e^a{}_\mu(X), \Gamma^a{}_{b\mu}(X), \Psi(X), \phi(X)$, a scalar field $\varphi(x)$ as well as $\overline{GA}(4,R)$-nonlinearly-realized SL(2,C) Dirac field $\xi(X)$.

The most general first-order gauge field Lagrangian takes the form

$$L_g = L_g(e^a{}_\mu, R^a{}_{\mu\nu}, R^a{}_{b\mu\nu}) = eL_g(R^a{}_{cd}, R^a{}_{bcd}).$$

If it contains at most second derivatives of the basic field variables, it consists of terms quadratic in torsions and curvatures and of the scalar curvature. Owing to the dimensional coupling constants, the scalar curvature and the $(\text{torsion})^2$ terms are not allowed by dilation invariance. We circumvent this fact by making use, besides the well known $\varphi^2 R$ terms, of a $\varphi^2(\text{torsion})^2$ term [15]. We make use of all possible torsion and curvature irreducible tensor operators with respect to the $\overline{GL}(4,R)$ group [16]. For anholonomized indices, there are two, respectively 4- and 20-dimensional, irreducible torsion pieces $R_{[acd]}$ and $R_{(a[c)d]}$, three, respectively 1-, 15- and 20-dimensional, irreducible Lorentz curvature pieces $R_{[abcd]}$, $R_{[ab[c]d]}$, and $R_{[a(b][c)d]}$, and two, respectively 15- and 45-dimensional, irreducible shear+dilation curvature pieces $R_{(a[b)cd]}$ and $R_{(ab[c)d]}$. As a result

$$L(e^a{}_\mu, \partial_\nu e^a{}_\mu, \Gamma^a{}_{b\nu}, \partial_\nu \Gamma^a{}_{b\mu}, \psi_A, \partial_\nu \psi_A, \phi_A, \partial_\nu \phi_A, \varphi, \partial_\nu \varphi, \xi, \partial_\nu \xi) =$$

$$= e\{\varphi^2 R - \frac{1}{4\alpha_1}\varphi^2 R_{[acd]}R^{[acd]} - \frac{1}{4\alpha_2}\varphi^2 R_{(a[c)d]}R^{(a[c)d]}$$

$$- \frac{1}{4\beta_1} R_{[abcd]}R^{[abcd]} - \frac{1}{4\beta_2} R_{[ab[c]d]} R^{[ab[c]d]} -$$

$$- \frac{1}{4\beta_3} R_{[a(b][c)d]} R^{[a(b][c)d]} - \frac{1}{4\kappa_1} R_{(a[b)cd]} R^{(a[b)cd]} -$$

$$- \frac{1}{4\kappa_2} R_{(ab[c)d]} R^{(ab[c)d]} \tag{11}$$

$$+ \bar{\psi} i g^{ab} \chi_a e_b{}^\mu D_\mu \psi + \frac{1}{2} g^{ab} e_a^\mu e_b^\nu ((D_\mu \phi)^+ (D_\nu \phi))$$

$$+ \frac{1}{2} g^{ab} e_a^\mu e_b^\nu \partial_\mu \varphi \partial_\nu \varphi + \bar{\xi} i \eta^{ab} \gamma_a e_b^\mu \Delta_\mu \xi$$

$$+ \mu \bar{\psi} \phi \psi - \lambda_\phi (\phi^+ \phi)^2 - \lambda (\phi^+ \phi) \varphi^2 - \lambda_\varphi \varphi^4\}$$

R is the General-affine scalar curvature, $\alpha_1, \alpha_2, \beta_1, \beta_2, \beta_3, \kappa_1, \kappa_2, \mu, \lambda_\phi, \lambda, \lambda_\varphi$ are dimensionless constants, D_μ is a $\overline{GL}(4,R)$ covariant derivative and Δ_μ is the $\overline{GL}(4,R)$ nonlinear covariant derivative for the SO(1,3) Dirac field ξ.

8. Symmetry Breaking

We now consider a breaking of the $\overline{GA}(4,R)$ which preserves the Poincaré symmetry. In the case of a Weinberg-Salam type theory, the Yang-Mills action is scale-invariant, so that one starts with a Lagrangian invariant w.r.t. a direct product of an internal U(2) group with the dilation group. Introducing for the Higgs-Goldstone field an explicit $(\text{mass})^2$ term which breaks scale-invariance, and using the wrong sign, one triggers a spontaneous breakdown of the internal gauge symmetry. Thus "spontaneous" breakdown of the internal symmetry is really achieved with the help of an explicit breaking of the "external" dilation group. In our case, however, the dilation symmetry is one of the key-stones of the $\overline{GA}(4,R)$ gauge field theory and the symmetry breaking has to be more subtle. We find that the deunitarizing automorphim is of crucial importance for the $\overline{SL}(4,R)$ symmetry breaking. Had we started with a manifield ϕ which transforms w.r.t. an $\overline{SL}(4,R)$ unirrep, we would not have a Lorentz symmetry preserving tadpole for this field since the representations of the Lorentz subgroup in the reduction are infinite-dimensional. However, for our A-deunitarized $\overline{SL}(4,R)$ representations the physical Lorentz subgroup is represented nonunitarily and our tensorial manifield ϕ can be chosen so that the lowest component transforms as a scalar (0,0) Lorentz representation.

We now list a few possible $\overline{GL}(4,R)$ symmetry breaking scenarios.
(i) We take both $\phi_A(x) = v_\phi \delta_{Ao} + \Xi_A(x)$, the index o in δ_{Ao} denoting the (0,0) Lorentz-scalar level and $\varphi(x) = v_\varphi + \xi(x)$, where $v_\phi = \langle 0 | \phi(x) | 0 \rangle \neq 0$ and $v_\varphi = \langle 0 | \varphi(x) | 0 \rangle \neq 0$. As a result there emerge mass terms for the manifields ψ, ϕ for φ and for the $\Gamma_{(ab)\mu}$, and $\Gamma^a{}_{a\mu}$ connection fields and Newton's constant is also generated.
(ii) We take only $\phi_A(x) = v_\phi \delta_{Ao} + \Xi_A(x)$, where $v_\phi = \langle 0 | \phi(x) | 0 \rangle \neq 0$. By doing this we break explicitly at the same time both the dilation and the $\overline{SL}(4,R)/\overline{SO}(1,3)$. We now get

$$v(\varphi) = \lambda v_\phi^2 \varphi^2 + \lambda_\varphi \varphi^4,$$

and provided $\lambda < 0$ we find a "spontaneous" nontrivial v.e.v. for the $\varphi(x)$ field, $v'_\varphi = (-\lambda/2\lambda_\phi)^{1/2} v_\phi$, and recover the results of case (i), however, now in terms of only one parameter v_ϕ.
(iii) Finally, we consider the most elegant possibility, namely to break explicitly only the dilation subgroup of the $\overline{GL}(4,R)$ group: $\varphi(x) = v_\varphi + \xi(x)$, $v_\varphi = \langle 0 | \varphi(x) | 0 \rangle \neq 0$, and we find for the Planck length, i.e. the Newton constant

$$\ell_{P\ell} = 1/(\sqrt{2}v_\varphi), \quad G_N = 1/(16\pi v_\varphi^2). \tag{12}$$

The tensorial manifield potential reads

$$v(\phi) = \lambda v_\varphi^2 \phi^2 + \lambda_\phi (\phi^2)^2 \ , \text{ where } \phi^2 = \Sigma_A \phi_A^* \phi^A \ , \quad A = (j_1, j_2).$$

This potential has a minimum provided

$$(\lambda v_\varphi^2 + 2\lambda_\phi (\Sigma_B \phi_B^* \phi^B))\ \phi^A = 0, \tag{13}$$

and if $\lambda < 0$ we are in a situation of a spontaneous symmetry breakdown of the $\overline{SL}(4,R)$ group. Our tensorial manifield transforms as $D^{pr}_{\overline{SL}4R}(c_2, d_2, e_2;\ (0,0))$ which does not contain in its decomposition an $\overline{SL}(3,R)$ scalar. On account of A, the only subgroup scalar is the Lorentz scalar $(0,0)$. Applying an appropriate $\overline{SL}(4,R)/\overline{SO}(1,3)$ transformation (within an irreducible space) we have $\Sigma_B \phi_B^* \phi^B = \phi_0 \phi^0$, and obtain

$$\phi_A(x) = v'_\phi \delta_{A0} + \Xi_A(x), \quad v'_\phi = (-\lambda/2\lambda_\phi)^{1/2} v_\varphi \tag{14}$$

The spinorial manifield mass becomes

$$M(\Psi) = \mu v'_\phi \tag{15}$$

$(0,0)$ is the lowest component of ϕ; at the next level we find three representations: $(2,0)$, $(0,2)$ and $(1,1)$. The first two of these cannot be obtained from $(0,0)$ via a single Lie algebra application, while $|(1,1)\rangle = Q_{(a,b)}|(0,0)\rangle$, i.e. $(1,1) \sim \overline{SL}(4,r)/\overline{SO}(1,3)$. Thus the $(1,1)$ components of ϕ are the Nambu-Goldstone fields which become the longitudinal components of the shear connections. By making use of a gauge transformation $\exp(-iQ^{(ab)}\ \Xi_{(ab)}(x))$, $\Xi_{(ab)} \sim (1,1)$ we find

$$\overline{\Gamma}_{(ab)\mu} = \Gamma_{(ab)\mu} - \partial_\mu \Xi_{(ab)}$$

The shear potential mass matrix is given by $\langle Q_{(ab)} v_\phi | Q_{(cd)} v_\phi \rangle$ and the $(Q_{(ab)})^B_A$ matrix elements which contribute to it correspond to $(0,0)\to(1,1)\to(0,0)$. From ref. [18] we find, using 3-j coefficients

$$M^2(\Gamma_{(ab)\mu}) = \frac{1}{4}(16 + e_2^2) \begin{pmatrix} 0 & 1 & 1 \\ 0 & -a & a \end{pmatrix}^2 \begin{pmatrix} 0 & 1 & 1 \\ 0 & -b & b \end{pmatrix}^2 \left(v'_\phi\right)^2 \tag{16}$$

where (in the spherical basis) $a,b = +1,0,-1$. For the dilation potential mass we find

$$M^2(\Gamma^a{}_{a\mu}) = v_\phi'^2 \ . \tag{17}$$

Assuming the dimensionless constants to be of order 1, we find that v'_ϕ is of the order of the Planck mass $M_{P\ell}$, and thus all nonzero-mass modes have masses of order $M_{P\ell}$. However, other values of κ_1, κ_2 could diminish the mass of the shear connections.

At low energies or at distances $\ell \gg \ell_{P\ell}$ the potentials $\Gamma_{(ab)\mu}$ and $\Gamma^a_{a\mu}$ do not contribute and we obtain (3). Moreover, on account of $v'_\phi \neq 0$ the global $\overline{GL}(4,R)$ symmetry too is broken down to Lorentz invariance. Thus when the macroscopic gravitational field is "turned off" we arrive at the Minkowski space-time of special relativity - and not at an affine space-time as in the unbroken $\overline{SL}(4,R)$ case [23]. The Principle of Equivalence is thus effectively fulfilled and we recover Einstein's Gravity over macroscopic distances. At the same time, the Planck scale is a "soft" feature, and horizons occur macroscopically only. At very short distances, the theory is scaleless and potentially renormalizable.

References

[1] Goroff and Sagnotti, Phys. Lett. B160 (1985) 81.

[2] Y. Ne'eman and T. Regge, Phys. Lett. B74 (1978) 54; Rivista d. Nuovo Cim. 1, N5 (ser. 3, 1978).

[3] J. Thierry-Mieg and Y. Ne'eman, Ann. Phys. (NY) 123 (1979) 247. Y. Ne'eman, E. Takasugi and J. Thierry-Mieg, Phys. Rev. D22 (1980) 2371.

[4] F. W. Hehl, P. v.d. Heyde, G. D. Kerlick and J. M. Nester, Rev. Mod. Phys. 48 (1976) 393.

[5] P. A. M. Dirac, Phys. Rev. 114 (1959) 924.

[6] K. Stelleand and P. C. West, Nucl. Phys. B145 (1978) 175. P. K. Townsend and P. v. Nieuwenhiuzen, Phys. Rev. D19 (1979) 3592.

[7] M. J. Bowick and S. G. Rajeev, Phys. Rev. Lett. 58 (1987) 535; J. Mickelsson, MIT preprint CTP 1448 (1987).

[8] V. O. Ogievetsky, Lett. Nuovo Cim. 8 (1973) 988.

[9] L. Smolin, Nucl. Phys. B247 (1984) 511.

[10] R. Kuhfuss and J. Nitsch, Gen. Rel. and Grav. 18 (1986) 1207.

[11] A. Komar, Phys. rev. D30 (1984) 305; P. G. Bergmann and A. Komar, J. Math. Phys. 26 (1985) 2030.

[12] J. W. Moffat, J. Math. Phys. 25 (1984) 347.

[13] F. W. Hehl, G. D. Kerlick and P. v.d. Heyde, Phys. Lett. 63B (1976) 446.

[14] F. W. Hehl, E. A. Lord and Y. Ne'eman, Phys. Lett. 71B (1977) 432; Phys. Rev. D17 (1978) 428.

[15] Y. Ne'eman and Dj. Šijački, Ann. Phys. (N.Y.) 120 (1979) 292; Proc. Nat. Acad. Sci. (USA) 76 (1979) 561; Dj. Šijački, Phy. Lett. 109B (1982) 435.

[16] Dj. Šijački in "Frontiers in Particle Physics '83", eds. Dj. Šijački et al (World Scientific, Singapore 1984), p. 382.

[17] Dj. Šijački and Y. Ne'eman, J. Math. Phys. 26 (1985) 2457.

[18] Dj. Šijački, "$\overline{SL}(n,R)$ Spinors for Particles, Gravity and Superstrings', Conference on Spinors in Physics and Geometry, Trieste 1986;
Ann. Isr. Phys. Sc. 3 (1980) 35.
[19] V. O. Ogievetsky and I. V. Polubarinov, JETP 48 (1965) 1625.
C. J. Isham, A. Salam and J. Strathdee, Ann. Phys. (N.Y.) 62 (1971) 98.
[20] Y. Ne'eman and Dj. Šijački, Int. J. Mod. Phys. A2 (1987),1655.
[21] Y. Ne'eman and Dj. Šijački, Phys. Lett. 157B (1985) 267 and 275.
[22] A. Cant and Y. Ne'eman, J. Math. Phys. 26 (1985) 3100.
[23] J. Mickelsson, Comm. Mat. Phys. 88 (1983) 551.

GAUGE THEORY FOR DIFFEOMORPHISM GROUPS

Peter W. Michor
Institut für Mathematik der Universität Wien
Strudlhofgasse 4
A-1090 Vienna
Austria

ABSTRACT: We consider fibre bundles without structure group and develop the theory of connections, curvature, parallel transport, (nonlinear) frame bundle, the gauge group and it's action on the space of connections and the corresponding orbit space. We describe the classifying space and the rudiments of characteristic class theory and we give several self-duality conditions.

0. INTRODUCTION

Gauge theory is usually formulated in the context of a principal fibre bundle with structure group a compact Lie group, or some associated vector or fibre bundle. We start here with a fibre bundle $p: E \to M$ with standard fibre S without structure group, i.e. with structure group the whole diffeomorphism group Diff(S) of the standard fibre S. A (generalized) connection will be just a projection on the vertical bundle and it's curvature will be given with the help of the Frölicher-Nijenhuis bracket. The local theory of this point of view has been developped very far by Marco Modugno (1986), but he never considered parallel transport, which in general exists only (fibrewise) locally. If the standard fibre S is compact, parallel transport is globally defined, and on each fibre bundle there are connections with global parallel transport. We also give a semi-local description of connections via 'Christoffel forms' which allows to decide whether a given (generalized) connection is induced from a principal one for a given G-structure on the bundle.

Then we construct the holonomy group of a complete connection and also the holonomy Lie algebra. The holonomy group is a subgroup of the diffeomorphism group of some fibre; the holonomy Lie algebra is desribed by the curvature and the parallel transport, and if it turns out to be finite dimensional we can show that the holonomy group is a finite dimensional Lie group, the bundle bears a G-structure and the connection is induced from a principal one.

Next we construct the (nonlinear) frame bundle of a fibre bundle, whose standard fibre now has to be compact. This is a principal fibre bundle with structure group the diffeomorphism group of the standard

K. Bleuler and M. Werner (eds.), Differential Geometrical Methods in Theoretical Physics, 345–371.

fibre S, which is a Lie group modelled on nuclear Fréchet spaces. We use the differential geometry of manifolds of mappings and diffeomorphism groups as developped in Michor (1980): No Sobolev spaces are used. Any connection on the bundle E lifts to a principal connection on this frame bundle and the usual calculus of principal connections holds true. By lifting we see that any principal connection on this frame bundle admits a parallel transport (S compact is essential here).

The gauge group of the frame bundle coincides with the group of all diffeomorphism of E which respect fibres and cover the identity on M. This is a Lie group modelled on nuclear (LF)-spaces now; it acts smoothly on the nuclear (LF)-manifold of all connections on E and we take a brief glance on the space of orbits. I expect that it is stratified into smooth manifolds, each one corresponding to a conjugacy class of holonomy groups in the gauge group. Usual moduli spaces should be subsets of this corresponding to holonomy groups whose connected component of the identity sits in the conjugacy class of a fixed compact Lie group in the gauge group.

The classifying space of Diff(S) for compact S can be chosen as the manifold of all submanifolds of a Hilbert space of type S (a sort of nonlinear Grassmannian). It carries a universal S-bundle and even a universal connection (given by othogonal projection): Any connection on an S-bundle is pullback of this universal connection. Diff(S) contains the group of diffeomorphisms preserving some volume density as a smooth deformation retract, so on each bundle $E \to M$ we may choose a fibrewise volume form and connections respecting these volume forms. For these data the Chern-Weil construction works and gives a characteristic class in the second cohomology of M with twisted coefficients. These results have been obtained by Gerd Kainz in his doctoral dissertation (1988).

Finally we describe one type of Yang-Mills functional depending on a Riemannian metric on the base space M and a fibrewise symplectic structure, and we give several types of self-duality conditions.

Lots of open problems remain: The (infinite dimensional) differential geometry of the holonomy groups, the Ambrose Singer theorem if the holonomy Lie algebra is infite dimensional, the moduli space and it's stratification and the exploitation of self-duality.

1. CONNECTIONS ON FIBRE BUNDLES

1.1. Let (E,p,M,S) be a fibre bundle: E, M, S are smooth finite dimensional second countable manifolds, $p: E \to M$ is smooth and each x in M has an open neighbourhood U such that $p^{-1}(U) =: E|U$ is fibre respectingly diffeomorphic to $U \times S$. S is called the <u>standard fibre</u>.

The vertical bundle of (E,p,M,S), called $V(E)$, is the kernel of $Tp: TE \to TM$. It is a sub vector bundle of $TE \to E$.

A <u>connection</u> on (E,p,M,S) is just a fibre linear projection $\Phi: TE \to VE$; so Φ is a 1-form on E with values in the vertical bundle, $\Phi \in \Omega^1(E; VE)$. The kernel of Φ is called the <u>horizontal subspace</u> $\ker\Phi$. Clearly for each u in E the mapping $T_u p: (\ker\Phi)_u \to T_{p(u)}M$ is a linear isomorphism, so $(Tp, \pi_E): \ker\Phi \to TM \times_M E$ is a diffeomorphism, fibre linear in the first coordinate, whose inverse $\chi: TM \times_M E \to \ker\Phi \to TE$ is called the <u>horizontal lifting</u>. Clearly the connection Φ can equivalently be described by giving a horizontal sub vector bundle of $TE \to E$ or by specifying the horizontal lifting χ satisfying $(Tp,\pi_E)\,\chi = Id$ on $TM \times_M E$.

1.2. The <u>curvature</u> of a connetion $\Phi \in \Omega^1(E;VE)$ is most conveniently expressed with the help of the Frölicher-Nijenhuis bracket: see the appendix. The Frölicher-Nijenhuis bracket of Φ with itself turns out to be $[\Phi,\Phi](X,Y) = 2\,\Phi[(Id - \Phi)X, (Id - \Phi)Y]$ for vector fields X, Y in $\mathfrak{X}(E)$. This formula shows that $[\Phi,\Phi]$ is the obstruction against integrability of the horizontal bundle $\ker\Phi$. We put $R := [\Phi,\Phi]$ in $\Omega^2(E, VE)$, a 2-form on E with values in the vertical bundle, and we call R the <u>curvature</u>. It's relation with the usual curvature on principal bundles is explained in section 2.

By the graded Jacobi identity we have $[R,\Phi] = \frac{1}{2}[[\Phi,\Phi],\Phi] = 0$; this is the form of the <u>Bianchi identity</u> in our setting.

An immediate consequence of the naturality of the Frölicher-Nijenhuis bracket (see appendix) is the following: Let $f: N \to M$ be a smooth mapping. Consider the pullback bundle (f^*E, f^*p, N, S) :

$$\begin{array}{ccc} f^*E & \xrightarrow{\;p^*f\;} & E \\ \downarrow{\scriptstyle f^*p} & & \downarrow{\scriptstyle p} \\ N & \xrightarrow{\;f\;} & M \end{array}$$

Since p^*f is fibrewise a diffeomorphism, $(f^*\Phi)_u(X) = T_u(p^*f)^{-1}\Phi\, T_u(p^*f)X$ is well defined for X in $T_u(f^*E)$ and a connection Φ on E and we get in this way a connection $f^*\Phi$ on f^*E. Clearly $f^*\Phi$ and Φ are p^*f-related and so their curvatures are also p^*f-related.

1.3. Let us suppose that a connection Φ on (E,p,M,S) has zero curvature. Then the horizontal bundle $\ker\Phi$ is integrable: each point u of E lies on a leaf $L(u)$ such that $T_v(L(u)) = (\ker\Phi)_v$ for each v in $L(u)$. Then $p|L(u)\colon L(u) \to M$ is locally a diffeomorphism. But in general it is neither surjective nor is it a covering of it's image.

1.4. Local description of a connection Φ on (E,p,M,S). Let us fix a fibre bundle atlas (U_α,ψ_α), i.e. (U_α) is an open cover of M and $\psi_\alpha\colon E|U_\alpha \to U_\alpha \times S$ is a diffeomorphism respecting all fibres over U_α. Then $\psi_\alpha\, \psi_\beta^{-1}(x,s) = (x,\psi_{\alpha\beta}(x,s))$, where $\psi_{\alpha\beta}\colon(U_\alpha \cap U_\beta) \times S \to S$ is smooth and $\psi_{\alpha\beta}(x,.)$ is a diffeomorphism of S for each x in $U_{\alpha\beta} := U_\alpha \cap U_\beta$. We will later consider the mappings $\psi_{\alpha\beta}\colon U_{\alpha\beta} \to \mathrm{Diff}(S)$; their differentiability will be a subtle question. $(\psi_{\alpha\beta})$ satisfies the usual cocycle condition for transition functions. Given a connection Φ on E we consider now the connection $(\psi_\alpha^{-1})^*\Phi$ in $\Omega^1(U_\alpha \times S,\, U_\alpha \times TS)$, which is described by $((\psi_\alpha^{-1})^*\Phi)(\xi_x,\eta_y) = -\Gamma^\alpha(\xi_x,y) + \eta_y$ for ξ_x in T_xU_α and η_y in T_yS, and where $(0_x,\, \Gamma^\alpha(\xi_x,y)) := -\,T(\psi_\alpha)\,\Phi\,(T(\psi_\alpha)^{-1}(\xi_x,0_y))$. We consider Γ^α as an element of $\Omega^1(U_\alpha,\mathfrak{X}(S))$, a 1-form on U_α with values in the infinite dimensional Lie algebra $\mathfrak{X}(S)$ of all vector fields on the standard fibre. The Γ^α are called the Christoffel forms of the connection Φ with respect to the bundle atlas (U_α,ψ_α). A short computation shows that they transform as follows: $T_y(\psi_{\beta\alpha}(x,.))\,\Gamma^\alpha(\xi_x,y) = \Gamma^\beta(\xi_x,\psi_{\beta\alpha}(x,y)) - T_x(\psi_{\beta\alpha}(.,y))\xi_x$. The curvature of the connection $(\psi_\alpha^{-1})^*\Phi$ is given by (using 1.2) $(\psi_\alpha^{-1})^*R = \frac{1}{2}(\psi_\alpha^{-1})^*[\Phi,\Phi] = \frac{1}{2}[(\psi_\alpha^{-1})^*\Phi,\, (\psi_\alpha^{-1})^*\Phi]$ and a short calculation yields $((\psi_\alpha^{-1})^*R)((\xi_1,\eta_1),(\xi_2,\eta_2)) = d\Gamma^\alpha(\xi_1,\xi_2) + [\Gamma^\alpha(\xi_1),\Gamma^\alpha(\xi_2)]$, where $d\Gamma^\alpha$ is the usual exteriour derivative of the 1-form Γ^α with values in $\mathfrak{X}(S)$ and the bracket is the one of $\mathfrak{X}(S)$. This is the Maurer-Cartan formula which in this setting appears only in the level of local description.

1.5. THEOREM (Parallel transport): Let Φ be a connection on a fibre bundle (E,p,M,S) and let $c:(a,b) \to M$ be a smooth curve with 0 in (a,b), $c(0) = x$. Then there is a neighbourhood U of $E_x \times \{0\}$ in $E_x \times (a,b)$ and a smooth mapping $Pt_c: U \to E$ such that:

1. $p(Pt(c,u_x,t)) = c(t)$ if defined and $Pt(c,u_x,0) = u_x$.
2. $\Phi(\frac{d}{dt} Pt(c,u_x,t)) = 0_{c(t)}$ if defined.
3. If $f: (\bar{a},\bar{b}) \to (a,b)$ is smooth with 0 in $(\bar{a},\bar{b})$, then $Pt(c,u_x,f(t)) = Pt(c\circ f, Pt(c,u_x,f(0)),t)$, if defined.
4. U is maximal for properties 1 and 2.
5. In a certain sense Pt depends smoothly on c, u_x and t.

1st proof: In local bundle coordinates $\Phi(\frac{d}{dt} Pt(c,u_x,t)) = 0$ is an ordinary differential equation, nonlinear, with initial condition $Pt(c,u_x,0) = u_x$. So there is a maximally defined local solution curve which is unique. All further properties are consequences of uniqueness.

2nd proof: Consider the pullback bundle $(c^*E, c^*p, (a,b),S)$ and the pullback connection $c^*\Phi$ on it. It's curvature is zero, since the horizontal bundle is 1-dimensional. By 1.3 the horizontal foliation exists and the parallel transport just follows a leaf. Map it back into E.

3rd proof: Consider a fibre bundle atlas (U_α,ψ_α) as in 1.4. Then $\psi_\alpha(Pt(c,\psi_\alpha^{-1}(x,y),t) = (c(t),\gamma(y,t))$ and $\gamma(y,t)$ is the evolution operator ('flow') of the time dependent vector field $\Gamma^\alpha(\frac{d}{dt}c(t),.)$ on S. Clearly local solutions exist and all properties follow. qed.

1.6. A connection Φ on (E,p,M,S) is called a <u>complete connection</u>, if the parallel transport Pt_c along any smooth curve $c: (a,b) \to M$ is defined on the whole of $E_{c(0)} \times (a,b)$. If the standard fibre is compact, then any connection on (E,p,M,S) is complete: this follows immediately from the third proof of 1.5. The following result is due to S. Halperin (1987):

THEOREM: Each fibre bundle (E,p,M,S) admits complete connections.

Proof: In the following lemma we will show that there are complete Riemannian metrics g on E and g_1 on M such that $p: E \to M$ is a Riemannian submersion. Let the connection Φ be just orthonormal projection $TE \to VE$ with respect to g. Then the horizontal lift mapping $\chi(.,u): T_{p(u)}M \to T_uE$ is an isometry, thus the arc lenght of $Pt(c,u,.)$ equals that of c and thus $\lim_{t\to s} Pt(c,u,t)$ exists if $c(s)$ is defined and we may continue.

1.7. LEMMA: Let (E,p,M,S) be a fibre bundle. Then there are complete Riemannian metrics g on E and g_1 on M such that $p: E \to M$ is a Riemannian submersion.

Proof: Let $\dim M = m$. Let (U_α, ψ_α) be a fibre bundle atlas as in 1.4. By topological dimension theory (see Nagata 1965) the open cover (U_α) of M admits a refinement such that any $m+1$ members have empty intersection. Let (U_α) itself have this property. Choose a smooth partition of unity (f_α) subordinated to (U_α). Then the sets $V_\alpha := \{x: f_\alpha(x) > \frac{1}{m+1}\} \subset U_\alpha$ are still an open cover of M since $\Sigma f_\alpha(x) = 1$ and at most m of the $f_\alpha(x)$ can be nonzero. Assume that each V_α is connected. Then we choose an open cover (W_α) of M such that $\overline{W_\alpha} \subset V_\alpha$.

Now let g_1 and g_2 be complete Riemannian metrics on M and S respectively (see Nomizu-Ozeki 1961 or J. Morrow 1970). Then $g_1|U_\alpha \times g_2$ is a metric on $U_\alpha \times S$ and we consider the metric $g := \Sigma f_\alpha\, \psi_\alpha{}^*(g_1|U_\alpha \times g_2)$ on E. We claim that g is a complete metric on E. So let B be a closed g-bounded set in E. Then $p(B)$ is g_1-bounded in M since $p: (E,g) \to (M,g_1)$ is a Riemannian submersion (see local considerations below), and has thus compact closure in M since g_1 is complete. Then $B'_\alpha := \overline{p(B)} \cap \overline{W_\alpha}$ is compact in V_α and clearly $B''_\alpha := p^{-1}(B'_\alpha) \cap B$ is closed and g-bounded and it suffices to show that each B''_α is compact since there are only finitely many of them and they cover B. Now we consider $B_\alpha := \psi_\alpha(B''_\alpha) \subset V_\alpha \times S$. A little thought shows that B_α is still bounded with respect to the metric $g_\alpha := (\psi_\alpha{}^{-1})^* g$. We obtain $(g_\alpha)_{(x,y)} = (g_1)_x \times (\sum_\beta f_\beta(x)\, \psi_{\beta\alpha}(x,.)^* g_2)_y$. Let $r := g_\alpha$-diameter of B_α in $V_\alpha \times S$, choose $R > (m+1)^{1/2}(r+1)$ and choose (x_0,y_0) in B_α. Let $D = D(y_0,R)$ be the open disc with center y_0 and g_2-radius R in S. Since g_2 is complete, D has compact closure in S. We claim that $B_\alpha \subset B'_\alpha \times D$ which then implies that B_α is compact and finishes the proof. Suppose not and let (x,y) be in $B_\alpha \setminus B'_\alpha \times D$. Then for any piecewise smooth curve $c = (c_1,c_2): [0,1] \to V_\alpha \times D$ from (x_0,y_0) to (x,y) we have g_2-lenght $(c_2) = \int_0^1 |c_2'(t)|_{g_2} dt \geq R$. But then

g_α-lenght $(c) = \int_0^1 |c'(t)|_{g_\alpha} dt =$

$$= \int_0^1 \Big(|c_1'(t)|^2_{g_1} + \sum_\beta f_\beta(c_1(t))\, (\psi_{\beta\alpha}(c_1(t),.)^* g_2)(c_2'(t),c_2'(t)) \Big)^{1/2} dt \geq$$

$\geq \int_0^1 (f_\alpha(c_1(t))\, g_2(c_2'(t),c_2'(t)))^{1/2} dt \geq (m+1)^{-1/2} \int_0^1 |c_2'(t)|_{g_2}\, dt \geq$
$\geq (m+1)^{-1/2}\, R > r+1.$

So (x,y) cannot be in B_α either. qed.

2. ASSOCIATED BUNDLES AND INDUCED CONNECTIONS

2.1. Let G be a Lie group which might be infinite dimensional like $\mathrm{Diff}(S)$. Let us consider a principal bundle (P,p,M,G) with structure group G: so there is a bundle atlas $(U_\alpha, \psi_\alpha : P|U_\alpha \to U_\alpha \times G)$ such that the transition functions $\psi_{\alpha\beta}: U_{\alpha\beta} \times G \to G$ are given by $\psi_{\alpha\beta}(x,g) = \tilde{\psi}_{\alpha\beta}(x).g$, where $\tilde{\psi}_{\alpha\beta}: U_{\alpha\beta} \to G$ are smooth. Let $r: P \times G \to G$ be the principal right action, let $\mathfrak{g}$ be the Lie algebra of G and let $\zeta: \mathfrak{g} \to \mathfrak{X}(P)$ be the fundamental vector field mapping, i.e. $\zeta(X)(u) = T_e(r_u).X$ for X in $\mathfrak{g}$ and u in P, where $r_u(g) = r(u,g)$. We also write $r^g: P \to P$ for $r^g(u) = r(u,g)$.

A connection Φ in $\Omega^1(P; VP)$ is called a principal connection, if $(r^g)^*\Phi = \Phi$ for all g in G, i.e. $T(r^g)^{-1}\Phi\, T(r^g) = \Phi$. Then the curvature $R = \frac{1}{2}[\Phi,\Phi]$ is also (r^g)-related to itself (see the appendix). Since $X \to \zeta(X)(u)$ is an isomorphism of $\mathfrak{g}$ onto V_uP, there is a unique ω in $\Omega^1(P;\mathfrak{g})$ such that $\Phi(\xi) = \zeta(\omega(\xi))$ for ξ in TP and by the usual properties of fundamental vector fields we have $(r^g)^*\omega = \mathrm{Ad}(g^{-1})\,\omega$ and clearly $\omega(\zeta_X) = X$, so ω is the usual description of a principal connection. On the other hand there is ρ in $\Omega^2(P;\mathfrak{g})$ such that $\zeta(\rho(\xi,\eta)) = R(\xi,\eta)$ for ξ,η in T_uP.

LEMMA: $\rho = -d\omega - \frac{1}{2}[\omega,\omega]_{\mathfrak{g}}$ (Maurer-Cartan formula).

Proof: Clearly both sides vanish on fundamental vector fields and coincide on horizontal vector fields. qed.

2.2. Now we describe the Christoffel forms Γ^α from 1.4 for a principal connection $\Phi = \zeta\omega$ with respect to a principal bundle atlas (U_α,ψ_α). Here $T(\psi_\alpha)\,\Phi\,(T(\psi_\alpha^{-1})\xi_x, 0_g) = (0_x, -\Gamma^\alpha(\xi_x,g))$ by 1.4, where Γ^α is in $\Omega^1(U_\alpha,\mathfrak{X}(G))$. Let $s_\alpha: U_\alpha \to E|U_\alpha$ be given by $s_\alpha(x) = \psi_\alpha^{-1}(x,e)$, and let $\omega_\alpha = s_\alpha^*\omega$ in $\Omega^1(U_\alpha, \mathfrak{g})$, the usual physicists description of a connection. Then it turns out that $\Gamma^\alpha(\xi_x,g) = -\,T_e(\lambda_g)\,\mathrm{Ad}(g^{-1})\,\omega_\alpha(\xi_x)$, where λ_g denotes left translation.

2.3. Now let ℓ: $G \times S \to S$ be a left action. Then the associated bundle $P[S]$ is given as the orbit space $\frac{G \times S}{G}$, where G acts on $P \times S$ by $(u,s)g = (ug, g^{-1}s)$. The differentiable structure is most conveniently (since G might be infinite dimensional) described by: choose a cocycle of transition functions $(\tilde{\psi}_{\alpha\beta}: U_{\alpha\beta} \to G)$ for P associated to a principal bundle atlas and let them act on S via ℓ. Then glue the the $U_\alpha \times S$ to get the associated bundle. We have a smooth map $q: P \times S \to P[S]$, which is itself a principal bundle projection, and $q_u: \{u\} \times S \to (P[S])_{p(u)}$ is a diffeomorphism for each u in P.

Note that TG is again a Lie group, $TG = \mathfrak{g} \circledS G$ is the semidirect product in the right trivialisation, where G acts on $\mathfrak{g}$ via Ad. Then (TP, Tp, TM, TG) is again a principal bundle, $T\ell$: $TG \times TS \to TS$ is again a left action, and $T(P[S,\ell]) = TP[TS, T\ell]$ is again an associated bundle. It's vertical bundle $V(P[S,\ell]) = P[TS, T\ell \circ (0_G \times Id_{TS}) =: T_2\ell]$ is associated to P itself. In particular $VP = V(P[G,\lambda]) = P[TG, T_2\lambda]$, and $VP/G = P[\mathfrak{g}, Ad]$. All this is easily checked by considering the different actions of the cocycles of transition functions.

2.4. Now we want to induce connections. Let $\Phi = \zeta\omega$ be a principal connection on P. We will induce a connection $\bar{\Phi}$ on the associated bundle $P[S]$ with the help of the following diagram:

$$\begin{array}{ccc} TP \times TS & \xrightarrow{\quad \Phi \times Id \quad} & TP \times TS \\ \downarrow Tq & & \downarrow Tq \\ T(P[S]) = TP[TS, T_2\ell] & \overset{\bar{\Phi}}{\dashrightarrow} & T(P[S]) \end{array}$$
$$T(P[S]) \to TM \leftarrow T(P[S])$$

Clearly $\Phi \times Id$ is TG-equivariant, so it induces $\bar{\Phi}$, which turns out to be a projection onto the vertical bundle, the induced connection. Note that $\Phi \times Id$ and $\bar{\Phi}$ are q-related, so $R \times 0 = \frac{1}{2}[\Phi \times Id, \Phi \times Id] = \frac{1}{2}[\Phi, \Phi] \times 0$ and $\bar{R} = \frac{1}{2}[\bar{\Phi}, \bar{\Phi}]$ are also q-related: The curvature of an induced connection is the induced curvature.

2.5. Now for the converse. Let $E = P[S]$ be an associated bundle. Let us suppose that the action ℓ: $G \times S \to S$ is infinitesimally effective, i.e. $\zeta: \mathfrak{g} \to \mathfrak{X}(S)$ is injective, where $\zeta(X)(s) = T_e(\ell^s).X$ for X in $\mathfrak{g}$ and s in S is the fundamental vector field mapping.

THEOREM: In this situation, a connection Φ in $\Omega^1(P[S], V(P[S]))$ is induced by a (necessarily unique) pricipal connection on P if and only if for some bundle atlas $(U_\alpha, \psi_\alpha: P[S]|U_\alpha \to U_\alpha \times S)$ belonging to the G-structure (so the transition functions have values in G and act on S via ℓ) the Christoffel forms Γ^α of Φ take values in $\mathfrak{X}_{fund}(S)$, the Lie algebra of fundamental vector fields.

Indication of proof: If the Christoffel forms have values in $\mathfrak{X}_{fund}(S)$ then $\Gamma^\alpha(\xi_x) = -\zeta\omega_\alpha(\xi_x)$ for unique ω_α in $\Omega^1(U_\alpha,\mathfrak{g})$, since $\zeta: \mathfrak{g} \to \mathfrak{X}_{fund}(S)$ is an isomorphism. These ω_α transform as required:

$\omega_\alpha = Ad(\tilde{\psi}_{\beta\alpha}{}^{-1})\omega_\beta + \tilde{\psi}_{\beta\alpha}{}^{-1}\, d\tilde{\psi}_{\beta\alpha}$. So they give a principal connection on P. The converse is a computation.

3. HOLONOMY GROUPS

3.1. Let (E,p,M,S) be a fibre bundle with complete connection Φ and let M be connected. We fix a base point x_0 in M and identify E_{x_0} with the standard fibre S. For each closed piecewise smooth curve $c: [0,1] \to M$ through x_0 the parallel transport $Pt(c,.,1) =: Pt(c,1)$ (pieced together over the smooth parts) is a diffeomorphism of S; all these diffeomorphisms form the group $Hol(\Phi,x_0)$, the <u>holonomy group</u> of Φ at x_0, a subgroup of the diffeomorphism group $Diff(S)$. If we consider only those piecewise smooth curves which are homotopic to zero we get a subgroup $Hol_0(\Phi,x_0)$, called the <u>restricted holonomy group</u> of Φ at x_0.

3.2. For a complete connection Φ let $\chi: TM \times_M E \to TE$ be the horizontal lifting and let R be the curvature. For any x in M and X_x in T_xM the horizontal lift $\chi(X_x) = \chi(X_x,.): E_x \to TE$ is a vector field along E_x. For X_x, Y_x in T_xM we consider $R(\chi X_x,\chi Y_x)$ in $\mathfrak{X}(E_x)$. Now choose any piecewise smooth curve c from x_0 to x, consider the diffeomorphism $Pt(c,1)$: $S = E_{x_0} \to E_x$ and the pullback $Pt(c,1)^* R(\chi X_x,\chi Y_x)$ in $\mathfrak{X}(S)$. Let us denote the closed linear subspace generated by all these vector fields (for all x in M, X_x, Y_x in T_xM and curves c from x_0 to x) in $\mathfrak{X}(S)$ with the compact C^∞-topology by the symbol $hol(\Phi,x_0)$ and let us call it the <u>holonomy Lie algebra</u> of Φ at x_0.

LEMMA: $hol(\Phi,x_0)$ is a Lie subalgebra of (S).

Proof: Let X be in $\mathfrak{X}(M)$ and let $Fl^{\chi X}_t$ denote the local flow of the horizontal lift χX of X. It restricts to parallel transport along any of the flow lines of X in M. Then we have that

$$\frac{d}{dt}\Big|_0 (Fl^{\chi X}_s)^* (Fl^{\chi Y}_t)^* (Fl^{\chi X}_{-s})^* (Fl^{\chi Z}_z)^* R(\chi U,\chi V)|E_{x_0} =$$
$$= (Fl^{\chi X}_s)^* [\chi Y, (Fl^{\chi X}_{-s})^* (Fl^{\chi Z}_z)^* R(\chi U,\chi V)]|E_{x_0} =$$
$$= [(Fl^{\chi X}_s)^* \chi Y, (Fl^{\chi Z}_z)^* R(\chi U,\chi V)]|E_{x_0}$$

is in $hol(\Phi,x_0)$. By the Jacobi identity and some manipulation the result follows. qed.

3.3. The theorem of Ambrose and Singer now admits the following generalisation.

THEOREM: Let Φ be a connection on the fibre bundle (E,p,M,S) with compact standard fibre S and let M be connected. Suppose that for some x_0 in M the holonomy Lie algebra $hol(\Phi,x_0)$ is finite dimensional. Then there is a principal bundle (P,p,M,G) with finite dimensional structure group G, an irreducible connection ω on it and a smooth action of G on S such that the Lie algebra $\mathfrak{g}$ of G equals the holonomy algebra $hol(\Phi,x_0)$, the fibre bundle E is isomorphic to the associated bundle $P[S]$ and Φ is the connection induced by ω. G is isomorphic to the holonomy group $Hol(\Phi,x_0)$. P and ω are unique up to isomorphism.

Proof: Let us identify E_{x_0} and S. Then $\mathfrak{g} := hol(\Phi,x_0)$ is a finite dimensional Lie subalgebra of $\mathfrak{X}(S)$, and since S is compact, there is a finite dimensional connected Lie group G_0 of diffeomorphisms of S with Lie algebra $\mathfrak{g}$.

Claim 1: G_0 equals $Hol_0(\Phi,x_0)$, the restricted holonomy group.

Let f be in $Hol_0(\Phi,x_0)$, then $f = Pt(c,1)$ for a piecewise smooth closed curve through x_0. Since parallel transport is essentially invariant under reparametrisation (1.5.3), we can replace c by $c\circ g$, where g is smooth and flat at each corner of c. So we may assume that c is smooth. Then c is homotopic to zero, so by approximation we may assume that there is a smooth homotopy $H: \mathbb{R}^2 \to M$ with $H_1|I = c$ and $H_0|I = x_0$. Then $Pt(H_t,1)=f_t$ is a curve in $Hol_0(\Phi,x_0)$ which is smooth as a mapping $\mathbb{R}\times S \to S$.

Claim 2: $(\frac{d}{dt} f_t)\circ f_t^{-1} =: Z_t$ is in $\mathfrak{g}$ for all t.

To prove claim 2 we consider the pullback bundle $H^*E \to \mathbb{R}^2$ with the induced connection $H^*\Phi$. It is sufficient to prove claim 2 there. Let $X = \frac{\partial}{\partial s}$ and $Y = \frac{\partial}{\partial t}$ be the constant vector fields on $\mathbb{R}^2$, so $[X,Y] = 0$. Then $Pt(c,s) = Fl^{\chi X}_s$ and so on. Put $f_{t,s} = Fl^{\chi X}_{-s}\ Fl^{\chi Y}_{-t}\ Fl^{\chi X}_s\ Fl^{\chi Y}_t : S \to S$, so $f_{t,1} = f_t$. Then

$$(\tfrac{d}{dt} f_{t,s})\, f_{t,s}^{-1} = -(Fl^{\chi X}_s)^* \chi Y + (Fl^{\chi X}_s)^* (Fl^{\chi Y}_t)^* (Fl^{\chi X}_{-s}) \chi Y \quad \text{in } \mathfrak{X}(S) \text{ and}$$

$$(\tfrac{d}{dt} f_t)\, f_t^{-1} = \int_0^1 \tfrac{d}{ds}((\tfrac{d}{dt} f_{t,s})\, f_{t,s}^{-1})ds = \int_0^1 (-(Fl^{\chi X}_s)^*[\chi X, \chi Y] +$$
$$+ (Fl^{\chi X}_s)^*[\chi X,\ (Fl^{\chi Y}_t)^* (Fl^{\chi X}_{-s})^* \chi Y] - (Fl^{\chi X}_s)^* (Fl^{\chi Y}_t)^* (Fl^{\chi X}_{-s})^*[\chi X, \chi Y]\ )\, ds.$$

Since $[X,Y] = 0$ we have $[\chi X, \chi Y] = \Phi[\chi X, \chi Y] = R(\chi X, \chi Y)$, and also we have

$$(Fl^{\chi X}_t)^* \chi Y = \chi((Fl^{\chi X}_t)^* \chi Y) + \Phi((Fl^{\chi X}_t)^* \chi Y) = \chi Y + \int_0^t \tfrac{d}{dt} \Phi(Fl^{\chi X}_t)^* \chi Y\, dt =$$
$$= \chi Y + \int_0^t \Phi(Fl^{\chi X}_t)^*[\chi X, \chi Y]\, dt = \chi Y + \int_0^t (Fl^{\chi X})^* R(\chi X, \chi Y)\, dt.$$

Thus all parts of the integrand above are in $\mathfrak{g}$ and so $(\frac{d}{dt} f_t)\, f_t^{-1}$ is in $\mathfrak{g}$ for all t and claim 2 is proved.

Now claim 1 can be shown as follows: There is a unique smooth curve $g(t)$ in G_0 satisfying $Z_t.g(t) = \frac{d}{dt} g(t)$ and $g(0) = Id$; via the action of G_0 on S $g(t)$ is a curve of diffeomorphisms on S, generated by the time dependent vector field Z_t, so $g(t) = f_t$ and $f = f_1$ is in G_0.

Step 3: Now we make $Hol(\Phi, x_0)$ into a Lie group which we call G, by taking $Hol_0(\Phi, x_0)$ as its connected component of identity. Then $Hol(\Phi, x_0)/G_0$ is a countable group, since the fundamental group $\pi_1(M)$ is countable (by Morse theory M is homotopy equivalent to a countable CW-complex).

Step 4: Construction of a cocycle of transition functions with values in G. Let $(U_\alpha,\ u_\alpha : U_\alpha \to \mathbb{R}^n)$ be a locally finite smooth atlas of M such that each $u_\alpha : U_\alpha \to \mathbb{R}^n$ is a diffeomorphism onto $\mathbb{R}^n$. Put $x_\alpha = u_\alpha^{-1}(0)$. Choose smooth curves $c_\alpha : [0,1] \to M$ with $c_\alpha(0) = x_0$, $c_\alpha(1) = x_\alpha$. For each x in U_α let $c^x_\alpha : [0,1] \to M$ be the smooth curve $t \to u_\alpha^{-1}(t.u_\alpha(x))$, then c^x_α connects x_α and x and the mapping $(x,t) \to c^x_\alpha(t)$ is smooth $U_\alpha \times [0,1] \to M$. Now we define a fibre bundle atlas $(U_\alpha, \psi_\alpha : E|U_\alpha \to U_\alpha \times S)$ by $\psi_\alpha^{-1}(x,s) = Pt(c^x_\alpha, 1)\, Pt(c_\alpha, 1)\, s$. Clearly ψ_α is smooth. Let us investigate the transition functions:

$$\begin{aligned}\psi_\beta\, \psi_\alpha^{-1}(x,s) &= (x,\ Pt(c_\alpha,1)^{-1}\, Pt(c^x_\alpha,1)^{-1}\, Pt(c^x_\beta,1)\, Pt(c_\beta,1)\, s) \\ &= (x,\ Pt(c_\beta.c^x_\beta.(c^x_\beta)^{-1}.(c_\beta)^{-1},4)\, s) \\ &=: (x,\ \psi_{\beta\alpha}(x)\, s), \text{ where } \psi_{\beta\alpha} : U_{\beta\alpha} \to G.\end{aligned}$$

Clearly $\psi_{\beta\alpha}: U_{\beta\alpha} \times S \to S$ is smooth and consequently (see section 4) $\psi_{\beta\alpha}: U_{\beta\alpha} \to \mathrm{Diff}(S)$ is smooth and also $U_{\beta\alpha} \to G$. $(\psi_{\alpha\beta})$ is a cocycle of transition functions and we use it to glue a principal bundle with stucture group G over M which we call (P,p,M,G). It is clear that $P \times S/G$ equals (E,p,M,S).

Step 5: Lifting the connection Φ to P. For this we have to compute the Christoffel symbols of Φ with respect to the bundle atlas of step 4. For this we need more sophisticated methods. Since it is heuristically obvious we omit this part of the proof. qed.

3.4. Remarks: I expect that theorem 3.3 is also true if S is not compact but the connection Φ is supposed to be complete. The problem here is to show that the holonomy Lie algebra consists of complete vector fields. By a theorem of Palais (1957) it suffices to find complete vector fields which generate the holonomy Lie algebra.

It would be interesting to find the relation between $\mathrm{Hol}(\Phi,x_0)$ and $\mathrm{hol}(\Phi,x_0)$ if the latter is not supposed to be finite dimensional (for complete connections). Se 4.6 below for some curious fact.

4. THE (NONLINEAR) FRAME BUNDLE OF A FIBRE BUNDLE

4.1. Let (E,p,M,S) be a fibre bundle. Let $(U_\alpha, \psi_\alpha: E|U_\alpha \to U_\alpha \times S)$ be a fibre bundle atlas as described in 1.4. Then $\psi_\alpha \psi_\beta^{-1}(x,s) = (x,\psi_{\alpha\beta}(x,s))$ for smooth mappings $\psi_{\alpha\beta}: U_{\alpha\beta} \times S \to S$ such that $\psi_{\alpha\beta}(x,.)$ is in Diff(S).

For second countable smooth manifolds Y, Z the space $C^\infty(Y,Z)$ is an infinite dimensional smooth manifold modelled on nuclear (LF)-spaces (see Michor 1980). In general we have $C^\infty(U_{\alpha\beta}, C^\infty(S,S)) \subseteq C^\infty(U_{\alpha\beta} \times S,S)$ with equality if and only if S is compact (see Michor 1980, 11.9). The image in the right hand side consists of these maps f which satisfy: for any compact set K in $U_{\alpha\beta}$ there is a compact set K' in S such that $f(x_1,s) = f(x_2,s)$ for all x_1, x_2 in the same connected component of K and for all s in $S \setminus K'$. So $\psi_{\alpha\beta}: U_{\alpha\beta} \to \mathrm{Diff}(S)$ is smooth if and only if (for $U_{\alpha\beta}$ connected) $\psi_{\alpha\beta}(U_{\alpha\beta})$ is contained in a left coset of the open and closed subgroup $\mathrm{Diff}_c(S)$ of diffeomorphisms with compact support. This is in principle true only for bundles which are "trivial near fibrewise infinity" or have " discrete structure group near fibrewise infinity".

To avoid these topological difficulties we will assume for the rest of this section that the standard fibre S is compact. Then $\psi_{\alpha\beta}: U_{\alpha\beta} \to \mathrm{Diff}(S)$ is smooth without any restrictions and $(\psi_{\alpha\beta}: U_{\alpha\beta} \to \mathrm{Diff}(S))$ form the cocycle of transition functions for (E,p,M,S).

4.2. Now we define the nonlinear frame bundle of (E,p,M,S) as follows. Let $\mathrm{Diff}\{S,E\} := \bigcup_x \mathrm{Diff}(S,E_x)$ where the union is over all x in M, and give it the differentiable structure which one gets by applying the functor $\mathrm{Diff}(S,.)$ to the cocyle of transition functions $(\psi_{\alpha\beta})$. Then the resulting cocycle of Diff{S,E} gives it the structure of a smooth principal bundle over M with structure group Diff(S). The principal action is just composition from the right.

4.3. Consider now the smooth action $\mathrm{Diff}(S) \times S \xrightarrow{ev} S$ and the associated bundle as in the following diagram:

$$\begin{array}{ccc} \mathrm{Diff}\{S,E\} \times S & \longrightarrow & \dfrac{\mathrm{Diff}\{S,E\} \times S}{\mathrm{Diff}(S)} = \mathrm{Diff}\{S,E\}[S, ev] \\ \downarrow ev & \nearrow & \\ E \cong & & \end{array}$$

Clearly the associated bundle gives us back the bundle (E,p,M,S).

4.4. Let Φ in $\Omega^1(E;VE)$ be a connection on E. We want to realise Φ as a principal connection on Diff{S,E}. For that we must get a good picture of the tangent space of Diff{S,E}. With the methods of Michor 1980 it is easy to see that $T\,\mathrm{Diff}\{S,E\} = \bigcup_{x \text{ in } M} \{ f \text{ in } C^\infty(S,TE|E_x): Tp\circ f = \text{one point in } T_xM \text{ and } \pi_E\circ f \text{ is in } \mathrm{Diff}(S,E_x)\}$. Given the connection Φ we consider $\omega(f) = T(\pi_E\circ f)^{-1}\circ\Phi\circ f : S \to TE \to VE \to TS$ for f in T Diff{S,E}. Then $\omega(f)$ is a vector field on S and ω: T Diff{S,E} $\to \mathfrak{X}(S)$ is fibre linear, so it is a one form on Diff{S,E} with values in the Lie algebra $\mathfrak{X}(S)$ of the structure group Diff(S) (with the negative of the usual bracket).

LEMMA: ω in $\Omega^1(\mathrm{Diff}\{S,E\}; \mathfrak{X}(S))$ is a principal connection, and the induced connection on $E = \mathrm{Diff}\{S,E\}[S,ev]$ coincides with Φ.

Proof: This follows directly from 2.5. But we also give direct proof. The fundamental vector field $\zeta(X)$ on Diff{S,E} for X in $\mathfrak{X}(S)$ is given by $\zeta(X)(g) = Tg\circ X$. Then $\omega(\zeta(X)(g)) = Tg^{-1}\circ\Phi\circ Tg\, X = \quad X$, since $Tg\circ X$ has vertical values. So Φ reproduces fundamental vector fields.

Let h be in Diff(S) and denote by r^h the principal right action. Then
$((r^h)^*\omega)(f) = \omega(T(r^h)\, f) = \omega(f\circ h) = T(\pi_E\circ f\circ h)^{-1}\circ\Phi\circ f\circ h = Th^{-1}\circ\omega(f)\circ h =$
$= Ad_{Diff(S)}(h^{-1})\,\omega(f).$ qed.

4.5. THEOREM: Let (E,p,M,S) be a fibre bundle with compact standard fibre S. Then connections on E and principal connections on Diff{S,E} correspond to each other bijectively, and their curvatures are related as in 2.4. Each principal connection on Diff{S,E} → M admits a smooth global parallel transport. The holonomy groups and restricted holonomy groups are equal as subgroups of Diff(S).

Proof: This follows directly from section 2. Each connection on E is complete since S is compact, and the lift to Diff{S,E} of its parallel transport is the (global) parallel transport of the lift of the connection, so the two last assertions follow. qed.

4.6 Remark on the holonomy Lie algebra: Let M be connected, let $\rho = -d\omega - \frac{1}{2}[\omega,\omega]_{\mathfrak{X}(S)}$ be the usual $\mathfrak{X}(S)$-valued curvature of the lifted connection ω on Diff{S,E}. Then we consider the $\mathbb{R}$-linear span of all elements $\rho(\xi_f,\eta_f)$ in $\mathfrak{X}(S)$, where ξ_f, η_f in T_fDiff{S,E} are arbitrary (horizontal) tangent vectors, and we call this span hol(ω). Then by the Diff(S)-equivariance of ρ the vector space hol(ω) is an ideal in the Lie algebra $\mathfrak{X}(S)$.

LEMMA: Let $f: S \to E_{x_0}$ be any diffeomorphism in $\mathrm{Diff}\{S,E\}_{x_0}$. Then $f_*: \mathfrak{X}(S) \to \mathfrak{X}(E_{x_0})$ induces an isomorphism between hol(ω) and the $\mathbb{R}$-linear span of all $g^*R(\chi X,\chi Y)$, X,Y in T_xM, x in M and $g: E_{x_0} \to E_x$ any diffeomorphism.

The proof is obvious. Note that the closure of $f_*(\mathrm{hol}(\omega))$ is (a priori) larger than $\mathrm{hol}(\Phi,x_0)$ of section 3. Which one is the right holonomy Lie algebra?

5. GAUGE THEORY

We fix the setting of section 4. In particular S is supposed to be compact.

5.1. With some abuse of notation we consider the bundle

$\mathrm{Diff}\{E,E\} := \bigcup_{x\in M} \mathrm{Diff}(E_x,E_x)$, which bears the smooth structure described by the cocycle of transition functions $\mathrm{Diff}(\psi_{\alpha\beta}^{-1},\psi_{\alpha\beta}) = (\psi_{\alpha\beta})_* (\psi_{\beta\alpha})^*$, where $(\psi_{\alpha\beta})$ is a cocycle of transition functions for (E,p,M,S).

LEMMA: The assiciated bundle Diff{S,E}[Diff(S), conj] is isomorphic to the fibre bundle Diff{E,E}.

Proof: The mapping $A: \mathrm{Diff}\{S,E\} \times \mathrm{Diff}(S) \to \mathrm{Diff}\{E,E\}$, given by $A(f,g) = f\circ g\circ f^{-1}: E_x \to S \to S \to E_x$ for f in $\mathrm{Diff}(S,E_x)$, is Diff(S)-invariant, so it factors to a smooth mapping $\mathrm{Diff}\{S,E\}[\mathrm{Diff}(S)] \to \mathrm{Diff}\{E,E\}$. It is bijective and admits locally (over M) smooth inverses, so it is a diffeomorphism. qed.

5.2. The <u>gauge group</u> Gau(E) of the bundle (E,p,M,S) is by definition the space of all principal bundle automorphisms of (Diff{S,E},p,M,Diff(S)) which cover the identity on M. The differential geometry for the infinite dimensional spaces involved is given in (Michor 1980), so the usual reasoning gives that Gau(E) equals the space of smooth sections of the associated bundle Diff{S,E}[Diff(S),conj], which by 3.1 equals the space of sections of the bundle Diff{E,E} → M. We equip it with the topology and differentiable structure as a space of smooth sections: since only the image space is infinite dimensional, this makes no difficulties. Since the fibre S is compact we see from a local (on M) application of the exponential law (Michor 1980, 11.9) that $\Gamma(\mathrm{Diff}\{E,E\} \to M) \hookrightarrow \mathrm{Diff}(E)$ is an embedding of a splitting closed submanifold.

THEOREM: The gauge group Gau(E) = Γ(Diff{E,E}) is a splitting closed subgroup of Diff(E), if S is compact. It has an exponential map (which is not surjective on any neighbourhood of the identity). Its Lie algebra consists of all vertical vector fields with compact support on E, with the negative of the usual Lie bracket.

5.3. Remark: If S is not compact, then we may circumvent the nonlinear frame bundle, and we may define the gauge group Gau(E) directly as the splitting closed subgroup of Diff(E) which consists of all fibre respecting diffeomorphisms of E which cover the identity on M. Contrary to the finite dimensional case with noncompact structure group G Gau(E) is then not discrete. Its Lie algebra consists of all vertical vector fields on

E with compact support. For simplicity's sake we stick to compact fibre in this paper.

5.4. The <u>space of connections</u> or gauge potentials. Let $J^1(E)$ be the bundle of 1-jets of sections of $E \to M$. Then $J^1E \to E$ is an affine bundle, in fact $J^1E = \{\ell$ in $L(T_xM, T_uE)$: $Tp \circ \ell = Id_{T_xM}$, u in E with $p(u) = x\}$. Then a section γ in $\Gamma(J^1E)$ is just a horizontal lift mapping $TM \times_M E \to TE$ which is fibre linear over E. By the discussion of 1.1 γ then describes a connection and we may view $\Gamma(J^1E)$ as the space of connections of E. The next result follows easily from (Michor 1980, §11).

THEOREM: The action of the gauge group Gau(E) on the space of connections $\Gamma(J^1E)$ is smooth.

5.5. We will now give a different description of the action . We view now a connection Φ as always as a linear fibre projection Φ: $TE \to VE$, so $Conn(E) := \{\Phi$ in $\Omega^1(E;TE)$: $\Phi \circ \Phi = \Phi$, $Im(\Phi) = VE\}$, which is a closed splitting affine subspace of $\Omega^1(E;TE)$. The natural isomorphism Conn(E) = $\Gamma(J^1E)$ is also one for the topology and the differentiable structures. Then the action of f in Gau(E)$\subseteq$Diff(E) on a connection Φ is given by $f_*\Phi = (f^{-1})^*\Phi = Tf \circ \Phi \circ Tf^{-1}$. Now it is very easy to describe the infinitesimal action. Let X be a vertical vector field with compact support on E and consider its global flow Fl^X_t. Then $\frac{d}{dt}|_0 (Fl^X_t)^*\Phi = [X,\Phi]$, the Frölicher-Nijenhuis bracket (see appendix). The tangent space of Conn(E) at Φ is $T_\Phi Conn(E) = \{\Psi$ in $\Omega^1(E;VE)$: $\Psi|VE = 0\}$. The infinitesimal orbit at Φ in Conn(E) is then $\{[X,\Phi], X$ in $\mathfrak{X}_c^{vert}(E)\}$.

The isotropy subgroup of a connection Φ is $\{f$ in Gau(E): $f_*\Phi = \Phi\}$. Clearly this is just the group of those f which respect the horizontal bundle kerΦ. The most interesting object is of course the orbit space Conn(E)/Gau(E). I am convinced that it is stratified into smooth manifolds, each one corresponding to a conjugacy class of holonomy groups in Diff(S). Those strata whose holonomy groups are (up to conjugacy) contained in a fixed compact group (like SU(2)) acting on S I expect to be diffeomorphic to the strata of usual gauge theory.

6. CLASSIFYING SPACE FOR Diff(S) AND A CHARACTERISTIC CLASS.

This section is based on the doctoral dissertation of Gerd Kainz (1988).

6.1. Let $\mathbb{R}^\infty = \oplus\mathbb{R}$ be the countable direct sum of copies of $\mathbb{R}$, the space of finite sequences, with the direct sum topology. Let S be a compact manifold. By a slight generalisation of (Michor 1980, §13) the space $Emb(S, \mathbb{R}^\infty)$ of smooth embeddings $S \to \mathbb{R}$ is an open submanifold of the smooth manifold $C^\infty(S, \mathbb{R}^\infty)$ and it is also the total space of a smooth principal bundle with structure group Diff(S), acting by composition from the right. The base space $B(S, \mathbb{R}^\infty) := Emb(S, \mathbb{R}^\infty)/Diff(S)$ is a smooth manifold, modelled on nuclear (LF)-spaces. Its topology is Lindelöf and the model spaces admit smooth partitions of unity, so also $B(S, \mathbb{R}^\infty)$ admits smooth partitions of unity. $B(S, \mathbb{R}^\infty)$ can be viewed as the space of all submanifolds of $\mathbb{R}^\infty$ diffeomorphic to S, a nonlinear analogue of the infinite dimensional Grassmanian.

Furthermore the total space $Emb(S, \mathbb{R}^\infty)$ is smoothly contractible. Ramadas (1982) constructed the continuous homotopy $A: \ell^2 \times I \to \ell^2$ through isometries given by $A_0 = Id$, $A_t(a_0,a_1,a_2,\dots) =$
$= (a_0,a_1,\dots,a_{n-2}, a_{n-1}\cos\theta_n(t), a_{n-1}\sin\theta_n(t), a_n\cos\theta_n(t), a_n\sin\theta_n(t),$
$a_{n+1}\cos\theta_n(t), a_{n+1}\sin\theta_n(t),\dots)$ for $\frac{1}{n+1} \leq t \leq \frac{1}{n}$, where $\theta_n(t) = n((n+1)t-1)\frac{\pi}{2}$. Then $A_{1/2}(a_0,a_1,a_2,\dots) = (a_0,0,a_1,0,a_2,0,\dots)$ is in ℓ^2_{even} and $A_1(a_0,a_1,\dots) = (0,a_0,0,a_1,0,a_2,0,\dots)$ is in ℓ^2_{odd}. This same homotopy is defined also on $\mathbb{R}^\infty$ and it turns out that it is even smooth there in the Frölicher-Kriegl calculus (Frölicher-Kriegl 1988). So given two embeddings e_1 and e_2 in $Emb(S, \mathbb{R}^\infty)$ we first deform e_1 through embeddings to e_1' in $Emb(S, \mathbb{R}^\infty_{even})$, and e_2 to e_2' in $Emb(S, \mathbb{R}^\infty_{odd})$, and then we connect them by $t.e_1' + (1-t).e_2'$.

By the general theory of classifying spaces $B(S, \mathbb{R}^\infty)$ is then a classifying space for Diff(S)-principal bundles over finite dimensional manifolds.

6.2. We will now give a more detailed description of the classifying process. Let us again consider the smooth action $Diff(S) \times S \xrightarrow{ev} S$ as in 3.3 and the associated fibre bundle $Emb(S, \mathbb{R}^\infty)[S,ev] = Emb(S, \mathbb{R}^\infty) \times_{Diff(S)} S$ which we call $E(S, \mathbb{R}^\infty)$, a smooth fibre bundle over $B(S, \mathbb{R}^\infty)$ with standard fibre S. Remembering the interpretation of $B(S, \mathbb{R}^\infty)$ as the nonlinear

Grassmannian, we van visualize $E(S, \mathbb{R}^\infty)$ as the "universal S-bundle" over $B(S, \mathbb{R}^\infty)$ as follows: $E(S, \mathbb{R}^\infty) = \{(N,x) \text{ in } B(S, \mathbb{R}^\infty) \times \mathbb{R}^\infty : x \in N\}$. Then $T\,E(S, \mathbb{R}^\infty)$ consists of all (N,ξ,x,v), where N is in $B(S, \mathbb{R}^\infty)$, ξ is a vector field along and normal to N in $\mathbb{R}^\infty$, x is in N and v is a tangent vector at x of $\mathbb{R}^\infty$ with $v^\perp = \xi(x)$. This follows from the description of the principal fibre bundle $\mathrm{Emb}(S, \mathbb{R}^\infty) \to B(S, \mathbb{R}^\infty)$ given in (Michor 1980, §13). Obviously the vertical bundle $V\,E(S, \mathbb{R}^\infty)$ consists of all (N,x,v), where N is in $B(S, \mathbb{R}^\infty)$, x is in N and $v \in T_xN$. The orthonormal projection $\mathbb{R}^\infty \to T_xN$ defines a connection $\Phi_0 : T\,E(S, \mathbb{R}^\infty) \to V\,E(S, \mathbb{R}^\infty)$, given by $\Phi_0(N,\xi,x,v) = (N,x,v - \xi(x)) = (N,x, \text{tangential component of } v)$.

THEOREM: $(E(S, \mathbb{R}^\infty), p, B(S, \mathbb{R}^\infty), S)$ is a universal bundle for S-bundles and Φ_0 is a universal connection, i.e. for each bundle (E,p,M,S) and connection Φ in $\Omega^1(E;VE)$ there is a smooth (classifying) map $f : M \to B(S, \mathbb{R}^\infty)$ such that (E,Φ) is isomorphic to $(f^*E(S, \mathbb{R}^\infty), f^*\Phi_0)$. Homotopic maps pull back isomorphic S-bundles and conversely (the homotopy can be chosen smooth). The pulled back connection stays the same if and only if $i(T_{(x,t)}H.(0_x,\frac{\partial}{\partial t}))R_0 = 0$ for all x in M and t in I, where H is the homotopy and R_0 is the curvature of Φ_0.

Proof: Choose a Riemannian metric g_1 on the vector bundle $VE \to E$, another one g_2 on M and consider $(Tp|\ker\Phi)^*g_2 \oplus g_1$, a Riemannian metric on E for which the horizontal and the vertical spaces are orthogonal. By the theorem of Nash there is an isometric embedding $h : E \to \mathbb{R}^N$ for N large enough. We then embed $\mathbb{R}^N$ into $\mathbb{R}$ and consider $f : M \to B(S, \mathbb{R}^\infty)$, given by $f(x) = h(E_x)$. Then

$$\begin{array}{ccc} E & \xrightarrow{\;(f,h)\;} & E(S, \mathbb{R}^\infty) \\ \downarrow & & \downarrow \\ M & \xrightarrow{\;f\;} & B(S, \mathbb{R}^\infty) \end{array}$$

is fibrewise a diffeomorphism, so $f^*E(S, \mathbb{R}^\infty) = E$. Since also $T(f,h)$ maps horizontal and vertical vectors to othogonal vectors, it pulls back Φ_0 to Φ.
Now let H be a continuous homotopy $M \times I \to B(S, \mathbb{R}^\infty)$. Then we may approximate H by smooth maps with the same H_0 and H_1, if they are smooth (see Bröcker-Jänich 1973, the infinite dimensionality of $B(S, \mathbb{R}^\infty)$ does not disturb). Then we consider the bundle $H^*E(S, \mathbb{R}^\infty) \to M \times I$, equipped with the connection $H^*\Phi_0$, whose curvature is H^*R_0. Let ∂_t be the vector field

∂_t on $M \times I$. Paralleltransport along the lines $t \to (x,t)$ with respect to $H^*\Phi_o$ is given by the flow of the horizontal lift $(H^*\chi_o)(\partial_t)$ of ∂_t. Let us compute its action on $H^*\Phi_o$:

$$\frac{d}{dt}(Fl_t^{(H^*\chi_o)(\partial_t)})^* H^* \Phi_o = \frac{d}{dt} H^*(Fl_t^{\chi_o(TH(0,\partial_t)})^* \Phi_o = H^*[\chi_o TH(o,\partial_t), \Phi_o]$$
$$= - H^* i(\chi_o\, TH(0,\partial_t))R_o \qquad \text{qed.}$$

6.3. THEOREM (Ebin-Marsden 1970): Let S be a compact orientable manifold, let μ be a positive volume form on S with total mass 1. Then $\mathrm{Diff}(S)$ splits topologically and smoothly , $\mathrm{Diff}(S) =$ $\mathrm{Diff}(S) = \mathrm{Diff}_\mu(S) \times \mathrm{Vol}(S)$, where $\mathrm{Diff}_\mu(S)$ is the Fréchet Lie group of all μ-preserving diffeomorphisms and $\mathrm{Vol}(S)$ is the space of all volumes of total mass 1 on S.

6.4. A consequence of theorem 6.3 is that the classifying spaces of $\mathrm{Diff}(S)$ and $\mathrm{Diff}_\mu(S)$ are homotopically equivalent. The space $B_1(S, \mathbb{R}^\infty)$ of all submanifolds of $\mathbb{R}^\infty$ of type S and volume 1 (in the volume form induced from the inner product on $\mathbb{R}^\infty$) is a closed submanifold of codimension 1 (by the Nash-Moser-Hamilton inverse function theorem) and clearly it is the classifying space for $\mathrm{Diff}_\mu(S)$-bundles. This can be seen as in 6.2; the total space is $\mathrm{Emb}_\mu(S, \mathbb{R}^\infty)$, the space of all embeddings $e: S \to \mathbb{R}^\infty$ with $\mathrm{vol}(e^*\langle\ ,\ \rangle) = \mu$. This gives a principal fibre bundle with structure group $\mathrm{Diff}_\mu(S)$. The associated bundle $\mathrm{Emb}_\mu(S, \mathbb{R}^\infty)[S,ev] =$ $=(\mathrm{Emb}_\mu(S, \mathbb{R}^\infty) \times S)/\mathrm{Diff}_\mu(S)$ can again be viewed as $\{(N,x)$ in $B_1(S, \mathbb{R}^\infty) \times \mathbb{R}^\infty$ with x in $N\}$ and it is the total space of a bundle $E_\mu(S, \mathbb{R}^\infty)$ over $B_1(S, \mathbb{R}^\infty)$ with fibre type S and a volume form of mass 1 on each fibre, which depends smoothly on the base.

6.5. Bundles with fibre volumes: Let (E,p,M,S) be a fibre bundle and let μ be a section of the real line bundle $\Lambda^n V^*E \to E$, where $n = \dim S$, such that for each x in M the n-form $\mu_x := \mu | E_x$ is a positive volume form of total mass 1. Such a form will be called a <u>fibre volume</u> on E. Such forms exist on any oriented bundle with compact oriented fibre type. We may plug n vertical vector fields X_i from $\Gamma(VE)$ into μ to get a function $\mu(X_1,\dots,X_n)$ on E. But we cannot treat μ as a differential form on E.

LEMMA: Let μ_o be a fixed volume form on S and let μ be a fibre volume on E. Then there is a bundle atlas (U_α,ψ_α) on E such that $(\psi_\alpha|E_x)^*\mu_o = \mu_x$ for all x in U_α.

Proof: Take any bundle atlas (U_α,ψ'_α). By theorem 6.3 there is $\varphi_{\alpha,x}$ in Diff(S) such that $(\varphi_{\alpha,x})^*(\psi'^{-1}_\alpha|\{x\}\times S)^*\mu_x = \mu_o$, and $\varphi_{\alpha,x}$ depends smoothly on x. Then $\psi_\alpha = (Id\times\varphi_{\alpha,x})\ \psi'_\alpha$ is the desired bundle chart. qed.

6.6. Let Φ be a connection on (E,p,M,S) and let χ: $TM\times_M E \to TE$ be its horizontal lift. Then for X in $\mathfrak{X}(M)$ the flow $Fl^{\chi X}_t$ of its horizontal lift respects fibres, so for the fibre volume μ on E the pullback $(Fl^{\chi X}_t)^*\mu$ makes sense and is again a fibre volume on E. We can now define a kind of Lie derivative by $\tilde{L}_{\chi X}\mu := \frac{d}{dt}|_0 (Fl^{\chi X}_t)^*\mu$.

Let j: VE $\to$ TE be the embedding of the vertical bundle and consider $j^*: \Omega^k(E) \to \Gamma(\Lambda^k V^*E)$. Likewise we consider $\Phi^*: \Gamma(\Lambda^k V^*E) \to \Omega^k(E)$. These are algebraic mappings, algebra homomorphisms for the wedge product, and satisfy $j^*\circ\Phi^* = Id$. $\Phi^*\circ j^*$ is a projection onto the image of Φ^*. With these operators we may write $\tilde{L}_{\chi X}\mu = j^*\, L_{\chi X}\, \Phi^*\,\mu = j^*\, i_{\chi X}\, d\, \Phi^*\,\mu$, where $L_{\chi X}$ is the usual Lie derivative. Thus $\tilde{L}_{\chi X}\mu$ is $C^\infty(M)$-linear in X.

We may also regard $\tilde{L}$ as a linear connection on the vector bundle $\bigsqcup_{x\text{ in }M}\Omega(E_x) \to M$, whose fibre is the nuclear Frechet space $\Omega(S)$.

<u>Definition</u>: The connection Φ is called μ-<u>respecting</u>, if $\tilde{L}_{\chi X}\mu = 0$ for all X in (M).

LEMMA: 1. A connection Φ is μ-respecting if and only if for some (any) bundle atlas (U_α,ψ_α) as in lemma 6.5 the corresponding Christoffel forms Γ^α take values in the Lie algebra $\mathfrak{X}_{\mu_o}(S)$ of divergence free vector fields ξ ($L_\xi\mu_o = 0$).

2. On E there exist many μ-respecting connections.

3. If Φ is μ-respecting and R is its curvature, then for X_x, Y_x in T_xM the form $i(R(\chi X_x,\chi Y_x))\mu_x$ is closed in $\Omega^{n-1}(E_x)$.

Proof: 1. Put $\Phi^\alpha := (\psi_\alpha^{-1})^*\Phi$, a connection on $U_\alpha\times S$. Its horizontal lift is given by $\chi^\alpha(X) = (X,\Gamma^\alpha(X))$. Then $\tilde{L}_{\chi^\alpha(X)}\mu_o = L_{\Gamma^\alpha(X)}\mu_o$ and this is zero for all X in $\mathfrak{X}(U_\alpha)$ if and only if Γ^α is in $\Omega^1(U_\alpha, \mathfrak{X}_{\mu_o}(S))$.

2. By 1 μ-respecting connections exist locally and since $\tilde{L}_{\chi X}\mu$ is $C^\infty(M)$-linear in X we may glue them via a partition of unity on M.

3. Choose local vector fields X, Y on M extending X_x and Y_x such that $[X,Y] = 0$. Then $[\chi X,\chi Y] = R(\chi X,\chi Y)$, which we denote $R(X,Y)$ for short (see the proof of 3.3). We then have for the embedding $j_x: E_x \to E$:

$$d(i_{R(X,Y)}\mu|E_x) = d\, j_x^*\, i([\chi X,\chi Y])\, \Phi^*\mu = j_x^*\, d\, i([\chi X,\chi Y])\, \Phi^*\mu =$$
$$= j_x^*\, L_{[\chi X,\chi Y]}\Phi^*\mu = j_x^*\, [L_{\chi X}, L_{\chi Y}]\Phi^*\mu = 0.$$
qed.

6.7. So we may consider the de Rham cohomology class $[i(R(X,Y))\mu|E_x]$ in $H^{n-1}(E_x)$ for X, Y in $\mathfrak{X}(M)$. $H^{n-1}(p) := \bigsqcup_{x \text{ in } M} H^{n-1}(E_x) \to M$ is a vector bundle, described by the cocycle of transition functions $U_{\alpha\beta} \to GL(H^{n-1}(S))$, $x \to H^{n-1}(\psi_{\beta\alpha}(x,.))$, which are locally constant. So $H^{n-1}(p) \to M$ admits a unique flat linear connection ∇ respecting the resulting discrete structure group, and the induced covariant exteriour derivative $d_\nabla: \Omega^k(M; H^{n-1}(p)) \to \Omega^{k+1}(M; H^{n-1}(p))$ defines the de Rham cohomology of M with values in the twisted coefficient domain $H^{n-1}(p)$. We may view $[i_R\mu]: X_x, Y_x \to [i(R(X_x,Y_x))\mu_x]$ in $H^{n-1}(E_x)$ as an element of $\Omega^2(M; H^{n-1}(p))$.

LEMMA: 1. If γ in $\Gamma(\Lambda^k V^*E)$ induces a section $[\gamma]: M \to H^k(p)$, then $\nabla_X[\gamma] = [\tilde{L}_{\chi X}\gamma]$.

2. If Φ is a μ-respecting connection on E then $d_\nabla[i_R\mu] = 0$.

Proof: Choose a bundle atlas (U_α,ψ_α) as in lemma 6.5. Then the parallel sections of the vector bundle $H^k(p)$ for the unique flat connection respecting the discrete structure group are exactly those which in each bundle chart $U_\alpha \times H^k(S)$ are given by the (locally) constant maps $U_{\alpha\beta} \to H^k(S)$. Obviously $\nabla_X[\gamma] = [\tilde{L}_{\chi X}\gamma]$ defines a connection, and in $U_\alpha \times \Omega^k(S)$ we have $\tilde{L}_{\chi^\alpha(X)}\gamma = \tilde{L}_{(X,\Gamma^\alpha(X))}\gamma = d\gamma(X) + L_{\Gamma^\alpha(X)}\gamma$, where d is the exteriour derivative on U_α of $\Omega^k(S)$-valued maps. But if $[\gamma]$ is parallel, so constant, then $L_{\Gamma^\alpha(X)}\gamma = d\, i_{\Gamma^\alpha(X)} + i_{\Gamma^\alpha(X)}0$ is exact anyhow, so $d\gamma(X)(x)$ is exact. So 1 follows.

2. Let X_0, X_1, X_2 in $\mathfrak{X}(M)$. Then we have

$$d_\nabla[i_R\mu](X_0,X_1,X_2) = \Sigma_{\text{cyclic}}\{\nabla_{X_0}([i_R\mu](X_1,X_2)) - [i_R\mu]([X_0,X_1],X_2)\}$$
$$= \Sigma_{\text{cyclic}}[\, \tilde{L}_{\chi X_0}(i_{R(\chi X_1,\chi X_2)}\mu) - i_{R(\chi[X_0,X_1],\chi X_2)}\mu\,]$$
$$= \Sigma_{\text{cyclic}}[\, i([\chi X_0,R(\chi X_1,\chi X_2)])\mu + i_{R(\chi X_1,\chi X_2)}\tilde{L}_{\chi X_0}\mu - \dots\,]$$
$$= [\Sigma_{\text{cyclic}}\, i([\chi X_0,R(\chi X_1,\chi X_2)] + 0 - R(\chi[X_0,X_1],\chi X_2))\mu] ,$$

where we used $[\tilde{L}_{\chi X}, i_Z] = i_{[\chi X, Z]}$ for a vertical vector field Z.
Now we need the Bianchi identity $[R,\Phi] = 0$. Writing out the global formula for it from the appendix for horizontal vector fields χX_i we get:
$0 = [R,\Phi](\chi X_0, \chi X_1, \chi X_2) = \Sigma_{cyclic}(-\Phi([R(\chi X_0, \chi X_1), \chi X_2]) - R([\chi X_0, \chi X_1], \chi X_2))$
From this it follows that $d_\nabla[i_R\mu] = 0$. qed.

6.8. Definition: So we may define the Kainz-class $k(E)$ of the bundle (E,p,M,S) with fibre volume as the class (for μ-respecting connection)

$$k(E) = [[i_R\mu]_{H^{n-1}(p)}]_{H^2(M;H^{n-1}(p))}.$$

LEMMA: The class $k(E)$ is independent of the choice of the μ-respecting connection Φ on E.

Proof: Let Φ, Φ' both be μ-respecting connections, let χ, χ' be their horizontal liftings. Put $\Psi := \chi' - \chi$: $TM \times_M E \to VE$, which is fibre linear over E, then $\chi_t := \chi + t\Psi = (1-t)\chi + t\chi'$ is a curve of horizontal lifts which preserve μ. In fact $L_{\Psi X}\mu = 0$ (ordinary Lie derivative) for any X in $\mathfrak{X}(M)$. Let Φ_t be the connection corresponding to χ_t and let R_t be its curvature. Then we have $R_t(X,Y)$ $(= R_t(\chi_t X, \chi_t Y)) = \Phi_t[\chi_t X, \chi_t Y] =$

$$= [\chi_t X, \chi_t Y] - \chi_t[X,Y] = R(X,Y) + t[\chi X, \Psi Y] + t[\Psi X, \chi Y] + t^2[\Psi X, \Psi Y] - t\Psi[X,Y].$$

$$\frac{d}{dt} R_t(X,Y) = [\chi X, \Psi Y] + [\Psi X, \chi Y] + 2t[\Psi X, \Psi Y] - \Psi[X,Y].$$

$$\frac{d}{dt} i(R_t(X,Y))\mu = i_{[\chi X, \Psi Y]}\mu - i_{[\chi Y, \Psi X]}\mu + 2t\, i_{[\Psi X, \Psi Y]}\mu - i_{\Psi[X,Y]}\mu.$$

$$= [\tilde{L}_{\chi X}, i_{\Psi Y}]\mu - [\tilde{L}_{\chi Y}, i_{\Psi X}]\mu + 2t[L_{\Psi X}, i_{\Psi Y}]\mu - i_{\Psi[X,Y]}\mu$$

$$= \tilde{L}_{\chi X}\, i_{\Psi Y}\mu - \tilde{L}_{\chi Y}\, i_{\Psi X}\mu - i_{\Psi[X,Y]}\mu + 2t(0 + d\, i_{\Psi X}\, i_{\Psi Y}\mu).$$

Now we use 6.8.1 to compute the differential of the 1-form $[i_\Psi\mu]$:

$$d_\nabla[i_\Psi\mu](X,Y) = \nabla_X[i_{\Psi Y}\mu] - \nabla_Y[i_{\Psi X}\mu] - [i_{\Psi[X,Y]}\mu]$$

$$= [\, \tilde{L}_{\chi X}\, i_{\Psi Y}\mu - \tilde{L}_{\chi Y}\, i_{\Psi X}\mu - i_{\Psi[X,Y]}\mu\,].$$

So we get $\frac{d}{dt}[i_{R_t}\mu] = d_\nabla[i_\Psi\mu]$ in the space $\Omega^2(M; H^{n-1}(p))$. qed.

6.9. There is another interpretation of the Kainz class $k(E)$: Consider the Serre spectral sequence of the fibre bundle $E \to M$. Its E_2-term is just the cohomology of M with twisted coefficients in $H^*(p)$. Then we start with the element in $H^0(M; H^n(p))$ given by the fibre volume μ and consider its second differential in $H^2(M; H^{n-1}(p))$. This is exactly the Kainz class. So it does not depend on the choice of the fibre volume μ, only on the orientation of the bundle it defines. I have not been able to find a pure differential geometric proof of this fact.

7. SELF DUALITY

7.1. Let again (E,p,M,S) be a fibre bundle with compact standard fibre S. Let ω be a fibrewise symplectic form on E, so that ω_x in $\Omega^2(E_x)$ is a symplectic form on the fibre E_x for each x in M and depends smoothly on x. So as in 6.5 we may plug two vertical vector fields X, Y into ω to get a function $\omega(X,Y)$ on E.

If $\dim S = 2n$, let $(\omega^n)_x = \omega_x \wedge \ldots \wedge \omega_x$ be the Liouville volume form of ω_x on E_x. Let us suppose that each fibre E_x has total mass 1 for this fibre volume: multiply ω by a suitable positive smooth function on M.

I do not know whether lemma 6.5 remains true for ω.

7.2. A connection Φ on E is again called ω-respecting, if $L_{\chi X}\omega :=$ $= \frac{d}{dt}|_0 (Fl_t^{\chi X})^* \omega = 0$ for all X in $\mathfrak{X}(M)$.

LEMMA: If $\dim H^2(S;\mathbb{R}) = 1$ then there are ω-respecting connections on E and they form a convex set in $\mathrm{Conn}(E)$.

Proof: Let us first investigate the trivial situation. So let $E = M \times S$, $\omega: M \to Z^2(S)$ is smooth and ω_x is a symplectic form on S for each x in M. Since $\int \omega_x{}^n = 1$ for each x and $\alpha \to \alpha^n$ is a covering map $H^2(S) = \mathbb{R} \to$ $\to H^{2n}(S) = \mathbb{R}$ (outside 0), we see that the cohomology class $[\omega_x]$ in $H^2(S)$ is locally constant, so $\omega: M \to \omega_{x_0} + B^2(S)$ if M is connected. Thus $d^M\omega: TM \to B^2(S)$ is a 1-form on M with values in $B^2(S)$. Let $G: \Omega^k(S) \to \Omega^k(S)$ be the Green operator of Hodge theory for some Riemann-metric on S and define Γ in $\Omega^1(M,\mathfrak{X}(S))$ by $i(\Gamma(X_x))\omega_x = - d^* G\, (d^M\omega)_x(X_x)$. Since ω_x is non degenerate, Γ is uniquely detemined by this proceedure and we have $L_{\Gamma(X_x)}\, \omega_x = d^S\, i_{\Gamma(X_x)}\, \omega_x + 0 = - d\, d^*\, G\, (d^M\omega)_x(X_x) =$

$$= - (d^M\omega)_x(X_x), \text{ since it is in } B^2(S).$$

Also $\tilde{L}_{\chi X}\omega = \tilde{L}_{(X,\Gamma(X))}\omega = (d^M\omega)(X) + L_{\Gamma(X)}$, so Γ is the Christoffel form of an ω-respecting connection and $(\tilde{L}_{\chi X}\omega)_x$ depends only on X_x.

So there are local solutions and and we may use partition of unity on M to glue the horizontal lifts to get a global ω-respecting connection. Likewise the last assertion follows. qed.

7.3. Let Φ be an ω-respecting connection on (E,p,M,S). Let $R = R(\Phi)$ be the curvature of Φ. Then $i(R(\chi X_x,\chi Y_x))\omega_x$ is in $\Omega^1(E_x)$ and $i(R(\chi X_x,\chi Y_x))\,\omega_x^n$ is in $\Omega^{2n-1}(E_x)$ and as in 6.7 we may prove that both are closed forms, so $[i(R(\chi X_x,\chi Y_x))\omega_x]$ is in $H^1(E_x)$ and the class of the second form is in $H^{2n-1}(E_x)$

Let us also fix an auxiliary Riemannian metric on the base manifold M and let us consider the associated $*$-operator $*: \Omega^k(M) \to \Omega^{m-k}(M)$, where $m = \dim M$. Let us suppose that the dimension if M is 4, let us consider $[i_R\omega]$ in $\Omega^2(M; H^1(p))$ and $*[i_R\omega^n]$ in $\Omega^2(M; H^{2n-1}(p))$. Both are closed forms on M (before applying $*$), since the proof of 6.8 applies without change. Let $D_x: H^{2n-1}(E_x) \to H^1(E_x)$ be the Poincaré-duality operator which induces a duality operator $D: H^{2n-1}(p) \to H^1(p)$ for the twisted coefficient domains.

Definition: An ω-respecting connection Φ is called self dual or anti self dual if $D * [i_R\omega^n] = [i_R\omega]$ or $= -[i_R\omega]$ respectively.

7.4. For the self duality notion of 7.3 there is also a Yang-Mills functional, if M is compact: Let Φ be any connection (not necessarily ω-respecting. For X_i in T_xM consider the following 4-form.

$\varphi(\Phi)(X_1,\dots,X_4) = \frac{1}{4}\,\mathrm{Alt} \int_{E_x} i(R(\chi X_1,\chi X_2)\omega_x \wedge_S *(i_R\omega^n)_x(X_3,X_4)$, where Alt is the alternator of the indices of the X_i. The Yang-Mills functional is then $F(\Phi) = \int_M \varphi(\Phi)$.

A typical example of a fibre bundle with compact symplectic fibres is the union of coadjoint orbits of a fixed type of a compact Lie group. Related to this are open subsets of Poisson manifolds. ω-respecting connections exist if all orbits have second Betti number 1. The Yang-Mills functional exists even without this requirement and also the self-duality requirement of 7.3 can be formulated without it.

7.5. Now we scetch another notion of self duality: We consider a fibre-wise contact structure on the fibre bundle (E,p,M,S) with compact standard fibre S. So let α_x be a contact structure on E_x for each x in M, depending smoothly on x. In more detail: α_x is in $\Omega^1(E_x)$ and $\alpha_x \wedge d\alpha_x^n$ is a positive volume form on E_x, where $\dim S = 2n+1$ now. A connection Φ on E is again called α-respecting, if $L_{\chi X}\alpha = 0$ for each vector field on M. If R is the curvature of an α-respecting connection then $i(R(\chi X,\chi Y))\alpha$ is fibrewise locally constant. If S is connected, then it is a closed 2-form on M with real coefficients and defines a cohomology class in $H^2(M; \mathbb{R})$. But it is not clear that α-respecting connections exist.

A. APPENDIX: THE FRÖLICHER NIJENHUIS BRACKET

We include here a short review of those properties of the Frölicher Nijenhuis bracket that we need. References are (Michor, 1987) and (Frölicher-Nijenhuis 1956).

A.1. The space $\Omega(M) = \oplus\, \Omega^k(M)$ of differential forms on a manifold M is a $\mathbb{Z}$-graded commutative algebra. A linear mapping $D: \Omega(M) \to \Omega(M)$ is called of degree k if $D(\Omega^h(M)) \to \Omega^{h+k}(M)$, and it is a graded derivation of degree k if $D(\varphi \wedge \psi) = D\varphi \wedge \psi + (-1)^{hk} \varphi \wedge D\psi$ for φ in $\Omega^h(M)$. The space Der $\Omega(M)$ = $= \oplus\, \mathrm{Der}_k \Omega(M)$ of all graded derivations is a graded Lie algebra with the graded commutator $[D_1, D_2] = D_1 \circ D_2 - (-1)^{k_1 k_2} D_2 \circ D_1$ as bracket.
A derivation D is called algebraic, if $D|\Omega^0(M) = 0$; then the derivation property says that it is of tensorial character, uniquely determinend by $D|\Omega^1(M)$, which can be viewed as a section of $\Lambda^{k+1} T^*M \times TM \to M$. If K denotes this section, we write $D = i(K)$ and we have:

LEMMA: For K in $\Omega^{k+1}(M;TM)$, φ in $\Omega^h(M)$ and X_i in $\mathfrak{X}(M)$ we have

$$i(K)\varphi\,(X_1, \ldots, X_{k+h}) = = \frac{1}{(k+1)!\,(h-1)!} \,\Sigma\, \mathrm{sign}\sigma\, \varphi(K(X_{\sigma 1}, \ldots, X_{\sigma(k+1)}), X_{\sigma(k+2)}, \ldots).$$

A.2. The exteriour derivative d is a graded derivation which is not algebraic. In view of the well known equation $L_X = d\, i_X + i_X\, d$ we put $L_K = L(K) := [i(K), d]$ for K in $\Omega^k(M;TM)$ and call it the Lie derivation along K. Then $L(Id_{TM}) = d$.

LEMMA: Any derivation D in $\mathrm{Der}_k \Omega(M)$ can be written uniquely in the form $D = L(K) + i(P)$ for K in $\Omega^k(M;TM)$ and P in $\Omega^{k+1}(M;TM)$. D is algebraic if and only if $K = 0$, and $P = 0$ if and only if $[D,d] = 0$.

A.3. Definition: For K_j in $\Omega^{k_j}(M;TM)$ we clearly have $[[L(K_1), L(K_2)], d] =$ $= 0$, so $[L(K_1), L(K_2)] = L([K_1, K_2])$ for a unique $[K_1, K_2]$ in $\Omega^{k_1+k_2}(M;TM)$, which is called the Frölicher Nijenhuis bracket of K_1, K_2. For vector fields it coincides with the Lie bracket and it gives a graded Lie algebra structure to $\Omega(M;TM) = \oplus \Omega^k(M;TM)$.

A.4. THEOREM: 1. For K in $\Omega^k(M;TM)$ and P in $\Omega^{p+1}(M;TM)$ we have

$$[L(K),i(P)] = i([K,P]) - (-1)^{kp}L(i(P)K).$$

2. For K_j in $\Omega^{k_j}(M;TM)$ and φ in $\Omega^q(M)$ we have

$$[\varphi\wedge K_1,K_2] = \varphi\wedge[K_1,K_2] - (-1)^{(q+k_1)k_2}L(K_2)\varphi\wedge K_1 + (-1)^{q+k_1}d\varphi\wedge i(K_1)K_2.$$

3. For X,Y in $\mathfrak{X}(M)$, φ in $\Omega^q(M)$, ψ in $\Omega(M)$ we have

$$[\varphi\otimes X,\psi\otimes Y] = \varphi\wedge\psi\otimes[X,Y] + \varphi\wedge L_X\psi\otimes Y - L_Y\varphi\wedge\psi\otimes X + (-1)^q(d\varphi\wedge i_X\psi\otimes Y + i_Y\varphi\wedge d\psi\otimes X).$$

A.5. THEOREM: For K in $\Omega^k(M;TM)$ and P in $\Omega^p(M;TM)$ we have:

$$\begin{aligned}
[K,P](X_1,\dots,X_{k+p}) &= \\
&= \frac{1}{k!\,p!}\sum \operatorname{sign}\sigma\,[K(X_{\sigma 1},\dots,X_{\sigma k}),P(X_{\sigma(k+1)},\dots,X_{\sigma(k+p)})] \\
&+ \frac{1}{k!\,(p-1)!}\sum \operatorname{sign}\sigma\,P([K(X_{\sigma 1},\dots,X_{\sigma k}),X_{\sigma(k+1)}],X_{\sigma(k+2)},\dots) \\
&+ \frac{(-1)^{kp}}{(k-1)!\,p!}\sum \operatorname{sign}\sigma\,K([P(X_{\sigma 1},\dots,X_{\sigma p}),X_{\sigma(p+1)}],X_{\sigma(p+2)},\dots) \\
&+ \frac{(-1)^{k-1}}{(k-1)!(p-1)!\,2!}\sum \operatorname{sign}\sigma\,P(K([X_{\sigma 1},X_{\sigma 2}],X_{\sigma 3},\dots),X_{\sigma(k+2)},\dots) \\
&+ \frac{(-1)^{(k-1)p}}{(k-1)!(p-1)!\,2!}\sum \operatorname{sign}\sigma\,K(P([X_{\sigma 1},X_{\sigma 2}],X_{\sigma 3},\dots),X_{\sigma(p+2)},\dots)
\end{aligned}$$

A.5. Let $f:M\to N$ be a smooth mapping between smooth second countable manifolds. Two vector valued differential forms K in $\Omega^k(M;TM)$ and K' in $\Omega^k(N;TN)$ are called f-related, if for all X_i in T_xM we have $K'_{f(x)}(T_xf.X_1,\dots,T_xf.X_k) = T_xf.K_x(X_1,\dots,X_k)$. If this is the case then the pullback operator $f^*: \Omega(N)\to\Omega(M)$ intertwines $i(K)$, $i(K')$ and also intertwines $L(K)$, $L(K')$.

LEMMA: If K_j and K'_j are f-related for $j = 1,2$, then their Frölicher Nijenhuis brackets $[K_1,K_2]$ and $[K'_1,K'_2]$ are also f-related.

A.6. If X is a vector field on M, then $(Fl^X_t)^*: \Omega^k(M;TM)\to\Omega^k(M;TM)$ makes sense locally.

LEMMA: $\frac{d}{dt}(Fl^X_t)^*K = (Fl^X_t)^*[X,K]$ for the Frölicher Nijenhuis bracket.

A.7. Let me finally remark that the Frölicher Nijenhuis bracket is not the only natural bilinear concomitant between vector valued differential forms. See Kolař-Michor, 1987, for the determination of all these concomitants. There are no surprising ones.

REFERENCES

M.C. Abbati, R. Cirelli, A. Mania, P. Michor: The action of the gauge group on the space of connections. J. Math. Physics.

Th. Bröcker, K. Jänich: Einführung in die Differentialtopologie. Springer-Verlag 1973.

M. De Wilde, P. Lecomte: Algebraic characterizations of the algebra of functions and the Lie algebra of vector fields of a manifold. Compositio Math. 45 (2) (1982), 199-205.

D.G. Ebin, J.E. Marsden: Groups of diffeomorphisms and the motion of an incompressible fluid. Ann. Math. 92 (1970), 102-163.

A. Frölicher, A. Kriegl: Linear spaces and differentiation theory. Pure and applied Math. J. Wiley 1988.

A. Frölicher, A. Nijenhuis: Theory of vector valued differential forms I. Indagationes Math. 18 (1956), 338-359.

S. Halperin: Personal communication, Grenoble, June 1987.

G. Kainz: A universal connection for fibre bundles, preprint 1984.

G. Kainz: Eine charakteristische Klasse für F-Faserbündel. Manuscript 1985.

G. Kainz: Dissertation, Wien 1988.

I. Kolar, P. Michor: Determination of all natural bilinear operators of the type of the Frölicher-Nijenhuis bracket. Proc. Winter school Geometry and Physics, Srni 1987, Suppl. Rendiconti Circolo Mat. di Palermo.

P. Michor: Manifolds of differentiable mappings. Shiva Math. Series 3, Orpington (GB) 1980.

P. Michor: Manifolds of smooth maps IV: Theorem of De Rham. Cahiers Top. Geo. Diff. 24 (1)(1983), 57-86.

P. Michor: Differentialgeometrie II, Lecture course, winter 1986/87, Vienna, Austria.

M. Modugno: An introduction to systems of connections. Preprint Firenze 1986.

J. Morrow: The denseness of complete Riemannian metrics. J. Diff. Geo. 4 (1970), 225-226.

J. Nagata: Modern dimension theory, North Holland 1965.

K. Nomizu, H. Ozeki: The existence of complete Riemannian metrics. Proc. AMS 12 (1961), 889-891.

R.S. Palais: A global formulation of the Lie theory of transformation groups. Mem. AMS 22 (1957).

T.R. Ramadas: On the space of maps inducing isomorphic connections. Ann. Inst. Fourier 32 (1982), 263-276.

QUANTUM FIELD THEORY IN CURVED SPACETIME

Bernard S. Kay
Institute for Theoretical Physics
University of Zurich
Schönberggasse 9
CH-8001 Zürich
Switzerland

ABSTRACT. I begin with an informal introduction to the subject of Quantum Field Theory in Curved Spacetime, indicating its status as an approximate theory, its basic physical effect, and its range of validity. I emphasize the importance of the Hawking effect, and of the fact that - while an approximation - the subject appears to admit a consistent mathematical and conceptual framework in its own right. After some brief historical and motivational remarks, I then outline such a suitable mathematical framework for the case of a linear field theory. Its principal elements are a suitable algebra and a suitable set of admissible states. The discussion of the latter incorporates recent work of R.M. Wald and the author. Evidence is given that this framework is complete, allows for the definition of a suitable energy-momentum tensor, and - through a result of R.M. Wald and the author - gives new insight into the Hawking effect. I end by stating two conjectures whose resolution may lead to a deeper understanding of the subject. I point out a close connection between these conjectures and a question posed by Fredenhagen and Haag in their recent axiomatic discussion of quantum gravity.

1. QUANTUM FIELD THEORY IN CURVED SPACETIME AND ITS RELATION TO THE REST OF PHYSICS

1.1. Introduction

I would like to talk about Quantum Field Theory in Curved Spacetime (QFT in CST). QFT in CST is - as I shall explain in a moment - a simple model theory which describes certain situations where both general relativity and quantum theory are important, but where a full quantum-gravity theory is not necessary. It's not intended to be an exact theory - but rather it's an approximate theory with its own particular range of validity. To indicate how it fits in with the rest of physics, I've drawn in figure 1 a simplified model of what might be called the modern physicist's Weltanschauung. We have a hierarchy of

K. Bleuler and M. Werner (eds.), Differential Geometrical Methods in Theoretical Physics, 373–393.

theories with a fundamental theory of everything at the top and then a downward-branching cascade of approximate theories - each of which may be considered (when appropriate conditions are satisfied) as an approximation to the theory immediately above it [A]*.

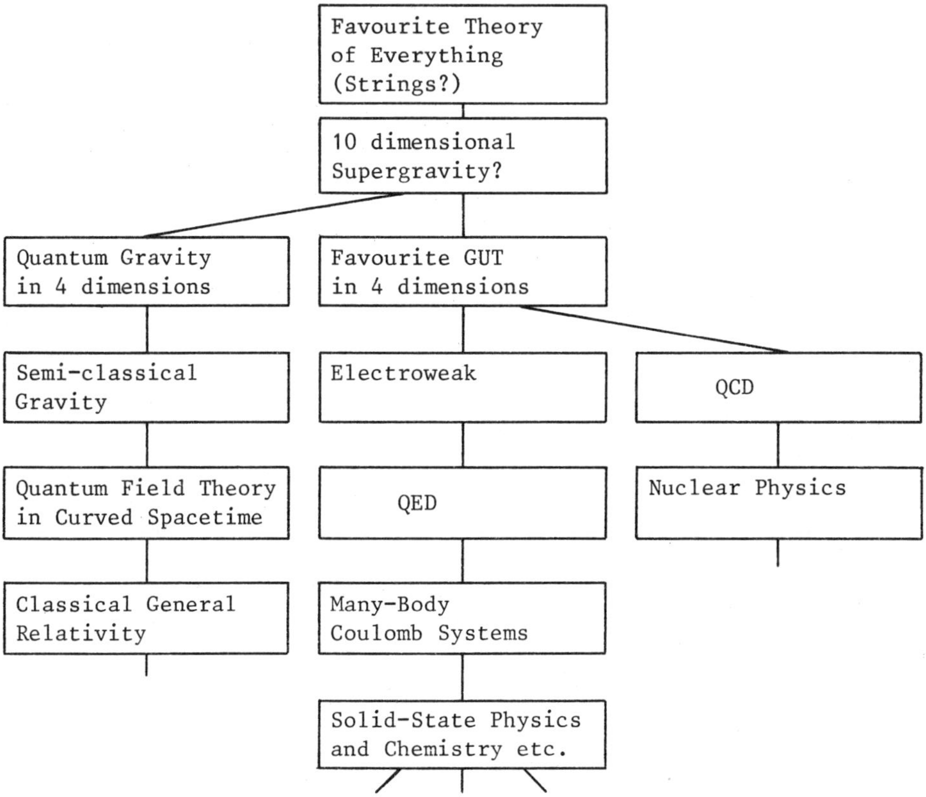

Fig. 1. "The Modern Physicist's Weltanschauung"

QFT in CST (strictly speaking Classical General Relativity with QFT in CST corrections) is located some way down the column where "gravitational physics" branches off from the rest of physics. I shall refer to fig. 1 again in a little while. First, let me begin to tell you about QFT in CST and about its range of validity.

1.2 The Basic Physical Effect of QFT in CST

As its name suggests, QFT in CST is a theory where one quantizes a

* The symbols [A], [B], etc. refer to the last section of this article entitled "Notes and Addenda", the symbols [1], [2] etc. to the references.

"matter field" in a classical (i.e. non-quantum), external (i.e. non-dynamical) gravitational field (i.e. curved spacetime). Its basic physical effect [B] is that a time-dependent gravitational field can create pairs of particles out of the vacuum. Thus, if, for example, one takes a heavy object (e.g. a ball of lead) and makes it vibrate, then, in principle, γ-rays, neutrinos and (if one could attain frequencies $\sim 10^{23}$ s^{-1}) electron-positron pairs etc. will be created out of the vacuum. In practice, the cross-sections and energies obtainable in the laboratory will, of course, be very small, and one has to look to extreme conditions comparable to those believed to have occurred in the early universe for important effects.

This pair-creation effect is closely analogous to the pair-creation effect one predicts if one quantizes e.g. a Dirac field in a time-dependent external electromagnetic field in flat spacetime [C].

1.3. The Range of Validity of QFT in CST

As for the range of validity of QFT in CST, one expects it to be a good approximation to quantum gravity for physical situations where the following conditions hold:

(1) (Since the gravitational field is assumed not to be quantized) Typical frequencies of sources (of the gravitational background) are very much less than the Planck frequency ($(c^5/G\hbar)^{1/2} \sim 10^{43}$ s^{-1}) and

(2) (Since the gravitational field is assumed not to be dynamical) A typical measure of the energy-momentum of the created particles is very much less than a typical measure of the energy-momentum of the sources.

1.4 The Hawking Effect

The most spectacular prediction of QFT in CST is undoubtedly the Hawking effect [1]. Hawking showed that QFT-in-CST-effects cause a (say spherically symmetric) black hole formed by collapse to radiate thermal radiation at a temperature $T = (8\pi M)^{-1}$ (in natural units of $k^{-1}(c^5\hbar/G)^{1/2}$) - inversely proportional to the mass M of the black hole. For example, a 10^{15} gram black hole (if a black hole with such a small mass could ever have been formed) would have a Hawking temperature equivalent to ~ 1 MeV and a luminosity of $\sim 10^{11}$ Watts. One sees from these figures that - for a black hole of this mass - quantum effects can certainly not be neglected. Nevertheless, (according to Conditions (1) and (2) above) quantum gravity effects should be unimportant and QFT in CST should be a good approximation for such a black hole (typical frequencies at this temperature are $\sim 10^{23}$ s^{-1} - 20 orders of magnitude below the Planck frequency) and should remain a good approximation for a long time [D]. One estimates that $\sim 10^8$ years will elapse before an appreciable fraction of the mass is lost [D]. (Of course, the question of what happens after a substantial amount of mass has been radiated away, and especially at the final stages of the presumed [D] "black hole evaporation" remains an outstanding challenge for any proposed theory of everything.)

Clearly a theory which is capable of predicting such an astonishing phenomenon is worthy of study, and I shall cite the Hawking effect as my first reason for studying QFT in CST.

1.5. QFT in CST as a Consistent Theory

I wish to also state a second reason for studying QFT in CST. To explain this, let me return for a moment to figure 1. In principle, the boxes in such a hierarchy may be classified into two categories: On the one hand, there are theories (let us call them "green" theories) which admit a consistent mathematical and conceptual framework in their own right. On the other hand, there are theories (let us call them "red" theories) which - if pushed too hard - reveal ambiguities or inconsistencies. To make full sense of a theory of the latter red type, one must ultimately appeal to some higher (green) level of the hierarchy. A priori, all conceivable colour schemes are possible. Thus, in a given hierarchy, a green theory may approximate a red theory, and a red theory a green theory quite independently of the colours of the other boxes. The only exception to this rule is that we would, of course, like our candidate for a fundamental theory of everything (assuming the hierarchy has a top!) to be coloured green.

It is not always an easy matter to decide which boxes should be coloured green and which red [E]. Nevertheless, I think everyone would agree that this is an important question to settle and that - when settled - the green boxes ought to have a privileged status in the scheme of things. I want to claim that QFT in CST has such a privileged status (and, in fact, appears to be the last green box in its branch of the hierarchy before the top). This is my second reason for studying QFT in CST: Viewed as a theory in its own right, QFT in CST [F] (at least when we model the matter by a linear field theory) appears to admit a consistent conceptual and mathematical framework. I believe it's interesting and worthwhile to get as clear as possible an understanding of this framework. Such an understanding may shed light on some of the less secure [A] higher levels of the hierarchy of theories.

1.6. The Aim of this Talk

The main aim of the rest of this talk will be (in Part 2) to describe such a suitable mathematical and conceptual framework for QFT in CST - as far as it is presently understood - and to make clear where open problems still lie. I shall take as my basic setting a globally hyperbolic spacetime [6] (M, g) and - as my matter-field model - the covariant Klein-Gordon equation [G]

$$(\Box_g + m^2)\phi = 0 \qquad (1)$$

I shall adopt as my basic ideology the algebraic approach to quantum field theory, and discuss first (in Section 2.1)

(i) the definition of a suitable field algebra

and then (in Section 2.2)

(ii) the definition of a suitable set of admissible states

Section 2.2 will be in many ways the main part of the talk, and incorporates recent work of R.M. Wald and the author. I will then give (in Section 2.3) three pieces of evidence that (i) and (ii) contain the essence of the theory. Finally - regarding the questions which are still open - I shall state (in Section 2.4) two conjectures whose resolution may lead to a deeper understanding of the subject. I shall also point out a close connection between these conjectures and a question posed by Fredenhagen and Haag in their recent axiomatic discussion of quantum gravity [15].

The account of the mathematical structure which I shall give in Section 2 - while brief - is rather complete. However I shall not have time to say as much as I would like about <u>why</u> I believe this is a good way to formulate things. Nor will there be time to illustrate <u>how</u> it provides all the necessary concepts for a physical interpretation. (Nor are my references in any way complete, although I do hope that the references in the references are.) To partially alleviate this situation, I would like to preface my account with a few historical and motivational remarks:

1.7. Some Historical and Motivational Remarks

The subject of QFT in CST goes back to work of Schrödinger (see the talk of L. Halpern at this conference [7]) who, already in the early 1930's, realized that a time-dependent gravitational field would give rise to pair creation. In subsequent decades, there was sporadic work both on the conceptual background and on cosmological applications and the subject had become quite popular by the late nineteen sixties and early seventies. However, the field only really took off with the announcement of the Hawking effect in 1974. Since then, there have been 13 years of intense activity. Several hundred papers have been (and continue to be) written by some hundred or so authors. The developments up to 1982 are summarized in the monograph of Birrell and Davies [8] (to which I refer for a detailed bibliography), and I know of four current projects to write further books emphasizing different aspects of the subject.

In the course of all this work, the subject has undergone several conceptual revolutions. <u>Most importantly</u>, we've learnt that the concepts "vacuum" and "particle" have only limited validity and to focus attention instead on local observables such as the field itself and its energy-momentum tensor. By learning to ask the right questions, we have seen that several initial apparent paradoxes disappear.

It was realized quite early on by several people that the algebraic approach to quantum field theory constitutes a particularly suitable mathematical framework for the subject since it focusses attention precisely on local observables, and clearly separates the issue of defining the theory from questions regarding possible preferred states which might deserve the name of "vacua" and "particles". During the last five years or so, such mathematical work has reached a stage where it is able to shed new light on some of the deeper aspects of the subject and especially on the Hawking effect. The framework I sketch below (and which is restricted to the case of linear fields) is closest to that given in recent work of Wald and myself [9] (to which I refer for more details and references) but (as [9]) it is also closely related to and in part influenced by earlier work of myself (see e.g. [10] and references therein and especially [11] - note also [5] and references therein), of Dimock [12] (see also [13] and references therein), of Sewell [4], and also of Haag Narnhofer and Stein [14] and Fredenhagen and Haag [15]. [14] discusses an axiomatic framework for interacting (i.e. non-linear) fields in curved spacetime, while [15] is an interesting attempt to incorporate the lessons learnt from a study of QFT in CST in a possible axiomatic framework for quantum gravity. (We shall mention [15] again in Section 2.4.)

By restricting ourselves here (and in [9]) to a linear model field theory, things are sufficiently simple as to allow a constructive approach. In fact, once the conceptual problems are sorted out, one can build on lessons which were learnt in the 1960's and 70's and earlier when Segal and Wightman and others discussed general mathematical frameworks for linear field theories. These lessons taught how questions about linear field theories may be reduced to questions about the underlying classical system together with the theory of quasi-free states and representations of the canonical commutation relations etc. (There is a vast literature on these matters which it is not possible to indicate here. See the references in the references - especially [10]) It should be emphasized, however, that QFT in CST raises genuinely new questions and, in particular, brings out an interesting interaction between the "old" functional analytic structures and the geometry of spacetime. In the light of these new curved-spacetime questions, the study of linear field theories remains of great interest: In fact, there are several unsolved problems in the non-linear case for which the linear case remains our only guide. Moreover, there are still interesting unsolved problems even in the linear case. (See the Conjectures in Section 2.4.)

2. SKETCH OF A CONCEPTUAL AND MATHEMATICAL FRAMEWORK FOR QFT IN CST

2.1. A Suitable Field Algebra

Given a globally hyperbolic spacetime (M, g) and the covariant Klein-

Gordon equation [G]

$$(\Box_g + m^2)\phi = 0 \qquad (1)$$

on (M, g), it's straightforward (here and elsewhere, we refer to [9] for for details and references -see also [10]) to construct a suitable *-algebra (with identity I) A generated by objects $\hat{\phi}(F)$, F real-valued in $C_0^\infty(M)$ (with the interpretation of "smeared quantum fields"

$$\hat{\phi}(F) = \int_M \hat{\phi}(x)F(x)|g|^{1/2}d^4x \;)$$

satisfying

(i) $\hat{\phi}(F)^* = \hat{\phi}(F)$

(ii) $((\Box_g + m^2)\hat{\phi})(F) \; (= \hat{\phi}((\Box_g + m^2)F)) = 0$

(iii) $[\hat{\phi}(F_1), \hat{\phi}(F_2)] = i\Delta(F_1, F_2)I$

where $\Delta(F_1, F_2)$ is the smeared (advanced minus retarded) fundamental solution to (1) - known (cf. [12]) to exist on globally hyperbolic spacetimes. (Formally, we have

$$\Delta(F_1, F_2) = \iint_{M\times M} \Delta(x, y)F_1(x)F_2(y)|g(x)|^{1/2}|g(y)|^{1/2}d^4x d^4y \quad (2))$$

It is important to note that the algebra A thus defined contains smeared versions of products of fields at different points such as $\hat{\phi}(x)\hat{\phi}(y)$ (they are just products of smeared fields), but not of objects like products of fields at the same point $\hat{\phi}(x)^2$ or $\hat{T}_{\mu\nu}(x)$. (For a discussion of such composite fields, see Note [I].)

2.2. A Suitable Set of Admissible States on A

2.2.1. We now turn to the question: What is a suitable definition for the set of "admissible states" on A? First we recall the definition of a state and some other basic notions from the algebraic approach to QFT:

1) A state ω is defined to be a positive (i.e.

$$\omega(A^*A) \geq 0 \quad \text{for all A in } A \qquad (3)\;)$$

normalized (i.e. $\omega(I) = 1$) linear (i.e. $\omega(\lambda A + \mu B) = \lambda\omega(A) + \mu\omega(B)$ etc.) functional on A, and is defined by specifying all its smeared n-point functions $\omega(\hat{\phi}(F_1)\ldots.\hat{\phi}(F_n))$.

2) The GNS Construction: Given a *-algebra A together with a state ω, one can always find a representation ρ of A as operators on a Hilbert

space H and a cyclic vector Ω in H such that

$$\omega(A) = \langle\Omega|\rho(A)\Omega\rangle$$

for all A in A. (The operators $\rho(A)$ will typically be unbounded, but this is not an important point here - see also the discussion in [15])

3) The Folium of a State: It will be convenient to use the notion (taken from [15]) of the folium of a state: Given an algebra A and a state ω as above, the folium of ω is defined to consist of the set of all states $\omega_\sigma(\cdot)$ defined by

$$\omega_\sigma(A) = \mathrm{tr}(\sigma\rho(A))_H$$

where σ ranges over the density matrices (i.e. trace-class operators with unit trace) on the GNS Hilbert space H, and ρ is the GNS representation belonging to (A, ω). (Notice that ω belongs to its own folium: To see this, take σ to be the projector onto the GNS cyclic vector Ω.)

The set of all states on A turns out to be much "too large". For example it contains states with "infinite" or undefinable energy-density (see Note [I]). We shall restrict this set in two steps [H]:

2.2.2. First Step: For a linear field theory, it seems reasonable to restrict attention to the union of the folia of some suitable set of quasi-free states. Recall (for simplicity we assume as part of the definition that the one-point function vanishes) that the quasi-free states are defined by the property that all their n-point functions are determined by the two-point function according to the combinatorial rules for Gaussian processes. For example, the four-point function is given in terms of the two-point function by

$$\omega(\phi(F_1)\phi(F_2)\phi(F_3)\phi(F_4)) = \omega(\phi(F_1)\phi(F_2))\omega(\phi(F_3)\phi(F_4))$$
$$+ \omega(\phi(F_1)\phi(F_3))\omega(\phi(F_2)\phi(F_4)) + \omega(\phi(F_1)\phi(F_4))\omega(\phi(F_2)\phi(F_3))$$

One can show that such quasi-free states are completely characterized by their (smeared) anticommutator functions $G(F_1, F_2)$:

$$G(F_1, F_2) = \omega(\phi(F_1)\phi(F_2) + \phi(F_2)\phi(F_1)) \qquad (4)$$

(Formally given in terms of a symmetrized two-point function $G(x, y)$ via - cf. (2) -

$$G(F_1, F_2) = \iint_{M\times M} G(x, y)F_1(x)F_2(y)|g(x)|^{1/2}|g(y)|^{1/2}d^4x\,d^4y \quad (5))$$

Each such G will automatically be a weak bisolution in the sense that

$$G((\Box_g + m^2)F_1, F_2) = G(F_1, (\Box_g + m^2)F_2) = 0 \qquad (6)$$

One can show that the positivity condition (3) for states implies the corresponding positivity conditions for G's:

$$\text{(i)} \quad G(F, F) \geqq 0 \quad \text{for all } F \text{ in } C_0^\infty(M) \qquad (7)$$

$$\text{(ii)} \quad \Delta(F_1, F_2) \leqq G(F_1, F_1)^{1/2} G(F_2, F_2)^{1/2} \text{ for all } F_1, F_2 \text{ in } C_0^\infty(M) \qquad (8)$$

Conditions (6), (7), (8) characterize G's which come from quasi-free states. Moreover, to every bilinear function G on $C_0^\infty(M)$ satisfying (6), (7), (8), there is a quasi-free state ω on A.

The (union of the folia of the) set of all quasi-free states remains too large. (It still contains states with infinite energy density.) But - within this class of states - we know how to impose a further "admissibility condition":

2.2.3. Second Step: This condition, which is motivated in part by the equivalence principle, and in part by the problem of renormalizing the energy-momentum tensor (see Note [I]) might be summarized by the demand that "the short-distance behaviour of our states should be what it ought to be". In more detail, the idea - which has a long history (see especially [16]) - is to demand that our quasi-free states are those whose two-point functions (we shall now work with half the anticommutator function $\frac{1}{2}G$) have the "Hadamard form" (for pairs of points x, y in a common convex normal neighbourhood [6] so that s below is well-defined - see Point (3) below):

$$(\tfrac{1}{2}G)(x, y) = H(x, y) + (\tfrac{1}{2}G)_{\text{smooth}}(x, y)$$

where $(\frac{1}{2}G)_{\text{smooth}}$ is a smooth two-point function which depends on the state, while H(x, y) is a universal (i.e. geometry-dependent, but state-independent) singular (on the light cone) piece taking the form:

$$H = -\frac{1}{(2\pi)^2}\left(\frac{U}{s} + V \ln s\right) \qquad (9)$$

where s is the square of the geodesic distance between x and y and U and V are smooth two-point functions fixed by demanding that U(x, x) = 1 and that

$$((\Box_g^x + m^2)H)(x, y)$$

(or equivalently the same expression with the Laplace-Beltrami operator acting on the y variable) be a smooth two-point function of x and y (i.e. a "Hadamard parametrix").

To make this idea into a precise definition, we need to deal with three points:

1) To fix U and V by the above demand, one typically takes U to be the "van Vleck-Morette determinant" and fixes V by recursion relations for the coefficients v_n in the power series

$$V = \sum_{n=0}^{\infty} v_n s^n \qquad (V,\ v_n \text{ each functions of x and y})$$

in s. The resulting power series is then usually only asymptotic.

2) To unambiguously translate the above idea into a definition of the smeared two-point function, one needs to impose the correct "ε-prescription" [H]:

$$\underline{\text{Replace}}\ s\ \underline{\text{by}}\ s - 2i\varepsilon(t(x) - t(y)) - \varepsilon^2$$

(where t is some choice of time coordinate which increases towards the future) on the right hand side of equation (9) and define H to be the real part of the resulting expression. (The imaginary part then gives the Hadamard expansion of half the commutator function $\frac{1}{2}\Delta$!) The limit as ε tends to zero is then to be taken after the integration (cf. (2), (5)) with the test functions.

3) (This is a crucial point which we shall re-examine in Section 2.4.) The Hadamard condition as stated above only determines the short-distance singularity "locally" - i.e. for pairs of points which lie in a common convex normal neighbourhood [6]. On the face of it, it still seems to allow the - presumably undesirable - possibility of non-local singularities [J] in the sense that for fixed x, there could be a critical value $y_{critical}$ of y (say spacelike separated from y and such that x and $y_{critical}$ do not lie in a single convex normal neighbourhood) such that G(x, y) blows up as $y \to y_{critical}$. One needs a global notion of Hadamard state which rules out such divergencies.

Robert Wald and myself have recently given (see [9] for details) a suitable definition of the notion of a global Hadamard state which takes into account the above three points.

2.2.4. <u>To summarize</u>: We define the set of admissible states to be the union of the folia of the quasi-free globally Hadamard states.

2.3. Evidence that the Definition of the Admissible States is a Good One

I now list three pieces of evidence that our definition of the set of admissible states is a good one:

1) First, one can show [9] that - on any globally hyperbolic

spacetime - there exist "many" quasi-free globally Hadamard states [K].

2) Second, for these states, one can define the expectation value of a suitable renormalized energy-momentum tensor (see Note [I] for more details).

3) Finally, for the quantized equation (1) on the Kruskal spacetime, Robert Wald and I have shown [9] (see also [17, 18]) that there is an "essentially unique" [L] Schwarzschild-isometry-invariant quasi-free globally Hadamard state. Moreover, this state, when restricted to an exterior Schwarzschild wedge and viewed with respect to Schwarzschild time-evolution, is a thermal equilibrium (i.e. KMS) state at the Hawking temperature $T = (8\pi M)^{-1}$ (cf. Section 1.4). For further discussion, see Note [L].

2.4. Two Conjectures

To end this talk, I would like to state two conjectures whose resolution will, I believe, deepen our understanding of QFT in CST. I shall also point out a connection between these conjectures and a question which Fredenhagen and Haag raise in their axiomatic program for quantum gravity [15].

First a definition: For any bounded region R, we define A_R to be the subalgebra of A generated by fields $\hat{\phi}(F)$ smeared with test functions F supported in R.

Conjecture 1. (see also Note[I]) Given any bounded open region R, there is a preferred (up to unitary equivalence) Hilbert space representation (say ρ_R of A_R on H_R) such that the restriction of each globally Hadamard state ω to A_R arises from a density matrix (say σ_ω^R on H_R):

$$\omega(A^R) = \mathrm{tr}(\sigma_\omega^R \rho_R(A^R))_{H_R} \quad \text{for all } A^R \text{ in } A_R$$

In other words, on any bounded open region, the globally Hadamard states define a unique folium.

The reason one expects this conjecture to be true is that the global Hadamard condition ensures that the ultra-violet behaviour is the same for all the states in question, while the assumption of a bounded region prevents any infra-red problems.

My second conjecture concerns the distinction between the local and global Hadamard notions explained in Section 2.2.3. I said there that, on the face of it, it seems that it might be possible for a state to be locally Hadamard, but still to have "non-local" singularities and hence not to be globally Hadamard. I now want to claim that, in actual fact, this cannot happen:

Conjecture 2.[M] Amongst the class of quasi-free states on A, the local and global Hadamard notions are equivalent.

Equivalently, reformulating things in terms of weak bisolutions i.e. of bilinear maps G from $C_0^\infty(M)\times C_0^\infty(M)$ to $\mathbb{R}$ satifsying (6) - cf. Sect. 2.2 :

Reformulation of Conjecture 2. Within the class of weak bisolutions G, the local Hadamard condition together with the positivity conditions (7), (8) imply the global Hadamard condition.

While I do not have time to present the evidence for its truth (see however Note [J]), I would like to emphasize that crucial to this conjecture are positivity and the precise factor $(1/2\pi)^2$ in the Hadamard condition (9): It is easy to find examples of weak bisolutions which are locally Hadamard but not globally Hadamard. But, the claim is that such bisolutions never arise from states - i.e. they violate the positivity conditions (7), (8). It is also possible to give examples of "pseudo-Hadamard" states which have non-local singularities - where by "pseudo-Hadamard" I mean that the factor of $(1/2\pi)^2$ in (9) is replaced by some strictly larger number c. The claim is that, for $c = 1/(2\pi)^2$, such non-local singularities are impossible. (Positivity prevents $c < 1/(2\pi)^2$.) This point is well illustrated by the de Sitter invariant states in de Sitter space [20].

I would like to draw attention to an interesting connection which seems to hold between this conjecture and a question which Fredenhagen and Haag pose as playing a central rôle in their axiomatic discussion of quantum gravity [15]. Their question might be paraphrased with the words:

"Are the laws of quantum gravity local?"

I will not pause to explain precisely what is meant by this question in its most general context. What will be important for us here is that one can ask an analogous question:

"Are the laws of QFT in CST local?"

in the framework of QFT in CST (- and that, hopefully, any progress in answering this latter question will shed some light on the former deeper question concerning quantum gravity.)

To explain both the meaning of this latter question and the connection with Conjecture 2, suppose first that the conjecture were not true: In this case, we could have two different quasi-free locally Hadamard states - one of which is globally Hadamard say, and one of which has non-local singularities. By reasoning similar to that we gave to justify belief in Conjecture 1, one expects that both these states would locally (i.e. in sufficiently small convex normal neighbourhoods about each point) define the same folium. However, because of the non-local singularities in the second state, one expects there to be "large" bounded open regions for which they define different folia. One therefore expects there to exist pairs of states which, while they both lie in the folia determined by the (globally) Hadamard states when restricted to sufficiently small regions, both lie in different folia when restricted to sufficiently large bounded regions. We would express this situation (in the spirit of [15]) by saying "The laws of QFT in CST are not local.".

On the other hand, suppose that Conjecture 2 were true (in the stronger version explained in Note [M]). Then, there would seem to be a chance that, if, for any (arbitrarily fine) covering of M by convex normal neighbourhoods N, a pair of states both lie in the folium determined by the (globally) Hadamard states on each region N, then they must both lie in the folium determined by the (globally) Hadamard states on any bounded open region. In this case, we would say that the laws of QFT in CST were local.

3. NOTES AND ADDENDA

A. A more accurate representation of the modern physicist's Weltanschauung might consist of a collection of competing alternative such hierarchies - each similar to fig. 1 in its lower levels, but with various competing arrangements of alternative theories occupying the upper levels. That the - until recently "best bet" - scenario selected for fig. 1 has already acquired - in the light of developments in string theories during the last few months - a rather old-fashioned look to it, provides a forceful illustration of this point. To qualify and complete this philosophy, we should add that sometimes not only downward branching, but also upward branching is possible. Moreover, each hierarchy of theories should be backed up by a network of "ideologies"/"methodologies" with boxes labelled "classical mechanics", "quantum mechanics", "thermodynamics" etc.

B. The statements in Section 1.2 need qualifying. Indeed the very notions of "vacuum" and "particles" have only a limited range of validity in a general curved spacetime context - see Section 1.7. This problem may be illustrated by the external-field-in-flat-spacetime analogy mentioned in Section 1.2 if one considers an external field which is everywhere strong and time-dependent.

C. While the analogy with flat-spacetime external field problems goes quite far, QFT in CST has interesting new features which transcend this analogy. (I have in mind here the Hawking effect [1] as well as the related Rindler-Fulling-Unruh story - see [2,3,4,5].) In so far as these new features have a single origin, it seems to be the new status which the notion of time acquires when one passes from flat spacetime - the kinematical arena of special relativity - to curved spacetime - the kinematical arena of general relativity. (Ironically, the Rindler-Fulling-Unruh story has to do with flat spacetime but from a curved-spacetime viewpoint!) Note that an even more drastic change of status in the notion of time occurs when one passes from the context of a fixed curved spacetime to the context of a dynamical gravitational field. It is tempting to speculate that there is a theory of quantum gravity waiting to be discovered which really does justice to this latter fully generally relativistic notion of time and which holds in store yet further surprises which transcend even the surprise of the Hawking effect!

D. Following Hawking [1], one estimates - by naively combining his formula with conservation of energy - a mass loss according to

$$\frac{dM}{dt} \cong c(\text{Area})(T^4) = c.16\pi M^2(8\pi M)^{-4}$$

(c a small number). This extrapolates to predict total evaporation with a lifetime ~ $((8\pi)^3/6c)(M/\text{Planck Mass})^3$ Planck times.

E. At the moment the informed hope/best bet seems to be that QCD may be tentatively coloured green since it has a chance of being constructable as an asymptotically free renormalizable quantum field theory. On the other hand, QED should probably be coloured red. One might hope for an intermediate level above QED but below strings which could be coloured green but, in the scheme depicted in fig. 1, the present feeling seems to be that all these intermediate levels have to be coloured red.

F. Included in our green box labelled "QFT in CST" would be the definition and properties of a suitable energy-momentum tensor operator - see Note [I]. The subsequent back-coupling of this energy-momentum tensor operator to the gravitational field (be it through some expectation value or matrix element as in semiclassical gravity, or "directly" as in quantum gravity) is to be regarded, however, as belonging to another (most likely red) level of the hierarchy - see fig. 1.

G. We assume a real scalar field ϕ. Most of the discussion in Section 2 goes through equally well if equation (1) is replaced by the massless wave equation or if the mass term is replaced by a scalar function (such as R/6) or even if (in the case of a complex field) we add an external electromagnetic field etc. Similarly, one could treat (with suitable replacements of CCRs by CARs etc.) the covariant Dirac equation or Maxwell equations etc. Some aspects of the theory of massless fields in CST require special treatment because of infra-red problems.

H. The admissibility condition outlined in Sections 2.2.2 and 2.2.3 - together with Conjecture 1 in Section 2.4 and the content of Note [I] - constitutes a sharper (in the light of the greater precision now available [9] in the notion of "Hadamard state") version of ideas which were first sketched by the author in [11]. To some extent, the details of the admissibility condition outlined here should be regarded as tentative and provisional. In fact, subsequently to (and independently of) [11],

Haag et al [14] proposed a related admissibility condition in the context of a possible axiomatic framework for not-necessarily-linear QFT in CST. Their condition takes the form of a short-distance scaling limit criterion (say SLC) on all the n-point functions of the states, and is in turn derived as a necessary condition for what they call the "principle of local definiteness" - PLD (see [14]). Note that Conjecture 1 here says essentially that, in the linear case, our "Hadamard" admissibility condition implies PLD. Note also that Haag et al show that SLC is not sufficient for PLD. It would be interesting to understand better the interrelationships between PLD, SLC and "Hadamard" (in the linear case). Presumably Hadamard is stronger than SLC (and for several issues (cf. Notes [I] and [L]) more detailed or stronger results are known in the linear case with the Hadamard assumption.) However, it is conceivable that SLC - together with the statement that the non-smooth part of the state depends only on the local geometry - implies Hadamard. Note in this connection that, in the presence of the field equation (6), fixing the correct leading singularity for the 2-point function may force the correct logarithmic term (see (9)). I thank R. Wald and G. Gonnella for discussions related to this point. Finally, let me note that the importance and significance of the epsilon-prescription (Point (2) in Section 2.2.3) was first clarified in the context of SLC in [14] where it was stressed that this epsilon prescription may be viewed as a local remnant of the spectral condition. (The full content of the spectral condition fails - of course - to generalize to non-globally-stationary spacetimes.)

<u>I</u>. The definition of composite operators may be illustrated by the problem of defining $\hat{\phi}^2(x)$. As noted in Section 2.1, such an object does not belong to A - even after smearing. Nevertheless, one can make sense of the expectation value of a renormalized $\hat{\phi}^2(x)$ in any of our admissible states ω by defining:

$$\omega(\hat{\phi}^2_{ren}(x)) = \lim_{y\to x} \{\omega(\hat{\phi}(x)\hat{\phi}(y)) - H(x, y)\} \qquad (I1)$$

where H(x, y) is the Hadamard parametrix (9). There are two points to be noted here. The first is that we are defining $\hat{\phi}^2{}_{ren}(x)$ only through its expectation values in the admissible states. A priori, we can talk about the "object itself" only if we think of it as the functional (states $\to$ $\mathbb{R}$-valued functions on M)

$$\omega \mapsto \omega(\hat{\phi}^2_{ren}(\cdot))$$

on the set of admissible states. (See however Conjecture 1A below.) A second crucial point is that the Hadamard parametrix is not uniquely determined (since, under changes of length scale - say from L to L′ - the division between "logarithmically singular part" and "finite part" changes: $ln(s/L^2) = ln(s/L'^2) + ln(L'^2/L^2)$) for a generic spacetime. In consequence, $\hat{\phi}^2{}_{ren}(x)$ is subject - in general - to the <u>finite renor-</u>

malization ambiguity

$$\hat{\phi}^2{}_{\text{ren}}(x) \to \hat{\phi}^2_{\text{ren}}(x) + cI$$

which may sometimes be fixed by a suitable renormalization condition. E.g. in the ground state [22,24] on a suitably asymptotically flat stationary spacetime, it is reasonable to demand that $\omega(\hat{\phi}^2_{\text{ren}}(x)) \to 0$ at spatial infinity. In special cases (such as conformally coupled matter in conformally flat spacetimes) the logarithmic divergence is absent and the ambiguity need not arise in the first place - a situation reflected in certain "analytic regularization schemes" by the apparent lack of need for any subtraction at all - but the physical significance of this fact is somewhat obscure and I shall term such cases "flukes". Note that the formula (I1) may be regarded as one particular manifestation ("point-splitting regularization and renormalization") of a general scheme:

$$\omega(\hat{\phi}^2_{\text{ren}}(x)) = \lim_{\Lambda\to\infty} \{\omega(\hat{\phi}^2_{\Lambda}(x)) - A(\Lambda)\} \tag{I2}$$

where Λ denotes some generic regularization procedure (or cutoff) and $A(\Lambda)$ is some suitable (state-independent) divergent function of Λ (ambiguous up to

$$A(\Lambda) \to A(\Lambda) + \alpha(\Lambda) \tag{I3}$$

where $\alpha(\Lambda)$ has a finite limit as $\Lambda \to \infty$) which is chosen such that, as the regularization is removed ($\Lambda \to \infty$) one obtains a finite result (again for any admissible state). It appears that this method is successful and gives the same finite result (up to the renormalization ambiguity) for any one of a wide range of regularization schemes - though in the present philosophy, the point-splitting scheme has a privileged status.

In the case of the energy-momentum tensor, a similar philosophy to the above applies except that one must now deal not only with products of fields but also with products of derivatives of fields at the same point. Thus one replaces eq. (I1) by an expression which involves suitable derivatives both of the two-point function and of the Hadamard parametrix. Another new feature is that the scheme must be modified so that the resulting $\omega(\hat{T}_{\mu\nu\text{ren}}(x))$ is covariantly conserved. A key paper on this point-splitting approach to the energy-momentum tensor is [16]. In other regularization schemes, it is often convenient to work in terms of a regularized action $S(g_{\mu\nu}, \hat{\phi})$ schematically (cf. eq. (I1)) defining a covariantly conserved $\omega(\hat{T}_{\mu\nu\text{ren}}(x))$ by

$$\omega(\hat{T}_{\mu\nu\text{ren}}(\cdot)) = \frac{2}{|g|^{1/2}} \lim_{\Lambda\to\infty} \omega\Big\{\frac{\delta}{\delta g'^{\mu\nu}}\Big(S^{\Lambda}(g'_{\mu\nu}, \hat{\phi}) - A_1(\Lambda)[\textstyle\int |g'|^{1/2}]^{\Lambda} I - A_2(\Lambda)[\int |g'|^{1/2} R']^{\Lambda} I - A_3(\Lambda)[\int |g'|^{1/2} R'^2]^{\Lambda} I - A_4(\Lambda)[\int |g'|^{1/2} C'^2_{\alpha\beta\gamma\delta}]^{\Lambda} I\Big)\Big\} \tag{I4}$$

(where it is to be understood that $\acute{g}_{\mu\nu}$ is set equal to the metric $g_{\mu\nu}$ on the given spacetime after the functional derivative is performed.) Here one needs to subtract the four "local geometrical terms" indicated, each with suitably divergent (in Λ) (state-independent) coefficients $A_i(\Lambda)$ (ambiguous up to $A_i(\Lambda) \to A_i(\Lambda) + \alpha_i(\Lambda)$ where $\alpha_i(\Lambda)$ converges) and the resulting $\hat{T}_{\mu\nu ren}$ is - in consequence - ambiguous up to corresponding "local geometrical terms"

$$\hat{T}_{\mu\nu ren} \to \hat{T}_{\mu\nu ren} + c_1 g_{\mu\nu} I + c_2 G_{\mu\nu} I + c_3 \overset{1}{H}_{\mu\nu} I + c_4 \overset{2}{H}_{\mu\nu} I.$$

(For a review of the computational state of the art using various regularization schemes, see [8]). The c_2, c_3, c_4-ambiguities in the Einstein tensor and "higher derivative" terms will - of course - disappear if the background spacetime satisfies Einstein's vacuum field equations. The c_1-ambiguity can also sometimes be fixed. E.g. in the ground state on a suitably asymptotically flat stationary spacetime, one would require $\omega(\hat{T}_{\mu\nu ren}(x))$ to go to zero at spatial infinity. (There are also "fluke" cases where the ambiguity never seems to arise just as discussed above for ϕ^2 - cf. [21].) One can also regard this renormalization ambiguity as being fixable by fixing the (four) coupling constants in a version of semiclassical gravity in which one adds two higher-derivative terms and a cosmological constant to the usual Einstein-tensor term before equating the resulting sum to some suitable expectation value or matrix element of the energy momentum tensor operator. (There are several possible versions of such a theory - each with its own distinct interpretation and each with its own difficulties - but a discussion of this would take us beyond the scope of this paper. See Note [F].)

So far, $\hat{T}_{\mu\nu ren}(x)$ has to be thought of (as we mentioned in the case of $\phi^2_{ren}(x)$) as a functional on the set of admissible states. I now want to conjecture that one can actually recover a good notion of a (smeared) energy-momentum tensor operator in the usual Hilbert-space sense, by adjoining to Conjecture 1 of Section 2.4 the following:

<u>Conjecture 1A.</u> Given any bounded open region R, then there is a (densely defined, unbounded) operator $\hat{T}^R_{\mu\nu}(F^{\mu\nu})$ (in a suitable sense affiliated to the von-Neumann algebra generated by $\rho_R(\mathcal{A}_R)$) on $\mathcal{H}_R$ such that, for all admissible states ω:

$$\int \omega(\hat{T}_{\mu\nu ren}(x)) F^{\mu\nu}(x) |g(x)|^{1/2} d^4x = \mathrm{tr}(\sigma^R_\omega \hat{T}^R_{\mu\nu}(F^{\mu\nu}))_{\mathcal{H}_R}$$

where $F^{\mu\nu}$ is any smooth tensor field compactly supported in R and ρ_R, $\mathcal{A}_R$, $\mathcal{H}_R$, σ^R_ω are as in Conjecture 1 (and a similar statement holds for ϕ^2 etc.).

(Such $\hat{T}^R_{\mu\nu}(F^{\mu\nu})$, $\phi^{2R}(F)$ will of course inherit any renormalization ambiguity.) This conjecture (modulo full precision in the notion of "Hadamard state") was first made in [11]. The above philosophy was also foreshadowed in [21] which discussed the definition of a renormalized $\hat{T}_{\mu\nu}(x)$ for the special case of locally flat spacetimes.

Finally, note that, if one accepts Conjectures 1 (of Sect. 2.4) and 1A (above), the final effect of restricting the set of all states to the set of admissible states will have been to effectively enlarge the algebra A (to something like the inductive limit of the local von Neumann algebras generated in the local representations ρ_R) so as to include composite objects such as the (spectral projections of the smeared) energy-momentum tensor. The set of admissible states on A will then be identifiable with something like the set of all (say normal) states on the resulting enlarged algebra.

J. The issue of non-local singularities appears first to have been raised by Najmi and Ottewill [19]. Note that the statements in [19] suggest that there do actually exist locally Hadamard states with non-local singularities - contrary to Conjecture 2 inSection 2.4. However, I believe the facts adduced towards this conclusion in [19] may be re-interpreted as evidence for Conjecture 2. To reconcile the apparently contradictory statements, I need to (and do) conjecture that the example of a Hadamard bisolution with non-local singularities $G^{(1)}$ given in [19] violates the positivity conditions (7), (8) of Section 2.2.2 here, and hence does not arise from any state (see also Note [M]).

K. Concerning Point (1) in Section 2.3: In fact it appears that almost all the "mode-splitting vacuum states" considered e.g. in [8] are quasi-free globally Hadamard states in our sense. In particular, one expects the natural ground states and KMS states which can be defined on (globally hyperbolic) stationary spacetimes (see [22, 10, 5]) to satisfy this condition. (Cf. [23, 24] which however used a notion of Hadamard state which lacked the full precision now attained in [9].) In general, the only exception to this rule is that a "mode-splitting vacuum state" fails to be Hadamard when it can be shown on "computational" grounds to be "singular" in the sense that e.g. its expectation value for the renormalized energy-momentum tensor diverges. E.g. presumably the Boulware and Unruh states are Hadamard on exterior Schwarzschild but fail to have Hadamard extensions to Kruskal because they are singular on the (in the case of Unruh, past) horizon. (These remarks are consistent with the uniqueness theorem [9] referred to in Point (3) in Section 2.3.)

L. The reason for the phrase "essentially unique" is that uniqueness is proven in [9] either on a certain ("large") subalgebra of the full algebra, or, on the full algebra but for states staisfying a certain ("mild") extra "no-zero-mode" condition. The full result in [9] concerns a large class of spacetimes with "bifurcate Killing horizons" which includes Kruskal, as well as Minkowski, de Sitter and many other examples. In the cases mentioned, the unique quasi-free globally Hadamard state presumably exists and is respectively the "Hartle-Hawking state", the Minkowski vacuum and the "Euclidean de Sitter vacuum". In some other cases (such as Kerr or Schwarzschild-de Sitter) we prove non-existence! (See [9] for full details, and [17, 18] for short summaries emphasizing different aspects of the work.) An obvious question in the light of the philosophy in this paper is whether this uniqueness result holds - not only amongst the set of quasi-free globally Hadamard states - but more generally amongst the set of all admissible states (i.e. amongst the union of the corresponding folia). This is a gap which I hope will soon be filled.

Finally, let me mention (cf. Note [H]) that Haag et al prove a related but weaker result in [14]. [14] considers states on *A* satisfying the KMS condition (with respect to Schwarzschild time evolution) at some temperature on an exterior Schwarzschild wedge, and shows that demanding that the state have an extension to all of Kruskal satisfying PLD everywhere forces the temperature to be the Hawking temperature.

M. There is a potential ambiguity in Sections 2.2.3, 2.3, 2.4 concerning the meaning of the phrase "(quasi-free) locally Hadamard state". This could either mean (a) that the state is globally Hadamard on every convex normal neighbourhood or (b) that, given the state, there exists a covering of *M* by convex normal neighbourhoods (maybe a very fine covering by very small such neighbourhoods) such that the state is (globally) Hadamard on each neighbourhood in the covering. The claim I wish to make is that Conjecture 2 also holds when one assumes the latter notion (b). It is this - stronger - form of the conjecture which has a close connection with the question "Are the laws of QFT in CST local?".

Note that the essential point of the conjecture is that positivity together with the local Hadamard notion implies the absence of non-local singularities. It may be that the notion of globally Hadamard implied does not have the full smoothness postulated in the definition of globally Hadamard given in [9]. I thank Robert Wald for a discussion on Conjecture 2 in which he raised this point.

The connection between Conjecture2 and the original question raised in [15] would be further clarified if the relationship between our Hadamard condition and SLC (see Note [H]) of [14], [15] were better understood.

ACKNOWLEDGMENTS

This work was supported in part by the Schweizerischer Nationalfonds.

I wish to thank Robert Wald for many useful discussions. I also thank Klaus Fredenhagen for a useful discussion.

Finally, I thank Giuseppe Gonnella, Tony Kakas and Urban Studer for carefully reading the manuscript.

REFERENCES

1. S.W. Hawking, Commun. Math. Phys. 43, 199 (1975)
2. S.A. Fulling, Phys. Rev. D7, 2850 (1973)
3. W.G. Unruh, Phys. Rev. D14, 870 (1976)
4. G.L. Sewell, Ann. Phys. (NY) 141, 201 (1982)
5. B.S. Kay, Commun. Math. Phys. 100, 57 (1985) (+ Erratum to appear)
6. R.M. Wald, General Relativity. Chicago University Press 1984
7. L. Halpern, In: Differential Geometric Methods in Theoretical Physics (Proceedings, Como 1987) ed. K. Bleuler. D. Reidel (to appear)
8. N.D. Birrell, P.C.W. Davies, Quantum Theory in Curved Space. Cambridge University Press 1982
9. B.S. Kay, R.M. Wald, Uniqueness of stationary non-singular quasi-free states in spacetimes with a bifurcate Killing horizon (to appear)
10. B.S. Kay, In: Differential Geometric Methods in Mathematical Physics (Proceedings, Clausthal 1980) ed. H.-D. Doebner, S.I. Andersson, H.R. Petry. Springer Lecture Notes in Mathematics 905 (1982)
11. B.S. Kay, Talk at the 10th International Conference on General Relativity and Gravitation (Padova, 1983). See "workshop chairman's report" by A. Ashtekar in the proceedings ed. by B. Bertotti, F. de Felice, A. Pascolini. D. Reidel 1984
12. J. Dimock, Commun. Math. Phys. 77, 219 (1980)
13. J. Dimock, B.S. Kay, Ann. Phys. (NY) 175, 366 (1987)
14. R. Haag, H. Narnhofer, U. Stein, Commun. Math. Phys. 94, 219 (1984)
15. K. Fredenhagen, R. Haag, Commun. Math. Phys. 108, 91 (1987)
16. R.M. Wald, Phys. Rev. D17, 1477 (1978)
17. B.S. Kay, R.M. Wald, In: Proceedings of the VII Italian National Congress on General Relativity and the Physics of Gravitation (Rapallo, 1986) eds. E. Massa, U. Bruzzo, R. Cianci. World Scientific Singapore 1987
18. B.S. Kay, R.M. Wald, In: Proceedings of the XV International Conference on Differential Geometric Methods in Theoretical Physics (Clausthal, 1986) ed. H.-D. Doebner. World Scientific, Singapore (to appear)
19. A.H. Najmi, A.C. Ottewill, Phys. Rev. D32, 1942 (1985)
20. B. Allen, Phys. Rev. D32, 3136 (1985)
21. B.S. Kay, Phys. Rev. D20, 3052 (1979)

22. B.S. Kay, Commun. Math. Phys. 62, 55 (1978)

23. S.A. Fulling, M. Sweeny, R.M. Wald, Commun. Math. Phys. 63, 257 (1981)

24. S.A. Fulling, F.J. Narcowich, R.M. Wald, Ann. Phys. (NY) 136, 243 (1981)

SUPERTHEORIES

Sergio Ferrara
CERN, Geneva, Switzerland
and
University of California, Los Angeles

1. INTRODUCTION

In the last decade the most fascinating development in the theory of elementary particle interactions has been the development of supersymmetric theories including internal and space-time gauge symmetries.[1] After it was established that supersymmetry, a symmetry exchanging bosons and fermions, can be consistently incorporated in Lagrangian quantum field theory, the discovery of supersymmetric theories of gravitation, called supergravities,[2] lead eventually to the recent major developments in superstring theories. Superstring theories,[3] since their revival occurred in 1984 with the anomaly cancellation mechanism of Green and Schwarz,[4] have been a rapidly changing field. The most stunning development which occurred in the last two years is probably the realization of an extremely rich variety of consistent superstring models living in space-time dimension less than (or equal) to ten.[5] Of particular interest are superstring theories directly formulated in four space-time dimensions with chiral fermions and N = 1 space-time supersymmetry.[6]

The present report is a brief review on some recent results in supergravity theories and their application to four-dimensional superstrings.

2. SUPERSYMMETRIC THEORIES

A supersymmetric theory is a theory in which the algebra of the space-time Poincaré group is enlarged by a grading representation, i.e., a set of fermionic generators $Q_\alpha{}^i$ where $\alpha = 1,\dots,4$ is a spinor index and i is a counting label. The basic anticommutation relation is[1]

$$\{Q_\alpha{}^i, Q_\beta{}^i\} = -2(\gamma^\mu C)_{\alpha\beta} P_\mu \delta^{ij} + (\text{central charges}) \qquad (2.1)$$

where C is the charge-conjugation matrix and P_μ is the space-time translation operator. A superalgebra with anticommutator (2.1) is called an N-extended supersymmetry algebra. In D = 4 space-time dimensions, massless particle multiplets are construed by a finite set

K. Bleuler and M. Werner (eds.), Differential Geometrical Methods in Theoretical Physics, 395–404.

of Poincaré representations with helicity states given by $\lambda, \ldots, \lambda - (N/2)$ where λ is the helicity of the Clifford vacuum of a given multiplet. Since, by PCT, each multiplet must contain states with opposite helicities, then it follows that the upper value of the helicity cannot be lower than N/4 for N even or (N+1)/4 for N odd. This gives the restriction on the possible number of N-extended algebras which can be realized in Lagrangian field theory with "consistent" interactions for the massless states. In theories not including gravity $N_{max} = 4$, in theories including gravity $N_{max} = 8$. The four-dimensional counting has its counterpart in the bound of the space-time dimension for which a supersymmetry algebra with particle multiplets up to spin one or two can exist. The bound $N_{max} = 8$ corresponds to the bound $D_{max} = 11$ while the bound $N_{max} = 4$ corresponds to $D_{max} = 10$.

It is worth mentioning that the upper critical dimension of superstrings corresponds precisely to D_{max}, i.e. the maximal space-time dimension in which combined (Yang-Mills) matter and gravity multiplets exist.

Superstring theories in which the massless modes only correspond to pure supergravity multiplets exist. They are called type II superstrings.[3] However, they do not correspond to D = 11 but rather to the two (inequivalent) versions of the N = 2, D = 10 super-Poincaré algebra.

What about the massive stringy modes? They are classified according to massive representations of the supersymmetry algebra. In this case, for D = 4, two kinds of multiplets exist: multiplets with vanishing central charges (see Eq. (2.1)) for which $J_{max} \geqslant N/2$ and those with non-vanishing central charges for which $J_{max} \geqslant N/4$ (or (N+1)/4 for odd N). In D = 4 space-time dimensions, the embedding of a gauge-Yang-Mills group, G, in the supersymmetry algebra requires that the generators T^A of G commute with the spinorial generators Q_α.[7] In Minkowski's space-time, there are only two exceptions to this rule. Superconformal supergravities[2] for which a U(N) group can be gauged. These theories, however, are non-unitary and require a gravity part of the Lagrangian which is quadratic in the Riemann curvature (more specifically in the Weyl tensor). Theories with a U(1) Abelian factor with a Fayet-Iliopoulos[8] term, in which case the gravitino is charged under the U(1) charge.[9]

For a non-Abelian gauge group $[T^A, Q_\alpha] = 0$, then it follows that particles and their superpartners belong to the same representation of G. Since a state with helicity $\lambda = +1/2$ is transformed into a state with helicity $\lambda = -1/2$ under the application of two spin-1/2 generators Q^1, Q^2, the only possibility of having supertheories with chiral fermions is to have a N = 1 theory. In this case the multiplet with $\lambda_{max} = 1/2$, in a complex representation of G, describes chiral fermions (with no mirror states).

For N = 1 supersymmetric theories in D = 4 dimensions, the massless representations are of the form

$$\left(\lambda, \lambda - \frac{1}{2} \right) \qquad \text{for } m = 0 \qquad (2.2)$$

$$J,\ J+\frac{1}{2}\ ,\ J-\frac{1}{2}\ ,\ J \quad \text{for } m \neq 0\ . \tag{2.3}$$

For $m = 0$ PCT conjugation adds states with helicities $1/2-\lambda$ and $-\lambda$ so one gets 4 states for $m = 0$ (arbitrary λ) and $4(2J+1)$ states for massive representations. The Wess-Zumino multiplet corresponds to $\lambda = 0$ (or 1/2) and $J = 0$. The Yang-Mills multiplet $\lambda = 1$ (or $-1/2$), the gravity multiplet to $\lambda = 2$ (or $-3/2$) and so on.

An important role in supersymmetric theories is played by the so called "auxiliary fields." They are the fields which must be added to the propagating modes in order to match the number of off-shell bosonic and fermionic degrees of freedom. This problem is completely solved for $N = 1$ and $N = 2$, $D = 4$ theories but is unresolved for $N = 4$, $D = 4$ theories or $D > 6$ higher dimensional theories.

The basic $N = 1$ multiplets of a $N = 1$ theory are the spin $(0^{\pm},1/2)$ multiplet[1,2]

$$z^i\ ,\ \chi_\alpha{}^i\ ,\ h^i \tag{2.4}$$

where z,h are complex scalars and χ is a Weyl spinor, i is a Yang-Mills index for a T-representation of G. The gauge multiplet

$$V_\mu{}^A\ ,\ \lambda_\alpha{}^A\ ,\ D^A \tag{2.5}$$

where λ is the gaugino and D is a real auxiliary field.

The (real) linear multiplet[10]

$$\phi,\ b_{\mu\nu},\ \psi_\alpha \tag{2.6}$$

when ϕ is real, $b_{\mu\nu}$ is a gauge antisymmetric tensor field (with gauge invariance $\delta b_{\mu\nu} = \partial_\mu\xi_\nu - \partial_\nu\xi_\mu$) and ψ_α is a Weyl spinor.

The gravity multiplet

$$e_{a\mu},\ \psi_\mu\ ,\quad \text{(auxiliary fields)} \tag{2.7}$$

where $e_{a\mu}$, ψ_μ are the vierbein and gravitino and the auxiliary fields have to be specified yet.

These multiplets have a "superfield representation" in superspace (x_μ,θ_α). The Wess-Zumino multiplet is called S^i, the linear multiplet L and the vector multiplet V^A. Multiplication rules of these multiplets exist which allow the construction of superinvariant Lagrangians.[1] In the case of superstring theories, there is always a "linear multiplet"[11] (2.6) which corresponds to the dilaton, the antisymmetric tensor and their fermionic superpartner. This multiplet, like the gravity multiplet, is "universal" in the sense that it is present in any consistent superstring model. The Wess-Zumino and vector multiplets are instead model dependent, i.e. they will depend on the particular Yang-Mills group and on the massless modes of the particular superstring model one is considering.

3. SUPERGRAVITY THEORIES REVISITED

In pure N = 1 supergravity an important role is played by its superconformal extension.[12] In fact, different auxiliary field formulations of N = 1 Poincaré supergravity (the gauge theory of the super-Poincaré group) get unified in the superconformal framework. In the present review, we will confine our discussion to the "minimal" formulation of N = 1 supergravity.

The N = 1 superconformal algebra can be described in terms of three basic elementary gauge fields[12]

$$e_{a\mu},\ \psi_\mu,\ A_\mu \tag{3.1}$$

where A_μ is an extra U(1) connection of the R-symmetry contained in the N = 1 superconformal algebra.[12]

The fields given in (3.1) form a complete supermultiplet as it can be checked by counting the bosonic and fermionic degrees of freedom (and by subtracting the gauge degrees of freedom). The super-Poincaré multiplets are obtained by introducing a conformal compensator which breaks the superconformal group to the super-Poincaré group, by fixing a particular conformal gauge.[13]

For a chiral compenmsator S_0 we get the old-minimal set of auxiliary fields[14]

$$S,\ P,\ A_\mu \tag{3.2}$$

where

$$S_0 = (1,\ 0,\ u = S - iP) \tag{3.3}$$

after gauge fixing. Note that A_μ has now four independent components since the U(1) gauge is also broken by the S_0 gauge fixing. For a linear compensator L, we get the new-minimal set of auxiliary fields[15]

$$A_\mu,\ b_{\mu\nu} \tag{3.4}$$

(with gauge invariances $\delta A_\mu = \partial_\mu\phi$, $\delta b_{\mu\nu} = \partial_\mu\xi_\nu - \partial_\nu\xi_\mu$) where

$$L = (1,\ 0,\ b_{\mu\nu}) \tag{3.5}$$

after gauge fixing. These different choices of compensators correspond to a different set of Poincaré supercurvatures of the corresponding superspace geometry.

In the old minimal formulation, we have the following three basic curvature multiplets[16]

$$W_{\alpha\beta\gamma},\ E_{\alpha\dot\alpha},\ R \tag{3.6}$$

where $W_{\alpha\beta\gamma}$ is the (superconformal) (chiral) Weyl multiplet which is "universal" since it does not depend on the conformal compensators. This multiplet contains the Weyl tensor in its θ-component and its square contains the Hirzenbruch signature density $R_{\mu\nu}{}^{ab}R^{\mu\nu cd}\epsilon_{abcd}$ in its θ^2 component.

The Ricci multiplet is a real multiplet which contains in its $\theta\sigma_\mu\bar{\theta}$ component the Ricci tensor and R is a chiral multiplet which contains in its $\theta\theta$ component the scalar curvature. These multiplets are related by Bianchi-identities as follows:[16]

$$\nabla^\alpha W_{\alpha\beta\gamma} = \nabla_\beta{}^{\dot{\tau}} E_{\gamma\dot{\tau}} + \nabla_\gamma{}^{\dot{\tau}} E_{\gamma\dot{\tau}} \quad , \quad (\bar{\nabla}_{\dot{\gamma}} W_{\alpha\beta\gamma} = 0) \tag{3.7}$$

$$\nabla^\alpha E_{\alpha\dot{\beta}} = \bar{\nabla}_{\dot{\beta}} \bar{R} \quad , \quad (\bar{\nabla}_{\dot{\beta}} R = 0) \tag{3.8}$$

Expressions of $E_{\alpha\beta}$ and R in terms of superderivative of the compensator multiplet S_0 are known. In the new-minimal formulation, the superspace geometry is described by the three following basic multiplets

$$W_{\alpha\beta\gamma},\ E_{\alpha\dot{\alpha}},\ W_\alpha \tag{3.9}$$

where the R multiplet has been replaced by a (chiral) field strength multiplet W_α which contains the scalar curvature in its θ component. These three multiplets are related by the following identities[17]

$$\nabla^\alpha W_\alpha = -\bar{\nabla}_{\dot{\alpha}} \bar{W}^{\dot{\alpha}} \tag{3.10}$$

$$W_\alpha = -i\, \sigma^a_{\alpha\dot{\alpha}} \bar{\nabla}^{\dot{\alpha}} E_a$$

The relation between $W_{\alpha\beta\gamma}$ and $E_{\alpha\dot{\alpha}}$ is the the same as in (3.7). Note that (3.10) implies that E_a is a linear multiplet in contrast with old-minimal supergravity. It is possible to define a (reducible) supermultiplet T_{ab} containing the full Riemann tensor whose decomposition in irreducible parts gives

$$T_{ab} = W_{ab} - \frac{1}{2}\sigma_{ab} W - \frac{1}{2}\epsilon_{abcd}\sigma_c \bar{\nabla} E_d \ . \tag{3.11}$$

In terms of this multiplet, the Gauss-Bonnet multiplet,[18] containing the R^*R and R^*R^* topological invariants, is simply given by[17]

$$T^*T = \bar{T}_{ab} T_{cd} \epsilon^{abcd} \ .$$

The W_α multiplet admits a gauge potential since from (3.10) it follows that

$$W_\alpha = \bar{\nabla}^2 \nabla_\alpha V_G \tag{3.12}$$

The V_G multiplet contains the A_μ U(1) connection in its $\theta\sigma_\mu\bar{\theta}$ component and the (supercovariant) scalar curvature in its $\theta^2\bar{\theta}^2$ component. In the superconformal framework, the pure supergravity Lagrangian are given by[13]

$$S_o \bar{S}_o \big|_D \qquad \text{old minimal} \tag{3.13}$$

$$-L \ln L\Big|_D \quad \text{new minimal} \tag{3.14}$$

where D is the superconformal D-density formula. In the Poincaré framework, they are given by

$$\det E\Big|_D \quad \text{old minimal} \tag{3.15}$$

$$V_G\Big|_D \quad \text{new minimal} \tag{3.16}$$

Eq. (3.14) is the Wess-Zumino result[19] that in old-minimal supergravity the supergravity Lagrangian is nothing but the superspace measure. Eq. (3.16) says instead that in the new-minimal approach pure supergravity is the Fayet-Iliopoulos term for the V_G gauge multiplet. Note that Eq. (3.16) is gauge invariant since the vector density of a chiral multiplet vanishes in new-minimal superspace (like in global supersymmetry). From the associated Bianchi identities of the curvature supermultiplets, we have seen that new and old minimal supergravity correspond to different superspace geometries. However, it turns out that in some cases, theories corresponding to these geometries turn out to be equivalent. This is possible whenever we can transform the chiral compensator S_0 into the linear compensator L via a (super)duality transformation. For this to be possible, the compensators must appear algebraically in the Lagrangian.[20] This is precisely the case for theories of gravity not including higher derivative terms.

Let us consider, for instance, a standard gauged Kahler model coupled to N = 1 supergravity[21]

$$\Phi(Z,\bar{Z}e^V)\, S_o\bar{S}_o\Big|_D + \mathrm{Re}\, f(Z)\, S_o^{\,3}\Big|_F + \mathrm{Re}\, f_{AB}(Z)\, W^A W^B\Big|_F \tag{3.17}$$

For such a model to be compatible with N = 1 new minimal supergravity, Φ and g must satisfy the following conditions[20]

$$n_i z_i \frac{\partial\phi}{\partial z_i} = n\bar{z}_i \frac{\partial\phi}{\partial z_i}\,, \quad n_i z_i \frac{\partial g}{\partial z_i} = 3g\,, \quad n_i z_i \frac{\partial f}{\partial z_i} = 0 \tag{3.18}$$

for some choices of n_i, the chiral weight of z_i. If this is the case, by sending $S_0 \to S_0 e^{-\Lambda}$, $Z \to Ze^{\Lambda}$, we get

$$\Phi(Z,\, \bar{Z}e^V,\, \Lambda+\bar{\Lambda})\, S_o\bar{S}_o\, e^{\Lambda+\bar{\Lambda}}\Big|_D + \text{terms independent of } \Lambda \tag{3.19}$$

which can be duality transformed by replacing $\Lambda + \bar{\Lambda} \to U$ (unconstrained real multiplet) and by adding the term LU with L linear.

4. FOUR-DIMENSIONAL SUPERSTRINGS

In the last two years a major advance in superstring theories has taken place with a deeper understanding of the constraints that local (world-sheet superconformal symmetry) and global (modular invariance) anomaly cancellation put on a particular superstring model. A large class of superstring theories have been constructed since then, many of them having chiral fermions and N = 1 space-time supersymmetry in four dimensions.[6]

It is not known whether this enormous class of superstring models can be considered as different string vacua of a unique higher dimensional theory like the ten-dimensional heterotic superstring.[22] A peculiar fact in superstring compactifications is the fact that new massless modes can be originated, in addition to the ordinary Kaluza-Klein modes. These modes can be regarded as "soliton" states in the bosonic construction or as ordinary states in the fermionic formulation when compactified coordinates are fermionized. The output of this is that the D = 10 heterotic superstring with gauge group SO(32) can have a four-dimensional vacuum with a larger gauge group SO(44).[23] This is the original thoroidal compactification discovered by Narain.[24] The gauge group SO(44) corresponds to a N = 4 supersymmetric theory whose point-field limit, in leading (α') order is given by N = 4 supergravity coupled to N = 4 Yang-Mills supermatter. By twisting the boundary conditions on the world-sheet in a way consistent with modular invariance, one can reduce both the gauge group and the number of supersymmetries.[6] The effect of the twisting is also the appearance of some new states (twisted sector) required by modular invariance. In the case of SO(44), the corresponding N = 2 and N = 1 theories have gauge groups of the form $SO(n_1) \times SO(n_2)$ for N = 2 and

$$\prod_{i=1}^{4} SO(n_i)$$

for N = 1.

The effective low-energy supergravity Lagrangians for these models have been constructed and an ansatz for the interactions of the massless twisted sector has also been given.[24] It can be shown that in all cases the Kahler potential[21]

$$K = -3 \log \phi \tag{4.1}$$

with ϕ as given by (3.17) and the superpotential and the f function satisfy Eq. (3.18), i.e. these theories can be formulated in the framework of new minimal supergravity.[15] Moreover, the coupling of the dilaton to the Yang-Mills curvature (field dependent gauge coupling)

$$\Phi F_{\mu\nu}^{2} \tag{4.2}$$

implies, by supersymmetry, a Green-Schwarz mechanism for the antisymmetric tensor field $b_{\mu\nu}$.[18,11] This means the replacement of $H_{\mu\nu\rho}$ with $H_{\mu\nu\rho} + a\Omega_{\mu\nu\rho}$ when $\Omega_{\mu\nu\rho}$ is the Yang-Mills Chern-Simons form. A similar mechanism occurs for the gravitational sector.

The relation between four-dimensional heterotic strings and different formulations of N = 1 supergravity has been recently clarified by looking at the structure of the gravitational supervertices. It turns out that the appropriate supergravity formulation of N = 1 4D-heterotic superstrings is the new minimal one.[26] The reason why this is the case is deeply rooted in the 2D superconformal algebra. It has been shown that in order to have space-time N = 1 supersymmetry, one needs a N = 2 superconformal symmetry on the world-sheet.[27]

The corresponding U(1) superconformal current

$$J(z) \tag{4.3}$$

can be used to construct a generator

$$R \sim \oint J(z) \tag{4.4}$$

which, when acting on physical vertices, generates an R-symmetry transformation. This R-symmetry can be promoted to a local symmetry since in the theory there is a vertex with the correct structure to correspond to a connection field for the R-symmetry:[26]

$$C^{\mu}(z,k) \sim J(z)\ \bar{\partial}X^{\mu}(\bar{z})\ \exp[ik\cdot X(z\bar{z}] \tag{4.5}$$

It is therefore natural to identify C_μ with the vertex operator of the A_μ auxiliary field gauge connection of new minimal supergravity. This identification can be proven by showing that the entire gravitational vertex, which transforms as the Einstein super-multiplet E_μ, fulfills the identities of new minimal supergravity:

$$\bar{D}^2 E_\mu = D^2 E_\mu = 0\ , \qquad D^\mu E_\mu = 0 \tag{4.6}$$

Note that in superfield language Eq. (4.5) gives the θ = 0 component of the E_μ multiplet. Eqs. (4.6) can be easily proven.

For isntance, for the θ = 0 component, we have

$$K^\mu C_\mu(z,k) = \frac{\partial}{\partial \bar{z}}\ [J(z)\ \exp(ik\cdot X)] \sim 0 \tag{4.7}$$

being a world-sheet total derivative which vanishes upon integration over a compact Riemann surface. Recently, direct superstring calculations have confirmed these results.[28]

5. CONCLUSIONS

The construction of vertex operators in string theories for all N = 1 supermultiplets, including the gravitational one, allows in principle to discuss string propagation in an arbitrary supersymmetric background, including the auxiliary fields. This may shed some light on the problem of supersymmetry breaking in superstring theories. These techniques have, in fact, been proven useful to show the possible occurrence of D-term breaking induced by string loop effects[29] in certain superstring models.

In order to have broken supersymmetry in a Minkowski background, a necessary condition is that the R-symmetry is broken, since the gravitino mass term

$$m_{3/2}\bar{\psi}_\mu \sigma^{\mu\nu} \psi_\nu \tag{5.1}$$

breaks this symmetry. Gaugino condensation[30] is precisely a mechanism which allows this phenomenon to occur. Whether more general stringy mechanisms are possible is one of the major problems to be solved to make superstring theories closer to reality.

REFERENCES

1. See, i.e., *Supersymmetry* (ed. S. Ferrara), World Scientific and North Holland (1987); *Supersymmetry and Supergravity* (ed. M. Jacob), North Holland and World Scientific (1986).
2. See, i.e., P. van Nieuwenhuizen, Phys. Rep. C68, 189 (1981).
3. See, i.e., M. Green and D. J. Gross, *Unified String Theories*, World Scientific (1985); M. B. Green, J. H. Schwarz, and E. Witten, Vols. I and II, *Cambridge Monographs on Mathematical Physics* (1987).
4. M. B. Green and J. H. Schwarz, Phys. Lett. 149B, 117 (1984).
5. K. Narain, Phys. Lett. 169B, 41 (1986); M. Mueller and E. Witten, Phys. Lett. 182B, 28 (1986).
6. H. Kawai, D. Lewellen, and S. Tye, Phys. Rev. Lett. 57, 1832 (1986); Nucl. Phys. B288, 1 (1987); I. Antoniadis, C. Bachas, and C. Kounnas, Nucl. Phys. B289, 477 (1987); D. Gepner, Princeton University Preprints 87-0370/87-0727 (1987).
7. S. Ferrara and B. Zumino, Nucl. Phys. B79, 413 (1974); A. Salam and J. Strathdee, Phys. Lett. 51B, 353 (1974).
8. P. Fayet and J. Iliopoulos, Phys. Lett. 251B, 461 (1974).
9. J. Bagger, Nucl. Phys. B211, 302 (1983).
10. S. Ferrara, J. Wess, and B. Zumino, Phys. Lett. 51B, 239 (1974); W. Siegel, Phys. Lett. 85B, 333 (1979); U. Lindstrom and M. Roĉek, Nucl. Phys. B222, 185 (1983).
11. S. Ferrara and M. Villasante, Phys. Lett. 186B, 85 (1987); S. Cecotti, S. Ferrara, and M. Villasante, Int. J. Mod. Phys. A2, 1839 (1987).
12. M. Kaku, P. K. Townsend, and P. van Nieuwenhuizen, Phys. Rev. D17, 3179 (1976).
13. T. Kugo and S. Uehara, Nucl. Phys. B226, 49 (1983); Progr. Theor. Phys. 73, 253 (1985).
14. S. Ferrara and P. van Nieuwenhuizen, Phys. Lett. 74B, 333 (1978); K. S. Stelle and P. C. West, Phys. Lett. 74B, 330 (1978).
15. M. F. Sohnius and P. C. West, Phys. Lett. 105B, 353 (1981).
16. See, e.g., B. Zumino in *Recent Developments in Gravitation*, Cargèse 1978 (eds. M. Levy and S. Deser), p. 405; J. Wess and J. Bagger in *Supersymmetry and Supergravity*, Princeton Series (1983).
17. S. Cecotti, S. Ferrara, M. Porrati, and S. Sabharwal, CERN-TH. 4786/87, UCLA/87/TEP/19; S. Ferrara, S. Sabharwal, and M. Villasante, UCLA/88/TEP/3.
18. S. Ferrara and B. Zumino, Nucl. Phys. B134, 301 (1978); P. S. Howe and R. W. Tucker, Phys. Lett. 80B, 138 (1978); P. K. Townsend and P. van Nieuwenhuizen, Phys. Rev. D19, 3592 (1979).
19. J. Wess and B. Zumino, Phys. Lett. 74B, 51 (1978).

20. S. Ferrara, L. Girardello, T. Kugo, and A. van Proeyen, Nucl. Phys. B223, 191 (1983).
21. E. Cremmer, B. Julia, J. Scherk, S. Ferrara, L. Girardello, and P. van Nieuwenhuizen, Nucl. Phys. B147, 105 (1979); E. Cremmer, S. Ferrara, L. Girardello, and A. van Proeyen, Nucl. Phys. B212, 413 (1983).
22. D. J. Gross, J. A. Harvey, E. Martinec, and R. Rohm, Phys. Rev. Lett. 54, 502 (1984).
23. K. S. Narain, M. H. Sarmadi, and E. Witten, Nucl. Phys. B279, 36 (1987).
24. K. S. Narain, Phys. Lett. 169B, 41 (1986).
25. S. Ferrara, L. Girardello, C. Kounnas, and M. Porrati, Phys. Lett. 192B, 368 (1987) and 194B, 358 (1978); I. Antoniadis, J. Ellis, E. Floratos, D. V. Nanopoulos, and T. Tomaras, Phys. Lett. 191B, 86 (1987).
26. S. Cecotti, S. Ferrara, and L. Girardello, Phys. Lett. 198B, 336 (1987).
27. W. Boucher, D. Friedan, and A. Kent, Phys. Lett. B172, 316 (1986); A. Sen, Nucl. Phys. B278, 289 (1986); B284, 423 (1987); L. Dixon, D. Friedan, and S. H. Shenker, Nucl. Phys. B282, 13 (1987); T. Banks, L. Dixon, D. Friedan, and E. Martinec, SLAC Preprint, SLAC-PUB-4377 (1987).
28. J. Lauer, D. Lüst, and S. Theisen, Max-Planck Preprint, MPI-PAE/Pth 86/87.
29. M. Dine, N. Seiberg, and E. Witten, Nucl. Phys. B289, 589 (1987); A. Atick, L. Dixon, and A. Sen, Nucl. Phys. B292, 109 (1987); M. Dine, I. Ichinose, and N. Seiberg, Nucl. Phys. B293, 253 (1987); A. Atick and A. Sen, SLAC-PUB-4292 (1987).
30. S. Ferrara, L. Girardello, and H. P. Nilles, Phys. Lett. 125B, 457 (1983); M. Dine, R. Rohm, N. Seiberg, and E. Witten, Phys. Lett. 156B, 55 (1985).

The Integrability Of N = 16 Supergravity

H.Nicolai
Institut für Theoretische Physik
Universität Karlsruhe

The maximally extended locally supersymmetric theory in two dimensions is N = 16 supergravity. This theory was recently constructed [1] by dimensional reduction of N = 16 supergravity in three dimensions [2]. On shell, it describes 128 bosonic and 128 fermionic degrees of freedom which are assigned to the two inequivalent spinor representations of SO(16). In three dimensions, their interactions are governed by an $E_8/SO(16)$ σ - model. In two dimensions, the nonlinearly realized E_8 invariance is enlarged to $E_9 = E_8^{(1)}$, the affine (Kac - Moody) extension of E_8. The emergence of infinite dimensional symmetries in two dimensions is related to the integrability of the theory: there exists a linear system whose compatibility conditions are equivalent to the equations of motion of the physical degrees of freedom [1]. More recently, an extension of this linear system has been discovered which also yields the equations of motion of some of the off-shell degrees of freedom [3]. This new system possesses a local N = 16 supersymmetry with transformation parameters ϵ^I (I = 1,...,16) subject to the condition

$$\gamma^\beta \gamma^\alpha D_\beta \epsilon^J = 0$$

In their bosonized form, the local N = 16 supersymmetry transformations become part of E_9, and thus all symmetries of the theory can be viewed as being contained in E_9 [3].

In view of the possible emergence of the hyperbolic algebra E_{10} (first conjectured by B.Julia [4]), it has been suggested that the N = 16 theory has a stringlike extension [1]. This is plausible because the algebra generated by all vertex operators of a string theory has a hyperbolic structure: the mass-shell conditions $\vec{k}^2 = 2, 0, -2, -4,...$ for the physical momenta coincide with the constraint $\vec{r}^2 \leq 2$, $\vec{r}^2 \in 2\mathbb{Z}$ obeyed by the "hyperbolic" roots $\vec{r}$ [5]. However, the precise structure and physical meaning of the E_{10} vertex operators is at present unknown.

[1] H.Nicolai, Phys.Lett. 194B (1987), 402; Karlsruhe preprint KA-THEP-87/10 (1987).
[2] N.Marcus and J.H.Schwarz, Nucl.Phys. B228 (1983), 145.
[3] H.Nicolai and N.P.Warner, in preparation.
[4] B.Julia, in "Superspace and Supergravity", C.U.P.,Cambridge (1981).
[5] V.G.Kac,"Infinite Dimensional Lie - Algebras",C.U.P., Cambridge (1985).
P.Goddard and D.I.Olive, in "Vertex Operators in Mathematics and Physics", Publications of MSRI Nr. 3, Springer, (1984).

K. Bleuler and M. Werner (eds.), Differential Geometrical Methods in Theoretical Physics, 405.

CONFORMAL INVARIANT SUPERSYMMETRIC THEORIES IN FOUR DIMENSIONS: A PRACTICAL APPLICATION OF BRS COHOMOLOGY[1,2]

C. LUCCHESI, O. PIGUET
Dép. de Physique Théorique
Université de Genève
CH - 1211 Genève 4

and K. SIBOLD
Max-Planck-Inst. für Physik
und Astrophysik
Werner-Heisenberg Inst. für Physik
D-8000 München 40

ABSTRACT. A criterion for conformal invariance in $N = 1$ supersymmetric gauge theories is established. Only one-loop order quantities are involved, thanks to the non-renormalization theorem for the chiral anomalies.

1. Introduction

In this talk we shall discuss the possibility of conformal invariance in four-dimensional field theory. Conformal invariance is generally broken by an anomaly, which expresses itself in a set of non-vanishing Callan-Symanzik β–functions. Examples of theories with vanishing conformal anomaly in four-dimensional space-time are known from a long time: these are the ultra-violet finite $N = 2$ and $N = 4$ extended supersymmetric gauge theories [1]. More recently, some massless gauge theories with $N = 1$ supersymmetry were argued to be finite [2], hence conformal invariant.

The requirement of complete finiteness is of course very strong: it requires the vanishing of all coupling constant renormalizations, i.e. of the β–functions, and also of the "wave function" renormalizations, i.e. of the anomalous dimensions. We shall weaken these conditions by demanding the vanishing of the β–functions only. This amounts to impose the condition of conformal invariance "on shell", which is certainly more physical than the condition of finiteness.

Our investigation results in a criterion [3,4] for the vanishing of the β–functions, to all orders of perturbation theory. This criterion involves only one-loop quantities; it applies to $N = 1$ supersymmetric gauge theories. Briefly stated: all the chiral anomalies of the model under consideration — of the gauge type and of the rigid $U(1)$ type — must vanish. Then, provided all coupling constants are certain functions of the gauge coupling constant [5], the β–functions vanish to all orders. The two key points are: (1) the non-renormalization theorem for the chiral anomalies [6,7] which allows one to decide of their vanishing through the calculation of one-loop graphs, and (2) a relation between conformal and chiral anomalies,

[1] Talk given by O. Piguet.
[2] Supported in part by the Swiss National Science Foundation.

K. Bleuler and M. Werner (eds.), Differential Geometrical Methods in Theoretical Physics, 407–415.

specific of supersymmetric theories [8].

The two points mentioned above provide the plan of the present talk, which is divided in two independent sections: *Section 2* sketches a proof of the non-renormalization theorem which relies on the BRS cohomological structure of any renormalizable gauge theory. We present it for an ordinary gauge theory in order not to obscure the argument with the introduction of the heavy superspace machinery. Although this theorem is considered in the literature as well-known and established [6], we think it useful to present here our proof [7] which, beyond to being quite general, offers some insights into its premices. *Section 3* is devoted to the application of the theorem in the proof of the criterion of superconformal invariance [4].

2. The Adler-Bardeen Theorem for the $U(1)$ Anomaly [7].

Let us consider a gauge theory in four-dimensional space-time, with a gauge group G assumed to be simple. The generalization to any compact Lie group may be found in the original paper [7]. The BRS transformations of the gauge connection A, of the Faddeev-Popov ghost fields c, with ghost number one, and of the matter fields ψ^R read

$$\begin{aligned} sA &= dc + [A, c] \quad , \\ sc &= c^2 \quad , \\ s\psi^R &= c_i T^i_{(R)} \psi^R \quad , \quad s\bar{\psi}_R = -\bar{\psi}_R T^i_{(R)} c_i \quad . \end{aligned} \tag{2.1}$$

The index R denotes both the field and the irreducible representation, with generators $T^i_{(R)}$, it belongs to. We use the matrix and differential form notation

$$A = -i\, A_{\mu i} \tau^i dx^\mu \quad , \quad c = c_i \tau^i \quad ,$$

$$[\tau^i, \tau^j] = i\, f^{ij}_{\ \ k} \tau^k \quad , \quad Tr\, \tau^i \tau^j = \delta^{ij} \quad . \tag{2.2}$$

The BRS operator s is nilpotent and anticommutes with the exterior derivative d:

$$s^2 = d^2 = sd + ds = 0 \quad . \tag{2.3}$$

The BRS–invariant classical action reads (matter fields are spinors: scalar fields may be included)

$$\begin{aligned} \Gamma_{\text{class}} = \int d^4x \Big[-\frac{1}{4g^2} Tr(F^{\mu\nu} F_{\mu\nu}) + i \sum_R \bar{\psi}_R (\not{\partial} - i \not{A}_i T^i_{(R)}) \psi^R \Big] + \\ + \text{gauge fixing} \quad , \end{aligned} \tag{2.4}$$

with (wedge symbol $\wedge$ suppressed)

$$F = -\frac{i}{2} F_{\mu\nu i} \tau^i dx^\mu dx^\nu = dA + AA \quad . \tag{2.5}$$

It is well known [9] that the theory is renormalizable, with BRS–invariance preserved, if the representation for the matter fields is chosen to be anomaly free, which we shall assume to be the case.

We suppose that the classical theory is invariant under a set of chiral transformations of the type

$$\delta_5 \psi^R = i\,\gamma^5 e^R_S \psi^S \quad , \quad \delta_5 \bar{\psi}_R = i\,\bar{\psi}_S \gamma^5 e^S_R \quad , \tag{2.6}$$

which commute with the BRS transformations. (We consider here a massless theory). The chiral charge matrix e is Hermitian, and $e^R_S = 0$ if R and S denote inequivalent irreducible representations. These invariances are preserved by renormalization and are expressed by the Ward identities

$$\delta_5 \Gamma = \int d^4x \, w_5(x) \Gamma = 0 \quad ,$$

$$w_5(x) = \delta_5 \psi^R \frac{\delta}{\delta \psi^R} + \delta_5 \bar{\psi}_R \frac{\delta}{\delta \bar{\psi}_R} \quad , \tag{2.7}$$

where $\Gamma(A, \psi, c, ...) = \Gamma_{\text{class}} + O(\hbar)$ is the "vertex functional", generating the one-particule-irreducible Green functions, and the functional differential operator $w_5(x)$ commutes with the BRS transformations.

The BRS–invariant axial current associated to the invariance under (2.6) is conserved classically (tree graphs approximation). In the quantum theory (loop graphs), its conservation law is broken by the anomaly. Let us first look for the most general form this anomaly can take. The space-time integral (2.7) being zero, its integrant is a total derivative:

$$w_5(x)\Gamma = \partial_\mu J^\mu(x) \quad . \tag{2.8}$$

$J^\mu(x)$ is an "insertion", otherwise denoted by $J^\mu(x) \cdot \Gamma$, the generating functional for the (one-particle-irreducible) Green functions with the insertion of the local field operator $J^\mu(x)$. It is BRS–invariant in the classical theory:

$$J^\mu(x) = e^R_S \bar{\psi}_R \gamma^\mu \gamma^5 \psi^S + O(\hbar) \quad . \tag{2.9}$$

The action of the BRS transformations (2.1) on insertions is given by a certain functional operator [7,9], which we shall also denote by s. It fulfills the properties (2.3). Since s commutes with the operator w_5, the right-hand-side of (2.8) is BRS–invariant:

$$s\partial_\mu J^\mu(x) = 0 \quad . \tag{2.10}$$

Denoting the Hodge dual of J^μ by J^0_3 (J^q_p is a p–form with ghost number q), we can write (2.10) as

$$dsJ^0_3 = 0 \quad , \tag{2.11}$$

where we have used the anticommutativity of d and s. The d–cohomology in the space of local polynomials being trivial [10], sJ^0_3 is an exact form:

$$sJ^0_3 = -\,dJ^1_2 \quad . \tag{2.12}$$

From the nilpotency of s and the anticommutativity of s and d follows that sJ_2^1 is also a closed, hence exact, form. Repeating the procedure we obtain the system of quantum *descent equations* [11]:

$$\begin{aligned}
sJ_3^0 + dJ_2^1 &= 0 \quad , \\
sJ_2^1 + dJ_1^2 &= 0 \quad , \\
sJ_1^2 + dJ_0^3 &= 0 \quad , \\
sJ_0^3 &= 0 \quad .
\end{aligned} \tag{2.13}$$

A special solution K_p^q of this system of equations is given by

$$\begin{aligned}
K_3^0 &= "Tr(dAA + \frac{2}{3}A^3)" \qquad \text{with} \quad dK_3^0 = "TrF^2" \quad , \\
K_2^1 &= -\,"Tr(dcA)" \quad , \\
K_1^2 &= "Tr(dcc)" \quad , \\
K_0^3 &= \frac{1}{3}Trc^3 \quad .
\end{aligned} \tag{2.14}$$

The quotation marks indicate that composite fields have to be renormalized and are therefore ambiguously defined. The last expression, however, is *finite*, hence unambiguous. This finiteness property holds in fact in the Landau gauge, and the insertion K_0^3 has a unique BRS–invariant extension to any gauge, the Landau gauge limit being finite.

Given a special solution (2.14) defined within a certain renormalization scheme, the general solution of the descent equations (2.13) is

$$\begin{aligned}
J_3^0 &= rK_3^0 + J_3^{0\ \mathrm{inv}} + dL_2^0 \quad , \\
J_2^1 &= rK_2^1 + dL_1^1 + sL_2^0 \quad , \\
J_1^2 &= rK_1^2 + sL_1^1 \quad , \\
J_0^3 &= rK_0^3 \quad ,
\end{aligned} \tag{2.15}$$

the coefficient r being gauge independent and uniquely defined. This solution follows from the triviality of the s–cohomology in the space of local functionals with positive ghost number. The insertions L_p^q and $J_3^{0\ \mathrm{inv}}$ are "integration constants": they correspond to all the possible ambiguities in the definition of the renormalized insertions. The non-trivial cohomology in the zero-ghost sector is represented by the BRS–invariant insertion $J_3^{0\ \mathrm{inv}}$. The latter is the BRS–invariant axial current associated to the chiral symmetries (2.6), and its anomalous conservation law is obtained by differentiating the first of Eqs. (2.15),

$$dJ_3^{0\ \mathrm{inv}} = -r"Tr\, F^2" + dJ_3^0 \quad , \tag{2.16}$$

and by going to the dual, Eq. (2.8) being taken into account:

$$\partial_\mu J^{\mu\ \mathrm{inv}} = -r"Tr\, F^{\mu\nu}\tilde{F}_{\mu\nu}" + w_5\Gamma \quad . \tag{2.17}$$

In this equation r is the *non-renormalized anomaly coefficient.*

Remark: the cohomologies, the derivation of the descent equations and of their solutions are straightforward in the classical approximation, where one is dealing with local field polynomials. This is in fact a special case of the more general results of Ref. [10]. One shows that all this remains valid in higher orders of perturbation theory thanks to the iterative character of the latter.

The proof of the non-renormalization of the anomaly coefficient r uses the Callan-Symanzik equation, satisfied by the vertex functional:

$$\mathcal{C}\Gamma \equiv \left[\kappa\partial_\kappa + \beta_g\partial_g + \sum_\lambda \beta_\lambda\partial_\lambda - \gamma_A N_A - \gamma^R_S N^S_R\right]\Gamma = 0 \quad , \tag{2.18}$$

where

$$N_A = \int d^4x\, A\,\delta_A \quad , \quad N^S_R = \int d^4x \left[\psi^S\,\delta_{\psi^R} + \bar\psi_R\,\delta_{\bar\psi_S}\right] \tag{2.19}$$

are the "counting operators"; κ — with the dimension of a mass — is some renormalization parameter, and the coupling constants λ describe the gauge–invariant self-couplings of the matter fields — if scalar fields are present. The Callan-Symanzik operator C commutes with the BRS operator s and the chiral operator w_5 (2.7):

$$[C, s] = 0 \quad , \quad [C, w_5(x)] = 0 \quad . \tag{2.20}$$

From the last commutation relation and from the dual of Eq. (2.8) follows that

$$dCJ^0_3 = 0 \quad , \tag{2.21}$$

hence

$$CJ^0_3 = d(...) \quad . \tag{2.22}$$

On the other hand the expressions CK^q_p obey the same descent equations (2.13) as K^q_p since C commutes with s. Moreover, K^3_0 being finite obeys the Callan-Symanzik equation $CK^3_0 = 0$ (without own anomalous dimension). Hence the general form of CK^q_p is found according to the general solution (2.15), in particular

$$CK^0_3 = \text{BRS–inv.} + d(...) \quad . \tag{2.23}$$

The substitution of the expression (2.15) for J^0_3 into Eq. (2.22) and the use of the last result (2.23) yields

$$(Cr)K^0_3 + \text{BRS–inv.} + d(...) = 0 \quad , \tag{2.24}$$

hence

$$Cr = \left[\beta_g\partial_g + \sum_\lambda \beta_\lambda\partial_\lambda\right] r = 0 \quad . \tag{2.25}$$

As a consequence of the last equation, the anomaly coefficient r is independent of the coupling constants [12]. This means that the only contribution to r is given through the computation of a *one-loop* triangle graph, with the result

$$r = -\frac{\hbar}{(4\pi)^2}\sum_R e^R_R\, T(R) \quad , \tag{2.26}$$

where the coefficients e^R_R are the diagonal elements of the chiral charge matrix defined in Eq. (2.6) and the Dynkin index $T(R)$ is defined by

$$Tr\left(T^i_{(R)}T^j_{(R)}\right) = \delta^{ij}T(R) \quad . \tag{2.27}$$

This is the content of the non-renormalization theorem of Adler and Bardeen [6]. The proof presented here [7] shows that the roots of the theorem are constituted by

(1) the relative cohomology for the coboundary operators s and d, i.e. the descent equations (2.13);

(2) the finiteness of the cocycle insertion Trc^3;

(3) the scaling behaviour of the theory as expressed by the Callan-Symanzik equation.

3. Supercurrent anomalies and vanishing β–functions [4].

We now consider the case of $N = 1$ supersymmetric gauge theories. The gauge and matter supermultiplets are formed by adjoining to the gauge field A_μ its fermionic superpartner λ_α, a complex Weyl spinor, and to the matter spinors ψ^R_α — which we take in their Weyl representation — the complex scalar fields A^R. The ghost fields too have superpartners — commuting Weyl spinors — which we do not write explicitly. Notations and conventions are as in Section 2, and the classical BRS–invariant action is

$$\Gamma_{\text{class}} = \int d^4x\Big[-\frac{1}{4g^2}Tr\,(F^{\mu\nu}F_{\mu\nu} + i\lambda\not{D}\bar{\lambda} + ...) + i\psi^R\not{D}\bar{\psi}_R + A^R\not{D}\bar{A}_R + ...+$$
$$+\lambda_{(RST)}(\psi^R\psi^S A^T + ...) + \text{conjugate}\Big] + \text{gauge fixing} \quad , \tag{3.1}$$

where the dots represent terms enforced by the requirement of supersymmetry, in particular contributions from auxiliary fields.

Beyond BRS and supersymmetry, the theory is always invariant under the chiral transformations

$$\delta_5 A_\mu = 0 \quad , \quad \delta_5\lambda_\alpha = -i\lambda_\alpha \quad ,$$
$$\delta_5 A^R = -i\frac{2}{3}A^R \quad , \quad \delta_5\psi^R_\alpha = i\frac{1}{3}\psi^R_\alpha \quad , \tag{3.2}$$

which obviously do not commute with the supersymmetry transformations $A_\mu \leftrightarrow \lambda_\alpha$, $A^R \leftrightarrow \psi^R_\alpha$. The theory may also be invariant under a set of chiral transformations of the type (2.6), labelled by a :

$$\delta_a(A_\mu, \lambda_\alpha) = 0 \quad ,$$
$$\delta_a(A^R, \psi^R_\alpha) = i\,e_a{}^R{}_S(A^S, \psi^S_\alpha) \quad , \tag{3.3}$$

which commute with supersymmetry. Note that both (3.2) and (3.3) commute with the BRS transformations.

Assuming that there is no gauge anomaly, i.e. choosing an "anomaly free" representation for the matter fields, one can show [8] that all these invariances, BRS included, are preserved by renormalization.

The BRS–invariant axial currents associated to the symmetries (3.2) and (3.3) are anomalous. The corresponding anomaly coefficients are not renormalized, i.e. they are exactly given by their one-loop values:

$$r_5 = \frac{3\hbar}{4(4\pi)^2}\left[-C_2(G) + \frac{1}{3}\sum_R T(R)\right] \quad , \tag{3.4}$$

$$r_a = \frac{\hbar}{2(4\pi)^2}\sum_R e_a{}^R{}_R T(R) \quad . \tag{3.5}$$

We have explicitly displayed the factor $\hbar$, corresponding to the first order in the loop expansion. The proof of this non-renormalization theorem goes through a generalization of the one sketched in the preceeding section.

We now want to relate these chiral anomalies to the ones of dilatation, i.e. to the β–functions and to the anomalous dimensions γ appearing in the Callan-Symanzik equation (c.f. (2.18))

$$\left[\kappa\partial_\kappa + \beta_g\partial_g + \sum_{R,S,T}\beta_{RST}\partial_{\lambda_{RST}} - \gamma_{\text{gauge}}N_{\text{gauge}} - \gamma^R_S N^S_R - \gamma_k\partial_{a_k}\right]\Gamma = 0 \quad , \tag{3.6}$$

where N_{gauge} and N^S_R are the counting operators for the gauge supermultiplet and for the matter fields, respectively (c.f. (2.19)); the a_k's form an infinite set of gauge paramaters, and the γ_k's describe their renormalization [8]. The relationship between chiral and dilatation anomalies in supersymmetric theories is provided by the existence of the "supercurrent" multiplet [8], which is a multiplet of supersymmetry containing, among other components, the BRS–invariant currents associated to the chiral symmetry (3.2) — an axial current — and to translation invariance —the energy-momentum tensor. The divergence of the axial current and the trace of the energy-momentum tensor are as well related by supersymmetry. Hence the chiral anomaly for (3.2) and the dilatation anomaly belong to the same supermultiplet: both are described in terms of the Callan-Symanzik functions β and γ. Now, the chiral anomaly coefficient r_5 (3.4) is not renormalized, whereas β and γ get in general contributions from all orders of perturbation theory. This apparent paradox is explained when one realizes that the axial current belonging to the supercurrent multiplet *differs* by a finite renormalization from the axial current whose divergence is not renormalized [3,4,13]. This statement can be made precise, and this results in the identity:

$$r_5 = \beta_g\left(\frac{1}{g^3} + x_g\right) + \sum_{R,S,T}\beta_{RST}x^{RST} + \sum_a \gamma_a r_a \quad . \tag{3.7}$$

Here x_g and x^{RST} are of order $\hbar$ at least, r_5 and r_a are the anomaly coefficients (3.4) and (3.5), of order $\hbar$ strictly. The γ_a's are certain linear combinations of the matter fields anomalous dimensions γ^R_S and of the γ_k's.

With the help of the identity (3.7) one can state and prove the following criterion for vanishing β–functions:

Theorem. The two conditions below are necessary and sufficient for the β–functions to vanish to all orders of perturbation theory:

(1) the chiral anomalies (3.4) and (3.5) are absent:

$$r_5 = r_a = 0 \quad . \tag{3.8}$$

(2) the coupling constants λ_{RST} describing the self-interactions of the matter fields are functions (power series) of the gauge coupling constant g

$$\lambda_{RST} = \lambda_{RST}(g) \quad . \tag{3.9}$$

Proof.[3,4] Whereas the proof of necessity is not straightforward and will be skipped here, the sufficiency of conditions (1) and (2) is simple to establish. The first condition renders Eq. (3.7) homogenous in the β–functions. The second implies that the reduction equations [5]

$$\beta_{RST} = \beta_g \frac{d\lambda_{RST}}{dg} \tag{3.10}$$

must hold. In other words, the functions $\lambda_{RST} = \lambda_{RST}(g)$ must be solutions of the reduction equations. Eq. (3.7) thus reads

$$0 = \beta_g \left(\frac{1}{g^3} + x_g + \frac{d\lambda_{RST}}{dg} x^{RST} \right) \quad , \tag{3.11}$$

and implies the vanishing of β_g since the term in parenthesis is invertible (in the perturbative sense, its lowest order value $1/g^3$ being non-zero). The vanishing of β_{RST} then follows from the reduction equations (3.10).

A very important feature of this criterion is that its first condition involves only one-loop quantities, due to the non-renormalization theorem. In contrast the second condition amounts to the existence to all orders of solutions $\lambda_{RST}(g)$ to the reduction equations. However it can be shown that, if there exists a solution at lowest order which is isolated and non-degenerate, then this solution extends uniquely to a power series solution valid to all orders.

The criterion for vanishing β–functions — hence for superconformal invariance — thus reduces to a set of simple conditions which may be tested on one-loop quantities. This extends to all loops a result known from a long time to be valid up to the two loops order [14].

The conditions of the criterion are of course fulfilled by the well-known finite $N = 2$ and $N = 4$ super-Yang-Mills theories [1], but also by genuine $N = 1$ theories with — physically more attractive — chiral representations for the matter fields [3,4]. Let us also recall the finite $N = 1$ models mentioned in the Introduction [2], for which the β–functions *and* the anomalous dimensions vanish.

References

[1] S. Mandelstam, Nucl. Phys. **B213** (1983) 149; L. Brink, O. Lindgren and B. Nilsson, Nucl. Phys. **B212** (1983) 401; P.S. Howe, K. Stelle and P.C. West, Phys, Lett. **B124** (1983) 55; P.S. Howe, K.S. Stelle and P.K. Townsend, Nucl. Phys. **B236** (1984) 125.

[2] D.I. Kazakov, Phys. Lett. **B179** (1986) 952; JINR preprint E2-86-816 (1986); X.D. Jiang and X.J. Zhou, Phys. Lett. **B197** (1987) 156.

[3] O. Piguet and K. Sibold, Int. J. Mod. Phys. **A1** (1986) 913; Phys.Lett. **B177** (1986) 373.

[4] C. Lucchesi, O. Piguet and K. Sibold, 'Vanishing β–functions in N=1 supersymmetric gauge theories', preprint CERN-TH 4860/87 — UGVA/DPT 1987/10-553; preprint UGVA/DPT 1987/10-554, to appear in Phys. Lett. **B**

[5] R. Oehme and W. Zimmermann, Commun. Math. Phys. **97** (1985) 569; R. Oehme, K. Sibold and W. Zimmermann, Phys. Lett. **B147** (1984) 115 and **B153** (1985) 142; R. Oehme, Progr. Theor. Phys. Suppl. **86** (1986) 215.

[6] S. Adler and W. Bardeen, Phys. Rev. **182** (1969) 1517.

[7] C. Lucchesi, O. Piguet and K. Sibold, Int. J. Mod. Phys. **A2** (1987) 385.

[8] O. Piguet and K. Sibold, *Renormalized Supersymmetry*, Birkhaüser Boston Inc., 1986.

[9] O. Piguet and A. Rouet, Phys. Reports **76** (1981) 1.

[10] M. Dubois-Violette, M. Talon and C.M. Viallet, Commun. Math. Phys. **102** (1985) 105; Ann. Inst. H. Poincaré **44** (1986) 103.

[11] R. Stora, 'Algebraic structures and topological origin of anomalies', in *Progress in Gauge Field Theory*, Edited by G. 't Hooft et al., Plenum Press (1984); L. Baulieu, Phys. Reports **129** (1985) 1.

[12] G. Bandelloni, C. Becchi, A. Blasi and R. Collina, Commun. Math. Phys. **72** (1980) 239.

[13] M.T. Grisaru, B. Milewski and D. Zanon, Nucl. Phys. **B266** (1986) 589.

[14] D.R.T. Jones and L. Mezincescu, Phys. Lett. **B138** (1984) 293; A.J. Parkes and P.C. West, Phys. Lett. **B138** (1984) 99.

SUPERMANIFOLDS, SUPERMANIFOLD COHOMOLOGY, AND SUPER VECTOR BUNDLES

U. BRUZZO
Department of Mathematics
University of Genoa
Via L. B. Alberti 4
16132 Genova
Italy

ABSTRACT. We show that a suitable choice of the category of supermanifolds allows the introduction of a well-behaved category of super vector bundles. In particular, such supermanifolds have a tangent bundle with a standard fibre that at each point can be identified with the tangent space at that point. We also discuss the cohomological classification of super line bundles. In order to make the paper as much self-contained as possible, we give a brief account of supermanifold cohomology.

1. Introduction

In this paper we wish to describe the basic elements of a theory of super vector bundles (SVB's) over supermanifolds. The guiding principle is that the category of SVB's over a supermanifold M whose function sheaf is $\mathcal{G}$ should be equivalent to the category of locally free graded $\mathcal{G}$-modules; this generalizes the situation occurring in the case of topological, smooth and holomorphic vector bundles [1-3] and provides automatically an SVB with all the good properties that one naturally expects: for instance, the specification of an SVB is equivalent to the assignment of transition functions, and all the fibres of an SVB are isomorphic to each other.

Our interest in the search for a satisfactory theory of SVB's and of the strictly related principal super fibre bundles [4,5] stems from supersymmetric field theory and superstring theory. For instance, superbundles are involved in the development of a variational calculus on fibred supermanifolds [6,7], in geometrical models of supergravity and supersymmetric field theory [8-11], in the construction of topologically non-trivial solutions of supersymmetric field theory [5], in several cohomological constructions related to the anomaly problem in supersymmetric field theory (see

K. Bleuler and M. Werner (eds.), Differential Geometrical Methods in Theoretical Physics, 417–440.

e.g. [12,13]), and in the study of super Riemann surfaces [14-19], which in turn are relevant to superstring theory.

Our approach to supermanifolds derives from the so-called geometric viewpoint, basically due to De Witt and Rogers [20,21]. In this approach supermanifolds are topological manifolds, modelled on the even part of a graded cartesian product of Grassmann algebras by means of atlases whose transition functions fulfil a suitable "supersmoothness" condition. These supermanifolds are to be contrasted with graded manifolds [22-24] (also called supermanifolds in the Russian literature) where one does not enlarge the set of points of the manifold, but rather considers a bigger function sheaf, enlarged so as to include "anti-commuting objects".

The delicate point in the "geometric" approach is the choice of the class of transition functions. The G^∞ functions originally introduced by Rogers [21] are not suitable, for their derivatives with respect to odd variables are not well defined, and the sheaf of derivations of G^∞ functions is not locally free [25,26]. This means for instance that tangent vector fields do not have component expressions in terms of local coordinates.

Rogers has recently introduced the class of GH^∞ functions [27], which are free from the above mentioned drawbacks; in particular, the sheaf of derivations is locally free. However, it is not possible to develop a satisfactory theory of SVB's within the category of GH^∞ supermanifolds [28], and, in particular, these supermanifolds do not have a good graded tangent space. The reason of this bad behaviour can be traced back to the fact that GH^∞ functions on a supermanifold M are maps $M \to B_L$, where B_L is the exterior algebra over $\mathbf{R}^L$, but the space of constant functions on a connected set is smaller than B_L. Indeed, the space $\mathcal{V}_x$ of values taken by all the GH^∞ functions at a fixed point $x \in M$ can be strictly smaller that B_L; moreover, the $\mathcal{V}_x$'s considered for variable x can be different. So a GH^∞ function is not a section of a suitable trivial bundle on M in any reasonable way, and the existence of a graded tangent bundle with a standard fibre is precluded, since its fibre at $x \in M$ should be isomorphic with $\mathcal{V}_x^{m+n}$ if M has dimension (m, n).

However, there exists a strictly related category of supermanifolds (that we call $\mathcal{G}$-supermanifolds) which do not have these defects; for such supermanifolds the theory of super vector bundles is very much alike the theory of holomorphic bundles on complex supermanifolds. It should be stressed that $\mathcal{G}$-supermanifolds are not a new structure, in that they fit into Rothstein's axiomatics for supermanifolds [26]; however, we think it is worth while to study them because they provide concrete examples of Rothstein's axiomatics, and because they are defined in terms of atlases; that is, $\mathcal{G}$-supermanifolds are closer to the "geometric" viewpoint. For a detailed comparison between $\mathcal{G}$-supermanifolds and Rothstein's supermanifolds see Ref. 28.

The plan of this paper is as follows. In Section 2 we review some algebraic material, give a brief introduction to sheaves, discuss the properties of GH^∞ functions, and introduce $\mathcal{G}$-supermanifolds. In Section 3 we define the concept of SVB and show the categorial equivalence between the SVB's over a $\mathcal{G}$-supermanifold M and the locally free graded modules over the structure sheaf $\mathcal{G}$ of M. We also discuss the relationship between the graded tangent bundle, introduced in terms of the sheaf of derivations of $\mathcal{G}$, and the graded tangent space. In Section 4 we recall from Refs. 29, 30 some basic facts about the cohomology of supermanifolds; in particular, we introduce the cohomology of the complex of "supersmooth" differential forms, i.e. a "super de Rham cohomology", and analyze its relationship to the ordinary de Rham cohomology of M. This will allow us to show that the Čech cohomology of the structure sheaf of a $\mathcal{G}$-supermanifolds is not trivial. Finally, in Section 5 we consider the SVB's of rank (1,0), namely, super line bundles, and relate their classification to the integer cohomology of the base supermanifold.

2. $\mathcal{G}$-supermanifolds

2.1 ALGEBRAIC PRELIMINARIES

Let B be a $\mathbf{Z}_2$-graded commutative algebra over a field F, i.e. B is a $\mathbf{Z}_2$-graded † F-vector space $B = B_0 \oplus B_1$ endowed with a structure of F-algebra such that

$$ab = (-1)^{rs}\, ba \in B_{r+s \bmod 2} \qquad \forall a \in B_r,\ b \in B_s\,.$$

For fixed integers m, n, the cartesian power B^{m+n} can be graded according to

$$B^{m+n} = B^{m,n} \oplus B^{\underline{m},\underline{n}}\,,\quad B^{m,n} = B_0^m \times B_1^n,\quad B^{\underline{m},\underline{n}} = B_1^m \times B_0^n\,.$$

A graded left B-module Σ is a left B-module with a gradation $\Sigma = \Sigma_0 \oplus \Sigma_1$ such that

$$ax \in \Sigma_{r+s} \qquad \forall a \in B_r,\ x \in \Sigma_s\,;$$

a graded right B-module is defined similarly. Any graded left B-module Σ can be turned into a graded right B-module, and viceversa, by letting

$$xa = (-1)^{|x|\,|a|}\, ax \quad \text{for all homogeneous } a \in B,\ x \in \Sigma\,,$$

where $||$ denotes the grading. Therefore one can use unambiguously the term "graded B-module". B^{m+n} is an example of graded B-module.

---

† Henceforth the prefix "$\mathbf{Z}_2$-" will be dropped.

A map between graded B-modules $f : \Sigma \to \Pi$ is said to be even (resp. odd) if $f(x) \in \Pi_r$ (resp. $f(x) \in \Pi_{r+1}$) whenever $x \in \Sigma_r$, $r = 0,1$. If $f : \Sigma \to \Pi$ is homogeneous, it is said to be graded B-linear (or a morphism of graded B-modules) if $f(ax) = (-1)^{|f|\,|a|}\, af(x)$ $\forall x \in \Sigma$ and for all homogeneous $a \in B_L$; if f is nonhomogeneous, f is said to be graded B-linear if both its even and odd parts are.

The graded tensor product $\Sigma \otimes_B \Pi$ of two graded B-modules is defined as the quotient of the module $B^{\Sigma \times \Pi}$ with respect to the submodule generated by all elements of the form

$$(ax + by, z) - a(x,z) - b(y,z), \qquad (x, aw + bz) - (-1)^{|a|\,|x|}\, a(x,w) - (-1)^{|b|\,|x|}\, b(x,z)$$

with $x, y \in \Sigma$, $w, z \in \Pi$, $a, b \in B$, all homogeneous. The grading of $\Sigma \otimes_B \Pi$ is such that the natural map $\Sigma \otimes_B \Pi \times \Pi^* \to \Sigma$ is even.

Finally, a graded B-module is said to be free of rank (m,n) if it is free of rank $m+n$ as a B-module, and any basis formed by homogeneous elements has m and n odd elements.

In the following B will be identified with either B_L, the real exterior algebra over $\mathbf{R}^L$, or C_L, the complex exterior algebra over $\mathbf{C}^L$, where L is a fixed integer. In any case the various modules B_L^{m+n}, $B_L^{m,n}$, etc. will be endowed with their vector space topology. Moreover, one has a splitting $B_L = \mathbf{R} \oplus N_L$, where N_L is the nilpotent ideal of B_L. The projections $\sigma : B_L \to \mathbf{R}$ and $\sigma : B_L \to N_L$, following De Witt, are called body and soul map respectively. One can also consider a body map $\sigma^{m,n} : B_L^{m,n} \to \mathbf{R}^m$ defined by letting $\sigma^{m,n}(x^1 \ldots x^m, y^1 \ldots y^n) = (\sigma(x^1) \ldots \sigma(x^m))$. Analogous splittings and maps are defined for C_L.

2.2 SHEAF THEORY

Given a topological space X, a correspondence assigning an abelian group $\mathcal{F}(U)$ to each open set U in X is said to be a sheaf of abelian groups on X if the following properties are verified:

(i) if V and U are open sets with $V \subset U$ there exists a restriction morphism $\rho^U_V : \mathcal{F}(U) \to \mathcal{F}(V)$; in particular, $\rho^U_U = id$.

(ii) if $W \subset V \subset U$ are inclusions of open sets, then $\rho^U_V \circ \rho^V_W = \rho^U_W$;

(iii) if $\{U_i,\, i \in I\}$ is a collection of open sets in X, $U = \bigcup_{i \in I} U_i$, and $s, t \in \mathcal{F}(U)$ are such that $\rho^U_{U_i}(s) = \rho^U_{U_i}(t)$ $\forall i \in I$, then $s = t$;

(iv) (gluing property) if $\{U_i,\, i \in I\}$ and U are as above, and a collection $\{s_i \in \mathcal{F}(U_i),\, i \in I\}$ is given such that $\rho^{U_i}_{U_i \cap U_j}(s_i) = \rho^{U_j}_{U_i \cap U_j}(s_j)$, then there exists a section $s \in \mathcal{F}(U)$ such that $\rho^U_{U_i}(s) = s_i$.

The elements of $\mathcal{F}(U)$ are called the sections of $\mathcal{F}$ over U. For all $x \in X$ the *stalk* $\mathcal{F}_x$ of $\mathcal{F}$ at x is the abelian group defined as the direct limit of the $\mathcal{F}(U)$'s over all open neighbourhoods U of x; its elements are called the *germs* of sections of $\mathcal{F}$ at x.

A sheaf morphism $\varphi : \mathcal{F} \to \mathcal{G}$ between two sheaves $\mathcal{F}$ and $\mathcal{G}$ on X is a collection $\{\varphi_U, U \text{ open in } X\}$ of group morphisms $\varphi_U : \mathcal{F}(U) \to \mathcal{G}(U)$ such that for all inclusions of open sets $V \subset U$ one has $\rho_V^U \circ \varphi_U = \varphi_V \circ \rho_V^U$. By a direct limit procedure, φ defines morphisms between the stalks, $\varphi_x : \mathcal{F}_x \to \mathcal{G}_x$. In particular, a sequence of sheaf morphisms

$$0 \to \mathcal{F} \to \mathcal{G} \to \mathcal{H} \to 0$$

is said to be exact if restricting the morphisms to the stalks the resulting sequence of abelian groups is exact.

Examples of sheaves are the sheaf of germs of smooth functions on a smooth manifold or of holomorphic functions on a complex manifold.

A related concept is that of *presheaf*, which is defined similarly to a sheaf, but without imposing the properties (iii) and (iv). It is easily seen that, given an abelian group G and a topological manifold X, the rule $U \to G$ for all open $U \subset X$ yields a presheaf (called the constant presheaf with stalk G) which is not a sheaf. On the other hand, starting from a presheaf $\mathcal{F}$ one can construct in a canonical way a sheaf $\tilde{\mathcal{F}}$, called the sheafification of $\mathcal{F}$, together with an injective morphism of presheaves $\mathcal{F} \to \tilde{\mathcal{F}}$ [31,32]. The constant sheaf on X with stalk an abelian group G (usually denoted again by G) is just the sheafification of the constant presheaf previously introduced.

2.3 GH^∞ FUNCTIONS

Following Rogers [27], we consider a ring of B_L-valued functions on $B_L^{m,n}$, called GH^∞ functions. If X is any smooth manifold, we let $\mathcal{C}_L(V)$ be the graded B_L-algebra of C^∞ B_L-valued functions on the open set $V \subset X$. Let L and L' be two positive integers such that $L' \leq L$, let U be an open set in $\mathbf{R}^m$, and define a map

$$Z_{L',L} : \mathcal{C}_{L'}(U) \to \mathcal{C}_L((\sigma^{m,0})^{-1}(U))$$

by letting

$$Z_{L',L}(f)(x^1 \dots x^m) = \sum_{k_1 \dots k_m = 0}^{L} \frac{1}{k_1! \dots k_m!} \partial_1^{k_1} \dots \partial_m^{k_m} f \mid_{\sigma^{m,0}(x^1 \dots x^m)} s(x^1)^{k_1} \dots s(x^m)^{k_m} .$$

$Z_{L',L}$ is an injective morphism of graded $B_{L'}$-algebras; we denote its image by $\widehat{\mathcal{GH}}[(\sigma^{m,0})^{-1}(U)]$. This set is to be identified with the graded algebra of GH^∞

functions of even variables on $(\sigma^{m,0})^{-1}(U)$. This means that GH^∞ functions of even variables are determined by their value for real arguments, their value at a non-real point x being given by a Taylor expansion around the body of x.

GH^∞ functions of both even and odd variables are defined on open sets $(\sigma^{m,n})^{-1}(U)$, where U is an open set in $\mathbf{R}^m$, as polynomials in the odd variables with GH^∞ functions of even variables as coefficients. Such functions form a graded $B_{L'}$ algebra, denoted by $\mathcal{GH}[(\sigma^{m,n})^{-1}(U)]$, and can be expressed as

$$F(x^1 \dots x^m, y^1 \dots y^n) = F_0(x^1 \dots x^m) + \sum_{\substack{k=1\dots n \\ 1\leq\alpha_1<\dots<\alpha_k\leq n}} F_{\alpha_1\dots\alpha_k}(x^1 \dots x^m)\, y^{\alpha_1}\dots y^{\alpha_k} \tag{2.1}$$

where $F_0,\ F_{\alpha_1\dots\alpha_k} \in \widehat{\mathcal{GH}}[(\sigma^{m,0})^{-1}(U)]$.

As far as the derivatives of GH^∞ functions are concerned, the basic idea is that they are determined by the development

$$F(x+h,\, y+k) = F(x,y) + \sum_{i=1}^{m} h^i \frac{\partial F}{\partial x^i}(x,y) + \sum_{\alpha=1}^{n} k^\alpha \frac{\partial F}{\partial y^\alpha}(x,y) + O(h,k)^2 \,. \tag{2.2}$$

Now, while Eq. (2.2) defines unambiguously the derivatives with respect to the even variables, those with respect to odd variables can be not defined if L' is too large, since the coefficient of the term $\partial F/\partial y^\alpha$ in Eq. (2.2) is nilpotent, and therefore one can add terms to $\partial F/\partial y^\alpha$ without altering the right-hand side of (2.2). However, if the condition

$$L - L' \geq n \tag{2.3}$$

is fulfilled, one can easily check that Eq. (2.2) defines *all* partial derivatives of F.

The G^∞ functions originally introduced by Rogers [21] are recovered by setting $L = L'$. In this case, unless one takes $n = 0$, the condition (2.3) is violated. The opposite case is $L' = 0$, which produces the so-called H^∞ functions, first considered by Batchelor [33] and De Witt [20]. It is often claimed that these functions are not suitable for modelling supersymmetric field theory, since by definition a H^∞ function of even variables is real when its argument is restricted to real values, which physically means that one is restricting the fields to space-time. In this sense the coefficients of the "superfield expansion" (2.1) are real, and therefore cannot describe physical particles with Fermi statistics, since the fields representing such particles must anticommute among them.

The previous discussion suggests to consider functions for which Eq. (2.3) holds with $L' > 0$. In the following, we shall understand that this condition holds.

We wish now to discuss the properties of GH^∞ functions. Let us introduce the sheaf $\mathcal{GH}$ of graded $B_{L'}$-algebras over $B_L^{m,n}$, whose sections over an open set $U \subset$

$B_L^{m,n}$ are the functions in $\mathcal{GH}[(\sigma^{m,n})^{-1}\,\sigma^{m,n}(U)]$ restricted to U. This seemingly strange definition is due to the fact that a GH^∞ functions is naturally defined on a whole "tube" over an open set in $\mathbf{R}^m$. The subsheaf $\widehat{\mathcal{GH}}$ of germs of GH^∞ functions of even variables can be characterized as follows:

$$\widehat{\mathcal{GH}}(U) = \{f \in \mathcal{GH}(U) \mid \frac{\partial f}{\partial y^\alpha} = 0, \quad \alpha = 1 \dots n\}.$$

The two sheaves are related by the isomorphism

$$\mathcal{GH} \simeq \widehat{\mathcal{GH}} \otimes_{B_{L'}} \Lambda[n, B_{L'}]$$

where $\Lambda[n, B_{L'}]$ is the exterior algebra over $B_{L'}$ with n generators.

DEFINITION 2.1 *The sheaf* $Der\,\mathcal{GH}$ *of graded derivations of* $\mathcal{GH}$ *is the sheafification of the presheaf over* $B_L^{m,n}$ *whose sections over an open* $U \subset B_L^{m,n}$ *are the morphisms of sheaves of graded* $B_{L'}$*-algebras* $D : \mathcal{GH}|U \to \mathcal{GH}|U$ *which satisfy the graded Leibniz rule, i.e.*

$$D(fg) = D(f)\,g + (-1)^{|f||D|}\,f\,D(g) \qquad \forall f,g \in \mathcal{GH}(V) \quad \text{for any open } V \subset U.$$

We recall that a rank (r,s) locally free graded $\mathcal{GH}$-module $\mathcal{E}$ is a sheaf on $B_L^{m,n}$ of $\mathcal{GH}$-modules with the property that any $x \in B_L^{m,n}$ has a neighbourhood U such that $\mathcal{E}|U$ is isomorphic with $(\mathcal{GH}|U)^{r+s}$.

PROPOSITION 2.1 *The sheaf* $Der\,\mathcal{GH}$ *over* $B_L^{m,n}$ *is a locally free graded* $\mathcal{GH}$*-module of rank* (m,n). *In particular,* $Der\,\mathcal{GH}(U)$ *is the graded* $\mathcal{GH}(U)$*-module generated by the derivations*

$$\left\{\frac{\partial}{\partial x^1}, \dots, \frac{\partial}{\partial x^m}, \frac{\partial}{\partial y^1}, \dots, \frac{\partial}{\partial y^n}\right\}$$

defined by Eq. (2.2).

Proof. The simple but rather technical proof was given in Ref. 28. □

So the sheaf $\mathcal{GH}$ has good properties as far as its behaviour on open sets is concerned. The problems arise when one tries to introduce a notion of *graded tangent space*. The most natural definition seems to be the following:

DEFINITION 2.2 *For any* $x \in B_L^{m,n}$, $T_x B_L^{m,n}$ *is the graded* B_L*-module of morphisms of graded* $B_{L'}$*-modules* $X : \mathcal{GH}_x \to B_L$ *satisfying the graded Leibniz rule*

$$X(fg) = X(f)\,\tilde{g} + (-1)^{|f||X|}\,\tilde{f}\,X(g) \qquad \forall f,g \in \mathcal{GH}_x\,,$$

where the tilde denotes evaluation of germs.

The point here is that the various $T_x B_L^{m,n}$, considered for variable x, are not all isomorphic. This is due to the fact that the various stalks $\mathcal{GH}_x$ are not isomorphic. So, by gluing together all the $T_x B_L^{m,n}$, one obtains a bundle without a standard fibre.

The reasons of this misbehaviour can be traced back to the fact that the space of values taken by all GH^∞ functions is different from point to point. This space $\mathcal{V}_x$ at a point $x \in B_L^{m,n}$ can be introduced by means of the exact sequence

$$0 \to \mathcal{M}_x \to \mathcal{GH}_x \to \mathcal{V}_x \to 0 \tag{2.4}$$

where $\mathcal{GH}_x$ is the stalk of $\mathcal{GH}$ at x, namely, the graded $B_{L'}$-algebra of germs of GH^∞ functions at x, and $\mathcal{M}_x$ is the maximal ideal of germs which vanish when evaluated at x. The fact that the $\mathcal{V}_x$'s for different x's can be different follows from Eq. (2.1): if $x \in U \subset B_L^{m,n}$ is real, and $f \in \mathcal{GH}(U)$, then $f(x) \in B_{L'}$; on the other hand, it is possible to choose x such that $B_{L'} \subset \mathcal{V}_x \subset B_L$ strictly.

It is obvious — since all we have said so far is of a local nature — that all this applies not not only to the case of $B_L^{m,n}$, but also to a generic GH^∞ supermanifold, intended as a topological space modelled on $B_L^{m,n}$ by means of atlases whose transition functions are GH^∞.

2.4 $\mathcal{G}$-SUPERMANIFOLDS

We shall presently show that a different choice of the structure sheaf on $B_L^{m,n}$ allows one to get rid of the above mentioned drawbacks. It is therefore convenient to define supermanifolds as topological manifolds whose structure sheaf is locally isomorphic with the newly introduced sheaf over $B_L^{m,n}$.

The new sheaf will be called $\mathcal{G}$ and is defined as

$$\mathcal{G} = \mathcal{GH} \otimes_{B_{L'}} B_L \,.$$

Its sections will be called $\mathcal{G}$-functions or supersmooth functions. The product whose action on homogeneous elements is defined by $(f \otimes a)(g \otimes b) = (-1)^{|a|\,|g|}\, fg \otimes ab$ makes $\mathcal{G}$ into a sheaf of graded commutative B_L-algebras. Of course there is an injection $\mathcal{GH} \to \mathcal{G}$, given by $f \mapsto f \otimes 1$; this map is not surjective, since $f \otimes a$ with f a GH^∞ function and $a \in B_L/B_{L'}$ is not in its image. Moreover we define a sheaf morphism $\delta : \mathcal{G} \to C_L$, where C_L is the sheaf of smooth B_L-valued functions on $B_L^{m,n}$, by letting

$$\delta(f \otimes a) = fa \,. \tag{2.5}$$

Finally, the sheaf $Der\,\mathcal{G}$ is defined on analogy of $Der\,\mathcal{GH}$ (see Definition 2.1).

PROPOSITION 2.2 *(i) The image of δ is the sheaf $\mathcal{G}^\infty$ of G^∞ functions on $B_L^{m,n}$. (ii) δ is injective when restricted to the subsheaf*

$$\hat{\mathcal{G}} = \widehat{\mathcal{G}\mathcal{H}} \otimes_{B_{L'}} B_L .$$

(iii) There is a sheaf isomorphism

$$Der\,\mathcal{G} \simeq Der\,\mathcal{G}\mathcal{H} \otimes_{B_{L'}} B_L .$$

Proof. (i) is obvious. In order to prove (ii), let $\hat{\mathcal{G}}^\infty$ be the sheaf of germs of G^∞ functions of even variables. For any open $U \subset B_L^{m,n}$, a function $f \in \hat{\mathcal{G}}^\infty(U)$ can be represented as

$$f = Z_{0,L}(\hat{f}^\mu)\,\beta_\mu \tag{2.6}$$

where the $\hat{f}^\mu$ are smooth real-valued functions on $\sigma^{m,n}(U)$, and μ is a multi-index ranging in the set M_L of strictly increasing sequences of integers with values between 1 and L. $\{\beta_\mu,\ \mu \in M_L\}$ is a basis of B_L obtained by letting

$$\beta_\mu = e_{\mu_1} \dots e_{\mu_k} \quad \text{if} \quad \mu = \{\mu_1 \dots \mu_k\},$$

$\{e_1 \dots e_L\}$ being a basis of $\mathbf{R}^L$. Then $\eta(f) = Z_{0,L}(\hat{f}^\mu) \otimes \beta_\mu$ defines a sheaf morphism $\hat{\mathcal{G}}^\infty \to \hat{\mathcal{G}}$ such that $\delta \circ \eta = \eta \circ \delta = id$; therefore η is an isomorphism. This also implies point (iii). □

The partial derivatives of sections of $\mathcal{G}$ are defined according to the rule

$$\frac{\partial}{\partial x^i}(f \otimes a) = \frac{\partial f}{\partial x^i} \otimes a,\ i = 1 \dots m; \qquad \frac{\partial}{\partial y^\alpha}(f \otimes a) = \frac{\partial f}{\partial y^\alpha} \otimes a,\ \alpha = 1 \dots n. \tag{2.7}$$

A straightforward consequence of Proposition 2.2 is the following

COROLLARY 2.1 *$Der\,\mathcal{G}$ is a locally free $\mathcal{G}$-module; for all open sets $U \subset B_L^{m,n}$, $Der\,\mathcal{G}(U)$ is generated over $\mathcal{G}(U)$ by the derivations*

$$\left\{ \frac{\partial}{\partial x^1}, \dots, \frac{\partial}{\partial x^m}, \frac{\partial}{\partial y^1}, \dots, \frac{\partial}{\partial y^n} \right\},$$

defined by Eq. (2.7). □

The constant sections of $\mathcal{G}$ over $U \subset B_L^{m,n}$ are defined as the sections $f \in \mathcal{G}(U)$ such that $\delta(f)$ is constant. If U is connected, the set of constant sections of $\mathcal{G}$ over U is just B_L. Composing δ with the evaluation of a G^∞ function at x one obtains an evaluation morphism $\sim$: $\mathcal{G}_x \to B_L$; there is an exact sequence of graded B_L-modules

$$0 \to \mathcal{M}_x \to \mathcal{G}_x \to B_L \to 0 . \tag{2.6}$$

where $\mathcal{M}_x$ is the ideal of germs in $\mathcal{G}_x$ which vanish when evaluated at $x \in B_L^{m,n}$.

DEFINITION 2.3 *(i) If U and V are open sets in $B_L^{m,n}$, a smooth map $\varphi : U \to V$ is said to be a $\mathcal{G}$-map if $\varphi^*(\mathcal{G}|V)$ is a subsheaf of $\mathcal{G}|U$.*

(ii) A Hausdorff, second countable topological space M is an (m,n)-dimensional $\mathcal{G}$-supermanifold if it admits an atlas $\mathcal{A} = \{(U_\alpha, \phi_\alpha) | \phi_\alpha : U_\alpha \to B_L^{m,n}\}$ such that the transition functions $\phi_\alpha \circ \phi_\beta^{-1}$ are $\mathcal{G}$-maps.

It should be noticed that any $\mathcal{G}$-supermanifold M also carries a structure of smooth manifold; in particular, $\mathcal{G}$-supermanifolds are paracompact. Given two $\mathcal{G}$-supermanifolds M, N, a smooth morphism $f : M \to N$ is said to be a $\mathcal{G}$-morphism if its coordinate expressions are $\mathcal{G}$-maps.

Since there is an injective but not surjective morphism $\mathcal{GH} \to \mathcal{G}$ of sheaves over $B_L^{m,n}$, any GH^∞ map is a $\mathcal{G}$-map, but the converse is false. Thus the category of GH^∞ supermanifolds is strictly included into the category of $\mathcal{G}$-supermanifolds.

The structure sheaf $\mathcal{G}^M$ of M is by definition the sheaf on M such that $\phi_\alpha^* : \mathcal{G}|\phi_\alpha(U_\alpha) \to \mathcal{G}^M|U_\alpha$ is a sheaf isomorphism for any α. If confusion cannot arise, $\mathcal{G}^M$ will also be denoted by $\mathcal{G}$.

The analysis of the properties of the graded tangent space to a $\mathcal{G}$-supermanifold is deferred to Section 3, after the introduction of super vector bundles.

2.5 EXAMPLES

We give some examples of $\mathcal{G}$-supermanifolds.

Example 2.3.1: Even supermanifolds. Whenever dim $M = (m, 0)$, we may assume $L = L'$; then the structure sheaf $\mathcal{G}$ coincides with the sheaf $\mathcal{G}^\infty$ of G^∞ functions on M (see Proposition 2.2).

Example 2.3.2. In particular, let us consider $M = \mathbf{R} \times S^1$ endowed with a structure of (1,0) dimensional $\mathcal{G}$-supermanifold. We take $L = L' = 2$; if $\{e_1, e_2\}$ is a basis of $\mathbf{R}^2$, B_L has a basis $\{1, e_1, e_2, e_1 e_2\}$. We choose two charts $(x, U_1 \times \mathbf{R})$ and $(w, U_2 \times \mathbf{R})$, where U_1 (U_2) is S^1 without the north pole (south pole). x and w are given in terms of $z \in \mathbf{R}$ and the stereographic angles θ, ϕ, respectively from the north and south pole, as follows:

$$x = e^z + \tan\tfrac{\theta}{2}\, e_1 e_2, \quad -\tfrac{\pi}{2} < \theta < \tfrac{\pi}{2};$$
$$w = e^{-2z}\,[e^z - \tan(\tfrac{\pi}{4} - \tfrac{\phi}{2})\, e_1 e_2], \quad -\tfrac{\pi}{2} < \phi < \tfrac{\pi}{2}.$$

It is easily shown that x and w are C^∞ diffeomorphisms and that the transition functions $x(w)$ and $w(x)$ are GH^∞.

Thus with this atlas M is a (1,0) dimensional GH^∞ supermanifold, and therefore is also a $\mathcal{G}$-supermanifold. In this case we have a sheaf isomorphism

$$\mathcal{G}^M \simeq \mathcal{GH}^M \otimes_{B_{L'}} B_L\,. \tag{2.7}$$

It should be noticed that if M is a $\mathcal{G}$-supermanifold but not a GH^∞ supermanifold, the previous isomorphism does not hold.

A direct calculation shows that a global GH^∞ function on M has the form

$$f = K + f^i e_i, \; i = 1,2,3,$$

where the constant K and the C^∞ functions f^i of z are real-valued.

Example 2.3.3. Let $M = T^2 \times \mathbf{R}^2$, with T^2 the two-dimensional torus, and set $L = 2$, $L' = 1$. If z, ξ are local real coordinates in T^2, and u, t real coordinates in $\mathbf{R}^2$, we put in M local $B_L^{1,1}$-valued coordinates $x = z + ue_1e_2$, $w = \xi e_1 + te_2$. A direct computation shows that the global GH^∞ functions on M have the form

$$f = \alpha + \gamma e_1 + [u\alpha' - t\mu] e_1 e_2 \tag{2.8}$$

where α, γ, and μ are periodic real valued functions of z and $'$ denotes differentiation. With this atlas M is a (1,1) dimensional GH^∞ supermanifold, and so also a $\mathcal{G}$-supermanifold. The sheaf isomorphism (2.7) holds again.

3. Super vector bundles and graded tangent space to a $\mathcal{G}$-supermanifold

After providing a natural definition of super vector bundle (SVB) over a $\mathcal{G}$-supermanifold with structure sheaf $\mathcal{G}$, we shall show that the category of such bundles is equivalent to the category of locally free graded $\mathcal{G}$-modules. The graded tangent bundle and graded tangent space are introduced and their relationship studied.

3.1 $\mathcal{G}$-SUPERBUNDLES

A triple (E, M, π), where E and M are $\mathcal{G}$-supermanifolds, and $\pi : E \to M$ is a surjective $\mathcal{G}$-morphism, is said to be a ***$\mathcal{G}$-superbundle***. In the following we shall only consider *locally trivial* $\mathcal{G}$-superbundles, in the sense that there are a cover $\{U_\alpha\}$ of M and $\mathcal{G}$-diffeomorphisms

$$\psi_\alpha : \pi^{-1}(U_\alpha) \to U_\alpha \times F \qquad \text{with } \mathrm{pr}_2 \circ \psi_\alpha = \pi,$$

where F is a $\mathcal{G}$-supermanifold, called the *standard fibre of E*.

DEFINITION 3.1 *A locally trivial $\mathcal{G}$-superbundle (E, M, π) with standard fibre F is said to be a super vector bundle of rank (r,s) if F is a free graded B_L-module of rank (r,s), and for all $x \in U_\alpha$, $\psi_\alpha|\pi^{-1}(x) : \pi^{-1}(x) \to F$ is an even morphism of graded B_L-modules.*

Without loss of generality, we may assume $F \equiv B_L^{r+s}$. Moreover, the fibre $\pi^{-1}(x)$ will also be denoted by E_x.

DEFINITION 3.2 *Given two SVB's E and G over $\mathcal{G}$-supermanifolds M and N respectively, a morphism $f : E \to G$ of SVB's is a $\mathcal{G}$-map which restricted to any E_x gives a morphism of graded B_L-modules of E_x into a fibre of G.*

One can easily verify that, choosing as objects the SVB's of a fixed rank (r,s) over a $\mathcal{G}$-supermanifold M, and as morphisms the morphisms between rank (r,s) SVB's over M, all the properties necessary for having a category are fulfilled. This category will be denoted by $\mathbf{SVB}^{r,s}(M)$.

As it happens in the ordinary theory, an SVB can be specified by means of its transition functions. These are maps with values in the general linear supergroup [34], which is defined as follows. The set of $(r+s) \times (r+s)$ matrices with entries in B_L, denoted by $gl(r+s)$, is a graded B_L-module of dimension $(r^2+s^2, 2rs)$. Its even part, denoted by $gl(r,s)$, is formed by matrices of the form

$$X = \begin{pmatrix} A & B \\ C & D \end{pmatrix}$$

where the matrices A and D are $r \times r$ and $s \times s$ respectively and have entries in $(B_L)_0$, while B and C are $r \times s$ and $s \times r$ respectively and have entries in $(B_L)_1$. The general linear supergroup $GL(r,s)$ is defined as the open subset of $gl(r,s)$ whose elements are invertible matrices; it is isomorphic with the module of even automorphisms of B_L^{r+s}. Moreover, it can be given the structure of a $\mathcal{G}$-supermanifold of dimension $(r^2+s^2, 2rs)$.

Now, given an SVB (E, M, π) of rank (r,s), with trivializing isomorphisms $\psi_\alpha : \pi^{-1}(U_\alpha) \to U_\alpha \times B_L^{r+s}$, we define its transition functions $g_{\alpha\beta} : U_\alpha \cap U_\beta \to GL(r,s)$ by letting

$$g_{\alpha\beta}(x)(\cdot) = \mathrm{pr}_2 \circ \psi_\alpha \circ \psi_\beta^{-1}(x, \cdot) \qquad \forall x \in U_\alpha \cap U_\beta \,.$$

The transition functions are by construction $\mathcal{G}$-maps, and the usual cocycle conditions are verified:

$$g_{\alpha\beta}(x) \cdot g_{\beta\gamma}(x) \cdot g_{\gamma\alpha}(x) = 1 \quad \forall x \in U_\alpha \cap U_\beta \cap U_\gamma \,. \tag{3.1}$$

Supposing one starts with a set of $\mathcal{G}$-maps $g_{\alpha\beta} : U_\alpha \cap U_\beta \to GL(r,s)$ satisfying the cocycle condition (3.1), the standard construction of a vector bundle from its transition functions can be adapted to the present case, thus showing the existence of an equivalence class of SVB's on M whose representatives have the given $g_{\alpha\beta}$'s as transition functions. The proof given in Ref. 3 is particularly suitable to this aim.

3.2 SUPER VECTOR BUNDLES AND LOCALLY FREE $\mathcal{G}$-MODULES

For a fixed $\mathcal{G}$-supermanifold with structure sheaf $\mathcal{G}$, let $\mathbf{GM}^{r,s}(\mathrm{G})$ be the category of locally free graded $\mathcal{G}$-modules of rank (r,s). The proof of the equivalence between the categories $\mathbf{GM}^{r,s}(\mathrm{G})$ and $\mathbf{SVB}^{r,s}(M)$ is again standard and proceeds as in the case of topological, smooth and holomorphic bundles [3]. For the reader's convenience, we sketch it here; further details can be found in Ref. 28.

If $\mathcal{E}$ is a locally free $\mathcal{G}$-module of rank (r,s), there are on M a cover $\{U_\alpha\}$ and a family of isomorphisms

$$\varphi_\alpha : \mathcal{E}|U_\alpha \to (\mathcal{G}|U_\alpha)^{r+s} . \tag{3.2}$$

In terms of these isomorphisms we define sheaf morphisms

$$h_{\alpha\beta} : (\mathcal{G}|U_\alpha \cap U_\beta)^{r+s} \to (\mathcal{G}|U_\alpha \cap U_\beta)^{r+s}$$

by letting $h_{\alpha\beta} = \varphi_\alpha \circ \varphi_\beta^{-1}$. Now, the condition

$$\widetilde{h_{\alpha\beta}(f)} = g_{\alpha\beta}(x) \cdot \tilde{f} \tag{3.3}$$

determines $\mathcal{G}$-maps

$$g_{\alpha\beta} : U_\alpha \cap U_\beta \to GL(r,s) \tag{3.4}$$

satisfying the cocycle condition (3.1). In this way an isomorphism class of SVB's over M is found. On the other hand, the sheaf of sections of an SVB E over M is a locally free graded $\mathcal{G}$-module. Routine checks show that the two processes are one the inverse of the other, and are well-behaved with respect to the morphisms. Therefore we have:

PROPOSITION 3.1 *The categories* $\mathbf{GM}^{r,s}(G)$ *and* $\mathbf{SVB}^{r,s}(M)$ *are equivalent.* □

The reader should notice that the above construction does not work in the case of GH^∞ supermanifolds; in particular, in that case Eq. (3.3) does not define the transition functions since $\tilde{f}$ does not belong to a free module.

For future use, we prove the following result.

PROPOSITION 3.2 *Let $\mathcal{E}$ be a locally free graded $\mathcal{G}$-module, and let E be an SVB over M corresponding to $\mathcal{E}$ according to Proposition 3.1. For all $x \in M$ there is a canonical isomorphism $E_x \simeq \mathcal{E}_x / \mathcal{M}_x \mathcal{E}_x$, where $\mathcal{M}_x$ is the ideal of $\mathcal{G}_x$ appearing in the exact sequence (2.6).*

Proof. Let $x \in U_\alpha \cap U_\beta$, and consider trivializing morphisms of $\mathcal{E}$ as in Eq. (3.2). They entail isomorphisms

$$\mathcal{E}_x / \mathcal{M}_x \mathcal{E}_x \simeq \mathcal{G}_x^{r+s} / \mathcal{M}_x \mathcal{G}_x^{r+s} \simeq B_L^{r+s}$$

where the last isomorphism follows from the exact sequence (2.6). Now, using the trivialization of E associated with the transition functions $g_{\alpha\beta}$ of Eq. (3.3), one can associate an element of E_x with any element of B_L^{r+s}. The resulting map is obviously an isomorphism, which is easily shown to be independent of the choice of trivializing morphisms φ_α. □

3.3 GRADED TANGENT BUNDLE AND TANGENT SPACE

In this Section we study the graded tangent bundle and graded tangent space to a $\mathcal{G}$-supermanifold. The results proved here do *not* hold in the case of G^∞ or GH^∞ supermanifolds. In accordance with the content of §3.2 and Corollary 2.1, the sheaf $Der\,\mathcal{G}$ of graded derivations of the structure sheaf $\mathcal{G}$ of an (m,n) dimensional $\mathcal{G}$-supermanifold M corresponds to a rank (m,n) SVB over M, that we call the *graded tangent bundle* to M and denote by TM. We also define the graded tangent space to M on the analogy of Definition 2.2:

DEFINITION 3.3 *For any $x \in M$, T_xM is the graded B_L-module of morphisms of graded B_L-modules $X : \mathcal{G}_x \to B_L$ satisfying the graded Leibniz rule*

$$X(fg) = X(f)\,\tilde{g} + (-1)^{|f||X|}\,\tilde{f}\,X(g) \qquad \forall f, g \in \mathcal{G}_x\,.$$

PROPOSITION 3.3 *For any $x \in M$ there is a canonical isomorphism $(TM)_x \simeq T_xM$.*

Proof. In accordance with Proposition 3.2, we can regard in a canonical way $(TM)_x$ as the quotient $(Der\,\mathcal{G})_x/\mathcal{M}_x(Der\,\mathcal{G})_x$. We also notice that, as a consequence of the definition of direct limit, any $D \in Der\,\mathcal{G}(U)$ may be restricted to a derivation of $\mathcal{G}_x$, provided that $x \in U$, so that there is a canonical morphism $(Der\,\mathcal{G})_x \to Der(\mathcal{G}_x)$. Then we may define a map $(TM)_x \to T_xM$ by letting

$$\bar{D} \mapsto X, \qquad X(f) = \widetilde{Df}\,,$$

where $D \in (Der\,\mathcal{G})_x$ is any representative of $\bar{D} \in (TM)_x$. An explicit computation shows that X fulfils the graded Leibniz rule and that the map $D \mapsto X$ is an isomorphism. □

Finally, the following proposition shows that the graded tangent space is a genuine generalization of the tangent space of ordinary differential geometry.

PROPOSITION 3.4 *The even part of T_xM is canonically isomorphic, as a real vector space, to the ordinary tangent space at x to the smooth manifold underlying M.*

Proof. Let $\psi(\cdot) = \{x^1(\cdot)\ldots x^m(\cdot), y^1(\cdot)\ldots y^n(\cdot)\}$ be a coordinate system around x, and let $\psi(x) = a$. It is a consequence of Corollary 2.1 and Proposition 3.3 that the elements $(X_1 \ldots X_{m+n})$ of T_xM defined by

$$X_i(f) = \widetilde{\frac{\partial f}{\partial x^i}}, \; i = 1 \ldots m, \quad X_{m+\alpha} = \widetilde{\frac{\partial f}{\partial y^\alpha}}, \; \alpha = 1 \ldots n \quad \forall f \in \mathcal{G}_x$$

are a basis of T_xM. If $X \in (T_xM)_0$, let $X = \sum_{A=1}^{m+n} b^A X_A$. Then

$$\left[\frac{d}{dt}\,\psi^{-1}(a+tb)\right]_{t=0}, \tag{3.5}$$

with t real number, defines a tangent vector to the smooth manifold M at x in the ordinary sense. It is easily seen that Eq. (3.5) establishes an isomorphism of vector spaces which is independent of the coordinate system ψ. □

4. Supermanifold cohomology

We give here a brief outline of cohomology of supermanifolds. We shall be concerned with the cohomology of the complex of supersmooth differential forms and with the cohomology of a supermanifold with values in its structure sheaf; in the latter case we shall use Čech cohomology. Even though the proof of some results in supermanifold cohomology [30] seems to require sheaf cohomology à la Grothendieck [31], those results can be expounded in terms of Čech cohomology, since in our case the base space is paracompact and therefore the two cohomologies are isomorphic [31].

4.1 SHEAF COHOMOLOGY

Let X be a a topological space with an open cover $\mathcal{U} = \{U_\alpha,\, \alpha \in J\}$, J being an ordered set, and let $\mathcal{F}$ be a sheaf of abelian groups on X. For a fixed set of indices $\alpha_0 \dots \alpha_p \in J$ define $U_{\alpha_0 \dots \alpha_p} = U_{\alpha_0} \cap \dots \cap U_{\alpha_p}$. The *Čech complex of the sheaf $\mathcal{F}$* is the sequence of abelian groups $C^p(\mathcal{U}, \mathcal{F}), p = 0, 1, 2, \dots$ whose p-th term is

$$C^p(\mathcal{U}, \mathcal{F}) = \prod_{\alpha_0 < \dots < \alpha_p} \mathcal{F}(U_{\alpha_0 \dots \alpha_p}).$$

A differential operator $\delta : C^p(\mathcal{U}, \mathcal{F}) \to C^{p+1}(\mathcal{U}, \mathcal{F})$, satisfying $\delta^2 = 0$, is defined as follows: if $f = \{f_{\alpha_0 \dots \alpha_p}\} \in C^p(\mathcal{U}, \mathcal{F})$, then

$$(\delta f)_{\alpha_0 \dots \alpha_{p+1}} = \sum_{k=0}^{p+1} (-1)^k f_{\alpha_0 \dots \hat{\alpha}_k \dots \alpha_{p+1}},$$

where the caret denotes omission. The *Čech cohomology of X with values in $\mathcal{F}$ with respect to the cover $\mathcal{U}$* is the cohomology of the differential complex $(C^*(\mathcal{U}, \mathcal{F}), \delta)$, i.e.

$$H^p(\mathcal{U}, \mathcal{F}) = \frac{\ker \delta : C^p(\mathcal{U}, \mathcal{F}) \to C^{p+1}(\mathcal{U}, \mathcal{F})}{\operatorname{im} \delta : C^{p-1}(\mathcal{U}, \mathcal{F}) \to C^p(\mathcal{U}, \mathcal{F})}.$$

The *Čech cohomology of X with values in $\mathcal{F}$* is the direct limit of the Čech cohomologies with values in $\mathcal{F}$ over all the open covers of X; the cohomology groups so obtained are denoted by $H^p(X,\mathcal{F})$. As a consequence of the definition of direct limit, there is a group morphism

$$H^p(\mathcal{U},\mathcal{F}) \to H^p(X,\mathcal{F}). \tag{4.1}$$

We state two results that we shall use later on. For proofs, and for a comprehensive treatment of Čech cohomology, the reader may refer to Ref. 31, while a quicker introduction to the argument is to be found in Ref. 32.

PROPOSITION 4.1 *If X is paracompact, a sufficient condition for the morphism (4.1) to be one-to-one is that for all non-void intersections $U_{\alpha_0\dots\alpha_q}$, one has*

$$H^p(U_{\alpha_0\dots\alpha_q},\mathcal{F}) = 0 \quad \textit{for} \quad p \geq 1.$$

□

PROPOSITION 4.2 *If X is paracompact, and*

$$0 \longrightarrow \mathcal{F} \longrightarrow \mathcal{G} \longrightarrow \mathcal{H} \longrightarrow 0$$

is an exact sequence of sheaves on X, there is a long exact sequence in Čech cohomology

$$0 \longrightarrow H^0(X,\mathcal{F}) \longrightarrow H^0(X,\mathcal{G}) \longrightarrow H^0(X,\mathcal{H}) \stackrel{\partial}{\longrightarrow} H^1(X,\mathcal{F}) \longrightarrow \dots$$

$$\dots \longrightarrow H^p(X,\mathcal{F}) \longrightarrow H^p(X,\mathcal{G}) \longrightarrow H^p(X,\mathcal{H}) \stackrel{\partial}{\longrightarrow} H^{p+1}(X,\mathcal{F}) \longrightarrow \dots$$

where the ∂'s are the so-called connecting morphisms. □

An important topological invariant of smooth manifolds is their de Rham cohomology. The de Rham cohomology $H^\star_{DR}(X)$ of a smooth manifold X is defined as the cohomology of the differential complex

$$\Omega^0(X) \stackrel{d}{\longrightarrow} \Omega^1(X) \stackrel{d}{\longrightarrow} \Omega^2(X) \stackrel{d}{\longrightarrow} \dots$$

where d is Cartan's exterior differential, and Ω^k is the sheaf of differential k-forms on X. It is a fundamental result in de Rham theory that this cohomology is isomorphic with the Čech cohomology of X with values in the constant sheaf $\mathbf{R}$:

$$H^p_{DR}(X) \simeq H^p(X,\mathbf{R}), \quad p \geq 0. \tag{4.2}$$

This result relies on the fact that, due to the existence of smooth partitions of unity on X, the cohomology of X with values in the sheaf Ω^k is trivial for any $k \geq 0$, i.e.

$$H^p(X, \Omega^k) = 0 \quad \forall p > 0,\ \forall k \geq 0 .$$

This is usually stated by saying that the sheaves Ω^k are *acyclic*.

4.2 COHOMOLOGY OF $\mathcal{G}$-SUPERMANIFOLDS

Let M be a $\mathcal{G}$-supermanifold, and let $Der^* \mathcal{G}$ be the B_L-dual of the sheaf $Der\,\mathcal{G}$ over M. The sheaf $\mathcal{G}^k$ of supersmooth k-forms on M is defined as

$$\mathcal{G}^k = \wedge_{\mathcal{G}} Der^* \mathcal{G} \quad \text{for } k \geq 1, \quad \mathcal{G}^0 \equiv \mathcal{G} .$$

Let us denote the pairing $Der\mathcal{G} \otimes_{\mathcal{G}} Der^* \mathcal{G} \to \mathcal{G}$ by $\rfloor$. We define an exterior differential $d : \mathcal{G} \to \mathcal{G}^1$ by letting

$$D \rfloor df = D(f) \quad \forall f \in \mathcal{G}(U),\ D \in Der\,\mathcal{G}(U), \quad \text{and for any open } U \subset M ;$$

this is extended to morphisms $\mathcal{G}^p \to \mathcal{G}^{p+1}$, $p \geq 1$, in the usual way, so that $d^2 = 0$. The supersmooth de Rham cohomology of M, denoted by $H^*_{SDR}(M)$, is the cohomology of the differential complex

$$\mathcal{G}(M) \xrightarrow{d} \mathcal{G}^1(M) \xrightarrow{d} \mathcal{G}^2(M) \to \dots .$$

Now, M has also a structure of smooth manifold, so that we can compare break $H^*_{SDR}(M)$ with the ordinary de Rham cohomology of S, $H^*_{DR}(M)$. Each $\mathcal{G}^p(M)$ injects into $\Omega^p(M) \otimes_{\mathbf{R}} B_L$, and there is a corresponding morphism in cohomology,

$$\lambda : H^p_{SDR}(M) \to H^p_{DR}(M) \otimes B_L , \quad p \geq 0 . \tag{4.3}$$

λ is obviously an isomorphism for $k = 0$, since

$$H^0_{SDR}(M) = \ker d : \mathcal{G} \to \mathcal{G}^1 \simeq (B_L)^N , \quad H^0_{DR}(M) \simeq \mathbf{R}^N ,$$

where N is the number of connected components of M (notice that also $H^0_{SDR}(M) \simeq H^0(M, B_L)$). In general, we have the following result.

PROPOSITION 4.3 *Let $q \geq 1$, and assume that $H^k(M, \mathcal{G}^p) = 0$ for $0 \leq p \leq q-1$ and $1 \leq k \leq q$. Then $H^k_{SDR}(M) \simeq H^k_{DR}(M) \otimes B_L \simeq H^k(M, B_L)$ for $1 \leq k \leq q$.*

Proof. The demonstration proceeds as in the sheaf-theoretic proof of the Čech-de Rham isomorphism [35]. The constant sheaf B_L on M can be injected into $\mathcal{G}$ by letting $a \mapsto 1 \otimes a$; moreover, let us define the sheaf of closed supersmooth k-forms, $\mathcal{Z}^k = \ker d : \mathcal{G}^k \to \mathcal{G}^{k+1}$, $k \geq 1$. The sequences

$$0 \to B_L \to \mathcal{G} \to \mathcal{Z}^1 \to 0, \qquad 0 \to \mathcal{Z}^k \to \mathcal{G}^k \to \mathcal{Z}^{k+1} \to 0 \tag{4.4}$$

are both exact as a consequence of the Poincaré Lemma for supersmooth forms, that we shall prove in the Appendix. Due to the acyclicity hypothesis, the long exact sequences induced in cohomology by the sequences (4.4) give

$$0 \to H^0(M, B_L) \to H^0(M, \mathcal{G}) \to H^0(M, \mathcal{Z}^1) \to H^1(M, B_L) \to 0$$

$$0 \to H^k(M, \mathcal{Z}^1) \to H^{k+1}(M, B_L) \to 0, \qquad 1 \leq k \leq q$$

$$0 \to H^0(M, \mathcal{Z}^p) \to H^0(M, \mathcal{G}^p) \to H^0(M, \mathcal{Z}^{p+1}) \to H^1(M, \mathcal{Z}^p) \to 0, \ 1 \leq p \leq q-1,$$

$$0 \to H^k(M, \mathcal{Z}^{p-1}) \to H^k(M, \mathcal{Z}^p) \to 0, \quad 1 \leq k \leq q, \ 1 \leq p \leq q-1.$$

These sequences in turn induce a chain of isomorphisms

$$\begin{aligned} H^k(M, B_L) &\simeq H^{k-1}(M, \mathcal{Z}^1) \simeq \dots \\ &\simeq H^1(M, \mathcal{Z}^{k-1}) \simeq \frac{H^0(M, \mathcal{Z}^k)}{\operatorname{im} H^0(M, \mathcal{G}^{k-1})} \\ &\equiv H^k_{SDR}(M). \end{aligned}$$

The proof now follows from the isomorphisms

$$H^k(M, B_L) \simeq H^k(M, \mathbf{R}) \otimes B_L \simeq H^k_{DR}(M) \otimes B_L.$$

□

We can use Proposition 4.3 two show that the sheaf $\mathcal{G}$ is not acyclic in general. Let us get back to Example 2.3.3; Eq. (2.8) asserts that

$$\mathcal{G}(M) = C^\infty(S^1) \otimes_{\mathbf{R}} B_L.$$

An explicit computation shows that

$$\mathcal{Z}^1(M) = \left[C^\infty(S^1) \otimes_{\mathbf{R}} B_L\right] \oplus \mathbf{R}\, e_1 e_2$$

$$\mathcal{B}^1(M) \equiv \operatorname{im} d : \mathcal{G} \to \mathcal{Z}^1 = \left[C_0^\infty(S^1) \otimes_{\mathbf{R}} B_L\right] \oplus \mathbf{R}\, e_1 e_2,$$

where $C^\infty(S^1)$ is the space of smooth functions on the circle and $C_0^\infty(S^1)$ denotes the space of functions in $C^\infty(S^1)$ whose integral over S^1 vanishes. $\mathcal{Z}^1(M)$ and $\mathcal{B}^1(M)$ are direct sums of graded B_L-modules; in particular, $\mathbf{R}\, e_1 e_2$ is a graded

B_L-module under the multiplication $a \cdot b = a\sigma(b)$, $a \in \mathbf{R}\, e_1 e_2$, $b \in B_L$. Taking the quotient one gets

$$H^1_{SDR}(M) = B_L\,.$$

On the other hand, one has

$$H^1_{DR}(M) \otimes B_L = B_L \oplus B_L$$

so that Proposition 4.3 implies $H^1(M, \mathcal{G}) \neq 0$.

The same result could be obtained with Example 2.3.2; indeed in that case we have

$$\mathcal{G}^1(M) = \mathcal{Z}^1(M) = [C^\infty(\mathbf{R}) \otimes N_L] \oplus [\mathbf{R} e_1 e_2], \quad \mathcal{B}^1(M) = C^\infty(\mathbf{R}) \otimes N_L$$

(with N_L the nilpotent ideal of B_L) whence

$$H^1_{SDR}(M) = \mathbf{R}\, e_1 e_2\,. \tag{4.5}$$

Since $H^1_{DR}(M) \otimes B_L = B_L$ we have again $H^1(M, \mathcal{G}) \neq 0$. It is interesting to note that the cohomology group (4.5) is not a free B_L-module.

The non-triviality of the cohomology of their structure sheaf is not a peculiarity of $\mathcal{G}$-supermanifolds: also Rogers' G^∞ and GH^∞, and, of course, Rothstein's supermanifolds have this property. On the contrary, the structure sheaf of a graded manifold is always acyclic [22].

The fact that in general the structure sheaf of a $\mathcal{G}$-supermanifold is not acyclic has obviously important consequences in the cohomological classification of SVB's, as we shall in the next Section in the case of super line bundles.

A particular class of supermanifolds whose structure sheaf is acyclic is given by the so-called De Witt supermanifolds. These are $\mathcal{G}$-supermanifolds that admit atlases whose open sets are homeomorphic to sets in $B_L^{m,n}$ of the type $U \times N_L^{m,n}$, U being an open set in $\mathbf{R}^m$; in other words, the coordinate maps are homeomorphisms into $B_L^{m,n}$ topologized with the coarsest topology such that the body map is continuous. These are the supermanifolds that so far have found applications in physics. The proof of the acyclicity of the structure sheaf of De Witt GH^∞ supermanifolds can be found in Ref. 30; the same proof works in the case of $\mathcal{G}$-supermanifolds.

5. Super line bundles

Let $C_L = B_L \otimes \mathbf{C}$. The theory of SVB's we have expounded in Section 3 works also when the standard fibre F is assumed to be C_L^{r+s}; in this case we shall speak of complex SVB's. We wish here to examine the rank (1,0) case, namely, we shall consider *complex super line bundles* (the rank (0,1) case is substantially identical since in both cases the standard fibre is C_L and the structure group is $(C_L)_0^\star$, the set of elements in $(C_L)_0$ with non-vanishing body). In this case the transition functions, for a fixed cover $\mathcal{U} = \{U_\alpha\}$, are $\mathcal{G}$-maps $g_{\alpha\beta} : U_\alpha \cap U_\beta \to (C_L)_0^\star$. Let us introduce the sheaves $\mathcal{I}$ and $\mathcal{I}^\star$ of $\mathcal{G}$-maps $M \to (C_L)_0$ and $M \to (C_L)_0^\star$ respectively. The set of transition functions of a complex super line bundle E over M determines a cocycle in $H^1(\mathcal{U}, \mathcal{I}^\star)$, and viceversa any cocycle in $H^1(\mathcal{U}, \mathcal{I}^\star)$ gives a class of equivalent sets of transition functions. Therefore, the set of isomorphism classes of complex super line bundles on a $\mathcal{G}$-supermanifold M may be identified with $H^1(M, \mathcal{I}^\star)$.

We wish to relate the classification of super line bundles over M to the integer cohomology of M. We consider the injection $j : \mathbf{Z} \to \mathcal{I}$ of the constant sheaf $\mathbf{Z}$ over M into $\mathcal{I}$ given by $j(z) = z \otimes 1 = 1 \otimes z$, and define a morphism $\exp : \mathcal{I} \to \mathcal{I}^\star$ as follows. We start by defining an exponential map $\exp : C_L \to C_L^\star$ by letting

$$\exp a = \sum_{k=0}^{\infty} \frac{(2\pi i a)^k}{k!} \tag{5.1}$$

where the series converges in the l^1 norm on the vector space C_L for all $a \in C_L$. Simple algebraic manipulations show that the sequence

$$0 \to \mathbf{Z} \to C_L \to C_L^\star \to 1 \tag{5.2}$$

is exact. The pointwise application of the map (5.1) defines a morphism $\exp : \mathcal{I} \to \mathcal{I}^\star$.

LEMMA 5.1 *The sequence*

$$0 \to \mathbf{Z} \xrightarrow{\ j\ } \mathcal{I} \xrightarrow{\ \exp\ } \mathcal{I}^\star \to 1 \tag{5.3}$$

is exact. □

The sequence (5.3), according to Proposition 4.2, induces a long exact sequence in Čech cohomology

$$\begin{aligned} \ldots \longrightarrow H^1(M, \mathbf{Z}) \longrightarrow H^1(M, \mathcal{I}) \longrightarrow & H^1(M, \mathcal{I}^\star) \\ & \xrightarrow{\ \delta\ } H^2(M, \mathbf{Z}) \longrightarrow H^2(M, \mathcal{I}) \longrightarrow \ldots \end{aligned} \tag{5.4}$$

where δ is the canonical connecting morphism. The element $\delta(E) \in H^2(M, \mathbf{Z})$ is the *obstruction class* of E.

Since in general δ is not injective, non-isomorphic super line bundles may have the same obstruction class; in this sense super line bundles behave like holomorphic bundles on complex manifolds. Indeed, we can define a Picard group

$$\mathrm{Pic}^0(M) \equiv \frac{H^1(M, \mathcal{I})}{\mathrm{im}\, H^1(M, \mathbf{Z})}$$

which classifies the non-trivial complex super line bundles whose obstruction class vanishes. For instance, the $\mathcal{G}$-supermanifolds of Examples 2.3.2 and 2.3.3 have a non-trivial Picard group.

However, if the base manifold M is De Witt, the sheaf $\mathcal{G}$ is acyclic, as we have discussed in Section 4.2, so that $\mathcal{I}$ is acyclic as well, and the exact sequence (5.4) implies that δ is one-to-one; in particular, $\mathrm{Pic}^0(M) = 0$. Thus if M is De Witt $H^2(M, \mathbf{Z})$ classifies faithfully the complex super line bundles over M.

A more detailed treatment of complex super line bundles and, more generally, of the cohomological classification of super vector bundles can be found in Refs. 36, 37.

Acknowledgments. The research on which this paper is based was done in collaboration with C. Bartocci, whom I wish to thank, and was partially supported by the Italian Ministry for Public Education through the research project "Geometria e Fisica", by the National Group for Mathematical Physics of the Italian National Research Council (CNR), and by the 'Consejo Superior de Investigación Científica' of Spain. I wish also to thank D. Hernández Ruipérez, J. Muñoz Masqué and J. M. Rabin for useful discussions.

Appendix: Poincaré lemma for supersmooth forms

We show that, if U is an open set in a $\mathcal{G}$-supermanifold M which is homeomorphic to an open ball in $B_L^{m,n}$, and $\omega \in \mathcal{G}^k(U)$ $(k \geq 1)$ is closed, i.e. $d\omega = 0$, then there is an $\eta \in \mathcal{G}^{k-1}(U)$ such that $\omega = d\eta$, i.e. ω is exact. We define a homotopy operator $K : \mathcal{G}^k(U) \to \mathcal{G}^{k-1}(U)$ as follows. If τ is a GH^∞ k-form on U (i.e. $\tau \in \mathcal{GH}^k(U)$), $(x^1(\cdot) \dots x^{m+n}(\cdot))$ are coordinates on U, and $\tau = dx^{A_k} \wedge \dots \wedge dx^{A_1}\, \tau_{A_1 \dots A_k}(x)$, with $x \in U$, and $A_i = 1 \dots m+n$, we let

$$K\tau = (-1)^k\, k\, dx^{A_{k-1}} \wedge \dots \wedge dx^{A_1}\, x^B \int_0^1 t^{k-1}\, \tau_{BA_1 \dots A_{k-1}}(tx)\, dt\,.$$

where, without loss of generality, we have assumed that the open ball U is centred in 0.

If $\omega \in \mathcal{G}^k(U)$ is of the type $\tau \otimes a$, with $\tau \in \mathcal{GH}^k(U)$ and $a \in B_L$, we set

$$K\omega = K\tau \otimes a\,. \tag{A.1}$$

Since as a consequence of our hypotheses $\mathcal{G}^k(U) \simeq \mathcal{GH}^k(U) \otimes_{B_{L'}} B_L$, Eq. (A.1) defines a morphism $K : \mathcal{G}^k(U) \to \mathcal{G}^{k-1}(U)$, which verifies $dK + Kd = 1$. Then it is sufficient to put $\eta = K\omega$ to get the result.

The case $k = 0$ must be treated separately. Since $\mathcal{G}(U) \simeq \hat{\mathcal{G}}(U) \otimes \Lambda[n, B_{L'}]$, any $f \in \mathcal{G}(U)$ is written as $f = f_\mu \otimes y^\mu$, with $\mu \in M_n$, $f_\mu \in \hat{\mathcal{G}}(U)$, and $y^\mu = y^{\mu_1} \dots y^{\mu_N}$ if $\mu = \{\mu_1 \dots \mu_N\}$. Then we have

$$0 = df = \sum_{i=1}^{m} dx^i \frac{\partial f_\mu}{\partial x^i} \otimes y^\mu + \sum_{\mu \in M_n} f_{\mu_1 \dots \mu_J \dots \mu_N} \otimes y^{\mu_1} \dots dy^{\mu_J} \dots dy^{\mu_N}\,.$$

This implies

$$f_\mu = 0 \quad \text{for} \quad \mu \neq \{0\}, \qquad \frac{\partial f_0}{\partial x^i} = 0\,,$$

i.e. f is constant.

References

1. Serre, J.-P.: *Faisceaux algèbriques cohérents.* Ann. Math. **61** (1950) 197-278.
2. Swan, R. G.: 'Vector bundles and projective modules.' *Trans. Amer. Math. Soc.* **105** (1962) 264-277.
3. Wells, R. O.: *Differential analysis on complex manifolds* (Springer-Verlag, New York, 1980).
4. Bruzzo, U., Cianci, R.: 'Mathematical theory of super fibre bundles.' *Class. Quantum Grav.* **1**, 213-226 (1984).
5. Bartocci, C., Bruzzo, U., Landi, G.: 'Chern-Simons forms on principal super fibre bundles.' Preprint S.I.S.S.A. 109/87/FM, Trieste 1987.
6. Bruzzo, U., Cianci, R.: 'Variational calculus on supermanifolds and invariance properties of superspace field theories.' *J. Math. Phys.* **28**, 786-791 (1987).
7. Bruzzo, U.: 'Lagrangian gauge theories on supermanifolds.' In *Differential geometric methods in Mathematical Physics*, Lect. Notes Math. **1251**, pp. 91-112 (Springer-Verlag, Berlin, 1987).
8. Yates, R. G.: 'Fibre bundles and supersymmetries.' *Commun. Math. Phys.* **76**, 255-258 (1980).
9. Awada, M. A.: 'Fibre bundles and supergravity.' *Commun. Math. Phys.* **91**, 53-64 (1983).

10. Bruzzo, U., Cianci, R.: 'Structure of supermanifolds and supersymmetry transformations.' *Commun. Math. Phys.* **95**, 393-400 (1984).
11. Bruzzo, U.: 'Geometry of rigid supersymmetry.' *Hadronic J.* **9**, 25-30 (1986).
12. Bonora, L., Pasti, P., Tonin, M.: 'Chiral anomalies in higher dimensional supersymmetric theories.' *Nucl. Phys. B* **286**, 150-174 (1987) and references therein.
13. Bruzzo, U., Landi G.: 'A new analysis of Weil triviality in supersymmetric gauge theories,' Preprint SU-4228-382 Syracuse University, April 1988, and Preprint 50/1988 Dip. di Matematica, Univ. di Genova.
14. Friedan, D., in *Proceedings of the Workshop on Unified String Theories,* ed. D. Gross and M. Green (World Scientific, Singapore, 1986).
15. Crane, L., Rabin, J. M.: 'Super Riemann surfaces: uniformization and Teichmüller theory.' *Commun. Math. Phys.* **113**, 601-624 (1988).
16. Hodgkin, L. A.: 'A direct calculation of super-Teichmüller space.' *Lett. Math. Phys.* **14**, 47-53 (1987); 'On metrics and super-Riemann surfaces.' *Lett. Math. Phys.* **14**, 177-184 (1987).
17. Batchelor, M., Bryant, P.: 'Graded Riemann surfaces.' *Commun. Math. Phys.* **114**, 243-255 (1988).
18. Le Brun, C., Rothstein, M.: 'Moduli of super Riemann surfaces.' Preprint Institute for Advanced Study, Princeton.
19. Giddings, S. B., Nelson, P.: 'Torsion constraints and super Riemann surfaces,' *Phys. Rev. Lett.* **59**, 2619 (1987).
20. De Witt, B.: *Supermanifolds* (Cambridge Univ. Press, London, 1984).
21. Rogers, A.: 'A global theory of supermanifolds.' *J. Math. Phys.* **21**, 1352-1365 (1980).
22. Kostant, B.: 'Graded manifolds, graded Lie theory, and prequantization', in *Differential geometric methods in Mathematical Physics,* Lect. Notes Math. **570**, pp. 177-306 (Springer-Verlag, Berlin, 1977).
23. Berezin, F. A., Lĕites, D. A.: 'Supermanifolds.' *Soviet Math. Dokl.* **16**, 1218-1222 (1975).
24. Lĕites, D. A.: 'Introduction to the theory of supermanifolds.' *Russian Math. Surveys* **35**, 1-64 (1980).
25. Boyer, C. P., Gitler, S.: 'The theory of G^{∞}-supermanifolds.' *Trans. Amer. Math. Soc.* **285**, 241-268 (1984).
26. Rothstein, M.: 'The axioms of supermanifolds and a new structure arising from them.' *Trans. Amer. Math. Soc.* **297**, 159-180 (1986).
27. Rogers, A.: 'Graded manifolds, supermanifolds and infinite-dimensional Grassmann algebras.' *Commun. Math. Phys.* **105**, 375-384 (1986).
28. Bartocci, C., Bruzzo, U.: 'Some remarks on the differential–geometric approach

to supermanifolds.' *J. Geom. Phys.* **4**, n. 3 (1987).

29. Bartocci, C., Bruzzo, U.: 'Cohomology of supermanifolds.' *J. Math. Phys.* **28**, 2363-2368 (1987).
30. Bartocci, C., Bruzzo, U.: 'Cohomology of the structure sheaf of real and complex supermanifolds.' Preprint 12/1987 Dipartimento di Matematica, Università di Genova.
31. Godement, R., *Théorie des faisceaux* (Hermann, Paris, 1964).
32. Hirzebruch, F.: *Topological methods in algebraic geometry* (Springer-Verlag, Berlin, 1966).
33. Batchelor, M.: 'Two approaches to supermanifolds.' *Trans. Amer. Math. Soc.* **258**, 257-270 (1980).
34. Rittenberg, V., Scheunert, M.: 'Elementary Construction of Graded Lie Groups.' *J. Math. Phys.* **19**, 713-718 (1978).
35. P. Griffiths, J. Harris. *Principles of algebraic geometry* (Wiley, New York, 1978).
36. Bartocci, C., Bruzzo, U.: 'Super line bundles.' Preprint 49/1988 Dip. di Matematica, Univ. di Genova.
37. Bruzzo, U., Hernández Ruipérez, D.: in preparation.

REPRESENTATIONS OF LIE SUPERALGEBRAS
AN INTRODUCTION

Manfred Scheunert
Department of Physics
University of Freiburg
Hermann-Herder-Straße 3
7800 Freiburg, W. Germany

ABSTRACT. An introductory overview over the theory of finite-dimensional representations of a Lie superalgebra is given. The most basic constructions with the representations of a Lie superalgebra are described, the enveloping algebra is introduced, and the induced representations are defined and then investigated for a special class of consistently Z - graded Lie superalgebras. On the example of the general linear Lie superalgebras the classical theory of roots and weights is generalized to the super case and it is shown how one is led to the so-called typical representations. Finally, the decomposition of the tensorial powers of a representation by means of the action of the symmetric group is briefly discussed.

1. INTRODUCTION

At present, the most profound results in the theory of Lie superalgebras are due to Kac. He has classified [1,2] the simple Lie superalgebras and he has shown [3,4] that a considerable part of the representation theory of semi-simple Lie algebras can be extended to the basic classical Lie superalgebras, provided one only considers the so-called typical irreducible (finite-dimensional) representations, i.e., irreducible representations which are "in general position". For example, the characters of these representations can be calculated (generalization of Weyl's classical formula), and any representation all of whose irreducible subquotients are typical, is completely reducible. Nevertheless, even the theory of finite-dimensional representations of the basic Lie superalgebras is far from complete. In particular, a satisfactory theory of the non-typical representations is still missing and the super analogue of the classical reduction of tensor spaces by means of symmetrization and contraction is only partly understood.

In the present work I am going to describe, under certain simplifying assumptions, some of the basic results in the representation theory of Lie superalgebras. After recalling a few elementary items in section 2, I shall introduce in section 3 the enveloping algebra of a Lie superalgebra and use it to discuss the so-called induced representations. In sec-

K. Bleuler and M. Werner (eds.), Differential Geometrical Methods in Theoretical Physics, 441–460.

tion 4 I shall show, on the example of the general linear Lie superalgebra gl(n,m), how the classical theory of roots and weights can be generalized to the super case and how one is led to the typical representations. Finally, in section 5 I shall briefly describe how the action of the symmetric group can be used to decompose the tensorial powers of the elementary gl(n,m) - module.

The present work is similar to an article which I have published in 1985 [5]. However, in contrast to this earlier work I intend here to provide an introductory overview over the subject without giving any details or proofs.

I close this introduction by collecting some conventions and notation as well as some basic definitions. For more details the reader is referred to reference [2] (see also [6]).

a) The base field will be any commutative algebraically closed field K of characteristic zero, for example, the field of complex numbers. Neither the algebras nor the vector spaces and modules are assumed to be finite-dimensional unless otherwise stated. The field of integers modulo 2 will be denoted by $Z_2 = Z/2Z$, its elements are $\bar{0}$ and $\bar{1}$, the residue classes of even or odd integers, respectively.

b) A Z_2 - *graded vector space* (superspace) is a vector space V endowed with a family $(V_\alpha)_{\alpha \in Z_2}$ of subspaces (the Z_2 - *gradation*) such that

$$V = V_{\bar{0}} \oplus V_{\bar{1}} .$$

The elements of $V_{\bar{0}}$ (resp. $V_{\bar{1}}$) are called *homogeneous of degree* $\bar{0}$ (resp. $\bar{1}$) or *even* (resp. *odd*).

A *subspace* V' of V is called *graded* if

$$V' = (V' \cap V_{\bar{0}}) \oplus (V' \cap V_{\bar{1}}) .$$

Let $W = W_{\bar{0}} \oplus W_{\bar{1}}$ be a second Z_2 - graded vector space and let L(V,W) denote the vector space of all linear mappings of V into W. Then L(V,W) admits a natural Z_2 - gradation defined by

$$L(V,W)_\gamma = \{g \in L(V,W) \mid g(V_\alpha) \subset W_{\alpha+\gamma} \text{ for all } \alpha \in Z_2\} ,$$

where $\gamma \in Z_2$.

c) A Z_2 - *graded algebra* (*superalgebra*) is an algebra A, endowed with a Z_2 - gradation

$$A = A_{\bar{0}} \oplus A_{\bar{1}}$$

such that

$$A_\alpha A_\beta \subset A_{\alpha+\beta} \quad \text{for all } \alpha, \beta \in Z_2 .$$

Homomorphisms of Z_2 - *graded algebras* are assumed to be *homogeneous of degree zero*.

d) Let L be a Z_2 - graded algebra whose multiplication is denoted by a pointed bracket; thus we have

$$L = \bigoplus_{\alpha \in Z_2} L_\alpha$$

and

$$\langle L_\alpha, L_\beta \rangle \subset L_{\alpha+\beta} \quad \text{for all} \quad \alpha,\beta \in Z_2 \ .$$

L is called a *Lie superalgebra* if the following identities are satisfied:

$$\langle A, B \rangle = -(-1)^{\alpha\beta} \langle B, A \rangle$$

super skew symmetry

$$(-1)^{\gamma\alpha} \langle A, \langle B, C \rangle\rangle + (-1)^{\alpha\beta} \langle B, \langle C, A \rangle\rangle + (-1)^{\beta\gamma} \langle C, \langle A, B \rangle\rangle = 0$$

super Jacobi identity

for all $A \in L_\alpha$, $B \in L_\beta$, $C \in L_\gamma$; $\alpha,\beta,\gamma \in Z_2$.

We note that this definition makes sense if Z_2 is replaced by Z (the ring of integers). An algebra of this type is called a *consistently* Z - *graded Lie superalgebra* [1,2]. Actually, many of the interesting Lie superalgebras are consistently Z - graded in a natural way and it may be quite helpful to take advantage of this finer structure.

e) An important class of Lie superalgebras is obtained as follows. Consider a Z_2 - graded vector space V. Obviously, L(V,V), endowed with its natural algebra structure and gradation, is an associative superalgebra. Let us introduce a new multiplication, the *supercommutator*, by

$$\langle A, B \rangle = A B - (-1)^{\alpha\beta} B A$$

for all $A \in L(V,V)_\alpha$ and $B \in L(V,V)_\beta$. Then L(V,V) is converted into a Lie superalgebra which is called the *general linear Lie superalgebra* of the Z_2 - graded vector space V and which will be denoted by $gl(V_{\bar{0}},V_{\bar{1}})$.

If V is finite-dimensional and if

$$\dim V_{\bar{0}} = n \quad , \quad \dim V_{\bar{1}} = m \ ,$$

it may be useful to choose a basis $(e_1, \ldots, e_n)$ of $V_{\bar{0}}$ and a basis $(e_{n+1}, \ldots, e_{n+m})$ of $V_{\bar{1}}$, to express the elements of $gl(V_{\bar{0}},V_{\bar{1}})$ in terms of their matrices with respect to the basis $(e_1, \ldots, e_{n+m})$ of V, and to transfer the structure of $gl(V_{\bar{0}},V_{\bar{1}})$ to the space of these matrices. The Lie superalgebra thus obtained is denoted by gl(n,m), its elements are usually written in their obvious block matrix form.

2. ELEMENTARY REPRESENTATION THEORY

Throughout this section L will denote an arbitrary Lie superalgebra.

Definition

A *graded representation* of a Lie superalgebra L in a Z_2 - graded vector space V is a homomorphism of L into the general linear Lie superalgebra $gl(V_{\bar{0}}, V_{\bar{1}})$ of V. A Z_2 - graded vector space which is endowed with a graded representation of L is called a *graded* L - *module*.

If ρ is the graded representation we consider and if $A \in L$, the representative $\rho(A)$ of A will frequently be denoted by A_V, and if $x \in V$ we shall even write $A \cdot x$ in place of $A_V x$.

Let V and W be two graded L - modules. An *even* linear mapping g of V into W which satisfies

$$A_W \circ g = g \circ A_V \quad \text{for all } A \in L$$

is called a *homomorphism* of the graded L - module V into the graded L - module W.

Concepts like graded submodules, graded quotient modules (with respect to graded submodules), direct sums of graded L - modules, graded irreducibility of a graded L - module,.... are defined in the obvious way.

Definition

Let V be a graded L - module and let $A \in L$ and $x \in V$. If

$$A_V x = 0$$

we say that x is *invariant* under (the action of) A or that A leaves the element x invariant.

We note that the set of all L - invariant elements of V, i.e., the set

$$V^L = \{x \in V | A \cdot x = 0 \text{ for all } A \in L\}$$

is a graded subspace of V. On the other hand, if y is a *homogeneous* element of V, the set of all elements A of L leaving y invariant, i.e., the set

$$\{A \in L | A \cdot y = 0\}$$

is a graded subalgebra of L.

Let us next describe some of the basic constructions with graded L - modules. All of these constructions are governed by the following *rule of thumb* which tells us how to generalize a classical (i.e., non-graded) multilinear relation to the graded case. We proceed in steps:

Consider first only objects which are homogeneous with respect to the pertinent Z_2 - gradation.

Fix an ordering of the objects appearing in the classical relation.

Add to each summand in the classical relation a factor of $(-1)^{\alpha\beta}$ whenever in this summand an object of degree α and an object of degree β have been interchanged (as compared with the fixed ordering).

Extend the resulting relation by multilinearity to arbitrary (i.e., not necessarily homogeneous) objects.

Obviously, the axioms for a Lie superalgebra and the definition of the supercommutator are chosen according to this rule.

Let us next consider the action of a Lie superalgebra on a space of linear mappings. Thus let V and W be two graded L - modules. We have already noted that the space L(V,W) of all linear mappings of V into W admits a natural Z_2 - gradation. It is easy to see that the equation

$$A\cdot g = A_W \circ g - (-1)^{\alpha\gamma} g \circ A_V ,$$

with $A \in L_\alpha$ and $g \in L(V,W)_\gamma$, defines on L(V,W) the structure of a graded L - module.

In particular, by choosing W = K, endowed with the gradation $K_{\bar{0}} = K$, $K_{\bar{1}} = \{0\}$ and with the zero action of L, we obtain the dual or contragredient L - module V* = L(V,K).

Returning to the general situation we note that a homogeneous linear mapping $g \in L(V,W)$ of degree γ is L - invariant if and only if

$$g(A\cdot x) = (-1)^{\gamma\alpha} A\cdot g(x) \quad \text{for all } A \in L_\alpha,\ x \in V .$$

Thus the mapping g is L - invariant if and only if it is L - superlinear in the obvious sense.

Let us next assume that V = W. Then we have

$$A\cdot g = \langle A_V, g\rangle$$

for all $A \in L$ and $g \in L(V,V)$ and the set of all L - invariant elements of L(V,V), i.e., the set

$$S = \{g \in L(V,V) \mid \langle A_V, g\rangle = 0 \text{ for all } A \in L\}$$

is a graded subalgebra of the associative superalgebra L(V,V). In this connection we mention the following generalized version of Schur's lemma.

Schur's lemma

Let V be a finite-dimensional graded L - module which is irreducible in the graded sense. Then the subalgebra S of L(V,V) defined above is equal to $K\cdot id_V$ or to $K\cdot id_V + K\cdot j$, where j is an odd linear mapping of V into itself with $j^2 = -id_V$.

The second alternative in Schur's lemma takes place if and only if the L - module V is not irreducible in the non-graded sense, i.e., if and only if V has a non-graded L - invariant subspace different from {0} and V. Obviously, in this case it follows that $\dim V_{\bar{0}} = \dim V_{\bar{1}}$. Of course, this latter equation is by no means characteristic for the second alternative. Actually, there are large classes of algebras (for example the algebras gl(n,m)) for which the second alternative does not occur at all.

A particularly important application of our rule of thumb is provided by the construction of tensor products of graded L - modules [6]. If $V_1, \dots, V_r$ are Z_2 - graded vector spaces, there exists a unique Z_2 - gradation on the tensor product $V_1 \otimes \dots \otimes V_r$ such that $x_1 \otimes \dots \otimes x_r$ is homogeneous of degree $\xi_1 + \dots + \xi_r$ whenever the elements $x_s \in V_s$ are homogeneous of degree ξ_s, for $s = 1,2,\dots,r$.

Now let $W_1, \dots, W_r$ be a second family of Z_2 - graded vector spaces and let $g_t \in L(V_t, W_t)$ be homogeneous of degree γ_t, for $t = 1,2,\dots,r$. Then there exists a unique linear mapping of $V_1 \otimes \dots \otimes V_r$ into $W_1 \otimes \dots \otimes W_r$, denoted by $g_1 \bar{\otimes} \dots \bar{\otimes} g_r$, such that

$$(g_1 \bar{\otimes} \dots \bar{\otimes} g_r)(x_1 \otimes \dots \otimes x_r) = \prod_{t>s} (-1)^{\gamma_t \xi_s} g_1(x_1) \otimes \dots \otimes g_r(x_r)$$

for all elements $x_s \in V_s$ which are homogeneous of degree ξ_s. Obviously, $g_1 \bar{\otimes} \dots \bar{\otimes} g_r$ is homogeneous of degree $\gamma_1 + \dots + \gamma_r$.

We extend the definition of $g_1 \bar{\otimes} \dots \bar{\otimes} g_r$ by multilinearity to arbitrary elements $g_t \in L(V_t, W_t)$ and call $g_1 \bar{\otimes} \dots \bar{\otimes} g_r$ the *graded tensor product of the mappings* $g_1, \dots, g_r$.

Consider next a third family $U_1, \dots, U_r$ of Z_2 - graded vector spaces and let $h_s \in L(U_s, V_s)$ be homogeneous of degree η_s, for $s = 1,2,\dots,r$. If, as before, the mappings $g_t \in L(V_t, W_t)$ are homogeneous of degree γ_t, we have

$$(g_1 \bar{\otimes} \dots \bar{\otimes} g_r) \circ (h_1 \bar{\otimes} \dots \bar{\otimes} h_r) = \prod_{t>s} (-1)^{\gamma_t \eta_s} (g_1 \circ h_1) \bar{\otimes} \dots \bar{\otimes} (g_r \circ h_r) .$$

Now suppose that $L_1, \dots, L_r$ are Lie superalgebras and that V_s is a graded L_s - module , for $s = 1,2,\dots,r$. Then the action of the product algebra $L_1 \times \dots \times L_r$ on $V_1 \otimes \dots \otimes V_r$ defined by

$$(A_1, \dots, A_r) \longrightarrow \sum_{s=1}^{r} \mathrm{id}_{V_1} \bar{\otimes} \dots \bar{\otimes} (A_s)_{V_s} \bar{\otimes} \dots \bar{\otimes} \mathrm{id}_{V_r}$$

for all $A_s \in L_s$, makes $V_1 \otimes \dots \otimes V_r$ into a graded $L_1 \times \dots \times L_r$ - module which is called the (exterior) *tensor product of the graded* L_s *- modules* V_s .

We note that if the graded L_s - modules V_s are finite-dimensional and irreducible in the graded sense, their graded tensor product is not necessarily graded - irreducible. For example, if r = 2 and if there exist mappings $j_s \in L(V_s, V_s)_{\bar{1}}$; s = 1,2, as described in Schur's lemma, then $j_1 \bar{\otimes} j_2$ is even and commutes with the action of $L_1 \times L_2$. A detailed study of this point has been carried out by Hochschild [7].

In the following we restrict our attention to the case

$L_1 = \ldots. = L_r = L$, with a fixed Lie superalgebra L. Then the "diagonal" action of L on $V_1 \otimes \ldots. \otimes V_r$ given by

$$A \longrightarrow \sum_{s=1}^{r} id_{V_1} \bar{\otimes} \ldots \bar{\otimes} A_{V_s} \bar{\otimes} \ldots \bar{\otimes} id_{V_r}$$

for all $A \in L$, defines on $V_1 \otimes \ldots. \otimes V_r$ the structure of a graded L-module, which is called the (interior) *tensor product of the graded* L*-modules* V_s.

The tensor product of graded L-modules is associative in the usual sense. The question of commutativity is somewhat more subtle. If π is a permutation of $\{1,2,\ldots.,r\}$, there exists a unique linear mapping

$$S_\pi : V_1 \otimes \ldots. \otimes V_r \longrightarrow V_{\pi^{-1}(1)} \otimes \ldots. \otimes V_{\pi^{-1}(r)}$$

such that

$$S_\pi(x_1 \otimes \ldots. \otimes x_r) = \varepsilon_r(\pi^{-1}; \xi_1, \ldots., \xi_r)\, x_{\pi^{-1}(1)} \otimes \ldots. \otimes x_{\pi^{-1}(r)}$$

for all homogeneous elements $x_s \in V_s$ of degree ξ_s, and S_π is an isomorphism of graded L-modules. The function ε_r is defined by

$$\varepsilon_r(\pi^{-1}; \xi_1, \ldots., \xi_r) = \prod (-1)^{\xi_i \xi_j} ,$$

where the product runs over all i,j with $i < j$ but $\pi(i) > \pi(j)$; it has the crucial property that for any two permutations π and τ of $\{1,2,\ldots.,r\}$ we have

$$\varepsilon_r(\pi\tau; \xi_1, \ldots, \xi_r) = \varepsilon_r(\pi; \xi_1, \ldots, \xi_r)\, \varepsilon_r(\tau; \xi_{\pi(1)}, \ldots, \xi_{\pi(r)})$$

for all $\xi_1, \ldots, \xi_r \in Z_2$. This relation implies that if $V_1, \ldots, V_r$ are all equal to a fixed graded L-module V, then $\pi \longrightarrow S_\pi$ is a representation of the symmetric group S_r of $\{1,2,\ldots, r\}$ by automorphisms of the graded L-module $V^{\otimes r} = V \otimes \ldots. \otimes V$, r factors.

Finally, let $V_1', \ldots, V_r'$ be a second family of graded L-modules and let $g_s \in L(V_s, V_s')$ be homogeneous of degree γ_s, for $s = 1,2,\ldots.,r$. Let S_π' be the mapping which is obtained if in the definition of S_π the modules V_s are replaced by V_s'. Then we have

$$S_\pi' \circ (g_1 \bar{\otimes} \ldots \bar{\otimes} g_r) = \varepsilon_r(\pi^{-1}; \gamma_1, \ldots, \gamma_r)\, (g_{\pi^{-1}(1)} \bar{\otimes} \ldots \bar{\otimes} g_{\pi^{-1}(r)}) \circ S_\pi .$$

The action of the symmetric group on tensor products is basic for the decomposition of tensorial powers of graded L-modules. A special case will be discussed in section 5.

3. THE ENVELOPING ALGEBRA INDUCED REPRESENTATIONS

As in the Lie algebra case, the (universal) enveloping algebra of a Lie superalgebra is a particularly useful tool in representation theory. In the present section we are going to discuss the basic properties of this algebra and then use it to construct the so-called induced representations [1,2,4].

If L is any Lie superalgebra, the *enveloping algebra* U(L) of L is defined to be the quotient algebra

$$U(L) = T(L)/J(L) ,$$

where T(L) is the tensor algebra of the vector space L and J(L) is the two-sided (graded) ideal of T(L) generated by all tensors of the form

$$A \otimes B - (-1)^{\alpha\beta} B \otimes A - \langle A, B \rangle$$

with $A \in L_\alpha$, $B \in L_\beta$; $\alpha, \beta \in Z_2$. It is easy to see that U(L) is a Z_2-graded associative algebra with a unit element. Moreover, it has the following basic properties whose analogues in the Lie algebra case are well-known.

a) The canonical mapping of L into U(L) = T(L)/J(L) is injective. This shows that L can be regarded as a graded subspace of U(L) and that for any two elements $A, B \in L$, the bracket $\langle A, B \rangle$ defined in L is nothing but the supercommutator of A and B, calculated in U(L).

b) The unit element and L generate the algebra U(L).

c) Every (graded) representation ρ of L in a (Z_2-graded) vector space V uniquely extends to a (graded) representation $\bar{\rho}$ of U(L) in V such that $\bar{\rho}(1) = id_V$.

d) Generalized Poincaré, Birkhoff, Witt theorem
Let $(E_i)_{i \in I}$ be a homogeneous basis of L (i.e., a basis consisting of homogeneous elements) and suppose that the index set I is totally ordered. Then the products

$$E_{i_1}^{n_1} E_{i_2}^{n_2} \dots . E_{i_r}^{n_r}$$

with $i_s \in I \; ; \; i_1 < i_2 < \dots . < i_r$

n_s integers $\geqslant 1$, $n_s = 1$ if E_{i_s} is odd,

form a basis of U(L). (The case r = 0 is allowed and yields the unit element.)

e) The Poincaré, Birkhoff, Witt theorem immediately implies the following corollary.

Corollary

Let L' be a graded subalgebra of L. Then:
I) The Z_2-graded algebra U(L') can be identified with the graded sub-

algebra of U(L) which is generated by L' and the unit element.
II) The left and right graded U(L') - modules U(L) have a homogeneous basis which can be constructed as in the Poincaré, Birkhoff, Witt theorem except that $(E_i)_{i\in I}$ now is a homogeneous basis of a graded subspace of L which is complementary to L'.

Obviously, the restriction to L of any graded representation of U(L) is a graded representation of L. Consequently, the property c) implies that the notions of a "graded representation of L" and a "graded representation of U(L)" and, moreover, those of a "graded L - module" and a "left graded U(L) - module" are all equivalent. It depends on the circumstances which language is to be preferred. The algebra U(L) has the advantage of being associative, however, in general, it is infinite-dimensional even if L has a finite dimension. In any case, the general techniques of the theory of associative algebras now are at our disposal. This fact will be utilized in the subsequent construction of the induced modules.

Let L be a Lie superalgebra and let L' be a graded subalgebra of L. We are going to describe a procedure through which any graded L'- module can be enlarged to become a graded module over L. Thus let W be a graded L'- module. According to the foregoing remarks W may be regarded as a left graded U(L') - module. Consequently, we can construct the tensor product

$$\overline{W} = U(L) \underset{U(L')}{\otimes} W$$

of the right graded U(L') - module U(L) with the left graded U(L') - module W. This tensor product can be defined to be the quotient of $U(L) \otimes W$ (the usual tensor product of vector spaces over K) modulo the (graded) subspace W' which is generated by all tensors of the form

$$X X' \otimes y - X \otimes X' \cdot y$$

with $X \in U(L)$, $X' \in U(L')$, and $y \in W$.

If $X \in U(L)$ and $y \in W$, the tensor product of X and y in $\overline{W}$ (i.e., the residue class of $X \otimes y$ modulo W') is usually denoted by $X \otimes y$, too. This will not lead to any confusion, since the tensor product $U(L) \otimes W$ over K will never appear in the sequel.

The space $\overline{W}$ has a natural Z_2 - gradation which is fixed by the property that $X \otimes y$ is homogeneous of degree $\xi + \eta$ whenever $X \in U(L)$ is homogeneous of degree ξ and $y \in W$ is homogeneous of degree η. Moreover, $\overline{W}$ has an obvious structure of a left graded U(L) - module which is specified by the equation

$$Y(X \otimes y) = (Y X) \otimes y$$

for all $X, Y \in U(L)$ and all $y \in W$. Endowed with the described Z_2 - gradation and L - module structure, $\overline{W}$ is called the *graded* L - *module induced from the graded* L'- *module* W.

The corollary to the Poincaré, Birkhoff, Witt theorem yields important information on $\overline{W}$, in particular, it shows how to construct a basis of $\overline{W}$. We do not want to go into detail but only mention two special consequences. First it follows that the mapping $y \longrightarrow 1 \otimes y$ of W into $\overline{W}$ is injective and, obviously, a homomorphism of graded L'-modules. In this sense, $\overline{W}$ may be considered as an enlargement of W. Secondly, let us assume that $L_{\overline{0}} \subset L'$. If W is finite-dimensional and if L' has a finite codimension m in L, it follows that

$$\dim \overline{W} = 2^m \dim W .$$

In particular, if L is finite-dimensional, we may apply this remark to $L' = L_{\overline{0}}$ and obtain a proof of the Ado theorem for Lie superalgebras: Any finite-dimensional Lie superalgebra has a faithful finite-dimensional graded representation.

Usually, it is not advantageous to construct representations of L by induction from $L_{\overline{0}}$. For example, it seems to be rather exceptional that a graded L-module induced from an irreducible $L_{\overline{0}}$-module is irreducible in the graded sense. In the following we shall consider an interesting special case in which better results can be obtained [5].

Let L be a finite-dimensional Lie superalgebra such that the $L_{\overline{0}}$-module $L_{\overline{1}}$ decomposes into the direct sum of two submodules L_{-1} and L_1 which satisfy

$$\langle L_{-1}, L_{-1} \rangle = \langle L_1, L_1 \rangle = \{0\} .$$

If we set $L_{\overline{0}} = L_0$, the foregoing assumptions are equivalent to the requirement that

$$L = L_{-1} \oplus L_0 \oplus L_1$$

be a consistently Z-graded Lie superalgebra (with $L_r = \{0\}$ if $r \neq 0, \pm 1$). We note that the Lie superalgebras gl(n,m), sl(n,m), sl(n,n)/center, osp(2,2r), the P-type Lie superalgebras, and the not centrally extended algebra of field theoretic supersymmetry are of this type.

Theorem

Let L be a finite-dimensional Lie superalgebra which satisfies the assumptions specified above.

a) If V is a finite-dimensional graded-irreducible L-module, then

$$V_0 = \{x \in V \mid A \cdot x = 0 \text{ for all } A \in L_{-1}\}$$

is an irreducible L_0-submodule of V. In particular, all its elements are homogeneous of the same degree.

b) Conversely, let W be a finite-dimensional irreducible L_0-module and let $\alpha \in Z_2$. Then there exists one and, up to isomorphism, only one finite-dimensional graded-irreducible L-module V such that the L_0-submodule V_0 is isomorphic to W and such that the elements of V_0 are homogeneous of degree α.

The L-module V specified in part b) of the theorem can be constructed as follows. Obviously,

$$L' = L_{-1} \oplus L_0$$

is a graded subalgebra of L. We introduce on W a Z_2 - gradation by

$$W_\alpha = W \quad , \quad W_{\alpha+\bar{1}} = \{0\}$$

and extend the representation of L_0 in W to a representation of L' in W by the requirement that

$$L_{-1} \cdot W = \{0\} .$$

Then W is a graded L'- module. The induced graded L - module

$$\overline{W} = U(L) \underset{U(L')}{\otimes} W$$

contains a unique maximal (proper) graded submodule $\overline{W}'$, and the quotient

$$V = \overline{W}/\overline{W}'$$

meets our requirements. Thus the construction of all finite-dimensional graded-irreducible L - modules has been reduced to the investigation of the induced L - modules $\overline{W}$ and of their maximal graded submodule $\overline{W}'$.

Remarks

1) The L_0- module $\overline{W}$ is canonically isomorphic to $(\bigwedge L_1) \otimes W$, where $\bigwedge L_1$ denotes the Graßmann algebra of L_1.

2) Using the results of reference [5] it can be shown that $\overline{W}'$ contains any proper (not necessarily graded) L - invariant subspace of $\overline{W}$. This implies that for the Lie superalgebras considered here the second alternative in Schur's lemma does not occur. On the other hand, if $\overline{W}$ is endowed with a Z - gradation which converts it into a Z - graded L - module [5], then $\overline{W}'$ is a Z - graded submodule of $\overline{W}$ [8].

4. WEIGHT MODULES OVER gl(n,m)

It is well-known that the algebraic theory of representations of a semisimple (or, more generally, of a reductive) Lie algebra is largely based on a careful investigation of the weights and weight spaces of the representations. In the present section I want to show how this approach can be generalized to the super case and how one is led to the typical representations. To reduce the technicalities to a minimum I shall only consider the example of the general linear Lie superalgebras, i.e., I shall set

$$L = gl(n,m) , \text{ with } n,m \geqslant 0 .$$

We have already noted that gl(n,m) is of the special type considered in section 3. More precisely, $gl(n,m)_0 = gl(n,m)_{\bar{0}}$ consists of the block diagonal matrices $\begin{pmatrix} A & 0 \\ 0 & D \end{pmatrix}$, whereas the elements of $gl(n,m)_{-1}$ and $gl(n,m)_1$ have the block matrix form $\begin{pmatrix} 0 & B \\ 0 & 0 \end{pmatrix}$ and $\begin{pmatrix} 0 & 0 \\ C & 0 \end{pmatrix}$, respectively.

In the following, the indices i,j,k,ℓ run through the values $1,2,\ldots,n+m$, unless otherwise stated, and we set

$$\sigma_i = \begin{cases} 1 & \text{if } 1 \leqslant i \leqslant n \\ -1 & \text{if } n+1 \leqslant i \leqslant n+m . \end{cases}$$

We begin by some basic definitions. Let E_{ij} denote the $(n+m)\times(n+m)$ matrix defined by

$$(E_{ij})_{k\ell} = \delta_{ik}\,\delta_{j\ell} .$$

Obviously, the E_{ij} form a homogeneous basis of $gl(n,m)$.

The subalgebra of all diagonal matrices in $gl(n,m)$ will be denoted by h. This is nothing but the standard Cartan subalgebra of the Lie algebra $gl(n,m)_0$. To construct the corresponding root space decomposition of $gl(n,m)$ let us define the linear forms ε_j on h by the requirement that $\varepsilon_j(H)$ be the jth diagonal element of the matrix H, for all $H \in h$, or, equivalently, that

$$\varepsilon_j(E_{ii}) = \delta_{ij} .$$

Then we have, for all $H \in h$,

$$\langle H, E_{ij} \rangle = (\varepsilon_i - \varepsilon_j)(H)\, E_{ij} .$$

In analogy with the Lie algebra case, we say that $\varepsilon_i - \varepsilon_j$; $i \neq j$, is a *root* of $gl(n,m)$ with respect to h and that E_{ij} is a corresponding *root vector*. The root $\varepsilon_i - \varepsilon_j$ is called even/odd depending on whether E_{ij} is even/odd, that is, if $\sigma_i \sigma_j = \pm 1$. The set Δ of all roots is called the *root system* of $gl(n,m)$ with respect to h,

$$\Delta = \{\varepsilon_i - \varepsilon_j \mid i \neq j\} .$$

As in the classical case, we choose a system Δ^+ of *positive roots*, namely

$$\Delta^+ = \{\varepsilon_i - \varepsilon_j \mid i < j\} .$$

Now let W be a graded $gl(n,m)$-module. For any linear form μ on h we define

$$W^\mu = \{x \in W \mid H\cdot x = \mu(H)x \text{ for all } H \in h\} .$$

Obviously, W^μ is a graded subspace of W. If $W^\mu \neq \{0\}$, we say that μ is a *weight* of the $gl(n,m)$-module W and that W^μ is the corresponding *weight space*. According to a classical result the sum of the subspaces

W^μ; $\mu \in h^*$, is direct. If this sum is equal to W we say that W is a *graded weight module* over gl(n,m) (with respect to h). The class of graded weight modules over gl(n,m) is closed under the formation of graded submodules, graded quotients, direct sums, and of graded tensor products. A particularly interesting class of graded weight modules is specified in the following definition.

Definition

A graded gl(n,m) - module W is called a *graded highest weight module* if there exist a homogeneous non-zero element $w \in W$ and a linear form Λ on h such that:

1) The element w generates the gl(n,m) - module W.

2) The element w is annihilated by the root vectors corresponding to the positive roots, i.e.,

$$E_{ij} \cdot w = 0 \quad \text{for } i < j .$$

3) $$H \cdot w = \Lambda(H)\, w \quad \text{for all } H \in h .$$

It is easy to see that W actually is a graded weight module. Moreover, we have

$$W^\Lambda = K w ,$$

and for any weight μ of W, the difference $\Lambda - \mu$ can be written as a sum of positive roots. This implies that the linear form Λ and the one-dimensional subspace Kw are uniquely determined. We call Λ the *highest weight* and the non-zero elements of W^Λ the *generators* of the highest weight module W (with respect to h and Δ^+).

Using the technique of induced modules we can show that for any linear form Λ on h there exist graded highest weight modules over gl(n,m) with highest weight Λ. In fact, let $\Lambda \in h^*$ and $\alpha \in Z_2$. The elements E_{ij} with $i \leqslant j$ span a graded subalgebra L" of gl(n,m) containing h. We convert K into a graded L" - module $K(\Lambda,\alpha)$ by the requirements that all its elements be homogeneous of degree α and that

$$E_{ij} \cdot 1 = 0 \quad \text{for } i < j$$

$$H \cdot 1 = \Lambda(H) \quad \text{for all } H \in h .$$

Then the induced gl(n,m) - module

$$M(\Lambda,\alpha) = U(gl(n,m)) \otimes_{U(L'')} K(\Lambda,\alpha)$$

is a graded highest weight module with highest weight Λ and its generators, proportional to $1 \otimes 1$, are homogeneous of degree α .

The so-called *Verma module* $M(\Lambda,\alpha)$ is maximal in the following sense. Let W be a highest weight module over gl(n,m) with highest weight Λ and let w be a generator of W which is homogeneous of degree α. Then there exists a unique homomorphism of the graded gl(n,m) - module $M(\Lambda,\alpha)$ onto the graded gl(n,m) - module W which maps $1 \otimes 1$ onto w. Thus W is isomorphic to a graded quotient of $M(\Lambda,\alpha)$.

The Verma module $M(\Lambda,\alpha)$ contains a proper graded submodule $N(\Lambda,\alpha)$ which includes all proper submodules of $M(\Lambda,\alpha)$. The quotient

$$I(\Lambda,\alpha) = M(\Lambda,\alpha)/N(\Lambda,\alpha)$$

is an irreducible graded gl(n,m) - module with highest weight Λ, whose generators are homogeneous of degree α. Any graded gl(n,m) - module with these properties is isomorphic to $I(\Lambda,\alpha)$.

Remark

Of course, for any $\Lambda \in h^*$ the two graded gl(n,m) - modules $M(\Lambda,\alpha)$; $\alpha \in Z_2$, are "essentially the same", they can be converted into each other by a simple "shift" of the Z_2 - gradation. The same holds for $I(\Lambda,\alpha)$.

It is easy to see that any finite-dimensional irreducible graded gl(n,m) - module is a highest weight module and hence isomorphic to some $I(\Lambda,\alpha)$. Conversely, let us determine the conditions on $\Lambda \in h^*$ under which $I(\Lambda,\alpha)$ is finite-dimensional. To solve this problem we use the gl(n,m) - modules induced from the irreducible graded modules over

$$L' = gl(n,m)_{-1} \oplus gl(n,m)_0$$

as discussed in section 3. Let $J(\Lambda)$ denote the irreducible $gl(n,m)_0$- module with highest weight Λ, trivially extended to a graded L'- module all of whose elements are homogeneous of degree α. Then the induced module

$$\bar{J}(\Lambda,\alpha) = U(gl(n,m)) \underset{U(L')}{\otimes} J(\Lambda)$$

is a graded gl(n,m) - module with highest weight Λ and its generators are homogeneous of degree α. Moreover, if $J(\Lambda)$ is finite-dimensional, then so is $\bar{J}(\Lambda,\alpha)$. This implies:

Theorem

The irreducible graded gl(n,m) - module $I(\Lambda,\alpha)$ is finite-dimensional if and only if the irreducible $gl(n,m)_0$- module $J(\Lambda)$ with highest weight Λ is finite-dimensional, i.e., if and only if the elements $\Lambda(E_{i,i} - E_{i+1,i+1})$ are integral $\geqslant 0$ for $1 \leqslant i < n$ and for $n+1 \leqslant i < n+m$.

The set of all $\Lambda \in h^*$ with the foregoing properties will be denoted by P_{++}. We stress that there is no condition on $\Lambda(E_{n,n} - E_{n+1,n+1})$.

Remark

The statement in the theorem above appears to be quite natural, hence one might think that it should hold under much more general circumstances. However, already for most of the orthosymplectic Lie superalgebras the analogous theorem is false.

Let us next discuss the reducibility of the finite-dimensional graded gl(n,m) - modules. To begin with, we recall that according to a theorem by Djoković and Hochschild [7,9] it is quite exceptional that a finite-dimensional Lie superalgebra L has all its finite-dimensional graded representations completely reducible: This is the case if and only if L is isomorphic to the direct product of a semi-simple Lie algebra and of some orthosymplectic algebras of the type osp(1,2r).

To motivate our procedure let us consider a finite-dimensional module V over a semi-simple Lie algebra L. In order to decompose V into irreducible submodules, we could proceed in the following two steps.

First we employ the Casimir operators of L and decompose V into a direct sum of L - submodules V_i in which every Casimir operator has only one eigenvalue. Since the eigenvalues of the Casimir operators characterize the (finite-dimensional) irreducible L - modules, the irreducible subquotients of each of the V_i are all isomorphic.

In the second step we consider each V_i separately. We choose a Cartan subalgebra h of L and notice that V_i is a weight module. Then the weight space decomposition of V_i with respect to h can be used to decompose V_i into irreducible L - submodules.

For the Lie superalgebra gl(n,m) neither of these two steps can be taken without a proviso.

Concerning the second step we point out that there exist finite-dimensional graded gl(n,m) - modules which are not weight modules. On the other hand, any finite-dimensional completely reducible graded gl(n,m) - module is a weight module. Consequently, in the following we shall restrict our attention to the finite-dimensional graded weight modules over gl(n,m). We remark that a finite-dimensional graded gl(n,m) - module is a weight module if and only if the central elements of the Lie algebra $gl(n,m)_0$ are represented by diagonalizable linear mappings. For $n,m \geqslant 1$ it would not be sufficient (even if $n \neq m$) to assume that this condition is satisfied for the central elements of gl(n,m) only.

Concerning the first (and more interesting) step we notice that the Casimir elements of gl(n,m) and their eigenvalues in a highest weight module are known explicitly [3-5,10-13]. Let us summarize the pertinent information.

A *Casimir element* of a Lie superalgebra L is an element of the enveloping algebra U(L) that supercommutes with all elements of U(L). The graded subalgebra of these elements of U(L) will be denoted by Z(L), the *supercenter* of U(L).

The supercenter Z(gl(n,m)) of U(gl(n,m)) is known:
Its elements are even and the special elements

$$C_r = \sum \sigma_{i_1}\sigma_{i_2}\ldots\sigma_{i_{r-1}} E_{i_r i_1} E_{i_1 i_2} \ldots . E_{i_{r-1} i_r} \quad ; \quad r \geqslant 1 ,$$

supplemented by the unit element, generate the algebra Z(gl(n,m)).

Let $C \in Z(gl(n,m))$ and let V be a graded highest weight module over gl(n,m) with highest weight Λ. Then the representative C_V of C in V is a scalar multiple of the identity,

$$C_V = \chi_\Lambda(C)\,\mathrm{id}_V ,$$

and the scalar factor $\chi_\Lambda(C)$ depends on C and Λ but not on the special graded highest weight module with highest weight Λ.

The mapping $\Lambda \longrightarrow \chi_\Lambda(C)$ is, for fixed $C \in Z(gl(n,m))$, a polynomial function on h*. Let ρ denote the half sum of the even positive roots minus the half sum of the odd positive roots, i.e.,

$$\rho = \frac{1}{2}\sum_{i<j} \sigma_i\sigma_j(\varepsilon_i - \varepsilon_j) .$$

It is advantageous to consider $\chi_\Lambda(C)$ as a function of $\lambda = \Lambda + \rho$ and to express λ in terms of the coordinates

$$\ell_i = \sigma_i\lambda(E_{ii}) .$$

Then we have

$$\chi_\Lambda(C) = \tilde{C}(\ell_1,\ldots,\ell_{n+m}) ,$$

where $\tilde{C}$ is a polynomial in $n+m$ indeterminates $x_1,\ldots,x_{n+m}$ and the substitution $x_i = \ell_i$ has been made.

The mapping $C \longrightarrow \tilde{C}$ is an injective homomorphism of the algebra Z(gl(n,m)) into the polynomial algebra $K[x_1,\ldots,x_{n+m}]$ (generalized *Harish - Chandra homomorphism*), and its image is known: It is generated (as an algebra) by 1 and the so-called supersymmetric power sums

$$P_r = \sum_{i=1}^{n+m} \sigma_i x_i^r \quad ; \quad r = 1,2,\ldots .$$

An intrinsic characterization of this image is known, too.

Let us now return to our main problem. We recall that P_{++} denotes the set of all $\Lambda \in h^*$ such that $I(\Lambda,\alpha)$ is finite-dimensional. In view of the aforementioned first step we determine those $\Lambda \in P_{++}$ such that, for any $\Lambda' \in P_{++}$ different from Λ, there exists a Casimir element C which "separates" Λ and Λ', i.e., such that

$$\chi_\Lambda(C) \neq \chi_{\Lambda'}(C) .$$

It is easy to see that Λ has this property if and only if

$$\ell_i \neq \ell_j \quad \text{for } 1 \leq i \leq n ;\ n+1 \leq j \leq n+m .$$

An element $\Lambda \in P_{++}$ with this property is called *typical* (in general position); a finite - dimensional irreducible graded gl(n,m) - module is said to be typical if its highest weight is typical.

Proceeding as sketched above, we now can prove [5]:

Theorem

Let V be a finite-dimensional graded weight module over gl(n,m) and let V_t be the sum of all typical (irreducible graded) submodules of V. The set of those graded submodules of V all of whose irreducible graded subquotients are non-typical has a unique maximal element V_n and V is the direct sum of V_t and V_n.

By definition, V_t is completely reducible (in the graded sense). Concerning the submodule V_n we only obtain the obvious information that V_n decomposes into a direct sum of indecomposable graded submodules and that in each indecomposable graded gl(n,m) - module any Casimir operator has only one eigenvalue.

In a slightly simplistic language the theorem may be formulated as follows: Typical (irreducible graded) subquotients split off from any finite-dimensional graded weight module over gl(n,m). This property characterizes the typical highest weights. In fact, if $\Lambda \in P_{++}$, the induced module $\overline{J}(\Lambda,\alpha)$ is irreducible (and hence isomorphic to $I(\Lambda,\alpha)$) if and only if Λ is typical [4,5]. If Λ is non-typical, then $\overline{J}(\Lambda,\alpha)$ is reducible but obviously an irreducible graded gl(n,m) - module with highest weight Λ cannot be split off.

The typical gl(n,m) - modules behave, from certain aspects, like the finite-dimensional irreducible modules over a semi-simple Lie algebra. For example, using classical techniques, Kac has proved for them a character formula à la Weyl [3,4]. On the other hand, let us note that the tensor product of two typical gl(n,m) - modules is not necessarily completely reducible.

To investigate the non-typical irreducible graded gl(n,m) - modules it seems to be promising to start from the modules $\overline{J}(\Lambda,\alpha)$ defined above. In fact one can try to construct suitable exact sequences of these modules (or of their duals), thus generalizing a technique which Rudakov [14] has employed to study the representations of the infinite-dimensional Lie algebras of Cartan type. Along these lines Bernstein and Leites [15] have obtained a character formula which was expected to hold for any finite-dimensional irreducible graded gl(n,m) - module. However, it is easy to see that this formula is wrong if $n,m \geqslant 2$, the trivial one-dimensional module being a counterexample. There have been several other attempts to derive a character formula for the non-typical irreducible graded gl(n,m) - modules, partly under certain additional assumptions. I do not go into detail here but refer to a recent letter by Hughes and King [16] in which the authors briefly review these attempts and propose a character formula which is conjectured to hold for any finite-dimensional irreducible graded sl(n,m) - module, with $n \neq m$. (Obviously, the finite-dimensional irreducible graded modules over gl(n,m) and sl(n,m) are closely related.)

The contents of this section show that the representation theory for gl(n,m) is still similar to that for reductive Lie algebras. For other simple or "almost simple" Lie superalgebras like the P - type algebras there do not exist any useful Casimir elements at all [13,17]. In such a case completely different techniques like those by Rudakov have to be employed.

5. DECOMPOSITION OF TENSORIAL POWERS

In the present final section I shall briefly describe how the action of the symmetric group can be used to decompose the tensorial powers of the elementary gl(n,m) - module.

Let V be a finite-dimensional Z_2 - graded vector space, considered as a graded $gl(V_{\bar{0}},V_{\bar{1}})$ - module in the obvious way. We set

$$\dim V_{\bar{0}} = n \quad , \quad \dim V_{\bar{1}} = m \, .$$

As shown in section 2 the tensorial powers $V^{\otimes r}$ of V have a canonical structure of a graded $gl(V_{\bar{0}},V_{\bar{1}})$ - module. Moreover, $V^{\otimes r}$ carries a natural representation $\pi \longrightarrow S_\pi$ of the symmetric group $\mathcal{S}_r$ such that the representatives S_π are automorphisms of the graded $gl(V_{\bar{0}},V_{\bar{1}})$ - module $V^{\otimes r}$.

In the non-graded case the relationship between the $\mathcal{S}_r$- and gl(V)-module structures of $V^{\otimes r}$ is even closer and, in particular, allows for a complete reduction of the gl(V) - module $V^{\otimes r}$ and for a classification of tensors according to their symmetry types, i.e., according to the representations of $\mathcal{S}_r$. It turns out that something analogous holds true in the super case as well. The following is a summary of results obtained by Sergeev [18] (see also [19]).

Let ρ_r be the canonical representation of $gl(V_{\bar{0}},V_{\bar{1}})$ in $V^{\otimes r}$, canonically extended to the enveloping algebra of $gl(V_{\bar{0}},V_{\bar{1}})$. The basic result, on which the more detailed investigation rests, reads as follows [18,19].

Theorem

Any linear mapping of $V^{\otimes r}$ into itself, which commutes with all the representatives S_π; $\pi \in \mathcal{S}_r$, has the form $\rho_r(X)$ with some element X of $U(gl(V_{\bar{0}},V_{\bar{1}}))$ or, more briefly : The commutant of the set $\{S_\pi | \pi \in \mathcal{S}_r\}$ is equal to $\rho_r(U(gl(V_{\bar{0}},V_{\bar{1}})))$.

This fact can be exploited in essentially the same way as in the classical case. It is well-known that the irreducible representations of $\mathcal{S}_r$ are classified by means of the partitions λ of r. Let S^λ denote the irreducible $\mathcal{S}_r$- module corresponding to λ and let W^λ denote the sum of all irreducible $\mathcal{S}_r$- submodules of $V^{\otimes r}$ which are isomorphic to S^λ. Then W^λ is a graded $gl(V_{\bar{0}},V_{\bar{1}})$ - submodule of $V^{\otimes r}$ and $V^{\otimes r}$ is the direct sum of the W^λ. The submodule W^λ is different from {0} if and only if $\lambda_{n+1} \leqslant m$. If this is the case the graded $gl(V_{\bar{0}},V_{\bar{1}})$ - module W^λ decomposes into the direct sum of $\dim S^\lambda$ irreducible graded $gl(V_{\bar{0}},V_{\bar{1}})$ - submodules which are all isomorphic to some irreducible graded $gl(V_{\bar{0}},V_{\bar{1}})$ - module V^λ. The highest weight of V^λ (identified with the n + m tuple of its standard classical coordinates) is equal to $(\lambda_1,\ldots,\lambda_n,\langle\lambda'_1 - n\rangle,\ldots,\langle\lambda'_m - n\rangle)$, where λ'_i is the length of the i th column in the diagram of λ and for all $\ell \in Z$ we set $\langle \ell \rangle = \max(\ell,0)$.

The multiplicities of the weights of V^λ can be characterized purely in terms of the representation theory of $\mathcal{S}_r$. Moreover, the tensor product of two such modules V^λ and V^μ can be decomposed into irreducible submodules by means of the Littlewood - Richardson rule.

Remark

According to classical representation theory it follows immediately from the theorem above that the $gl(V_{\bar{0}}, V_{\bar{1}})$ - module $V^{\otimes r}$ is completely reducible (in the graded sense). In the complex case, this fact can easily be derived by means of the theory of star representations of a Lie superalgebra (see section 2 of reference [20]).

As already mentioned the foregoing results are analogous to those in the non-graded case. Things are much less satisfactory if we consider mixed tensor products whose factors may be equal to V or V* or if b is a non-degenerate supersymmetric even bilinear form on V and $V^{\otimes r}$ is considered as a module over the corresponding orthosymplectic Lie superalgebra. In these cases we expect from the classical theory that in order to decompose the tensor products we also have to use contractions with respect to the canonical pairing of V* and V or with respect to b, respectively. Unfortunately, it turns out that the tensor products under consideration are not necessarily completely reducible. There is a multitude of publications on this and similar problems but, regrettably, I often have the impression that the results obtained are to be considered as useful educated conjectures rather than as rigorous mathematical theorems.

REFERENCES

1. V. G. Kac, *Adv. Math.* **26** (1977) 8
2. M. Scheunert, *The Theory of Lie Superalgebras,* Lecture Notes in Mathematics 716; Springer; Berlin, Heidelberg, New York (1979)
3. V. G. Kac, *Comm. Alg.* **5** (1977) 889
4. V. G. Kac, p. 597 in *Differential Geometrical Methods in Mathematical Physics II,* Bonn 1977; K. Bleuler, H. R. Petry, and A. Reetz eds.; Lecture Notes in Mathematics 676; Springer; Berlin, Heidelberg, New York (1978)
5. M. Scheunert, p. 421 in *Supersymmetry,* Bonn 1984; K. Dietz, R. Flume, G. von Gehlen, and V. Rittenberg eds.; Plenum; New York, London (1985)
6. M. Scheunert, *J. Math. Phys.* **24** (1983) 2658
7. G. Hochschild, *Ill. J. Math.* **20** (1976) 107
8. This last observation establishes the connection with reference [5].
9. D. Ž. Djoković and G. Hochschild, *Ill. J. Math.* **20** (1976) 134
10. V. G. Kac, *Proc. Natl. Acad. Sci. USA* **81** (1984) 645, Physics
11. A. M. Bincer, *J. Math. Phys.* **24** (1983) 2546
12. M. Scheunert, *J. Math. Phys.* **24** (1983) 2681
13. M. Scheunert, to appear in the proceedings of the 15 th International Conference on *Differential Geometric Methods in Theoretical Physics,* Clausthal 1986; World Scientific; Singapore
14. A. N. Rudakov, *Iz. Acad. Nauk SSSR, Ser. Mat.* **38**(4) (1974) 835
15. I. N. Bernstein and D. A. Leites, *C. R. Acad. Bulgare Sci.* **33**(8) (1980) 1049
16. J. W. B. Hughes and R. C. King, *J. Phys. A: Math. Gen.* **20** (1987) L 1047

17. M. Scheunert, *J. Math. Phys.* **28** (1987) 1180
18. A. N. Sergeev, *Funkts. Anal. Prilozhen.* **18**(1) (1984) 80
19. A. Berele and A. Regev, *Adv. Math.* **64** (1987) 118
20. M. Scheunert, W. Nahm, and V. Rittenberg, *J. Math. Phys.* **18** (1977) 146

CONCLUDING REMARKS

Michael Atiyah

Perhaps the dominant topic at this conference has been 2-dimensional quantum field theory in its various forms. In particular the structure of conformally invariant theories, involving the representation theory of the Virasoro algebra, has emerged as of particular importance. These ideas seem to be relevant to a surprisingly wide range of problems in both mathematics and physics. It is quite possible that we are seeing the beginning of a major new development in mathematics, comparable in its way to some of the great stages of the past

String theory has given a great impetus to all this work and I look forward to seeing where this leads us. However I should like to suggest that we do not ignore other avenues. For example the Donaldson theory of 4-dimensional manifolds is surely deserving of great attention and may yet have much to teach both mathematicians and physicists.

It is clear that we are living in exciting times and this conference has high-lighted some of the latest developments. Who knows what dramatic events still lie in store for us or what will be discussed at such meetings in a few years time?

K. Bleuler and M. Werner (eds.), Differential Geometrical Methods in Theoretical Physics, 461.

OBSERVATIONS OF TWO OF OUR BRIGHTEST STARS

Leopold Halpern*
Jet Propulsion Laboratory
California Institute of Technology
Pasadena, California 91109

ABSTRACT. The author who has been associated to E. Schroedinger at the University of Vienna 1956-59 and to P.A.M. Dirac at Florida State University 1974-84 discusses some of his experiences characterizing the two founders of modern physics and their outlook on science and various scientific problems.

Erwin Schroedinger and Paul Dirac differed widely in their character and modes of thinking. I knew both of them during their last years when hectic activity had given way to calmer contemplation and the widest and wisest outlook. A virtue common to them was to display no haughtiness. They possessed the simple dignity of the great man which Focker identified with Einstein. Arrogance was completely absent and detested by them if shown by others. One must however be profoundly autonomous and needs a fair amount of stubbornness to develop modern physics according to one's own ideas.

Dirac remarked to me that a lecture in which one is not particularly interested constitutes a "powerful sleeping pill" and he hardly ever hesitated to use this harmless stimulant.** At physics meetings one had usually reserved for him a seat in the front row and he gave thus a conspicuous example of how to make best use of the many hours one is compelled to spend at a crowded conference schedule. He adopted this habit early, (indeed he slept when Heisenberg gave his first introduction to quantum mechanics in Cambridge. He realized it's significance only later when his teacher suggested to study the publication.) Dirac did not do this in order to impress or offend anybody, nor would he have minded if during his own lecture even the whole audience had gone to sleep. After attending a lecture Dirac used

---

*NAS-NRC Resident Research Associate, 1976-1978, at the Jet Propulsion Laboratory, California Institute of Technology, 4800 Oak Grove Drive, Pasadena, California 91109. On leave of absence from Florida State University, Department of Physics, Tallahassee, Florida 32306.

**This was one of the very few stimulants he used. He neither smoked nor ever drunk alcohol and seldomly used coffee.

K. Bleuler and M. Werner (eds.), Differential Geometrical Methods in Theoretical Physics, 463–476.

to state an opinion at most in the form of a soft spoken question. Later he would make a dry remark with humorous flavour to me about it. Only if he was specially asked for his opinion would he state it calmly to the lecturer without argument.

An anecdote about Dirac tells that after one of his illustrious lectures, when questions were invited, somebody said that he had not understood certain features of the lecture. Dirac remained silent; when the chairman finally asked if he wanted to answer, Dirac replied that "a question was not asked -- this was a statement." This sharp distinction appears amusing but also arrogant. Dirac's lectures and explanations were indeed so clear that during the eleven years I spent with him, I very rarely had to ask him further questions -- but this anecdote seemed to me to be in contrast with his gentle modest manner.

I found out from him that the incident had indeed happened but the usual interpretation was wrong. In fact because the person continued after the statement at such great length in explanations that Dirac could not follow anymore, he did not imagine that this was still meant to be a question; in accord with his famous inequality "that there are always more people who like to talk then to listen" he thus remained silent.

I find it regrettable that this story is so frequently mistold by people who did not know or understand Dirac well enough and depict him thus as arrogant. Sometimes people also imitate his phrasing to unexperienced questioners -- but different from the original there is often another implicit statement in the question -- namely about the clearness of the lecture.) Dirac was aware that everybody's intelligence may prove creative in a different domain. When I told that a certain good car mechanic was hardly able to write, he remarked that this man's cleverness was directed towards a task that would be hard for himself to perform. He was however convinced that the very best brains are needed to achieve true progress in science.

Schroedinger saw it as his divine duty to instill the most thorough understanding into his audience. His presentations differed from the traditional norm at the University of Vienna by encouraging interruption for questions. The lectures reminded more of the dissertations of a philosopher in ancient Greece than that of a professor. Particularly impressive was his appearance in the seminars when routine presentations were transformed into highly spirited discussions. Even when severely handicapped by bronchitis and general weakness, Schroedinger would not save any effort to promote better understanding. This trait is also exemplified in his books which apart from giving a lucid introduction to a subject show so much wisdom conscientiousness and brilliance of presentation that reading them will be enjoyed throughout the ages. Schroedinger's intentions can well be characterized by his attitude towards the crackpots who sometimes approached him. He stressed to me that these people also need to be helped. He remarked however that with those who believe they are able to refute the theory of relativity by elementary arguments such help proves often futile.

Schroedinger's humanitarian attitude also becomes apparent when he was asked to comment on the benefits he expected from new discoveries. His concern for the environment and the health of every one is exemplary. One could only wish that more than one politician had chosen to consult with him.

Dirac never spoke a superfluous word. He had a marvelous capability to grasp and limit himself to the essential. This made his presentations so outstandingly clear. He told me that his thoughts arise from the subconscious in quite a different form than he used to express them finally to others.

Schroedinger was no doubt the most universal spirit I have come to know among scientists. He could lose himself into varied digressions during his lectures. The formation of his thoughts was apparently tied rather closely to the language. He mastered about 12 languages, wrote poetry in several of them and was very choosy in his idiomatic expressions. He used to illustrate his thoughts by transparent analogies and examples. Also his humor frequently took the form of these striking comparisons. Once when I tended to excuse lacking results in my research due to the difficulty and significance of the problem, he compared my attitude with that of a colleague who had returned empty handed from a tiger hunt and when teased argued "that for tigers this was not really so little." He also was able to characterize people outstandingly well and could predict their reactions. Once he urged me to stay for supper instead of preparing for a seminar discussion that was to take place in front of a visitor the next morning. He said this man would do most of the talking anyway and it turned out that way.

In contrast Dirac's humor condensed the most essential features of a situation, which often enough, without comment renders it ridiculous. His dry presentation of a joke was inevitably a success. No jokes appear in Dirac's lectures and writings, whereas Schroedinger's books abound with humorous analogies and digressions.

Dirac and Schroedinger learned early to express themselves in two languages. As a child Schroedinger visited relatives of his mother in England. Dirac's father, a French-Swiss language tutor in Bristol was rather tyrannic in his good intentions; he obliged his children to speak French at home and not even the most elementary request was granted to them unless they could express it well. Remarkably in the case of Dirac this seems to have contributed, perhaps even formed his individual abstract thinking. It proved difficult for the child, who grew up in an English environment to express himself in French. He thus adopted early the habit to limit his speech to the most essential and invent ways to obtain what he needed without requesting anybody's help. Schroedinger's father managed to create a good companionship with his son on their numerous excursions and journeys in particular to Tyrol. He obtained his elementary school training at home where apparently no teacher was able to spoil his desire for knowledge and understanding. His desire was to be a poet and philosopher. The circumstances led him however to a study of mathematics and physics and a thesis on an experimental subject. He expressed to me the opinion that a physicist should not be admitted to theoretical research before completing some experimental work.

He worked with Franz Exner in Vienna, who viewed a paradox in

the assumption of rigorously valid laws of motion for microscopic particles while macroscopic systems always are subject to statistical fluctuations. This doubt appears in Schroedinger's thinking throughout his life. It played a role in his earlier publications, his main discovery and also in his last paper: "Might Energy Be a Merely Statistical Concept."[1]

Another paradox with which Schroedinger was much concerned is the inevitable appearance of the continuum in physical description, whereas the points, which constitute it's elements, are devoid of physical meaning. He compared this appearance of the continuum, already at the birth of our science, with the legendary evil spell of the thirteenth fairy who was not invited to a prince's birthday party.[2] He hoped that the discrete character of the eigenvalues of wave mechanics would alleviate this situation but he found reason for doubts also in their exact physical validity.[3]

There was hardly another physicist who had given thought to the relationship of his science to all aspects of life and the mind as conscientiously as Schroedinger. He struggled with the problem of the separation of our own perception from the description of the physical world. The thorough discussion is presented in his poetic style in the books that appeared during his last decade.[2,4] In young age he acquired a profound knowledge of classical Greek philosophy and was also acquainted with Indian philosophy (the Upanisads). In the latter he saw a way to bridge the gap between mind and matter which Western thinking had not achieved satisfactorily. It is remarkable that he found it necessary to introduce mysticism into his picture of the world, apparently without having a mystical experience himself. One can hardly follow medieval Christian ideas or those of the Kabbalists without personally having a related religious experience. I think the same is true for the meditative experiences (which need not even be termed religious) from which the Upanisads originated. Neither Schroedinger nor Dirac were religious in the conventional sense* nor did they probably have occasion to practice meditation. If it were otherwise Schroedinger would not have remarked that he finds certain

---

*If either of them is identified as "religious" -- and even this has been done -- it should be qualified (least one wants to commit a falsification) that this religiosity was limited to the concern and compassion for the living and above all else to the "intellectual astonishment and it's awe of existence." I could not discover any trace of a religiosity in them, which claiming to posses full explanation for the latter, proceeds to justify activities leading to libel, the burning of books and people (or, to the least, refusing them a burial and to the worst promising them avoidance of disaster in a nuclear war.) Both Schroedinger and Dirac were aware that there is no inherent ethics in today's natural science. They were equally remote from aligning themselves with those, who claim to posses adequate knowledge from present (or even from past) scientific results for it's technocratic application to ends which ultimately curtail freedom and risk to invite disaster which is revealed only after subsequent findings show once more that our knowledge has been incomplete.

parts of the Bhagavad Gita "not particularly interesting."

I believe however that the intense contemplation and thinking about natural phenomena often brought Schroedinger and Dirac into a state of mind that may be akin to that resulting from meditation. This distinguishes them from many scientists who are unable to seek in nature much more than the manipulation of blind forces - somewhat like military people whose views are too narrow to recognize that features beyond military power really are crucial to the ultimate fate of the nations.

In more practical aspects Western scientific thought in this century has already been supplemented continuously: I don't believe that Ph. Lenard and his circle were that wrong in calling Einstein's thinking Jewish and even Kabbalistic -- they just completely misjudged it's value. During the last four decades important results in physics originated from modes of contemplation which are distinctly Asian. Whereas Schroedinger scrupulously analyzed every physical assumption to it's last consequences, Dirac had already been accustomed from his initial study of engineering to accept approximate description. The subsequent study of pure and applied mathematics, which he began due to lacking employment, introduced him equally well to it's rigorous mental constructions and their beauty. While continuing his studies in Cambridge he apparently lost further confidence in an exact description of nature after some disappointment with Bohr's theory. He adopted the mathematical beauty of a theory as a criterion of it's value. He later stressed "that the theory describing reality best ought to be the most beautiful -- because God made it that way." *

Following such esthetic considerations, which also imply the utmost of logical consistency he arrived at his famous equation from the motive of "pretty mathematics."[5]

Schroedinger and Dirac lacked interest and understanding in politics. Dirac became concerned with authoritarian governments when Stalin prevented his friend Kapitza to return from a visit to Russia. Dirac proved to be a true friend by visiting Kapitza as often as possible. After the war Kapitza came under severe pressure to work on the nuclear bombs by Beria. Refusing such collaboration, Kapitza emerged as one of the true heroes of our time.

Dirac was able to speak German well. He appreciated the contributions made by German physicists highly. But when the National Socialists displayed their activities he permanently stopped using the language. This may be a striking example how Hitler became the arch enemy and greatest destroyer of the influence of German language and culture that ever existed. It is heartbreaking to note that to this day people in some German speaking countries identify themselves still with the worst and most disastrous of what their region produced and remain indifferent towards the best. Dirac and Heisenberg were close friends. Dirac used to defend Heisenberg from accusations and suspicions.

* He used this expression to show that reality can result only from an awe inspiring source. The use of the word God does not imply a traditional religious image.

Schroedinger could not tolerate living in Berlin after Hitler came to power. He left his position there and accepted a professorship in Graz, Austria. His humanitarian views were incompatible with national socialism. He was outspoken about this even after the nazis controlled a part of the Austrian government and Austria's fate was sealed. Prof. W. Frank of the Dept of Mathematics, University of Salzburg recently found a newspaper article that reports on a lecture by Schroedinger which testifies to his opposition to the notion of national and racial superiority. Schroedinger was then fired and suffered severe duress; later the nazis approached him to publish a statement showing his enthusiasm for their regime. He told me that after much reflection he decided to make such a statement so enthusiastically that his friends abroad would recognize that it was not genuine. The conditions granted to him then improved. He was not prevented to visit with his wife an academy meeting in allied Italy. From there they fled to Belgium. He was then invited by De Valera to direct the newly established Center for Advanced Studies in Dublin where he stayed till 1956.

Dirac and Schroedinger had a considerable interest in cosmology which became dominant during their last years. Both considered the general theory of relativity as the best existing theory. They revered and admired Einstien. Einstein's burial was one of the few occasions when Dirac could not control his feelings and cried. During the Einstein centennial Dirac's state of health was already weak but he still accepted invitations to the many meetings all over the world, saying that he wanted to help the public realize how great a man Einstein was.

Schroedinger and Dirac proposed both generalizations of Einstein's theory. In light of such generalizations emerges often enough the senseless statement that not Einstein's general relativity but the theory of somebody else may be right. The truth of Einstein's theory is indisputably established. It is as correct as it ever could have been in Einstein's time. He was even centuries ahead in creating it. I consider it however equally likely that in relation to newer discoveries, made after it's creation it will finally have to undergo minor modifications the consequences of which are not yet observable. This in no way diminishes Einstein's achievement.

Schroedinger worked for many years in friendly competition with Einstein on a unified field theory of gravitation and electrodynamics. He recognized during the fifties that the theory could not fulfill his expectations and he abandoned it. He believed that general relativity well describes the situation in our solar system but he hesitated to apply it to extreme conditions. He had already seen before the most consistent classical constructions in physics shattered and it did not appear likely to him that the radical changes predicted by the theory at the Schwarzschild horizon need to be physically correct. He recognized the possibility of the passage through a horizon and had coined the comparison with Xeno's paradox of Achilleus and the turtle.[6] He recognized also the possibility of neutron stars. He had however also discovered already in 1939 the "alarming phenomenon of particle-antiparticle formation in matter free expanding universes. In 1959 he suggested to me: "Instead of assuming the validity of a theory under extreme conditions, remote from observability, it could prove more fruitful to analyze other of it's limitations -- in this case the question why the gravitational forces which are limited by the mass densities, are so weak. A new idea may be needed to explain this."

I consider this as one of the wisest suggestions made on the subject. In a recent article I have described how it influenced my research.[7]

Also Einstein was not convinced of the physical role of the horizon. Outstanding mathematical work has since shown that within the framework of classical general relativity there is no way to prevent the formation of horizons. This does of course not imply that these results need apply to physics. The decision of the question how close to the formation of a horizon the laws of general relativity remain valid is a difficult future task for astrophysical observation. The structure of physical research since Schroedinger's days underwent significant changes that are reminiscent of the change of production from the time of artists and artisans to the industrial revolution. The motives behind the propagation of certain fashionable research projects may be interesting to analyze nowadays. To me in any case the premature assumption of physics at or beyond the horizon appeared at variance with Schroedinger's principles. I objected, when the impressive mathematical arguments for complete collapse were presented at the second Symposium on Differential Geometry and Physics, pointing out the existence of a virtual part of Schroedinger's real pair formation I said: "There are more things between heaven and the black hole then dreamed of in the philosophy of relativist mathematicians."

One year later Hawking's paper on pair formation helped the real part of the above statement to recognition; but if the real part is assumed to be valid, the less conspicuous but more significant virtual part should also exist. I do not think that one can reliably obtain it from specific calculations and one may have to approach the problem in different ways.[7] The essence of Schroedinger's suggestion may in any case still prove of value in spite of the fact that "industrial exploitation" of others views has already progressed.

Dirac's famous speculation about a cosmological origin of dimensionless large numbers in physics originated already in the thirties. It is in a way a generalization of a suggestion by Eddington to a dependence on cosmic time. It is one of the most daring speculations ever made. (It could be the idea Schroedinger requested for the understanding of the weakness of the gravitational constant.) Dirac remained convinced of his Large Number's hypothesis all his life. He dedicated a large part of his last ten years working out it's consequences. He saw in general relativity the best existing theory and wanted his generalization to deviate as little as possible from it. He requested rigorously that every dimensionless large number in nature should vary with the appropriate power of the age of the universe. When in his last year physical data finally seemed to indicate that the gravitational constant does not vary with time he told me that he felt like Einstein when first experiments tended to contradict relativity: he remained convinced of his theory but he did not want to impose his conviction on the public.

Although Dirac raised the mathematical beauty of a theory to a criterium of it's value, he did not like the use of mathematics beyond the stringent requirements for the physical description; the same is true for Schroedinger. They considered mathematical orgies ridiculous if presented in connection with physics. They reveal a developed sense for the inessential, as Einstein once had said jokingly to Schroedinger.

Neither Schroedinger nor Dirac did believe that their most significant achievements in quantum theory had reached the final form. Dirac found mathematical correctness lacking in a theory that requires renormalization and he lost confidence in the Hamiltonian used.

They considered both the reduction of the wave packet as a feature subject to modification in future theory.

Schroedinger told me that among all the physicists who developed quantum mechanics he had the relatively best agreement on the interpretation with Dirac. I asked Dirac twenty years later about it and he said that he could agree with Schroedinger's statement, but the word relatively should be emphasized.

References

[1] E. Schroedinger Il Nuovo Cimento (10) **9** (1958) pp 162-70

[2] E. Schroedinger Science and Humanism Cambridge University Press (1952)

[3] E. Schroedinger 'The Meaning of Wave Mechanics' in Louis DeBroglie Physicien et Penseur p 16-32 Paris: Michel (1953), Reproduced in 'Collected Papers' **3** pp 693-710

[4] E. Schroedinger Mind and Matter Cambridge University Press (1958) My View of the World Cambridge University Press (1964)

[5] P.A.M. Dirac 'Pretty Mathematics' Int. J. Theor. Phys. (USA) **21** pp 603-5 (1982)

[6] E. Schroedinger Expanding Universes Cambridge University Press (1956) p 20

[7] L. Halpern 'Erwin Schroedinger's Views on Gravitational Physics during his Last Years at the University of Vienna and Some Research Ensuing from it Foundations of Physics **17** (1987) pp 1113-1130